P9-DNS-763

Human Anatomy Laboratory Manual
with Cat Dissections

Elaine N. Marieb, R.N., Ph.D.

Edison Community College
Holyoke Community College, Emeritus

SECOND EDITION

The Benjamin/Cummings Publishing Company, Inc.

Menlo Park, California • Reading, Massachusetts
New York • Don Mills, Ontario • Wokingham, U.K. • Amsterdam
Bonn • Paris • Milan • Madrid • Sydney • Singapore • Tokyo
Seoul • Taipei • Mexico City • San Juan, Puerto Rico

Executive Editor: Johanna Schmid
Sponsoring Editor: Lisa Moller
Assistant Editors: Natasha Banta, Thor Ekstrom
Production Editor: Lisa Weber
Art Supervisors: Kathleen Cameron, Carol Ann Smallwood
Photo Editor: Kathleen Cameron
Design Manager: Don Kesner
Composition and Film Manager: Lillian Hom
Composition and Film Assistant: Vivian McDougal
Permissions Editor: Marty Granahan
Manufacturing Supervisor: Merry Free Osborn
Marketing Manager: Nina Horne
Artists: Val Felts, Karl Miyajima, Kristin Otwell, Denise Schmidt
Copyeditor: Anita Wagner
Proofreader: Christine Sabooni
Indexer: William J. Richardson Associates, Inc.
Text Designer: Gary Head
Page Layout: Brenn Lea Pearson
Compositor/Film Preparation: GTS Graphics
Printer: Von Hoffmann Press
Cover Designer: Yvo Riezebos

Copyright © 1997 by The Benjamin/Cummings Publishing
Company, Inc.

All rights reserved. No part of this publication may be
reproduced, stored in a retrieval system, or transmitted in any
form or by any means, electronic, mechanical, photocopying,
recording, or any other media embodiments now known or
hereafter to become known, without the prior written
permission of the publisher. Manufactured in the United
States of America. Published simultaneously in Canada.

Many of the designations used by manufacturers and sellers
to distinguish their products are claimed as trademarks.
Where those designations appear in this book, and the
publisher was aware of a trademark claim, the designations
have been printed in initial caps or all caps.

The Author and Publisher believe that the lab experiments
described in this publication, when conducted in conformity
with the safety precautions described herein and according to
the school's laboratory safety procedures, are reasonably safe
for the students to whom this manual is directed. Nonetheless,
many of the described experiments are accompanied by some
degree of risk, including human error, the failure or misuse of
laboratory or electrical equipment, mismeasurement, spills of
chemicals, and exposure to sharp objects, heat, bodily fluids,
blood, or other biologics. The Author and Publisher disclaim
any liability arising from such risks in connection with any of
the experiments contained in this manual. If students have any
questions or problems with materials, procedures, or instructions on any experiment, they should *always* ask their instructor for help before proceeding.

The Benjamin/Cummings Series in Human Anatomy and Physiology

Textbooks

By R. A. Chase
The Bassett Atlas of Human Anatomy (1989)

By S. W. Langjahr and R. D. Brister
Coloring Atlas of Human Anatomy, second edition (1992)

By E. N. Marieb
Human Anatomy and Physiology, third edition (1995)

Human Anatomy and Physiology, Study Guide,
third edition (1995)

Human Anatomy and Physiology Laboratory Manual,
fourth edition (1996)
Cat Version, fifth edition (1996)
Fetal Pig Version, fifth edition (1996)

Essentials of Human Anatomy and Physiology,
fifth edition (1997)

The A & P Coloring Workbook: A Complete Study Guide,
fifth edition (1997)

By E. N. Marieb and J. Mallatt
Human Anatomy, second edition (1997)

Videotapes by California State University, Sacramento

By R. L. Vines and A. Hinderstein
Human Musculature (1989)

By R. L. Vines and University Media Services
The Human Cardiovascular System: The Heart (1995)
The Human Cardiovascular System: The Blood Vessels
(1995)

By R. L. Vines and J. Barrena
*Selected Actions of Hormones and Other
Chemical Messengers* (1994)

By R. L. Vines, R. Carter, and University Media Services
*The Human Nervous System: The Brain and Cranial
Nerves* (1996)
*The Human Nervous System: The Spinal Cord and Spinal
Nerves* (1996)

ISBN 0-8053-4057-2

4 5 6 7 8 9 10—VH—00 99 98

Credits appear following Appendix C.

Cover printed with soy-based inks.

The Benjamin/Cummings Publishing Company, Inc.
2725 Sand Hill Road
Menlo Park, CA 94025

Table of Contents

The Circulatory System

The Respiratory System

The Digestive System

The Urinary System

The Reproductive System

Review Sheets

Preface to the Instructor

Each new edition of this book and others like it brings a matured sensibility for the way teachers teach and students learn. This sensibility is achieved through years of teaching the subject and by listening to the suggestions of other instructors as well as those of students enrolled in multifaceted health-care programs. The second edition of *Human Anatomy Laboratory Manual* has been developed to facilitate the anatomy laboratory experience for both teachers and students.

As with the previous edition of this manual, this edition is intended for students in introductory human anatomy courses. This manual presents a wide range of anatomy laboratory experiences for students concentrating in nursing, physical therapy, occupational therapy, respiratory therapy, dental hygiene, pharmacology, health and physical education, as well as biology and premedical programs.

This manual studies anatomy of the human specimen in particular, but the cat and isolated animal organs are used in the dissection experiments.

ORGANIZATION

The organization and scope of this manual lend themselves to use in the one-term human anatomy course. The variety of anatomical studies enables instructors to gear their courses to specific academic programs, or to their own teaching preferences. Although the textbook *Human Anatomy, second edition* (Marieb and Mallatt, Benjamin/Cummings, 1997), provided the impetus for its preparation, the manual is largely based upon exercises developed for use independent of any textbook. It contains all the background discussion and terminology necessary to perform all manipulations effectively. This eliminates the need for students to bring a textbook into the laboratory.

The manual is comprehensive and balanced enough to be flexible, and carefully written so that students can successfully complete each of its 29 exercises with little supervision. The manual begins with an exercise on anatomical terminology and an orientation to the human body. The second exercise provides the necessary tools for studying the various body systems. Succeeding exercises on the microscope, the cell, and tissues lay the groundwork for a study of each body system from the cellular to the organ level. Exercises 6 through 29 explore the anatomy of the organ systems in detail.

Homeostasis is continually emphasized as a requirement for optimal health. Pathological conditions are viewed as a loss of homeostasis; these discussions can be recognized by the homeostasis imbalance logo ⚕ within the descriptive material of each exercise. This holistic approach encourages an integrated understanding of the human body.

NEW FEATURES

In an effort to create a laboratory resource to accompany *Human Anatomy* (the text), the first edition of this manual was prepared by using *Human Anatomy and Physiology Laboratory Manual, Cat Version, fourth edition,* as a basis, and deleting much of the physiological content. Although the first edition was a valiant effort given time restraints and has been well received by users, it did not quite meet the full requirements of a stand-alone anatomy course because it lacked some anatomical detail and was still a bit heavy on physiology experiments. Although I firmly believe that anatomy is best taught and understood from a functional anatomy bias, I have really tried to respond to reviewers' and users' feedback concerning aspects they deem most important. Hopefully the changes listed next will successfully fulfill these needs.

1. These lab exercises or topics containing heavy physiological content have been deleted: former Exercise 5 (The Cell—Transport Mechanisms and Cell Permeability) and 23 (General Sensation), the total blood cell counts and the cholesterol determination in the blood lab, the immunity discussion and tests in the Lymphatic System exercise, and parts of the exercises concerning the special senses.

2. Muscle anatomy has been "beefed up" and now contains study sections on muscles of the hand and foot. Likewise, bone anatomy, especially that concerning skull markings, has received greater emphasis in this new edition. Anatomical terminology has been updated throughout to correspond with that in *Human Anatomy, second edition.*

3. The presentation of the pulmonary and systemic vasculature has been totally revised in Exercise 24. For nearly every circulation and subcirculation discussed, there is a new two-page table containing pertinent text and a minimum of two pieces of art to illustrate the vascular pathways—one employing the typical anatomical

art and the second a "plumbing" type diagram which allows the pathway to be traced more clearly.

4. The colored photomicrographs on the inside of the front and back covers and the bifold insert of the first edition have been replaced with a sixteen-page full-color insert of atlases as detailed in items 5 through 7.

5. With the addition of a new ten-page Histology Atlas, the number of colored histology plates in this edition totals 60. Great care has been taken to select histology photomicrographs that will be most helpful to students and correspond closely with what they will view in the laboratory. While most tissues are stained with hematoxylin and eosin (H&E), a few depicted in the Histology Atlas have special stains, such as the photomicrograph of a pancreatic islet of Langerhans that has been differentially stained to distinguish between its glucagon-producing alpha cells and its insulin-secreting beta cells. For some organs studied, a low-power view appears first to orient the student to the tissue before presenting anatomical detail of the high-power view. Line drawings, corresponding exactly to the colored plates in the Histology Atlas, appear in appropriate exercises in conjunction with text instructing students to view the histologic preparations. This inclusion adds immeasurably to the utility and effectiveness of the Histology Atlas. Since these diagrams can be colored by the student to replicate the stained slides they are viewing, the line art will be an excellent student learning aid. The edges of the Histology Atlas are colored blue so that the Atlas can be quickly located.

6. The Cat Anatomy Atlas (formerly in the bifold) now includes an additional ventral cavity photograph for a total of six unique color photographs of cat organ dissections. The edges of the Cat Anatomy Atlas are pink for easy recognition.

7. Two color photographs of a Beauchene skull are also new to this edition. The Beauchene skull is an excellent learning tool for students because it clearly shows the relationship of the bones in the skull. The views shown highlight the three-dimensional quality and unique shape of each bone. The Beauchene Skull Photo Atlas is located between the Histology Atlas and the Cat Anatomy Atlas.

8. Five icons, which are visual mnemonics to alert students to special instructions or a lab manual feature, appear in this edition. Although the dissection tray and homeostatic imbalance icons were also used in the first edition, all five icons are described.

 The dissection tray icon appears at the beginning of lab activities to be conducted by the students.

The homeostasis imbalance icon directs the student's attention to conditions representing a loss of homeostasis.

A blood and body fluid icon appears when blood or other body fluids such as saliva or urine must be handled and signifies where special self-protective measures are to be taken.

A safety icon notifies students that specific safety precautions are to be observed when using certain equipment or conducting particular lab procedures (for example, gloves should be worn when handling preserved specimens).

The cooperative learning icon is seen before experiment heads or procedures where learning would be enhanced and/or time saved by having students work together (in pairs or teams).

9. Appendix B correlates some of the required anatomical laboratory observations with the corresponding sections of the A.D.A.M. Standard CD-ROM software (available for purchase from Benjamin/Cummings).

The A.D.A.M. "walking man" icon indicates that the A.D.A.M. software would enhance the study and comprehension of the laboratory topics in a given exercise.

10. A listing of barcodes corresponding to images on *Anatomy and PhysioShow: The Videodisc* (available to qualified adopters) is now provided in Appendix C.

The video icon signifies that specific videodisc images can be useful pedagogically to both instructors (for demonstration) and students (for self-study and review) in a given laboratory exercise.

11. The laboratory review sheets have been expanded and now include more labeling exercises, particularly for the muscular system anatomy exercise.

SPECIAL FEATURES

The following features, carefully developed in the first edition to provide students with an interesting and challenging experience in the human anatomy laboratory, have been retained:

- Learning objectives are clearly written and precede each exercise.
- A materials list precedes each exercise as well. A complete materials list also appears in the instructor's guide.
- Key terminology appears in boldface type, and each term is defined when introduced.
- Illustrations are plentiful, large, and of exceptional quality and clarity. Structures are consistently shown in their natural anatomical positions within the body.
- The realism of several aspects of anatomy is enhanced by the use of additional full-color photographs and illustrations. Color is used pedagogically; that is, the structures and systems are color-coded to facilitate identification. Several exercises are embellished with full color. The art is borrowed from the textbook associated with this manual, *Human Anatomy, second edition* (Marieb and Mallatt, Benjamin/Cummings, 1997), and the photos were carefully selected from the Stanford Bassett Collection, also featured in the Marieb and Mallatt textbook.
- All laboratory instructions and procedures incorporate the latest precautions as recommended by the Centers for Disease Control (CDC); these are reinforced by the laboratory safety procedures described on the back of the front cover and in the front section of the instructor's guide. These procedures can be easily photocopied and posted in the lab.
- Clinical information is scattered throughout the exercises to point out how systems behave when structural abnormalities occur. It is identifiable by the symbol ⚕ near the margin.
- Tear-out Laboratory Review Sheets, keyed to the exercises, require students to label diagrams and answer multiple-choice, short-answer, and essay questions. Every attempt has been made to achieve an acceptable balance between explanatory and recognition questions.
- An Instructor's Guide, complete with directions for lab set-up, comments on chapter exercises, and answers to the lab manual questions, is also available.

TEXT SUPPLEMENTS AND MULTIMEDIA RESOURCES

Instructors can use help in ordering supplies, anticipating pitfalls and problem areas, and locating and using appropriate audiovisual material. To this end, several supplements, all mentioned below, are now available free of charge to qualified adopters.

Instructor's Guide

The *Instructor's Guide* that accompanies the *Human Anatomy Laboratory Manual* contains a wealth of information for those teaching this course. The guide includes directions for lab setup and answers to the lab manual questions. The probable in-class time required for each lab is indicated by a clock icon. Other useful resources are the Trends in Instrumentation section that describes the latest laboratory equipment and technological teaching tools available.

Multimedia Resources

It has become clear that several new trends are impacting anatomy studies, namely declining popularity of animal dissection exercises, the increased use of multimedia (e.g., videodiscs, computers, etc.) in the laboratory and, hence, the subsequent desire for more computerized exercises.

Among the specific changes newly implemented to address these trends are the following multimedia products.

- *Anatomy and PhysioShow: The Videodisc*
- Five videotapes developed at California State University, Sacramento, (available to qualified adopters)
  The Human Nervous System: The Brain and Cranial Nerves and *The Human Nervous System: The Spinal Cord and Spinal Nerves* videotapes by R. L. Vines, R. Carter, and University Media Services (1996).
  The Human Cardiovascular System: The Heart and *The Human Cardiovascular System: The Blood Vessels* videotape by R. L. Vines and University Media Services (1995).
  Human Musculature videotape by R. L. Vines and A. Hinderstein (1989)

A.D.A.M. Software

Available for purchase from The Benjamin/Cummings Publishing Company to enhance student learning are:

- A.D.A.M. Standard
- A.D.A.M.-Benjamin/Cummings Cardiovascular Physiology Module by Elaine Marieb and Marvin Branstrom
- A.D.A.M.-Benjamin/Cummings Muscle Physiology Module by Marvin Branstrom

ACKNOWLEDGMENTS

I wish to thank the following reviewers for their contributions to this edition: William Brothers (San Diego Mesa College), Caitlin Goodwin (University of Rhode Island), Jane Aloi (Saddleback College), and Gail Patt (Boston University).

My continued thanks to my colleagues at Benjamin/Cummings who helped me in the production of this edition, especially Lisa Moller, my new sponsoring editor, and Johanna Schmid, my "old" one who "guardian angeled" this work until Lisa was on board. Kudos also to Lisa Weber, production editor for this edition. Her precise approach to her work and her easygoing nature have made my job easy. Also not to be forgotten are Natasha Banta, assistant editor, for her diligent work; Thor Ekstrom, the assistant editor who handled hundreds of details (and "flogged" me daily to get this preface in); Thomas Viano, assistant editor, who helped with videodisc and A.D.A.M. correlations; and Kathleen Cameron, who is responsible for the photo research.

"Off campus" but equally important are Yvo Riezebos who conjured up the cover idea; Valerie Felts, Karl Miyajima, Kristin Otwell, and Denise Schmidt, who drew the new art pieces for this edition; and Anita Wagner, the very competent copy editor who is now an old hand at working on my books. Thanks also to Carlene Tonini, who catalogued the new barcodes in Appendix C and the A.D.A.M. correlations in Appendix B. Rose Leigh Vines deserves a hand for the significant roles she and the team from Sacramento State played in producing the excellent new videotapes on the human nervous system.

Finally, since preparations for the next edition begin well in advance of its publication, I invite users of this edition to send me their comments and suggestions for subsequent editions.

Elaine N. Marieb
Department of Science
Edison Community College
Collier County Campus
7007 Lely Cultural Parkway
Naples, FL 33962-8977

A Word to the Student

Hopefully, your laboratory experiences will be exciting times for you. Like any unfamiliar experience, it really helps if you know in advance what to expect and what will be expected of you.

LABORATORY ACTIVITIES

The laboratory exercises in this manual are designed to help you gain a broad understanding of anatomy. So, you can anticipate that you will be examining models, dissecting isolated animal organs, and using a microscope to look at tissue slides (anatomical approaches). You will also investigate a limited selection of physiological phenomena (conduct visual tests, plot sensory receptor locations, etc.) to make your anatomy studies more meaningful.

Because some students question the use of animals in the laboratory setting, their concerns need to be addressed. Be assured that the commercially available major dissection animals and preserved organ specimens used in the A&P labs are *not* harvested from animals raised specifically for dissection purposes. Instead, cats brought to the SPCA or other animal shelters for humane extermination are sent to qualified biological supply houses, where they are prepared for laboratory use according to USDA guidelines. Organs that are of no use to the meat packing industry (such as the brain, heart, or lungs) are sent from slaughterhouses to biological supply houses for preparation.

A.D.A.M. CORRELATIONS

If the A.D.A.M. CD-ROM software is available for your use, Appendix B correlating the various laboratory topics with specific frames of the A.D.A.M. software will provide an invaluable resource to help you in your studies.

BARCODES

The barcodes found in Appendix C will allow you to access topic-related images from the new *Anatomy and PhysioShow: The Videodisc* as you work in the laboratory (provided your classroom has the appropriate equipment).

ICONS/VISUAL MNEMONICS

I have tried to make this manual very easy for you to use, and to this end, seven different icons (visual mnemonics) are described below:

 The **dissection tray icon** appears at the beginning of lab activities. Since most exercises have some explanatory background provided before the experiment(s), this visual cue alerts you that your lab involvement is imminent.

The **blood/body fluid icon** appears where blood or other body fluids (saliva, etc.) must be handled, and it signifies that you should take special measures to protect yourself.

The **safety icon** alerts you to special precautions in handling lab equipment or conducting certain procedures, e.g., wearing gloves when handling preserved specimens.

The **cooperative learning icon** is seen before procedures where your learning will be enhanced (or time saved) if you work with a partner (or team).

The **homeostasis imbalance icon** appears where a clinical disorder is described to indicate what happens when there is a structural abnormality or physiological malfunction, i.e., a loss of homeostasis.

The **A.D.A.M. walking man icon** alerts you where the use of A.D.A.M. would enhance your laboratory experience.

The **video icon** indicates that there are images on *Anatomy and PhysioShow: The Videodisc* that correlate with laboratory observations you are being asked to make.

HINTS FOR SUCCESS IN THE LABORATORY

With the possible exception of those who have photographic memories, most students can use helpful hints and guidelines to ensure that they have successful lab experiences.

1. Perhaps the best bit of advice is to attend all your scheduled labs and to participate in all the assigned exercises. Learning is an *active* process.
2. Scan the scheduled lab exercise and the questions in the review section in the back of the manual that pertain to it *before* going to lab.
3. Be on time. Most instructors explain what the lab is about, pitfalls to avoid, and the sequence or format to be followed at the beginning of the lab session. If you are late, not only will you miss this information, you will not endear yourself to the instructor.
4. Review your lab notes after completing the lab session to help you focus on and remember the important concepts.
5. Keep your work area clean and neat. This reduces confusion and accidents.
6. Assume that all lab chemicals and equipment are sources of potential danger to you. Follow directions for equipment use and observe the laboratory safety guidelines provided inside the front cover of this manual.
7. Keep in mind the real value of the laboratory experience—a place for you to observe, manipulate, and experience "hands on" activities that will dramatically enhance your understanding of the lecture presentations.

I really hope that you enjoy your anatomy laboratories and that this lab manual makes learning about intricate structures and functions of the human body a fun and rewarding process. I'm always open to constructive criticism and suggestions for improvement in future editions. If you have any, please write to me.

Elaine N. Marieb
Department of Science
Edison Community College
Collier County Campus
7007 Lely Cultural Parkway
Naples, FL 33962-8977

The Language of Anatomy

OBJECTIVES

1. To describe the anatomical position verbally or by demonstration.
2. To use proper anatomical terminology to describe body directions, planes, and surfaces.
3. To name the body cavities and indicate the important organs in each.

MATERIALS

Human torso model (dissectible)
Human skeleton
Demonstration: sectioned and labeled kidneys (three separate kidneys uncut or cut so that (a) entire, (b) transverse section, and (c) longitudinal sectional views are visible)

See Appendix B, Exercise 1 for links to A.D.A.M. Standard.

See Appendix C, Exercise 1 for links to *Anatomy and PhysioShow: The Videodisc.*

Most of us have a natural curiosity about our bodies. This fact is amply demonstrated by infants, who early in life become fascinated with their own waving hands or their mother's nose. The study of the gross anatomy of the human body elaborates on this fascination. Unlike the infant, however, the student of anatomy must learn to identify and observe the dissectible body structures formally. The purpose of any gross-anatomy experience is to examine the three-dimensional relationships of body structures—a goal that can never completely be achieved by using illustrations and models, regardless of their excellence.

When beginning the study of any science, the student is often initially overcome by jargon unique to the subject. The study of anatomy is no exception. But without this specialized terminology, confusion is inevitable. For example, what do *over, on top of, superficial to, above,* and *behind* mean in reference to the human body? Anatomists have an accepted set of reference terms that are universally understood. These allow body structures to be located and identified with a minimum of words and a high degree of clarity.

This exercise presents some of the most important *anatomical terminology* used to describe the body and introduces you to basic concepts of **gross anatomy,** the study of body structures visible to the naked eye.

ANATOMICAL POSITION

When anatomists or doctors refer to specific areas of the human body, they do so in accordance with a univer-

sally accepted standard position called the **anatomical position.** It is essential to understand this position, because much of the body terminology employed in this book refers to this body positioning, regardless of the position the body happens to be in. In the anatomical position the human body is erect, with feet together, head and toes pointed forward, and arms hanging at the sides with palms facing forward (Figure 1.1).

● Assume the anatomical position, and note that it is not particularly comfortable. The hands are held unnaturally forward rather than hanging partially cupped toward the thighs.

SURFACE ANATOMY

Body surfaces provide visible landmarks for study of the body.

Anterior Body Landmarks

Note the following regions in Figure 1.2a:

Abdominal: pertaining to the anterior body trunk region inferior to the ribs
Antebrachial: pertaining to the forearm
Antecubital: pertaining to the anterior surface of the elbow
Axillary: pertaining to the armpit

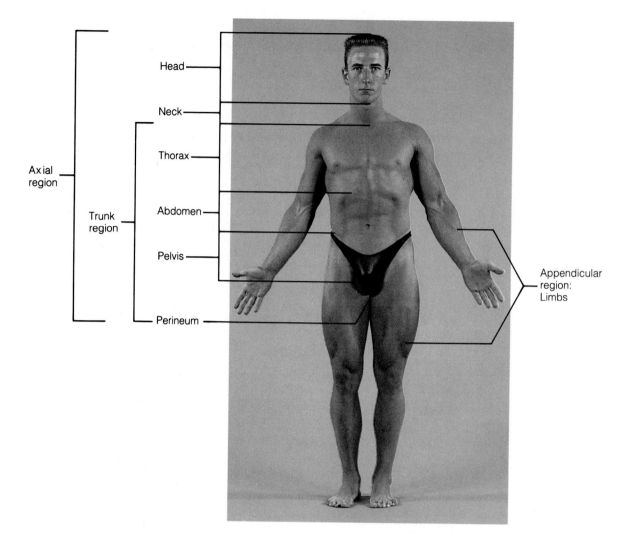

F1.1

Anatomical position.

Brachial: pertaining to the arm
Buccal: pertaining to the cheek
Carpal: pertaining to the wrist
Cervical: pertaining to the neck region
Coxal: pertaining to the hip
Crural: pertaining to the leg
Deltoid: pertaining to the roundness of the shoulder caused by the underlying deltoid muscle
Digital: pertaining to the fingers or toes
Femoral: pertaining to the thigh
Frontal: pertaining to the forehead
Hallux: pertaining to the great toe
Inguinal: pertaining to the groin
Mammary: pertaining to the breast
Manus: pertaining to the hand
Mental: pertaining to the chin

Nasal: pertaining to the nose
Oral: pertaining to the mouth
Orbital: pertaining to the bony eye socket (orbit)
Palmar: pertaining to the palm of the hand
Patellar: pertaining to the anterior knee (kneecap) region
Pedal: pertaining to the foot
Pelvic: pertaining to the pelvis region
Peroneal: pertaining to the side of the leg
Pollex: pertaining to the thumb
Pubic: pertaining to the genital region
Sternal: pertaining to the region of the breastbone
Tarsal: pertaining to the ankle
Thoracic: pertaining to the chest
Umbilical: pertaining to the navel

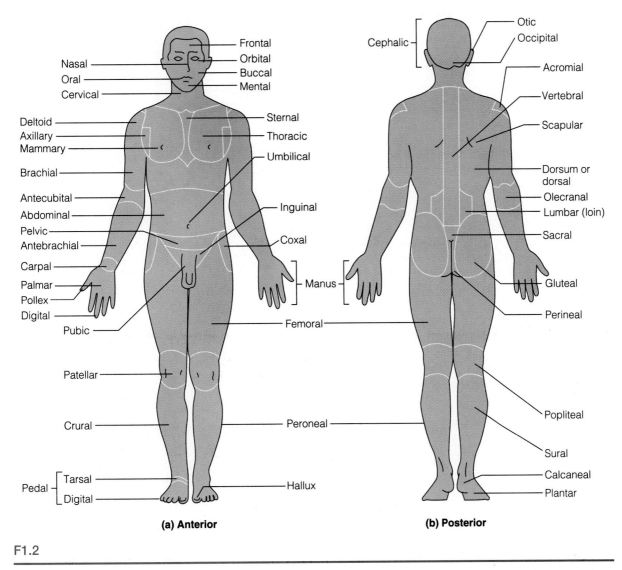

(a) Anterior   **(b) Posterior**

F1.2

Surface anatomy. (a) Anterior body landmarks. (b) Posterior body landmarks.

Posterior Body Landmarks

Note the following body surface regions in Figure 1.2b:

Acromial: pertaining to the point of the shoulder
Calcaneal: pertaining to the heel of the foot
Cephalic: pertaining to the head
Dorsum: pertaining to the back
Gluteal: pertaining to the buttocks or rump
Lumbar: pertaining to the area of the back between the ribs and hips; the loin
Occipital: pertaining to the posterior aspect of the head or base of the skull
Olecranal: pertaining to the posterior aspect of the elbow
Otic: pertaining to the ear
Perineal: pertaining to the region between the anus and external genitalia

Plantar: pertaining to the sole of the foot
Popliteal: pertaining to the back of the knee
Sacral: pertaining to the region between the hips (overlying the sacrum)
Scapular: pertaining to the scapula or shoulder blade area
Sural: pertaining to the calf or posterior surface of the leg
Vertebral: pertaining to the area of the spinal column

Locate the anterior and posterior body landmarks on yourself, your lab partner, and a torso model before continuing.

BODY ORIENTATION AND DIRECTION

Study the terms below, referring to Figure 1.3. Notice that certain terms have a different connotation for a four-legged animal than they do for a human.

Superior/inferior (*above/below*): These terms refer to placement of a structure along the long axis of the body. Superior structures always appear above other structures. For example, the nose is superior to the mouth, and the abdomen is inferior to the chest.

Anterior/posterior (*front/back*): In humans the most anterior structures are those that are most forward — the face, chest, and abdomen. Posterior structures are those toward the backside of the body. For instance, the spine is posterior to the heart.

Medial/lateral (*toward the midline/away from the midline or median plane*): The ear is lateral to the nose; the sternum (breastbone) is medial to the ribs.

The terms of position described above depend on an assumption of anatomical position. The next four term pairs are more absolute. Their applicability is not relative to a particular body position, and they consistently have the same meaning in all vertebrate animals.

Cephalad/caudal (*toward the head/toward the tail*): In humans these terms are used interchangeably with *superior* and *inferior*. But in four-legged animals they are synonymous with *anterior* and *posterior* respectively.

Dorsal/ventral (*backside/belly side*): These terms are used chiefly in discussing the comparative anatomy of animals, assuming the animal is standing. *Dorsum* is a Latin word meaning "back." Thus, *dorsal* refers to the backside of the animal's body or of any other structures; e.g., the posterior surface of the leg is its dorsal surface. The term *ventral* derives from the Latin term *venter*, meaning "belly," and always refers to the belly side of animals. In humans the terms *ventral* and *dorsal* are used interchangeably with the terms *anterior* and *posterior*, but in four-legged animals *ventral* and *dorsal* are synonymous with *inferior* and *superior* respectively.

Proximal/distal (*nearer the trunk or attached end/farther from the trunk or point of attachment*): These terms are used primarily to locate various areas of the body limbs. For example, the fingers are distal to the elbow; the knee is proximal to the toes.

Superficial/deep (*toward or at the body surface/away from the body surface or more internal*): These terms locate body organs according to their relative closeness to the body surface. For example, the lungs are deep to the rib cage, and the skin is superficial to the skeletal muscles.

 Before continuing, use a human torso model, a skeleton, or your own body to specify the relationship between the following structures. Use the correct anatomical terminology:

1. The wrist is _____ to the hand.

2. The trachea (windpipe) is _____ to the spine.

3. The brain is _____ to the spinal cord.

4. The kidneys are _____ to the liver.

5. The nose is _____ to the cheekbones.

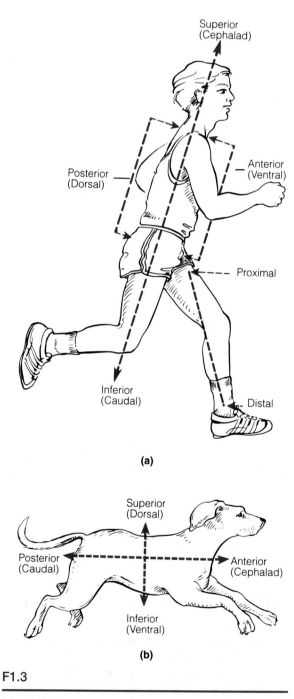

(a)

(b)

F1.3

Anatomical terminology describing body orientation and direction. (a) With reference to a human. **(b)** With reference to a four-legged animal.

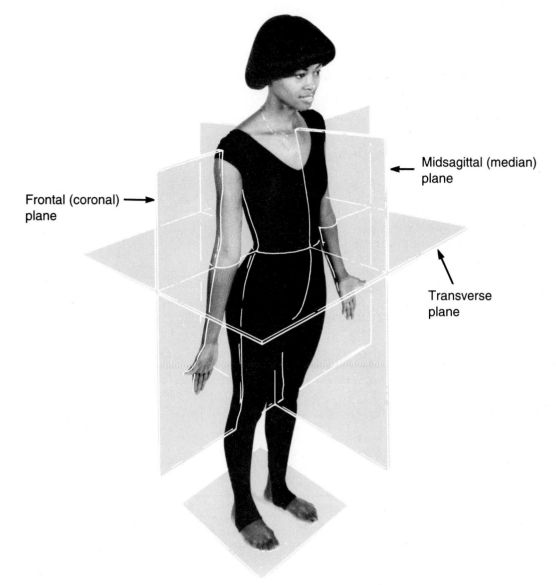

Frontal (coronal) plane

Midsagittal (median) plane

Transverse plane

F1.4

Planes of the body.

BODY PLANES AND SECTIONS

The body is three-dimensional and, in order to observe its internal structures, it is often helpful and necessary to make use of a **section,** or cut. When the section is made through the body wall or through an organ, it is made along an imaginary surface or line called a **plane.** Anatomists commonly refer to three planes (Figure 1.4) or sections which lie at right angles to one another.

Sagittal plane: A plane that runs longitudinally, dividing the body into right and left parts, is referred to as a sagittal plane. If it divides the body into equal parts, right down the median plane of the body, it is called a **midsagittal,** or **median, plane.** All other planes are referred to as **parasagittal planes.**

Frontal plane: Sometimes called a **coronal plane,** the frontal plane is a longitudinal plane that divides the body (or an organ) into anterior and posterior parts.

Transverse plane: A transverse plane runs horizontally, dividing the body into superior and inferior parts.

When organs are sectioned along the transverse plane, the sections are commonly called **cross sections.**

As shown in Figure 1.5, a sagittal or frontal plane section of an organ provides quite a different view than a transverse section.

Go to the demonstration area and observe the transversely and longitudinally cut organ specimens. Pay close attention to the different details of structure in the samples.

BODY CAVITIES

The body has two sets of cavities, which provide different degrees of protection to the organs within them (Figure 1.6).

5

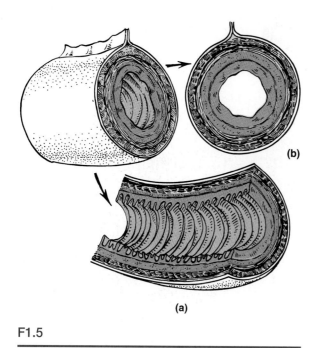

F1.5

Segment of the small intestine. (**a**) Cut longitudinally.
(**b**) Cut transversely.

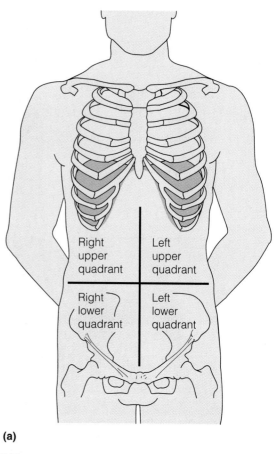

(**a**)

F1.7

Abdominopelvic surface and cavity. (**a**) The four
quadrants. (**b**) Nine regions delineated by four planes. The
superior horizontal plane is just inferior to the ribs; the
inferior horizontal plane is at the superior aspect of the hip
bones. The vertical planes are just medial to the nipples.
(**c**) Anterior view of the abdominopelvic cavity showing
superficial organs.

Dorsal Body Cavity

The dorsal body cavity can be subdivided into the **cranial cavity,** in which the brain is enclosed within the
rigid skull, and the **spinal cavity,** within which the delicate spinal cord is protected by the bony vertebral column. Because the cord is a continuation of the brain,
these cavities are continuous with each other.

Ventral Body Cavity

Like the dorsal cavity, the ventral body cavity is subdivided. The superior **thoracic cavity** is separated from
the rest of the ventral cavity by the dome-shaped diaphragm. The heart and lungs, located in the thoracic
cavity, are afforded some measure of protection by the
bony rib cage. The cavity inferior to the diaphragm is
often referred to as the **abdominopelvic cavity,** since
there is no further physical separation of the ventral cav-

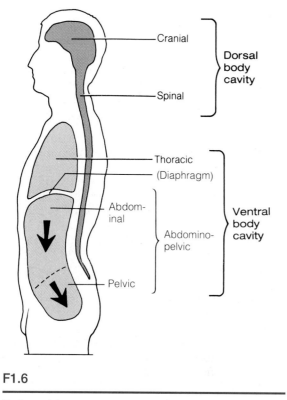

F1.6

Body cavities. Arrows indicate the angle of the relationship
between the abdominal and pelvic cavities.

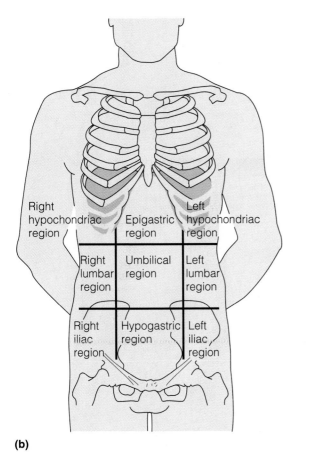

Right hypochondriac region

Epigastric region

Left hypochondriac region

Right lumbar region

Umbilical region

Left lumbar region

Right iliac region

Hypogastric region

Left iliac region

(b)

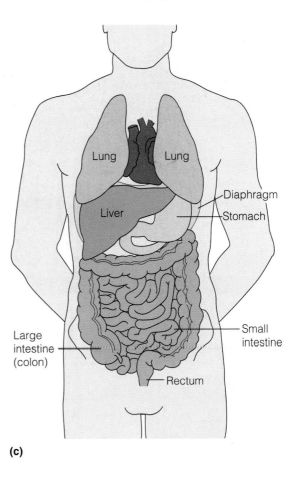

Lung

Lung

Diaphragm

Liver

Stomach

Large intestine (colon)

Small intestine

Rectum

(c)

F1.7 (*continued*)

ity. Some prefer to subdivide the abdominopelvic cavity into a superior **abdominal cavity,** which houses the stomach, intestines, liver, and other organs, and an inferior **pelvic cavity,** partially enclosed by the bony pelvis and containing the reproductive organs, bladder, and rectum. Notice in Figure 1.6 that the abdominal and pelvic cavities are not continuous with each other in a straight plane but that the pelvic cavity is tipped away from the perpendicular.

ABDOMINOPELVIC QUADRANTS AND RE-GIONS   Because the abdominopelvic cavity is quite large and contains many organs, it is helpful to divide it up into smaller areas for discussion or study. The scheme used by most physicians and nurses divides the abdominal surface (and the abdominopelvic cavity deep to it) into four approximately equal regions called **quadrants.** These quadrants are named according to their relative position—that is, *right upper quadrant, right lower quadrant, left upper quadrant,* and *left lower quadrant* (see Figure 1.7a).

Another scheme, commonly used by anatomists, divides the abdominal surface and abdominopelvic cavity into nine separate regions by four planes, as shown in Figure 1.7b. Although the names of these nine regions

are unfamiliar to you now, with a little patience and study they will become easier to remember. As you read through the descriptions of these nine regions below and locate them in the figure, notice the organs they contain by referring to Figure 1.7c.

Umbilical region: the centermost region, which includes the umbilicus
Epigastric region: immediately above the umbilical region; overlies most of the stomach
Hypogastric region: immediately below the umbilical region; encompasses the pubic area
Iliac regions: lateral to the hypogastric region and overlying the hip bones
Lumbar regions: between the ribs and the flaring portions of the hip bones
Hypochondriac regions: flanking the epigastric region and overlying the lower ribs

Locate the regions of the abdominal surface on a torso model and on yourself before continuing.

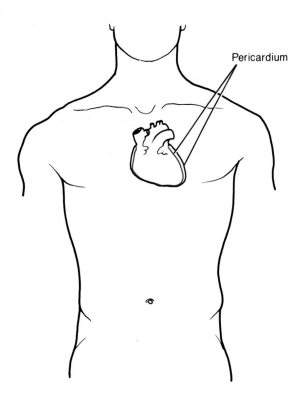

F1.8

Serous membranes. The visceral layer is shown in color; the parietal layer is shown in black.

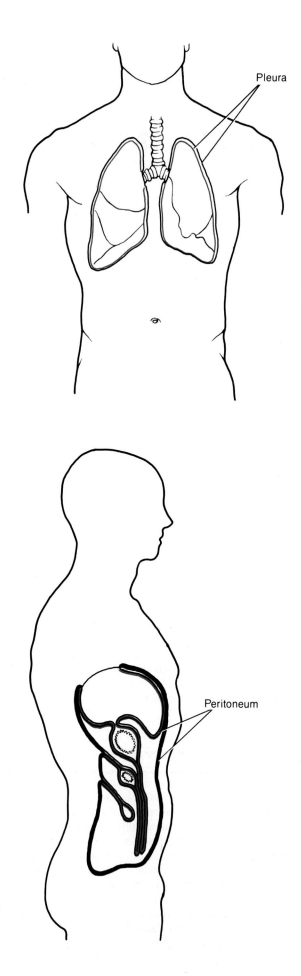

SEROUS MEMBRANES OF THE VENTRAL BODY CAVITY   The walls of the ventral body cavity and the outer surfaces of the organs it contains are covered with an exceedingly thin, double-layered membrane called the **serosa,** or **serous membrane.** The part of the membrane lining the cavity walls is referred to as the **parietal serosa,** and it is continuous with a similar membrane, the **visceral serosa,** covering the external surface of the organs within the cavity. These membranes produce a thin lubricating fluid that allows the visceral organs to slide over one another or to rub against the body wall without friction. Serous membranes also compartmentalize the various organs so that infection of one organ is prevented from spreading to others.

    The specific names of the serous membranes depend on the structures they envelop. Thus the serosa lining the abdominal cavity and covering its organs is the **peritoneum,** that enclosing the lungs is the **pleura,** and that around the heart is the **pericardium** (Figure 1.8).

Organ Systems Overview

OBJECTIVES

1. To name the human organ systems and indicate the major functions of each.
2. To list two or three organs of each system, and categorize the various organs by organ system.
3. To identify these organs in a dissected rat or on a dissectible human torso model or human cadaver.
4. To identify the correct organ system for each organ when presented with a list of organs (as studied in the laboratory).

MATERIALS

Freshly killed or preserved rat predissected by instructor as a demonstration or for student dissection (one for every two to four students), or dissected human cadaver
Dissecting pans and pins
Scissors
Forceps
Twine
Disposable plastic gloves
Human torso model (dissectible)

See Appendix B, Exercise 2 for links to A.D.A.M. Standard.

The basic unit or building block of all living things is the **cell.** Cells fall into four different categories according to their structures and functions. Each of these corresponds to one of the four **tissue** types: epithelial, muscular, nervous, and connective. An **organ** is a structure composed of two or more tissue types that performs a specific function for the body. For example, the small intestine, which digests and absorbs nutrients, is composed of all four tissue types. An **organ system** is a group of organs that act together to perform a particular body function. For example, the organs of the digestive system work together to ensure that food moving through the digestive system is properly broken down and that the end products are absorbed into the bloodstream to provide nutrients and fuel for all the body's cells. In all, there are eleven organ systems, which are described in Table 2.1. The lymphatic system also encompasses a *functional system* called the immune system, which is composed of an army of mobile *cells* that act to protect the body from foreign substances. Read through this summary before beginning the rat dissection.

RAT DISSECTION

Now you will have a chance to observe the size, shape, location, and distribution of the organs and organ systems. Many of the external and internal structures of the rat are quite similar in structure and function to those of the human, so a study of the gross anatomy of the rat should help you understand your own physical structure.

The following instructions have been written to complement and direct the student's dissection and observation of a rat, but the descriptions for organ observations from procedure 4 (p. 12) apply as well to super-

ficial observations of a previously dissected human cadaver. In addition, the general instructions for observing external structures can easily be extrapolated to serve human cadaver observations.

Note that four of the organ systems listed in Table 2.1 will not be studied at this time (integumentary, skeletal, muscular, and nervous), as they require microscopic study or more detailed dissection.

External Structures

1. If your instructor has provided a predissected rat, go to the demonstration area to make your observations. Alternatively, if you and/or members of your group will be dissecting the specimen, obtain a preserved or freshly killed rat (one for every two to four students), a dissecting pan, dissecting pins, scissors, forceps, and disposable gloves and bring them to your laboratory bench.

2. Don the gloves before beginning your observations. This precaution is particularly important when handling freshly killed animals, which may harbor internal parasites.

3. Observe the major divisions of the animal's body—head, trunk, and extremities. Compare these divisions to those of humans.

Oral Cavity

Examine the structures of the oral cavity. Identify the teeth and tongue. Observe the extent of the hard palate (the portion underlain by bone) and the soft palate (immediately posterior to the hard palate, with no bony support). Notice that the posterior end of the oral cavity leads into the throat, or pharynx. The pharynx is a passageway used by both the digestive and respiratory systems.

TABLE 2.1   Overview of Organ Systems of the Body

Organ system	Major component organs	Function
Integumentary (Skin)	Epidermal and dermal regions; cutaneous sense organs and glands	· Protects deeper organs from mechanical, chemical, and bacterial injury, and desiccation (drying out) · Excretes salts and urea · Aids in regulation of body temperature · Produces vitamin D
Skeletal	Bones, cartilages, tendons, ligaments, and joints	· Body support and protection of internal organs · Provides levers for muscular action · Cavities provide a site for blood cell formation
Muscular	Muscles attached to the skeleton	· Primary function is to contract or shorten; in doing so, skeletal muscles allow locomotion (running, walking, etc.), grasping and manipulation of the environment, and facial expression · Generates heat
Nervous	Brain, spinal cord, nerves, and sensory receptors	· Allows body to detect changes in its internal and external environment and to respond to such information by activating appropriate muscles or glands · Helps maintain homeostasis of the body via rapid transmission of electrical signals
Endocrine	Pituitary, thyroid, parathyroid, adrenal, and pineal glands; ovaries, testes, and pancreas	· Helps maintain body homeostasis, promotes growth and development; produces chemical "messengers" (hormones) that travel in the blood to exert their effect(s) on various "target organs" of the body
Cardiovascular	Heart, blood vessels, and blood	· Primarily a transport system that carries blood containing oxygen, carbon dioxide, nutrients, wastes, ions, hormones, and other substances to and from the tissue cells where exchanges are made; blood is propelled through the blood vessels by the pumping action of the heart · Antibodies and other protein molecules in the blood act to protect the body
Lymphatic	Lymphatic vessels, lymph nodes, spleen, thymus, tonsils, and scattered collections of lymphoid tissue	· Picks up fluid leaked from the blood vessels and returns it to the blood · Cleanses blood of pathogens and other debris · Houses lymphocytes that act via the immune response to protect the body from foreign substances (antigens)
Respiratory	Nasal passages, pharynx, larynx, trachea, bronchi, and lungs	· Keeps the blood continuously supplied with oxygen while removing carbon dioxide
Digestive	Oral cavity, esophagus, stomach, small and large intestines, rectum, and accessory structures (teeth, salivary glands, liver, and pancreas)	· Breaks down ingested foods to minute particles, which can be absorbed into the blood for delivery to the body cells · Undigested residue removed from the body as feces
Urinary	Kidneys, ureters, bladder, and urethra	· Rids the body of nitrogen-containing wastes (urea, uric acid, and ammonia), which result from the breakdown of proteins and nucleic acids by body cells · Maintains water, electrolyte, and acid-base balance of blood
Reproductive	Male: testes, scrotum, penis, and duct system, which carries sperm to the body exterior	· Provides germ cells (sperm) for perpetuation of the species
	Female: ovaries, uterine tubes, uterus, and vagina	· Provides germ cells (eggs) and the female uterus houses the developing fetus until birth

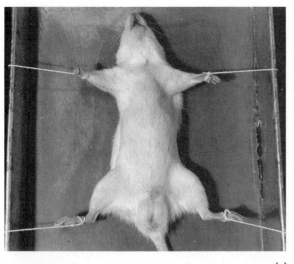

(a)

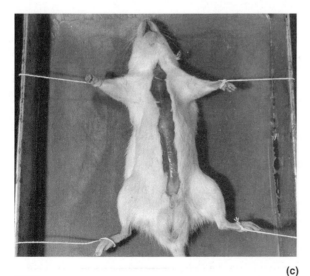

(b)

Ventral Body Cavity

1. Pin the animal to the wax of the dissecting pan by placing its dorsal side down and securing its extremities to wax. If the dissecting pan is not waxed, you will need to secure the animal with twine as shown in Figure 2.1a. (Some may prefer this method in any case.) Obtain the roll of twine. Make a loop knot around one upper limb, pass the twine under the pan, and secure the opposing limb. Repeat for the lower extremities.

2. Lift the abdominal skin with a forceps, and cut through it with the scissors (Figure 2.1b). Close the scissor blades and insert them under the cut skin. Moving in a cephalad direction, open and close the blades to loosen the skin from the underlying connective tissue and muscle. Once this skin-freeing procedure has been completed, cut the skin along the body midline, from the pubic region to the lower jaw (Figure 2.1c). Make a lateral cut about halfway down the ventral surface of each limb. Complete the job of freeing the skin with the scissor tips, and pin the flaps to the tray (Figure 2.1d). The underlying tissue that is now exposed is the skeletal musculature of the body wall and limbs. It allows voluntary body movement. Notice that the muscles are packaged in sheets of pearly white connective tissue (fascia), which protect the muscles and bind them together.

3. Carefully cut through the muscles of the abdominal wall in the pubic region, avoiding the underlying organs. Remember, *to dissect* means "to separate"—not mutilate! Now, hold and lift the muscle layer with a forceps and cut through the muscle layer from the pubic

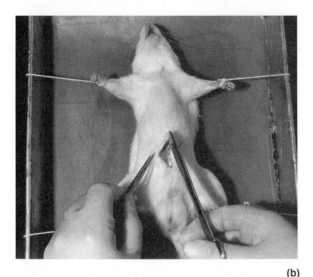

(c)

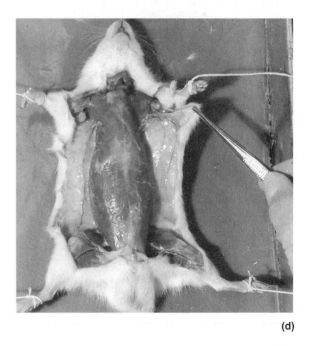

(d)

F2.1

Rat dissection: Securing for dissection and the initial incision. (a) Securing the rat to the dissection tray with twine. (b) Using scissors to make the incision on the median line of the abdominal region. (c) Completed incision from the pelvic region to the lower jaw. (d) Reflection (folding back) of the skin to expose the underlying muscles.

region to the bottom of the rib cage. Make two lateral cuts through the rib cage (Figure 2.2). A thin membrane attached to the inferior boundary of the rib cage should be obvious; this is the **diaphragm,** which separates the thoracic and abdominal cavities. Cut the diaphragm away to loosen the rib cage. You can now lift the ribs to view the contents of the thoracic cavity.

4.   Examine the structures of the thoracic cavity, starting with the most superficial structures and working deeper. As you work, refer to Figure 2.3, which shows the superficial organs.

Thymus: an irregular mass of glandular tissue overlying the heart.

Push the thymus to the side to view the heart.

Heart: median oval structure enclosed within the pericardium (serous membrane sac).
Lungs: flanking the heart on either side.

Now observe the throat region.

Trachea: tubelike "windpipe" running medially down the throat; part of the respiratory system.

Follow the trachea into the thoracic cavity; note where it divides into two branches. These are the bronchi.

Bronchi: two passageways that plunge laterally into the tissue of the two lungs.

To expose the esophagus, push the trachea to one side.

Esophagus: a food chute; the part of the digestive system that transports food from the pharynx (throat) to the stomach.

Follow the esophagus through the diaphragm to its junction with the stomach.

Stomach: a C-shaped organ important in food digestion and temporary food storage.

5.   Examine the superficial structures of the abdominopelvic cavity. Beginning with the stomach, trace the rest of the digestive tract.

Small intestine: connected to the stomach and ending just before a large saclike cecum.
Cecum: the initial portion of the large intestine.
Large intestine: a large muscular tube coiled within the abdomen.

Follow the course of the large intestine to the rectum, which is partially covered by the urinary bladder.

Rectum: terminal part of the large intestine; continuous with the anal canal.
Anus: the opening of the digestive tract (anal canal) to the exterior.

Now lift the small intestine with the forceps to view the mesentery.

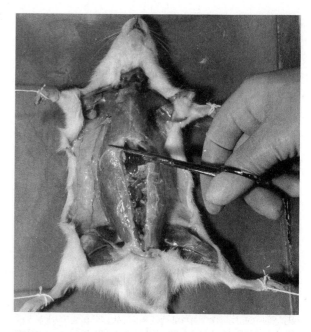

F2.2

Rat dissection: Making lateral cuts at the base of the rib cage.

Mesentery: an apronlike serous membrane; suspends many of the digestive organs in the abdominal cavity. Notice that it is heavily invested with blood vessels and, more likely than not, riddled with large fat deposits.

Locate the remaining abdominal structures.

Pancreas: a diffuse gland; rests dorsal to and in the mesentery between the first portion of the small intestine and the stomach.
Spleen: a dark red organ curving around the left lateral side of the stomach; considered part of the lymphatic system and often called the red blood cell graveyard.
Liver: large and brownish red; the most superior organ in the abdominal cavity, directly beneath the diaphragm.

6.   To locate the deeper structures of the abdominopelvic cavity, cut through the superior margin of the stomach and the distal end of the large intestine and lay them aside. (Refer to Figure 2.4 [p. 14] as you work.)

Examine the posterior wall of the abdominal cavity to locate the two kidneys.

Kidneys: bean-shaped organs; retroperitoneal (behind the peritoneum).
Adrenal glands: large glands that sit astride the superior margin of each kidney; considered part of the endocrine system.

Carefully strip away part of the peritoneum and attempt to follow the course of one of the ureters to the bladder.

Ureter: tube running from the indented region of a kidney to the urinary bladder.

Urinary bladder: the sac that serves as a reservoir for urine.

7.   In the midline of the body cavity lying between the kidneys are the two principal abdominal blood vessels. Identify each.

Inferior vena cava: the large vein that returns blood to the heart from the lower regions of the body.

Descending aorta: deep to the inferior vena cava; the largest artery of the body; carries blood away from the heart down the midline of the body.

8.   Only a cursory examination of reproductive organs will be done. First determine if the animal is a male or female. Observe the ventral body surface beneath the tail. If a saclike scrotum and a single body opening are visible, the animal is a male. If three body openings are present, it is a female. (See Figure 2.4.)

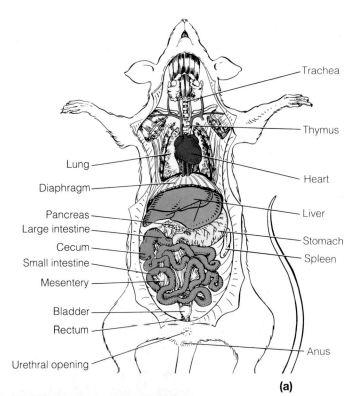

(a)

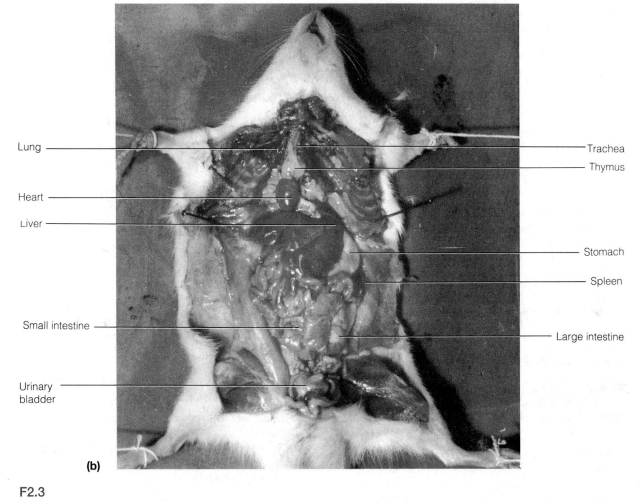

(b)

F2.3

Rat dissection: superficial organs of the thoracic and abdominal cavities. (a) Diagrammatic view. (b) Photograph.

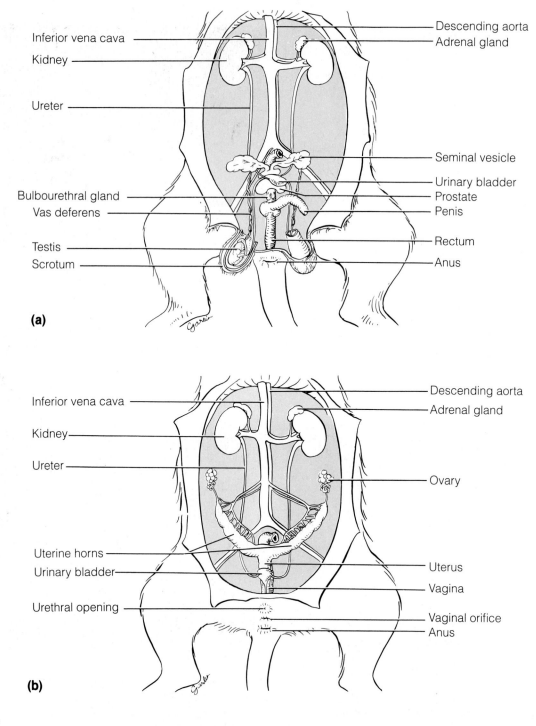

F2.4

Rat dissection: deeper organs of the abdominal cavity and the reproductive structures. (a) Male. **(b)** Female.

MALE ANIMAL   Make a shallow incision into the **scrotum.** Loosen and lift out the oval **testis.** Exert a gentle pull on the testis to identify the slender **vas deferens,** or sperm duct, which carries sperm from the testis superiorly into the abdominal cavity and joins with the urethra. The urethra runs through the penis of the male and carries both urine and sperm out of the body. Identify the **penis,** extending from the bladder to the ventral body wall. Figure 2.4a indicates other glands of the male reproductive system, but they need not be identified at this time.

FEMALE ANIMAL   Inspect the pelvic cavity to identify the Y-shaped **uterus** lying against the dorsal body wall and beneath the bladder (Figure 2.4b). Follow one of the uterine horns superiorly to identify an **ovary,** a small oval structure at the end of the uterine horn. (The rat uterus is quite different from the uterus of a human female, which is a single-chambered organ about the size and shape of a pear.) The inferior undivided part of the rat uterus is continuous with the vagina, which leads to the body exterior. Identify the **vaginal orifice** (external vaginal opening).

9. When you have finished your observations, store or dispose of the rat according to your instructor's directions. Wash the dissecting pan and tools with laboratory detergent. Dispose of the gloves. Then wash and dry your hands before continuing with the examination of the torso model.

EXAMINING THE HUMAN TORSO MODEL

Examine a human torso model to identify the organs listed below. (Note: If a torso model is not available, Figure 2.5 may be used for this part of the exercise.)

Dorsal body cavity: brain, spinal cord
Thoracic cavity: heart, lungs, bronchi, trachea, esophagus, diaphragm
Abdominopelvic cavity: liver, stomach, pancreas, spleen, small intestine, large intestine, rectum, kidneys, ureters, bladder, adrenal gland, descending aorta, inferior vena cava

As you observe these structures, locate the nine abdominopelvic areas studied earlier and determine which organs would be found in each area.

1. Umbilical region: _____

2. Epigastric region: _____

3. Hypogastric region: _____

4. Right iliac region: _____

5. Left iliac region: _____

6. Right lumbar region: _____

7. Left lumbar region: _____

8. Right hypochondriac region: _____

9. Left hypochondriac region: _____

Would you say that the shape and location of the human organs are similar or dissimilar to those of the rat?

_____

Assign each of the organs just identified to one of the organ system categories below.

Digestive: _____

Urinary: _____

Cardiovascular: _____

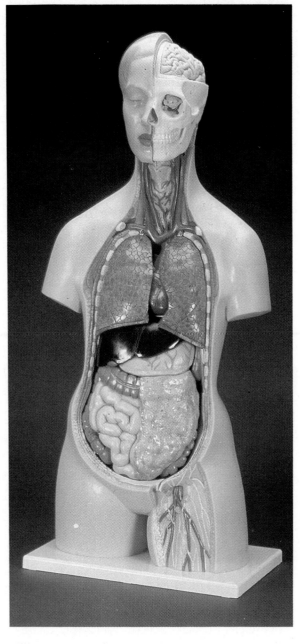

F2.5

Human torso model.

Reproductive: _____

Respiratory: _____

Lymphatic: _____

Nervous: _____

The Microscope

| OBJECTIVES | MATERIALS |

OBJECTIVES

1. To identify the parts of the microscope and list the function of each.

2. To describe and demonstrate the proper techniques for care of the microscope.

3. To define *total magnification* and *resolution*.

4. To demonstrate proper focusing technique.

5. To define *parfocal*, *field*, and *depth of field*.

6. To estimate the size of objects in a field.

MATERIALS

Compound microscope
Millimeter ruler
Prepared slides of the letter *e* or newsprint
Immersion oil
Lens paper
Prepared slide of grid ruled in millimeters (grid slide)
Prepared slide of 3 crossed colored threads
Clean microscope slide and coverslip
Toothpicks (flat-tipped)
Physiologic saline in a dropper bottle
Methylene blue stain (dilute) in a dropper bottle
Filter paper
Forceps
Beaker containing fresh 10% household bleach solution for wet mount disposal
Disposable autoclave bag

Note to the Instructor: The slides and coverslips used for viewing cheek cells are to be soaked for 2 hours (or longer) in 10% bleach solution and then drained. The slides, coverslips, and disposable autoclave bag (containing used toothpicks) are to be autoclaved for 15 min at 121°C and 15 pounds pressure to insure sterility. After autoclaving, the disposable autoclave bag may be discarded in any disposal facility and the glassware washed with laboratory detergent and reprepared for use. These instructions apply as well to any blood-stained glassware or disposable items used in other experimental procedures.

With the invention of the microscope, biologists gained a valuable tool to observe and study structures (like cells) that are too small to be seen by the unaided eye. As a result, many of the theories basic to the understanding of biologic sciences have been established. This exercise will familiarize you with the workhorse of microscopes—the compound microscope—and provide you with the necessary instructions for its proper use.

CARE AND STRUCTURE OF THE COMPOUND MICROSCOPE

The **compound microscope** is a precision instrument and should always be handled with care. At all times you must observe the following rules for its transport, cleaning, use, and storage:

- When transporting the microscope, hold it in an upright position with one hand on its arm and the other supporting its base. Avoid jarring the instrument when setting it down.

- Use only special grit-free lens paper to clean the lenses. Clean all lenses before and after use.

- Always begin the focusing process with the lowest-power objective lens in position, changing to the higher-power lenses as necessary.

- Never use the coarse adjustment knob with the high-power or oil immersion lenses.

- A coverslip must always be used with temporary (wet mount) preparations.

- Before putting the microscope in the storage cabinet, remove the slide from the stage, rotate the lowest-power objective lens into position, and replace the dust cover.

- Never remove any parts from the microscope; inform your instructor of any mechanical problems that arise.

1. Obtain a microscope and bring it to the laboratory bench. (Use the proper carrying technique!) Compare your microscope with the illustration in Figure 3.1 and identify the following microscope parts:

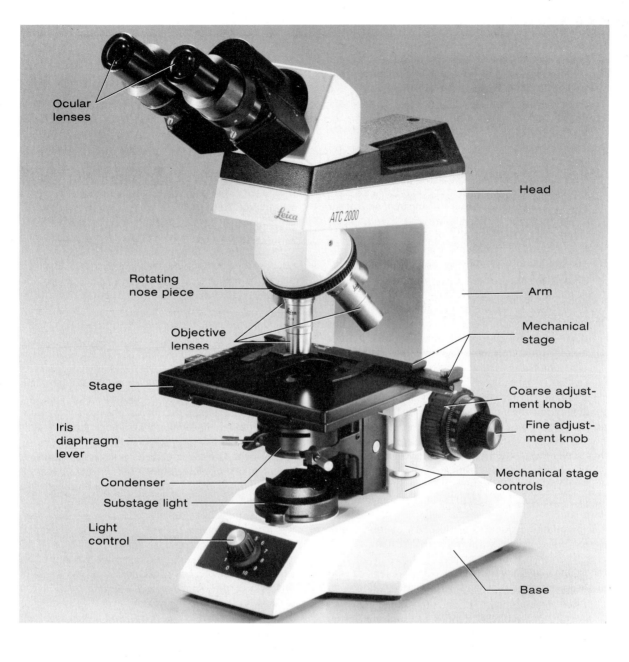

F3.1

Compound microscope and its parts.

Base: supports the microscope. (Note: Some microscopes are provided with an inclination joint, which allows the instrument to be tilted backward when viewing dry preparations.)

Substage light (or *mirror*): located in the base. In microscopes with a substage light source, the light passes directly upward through the microscope. If a mirror is used, light must be reflected from a separate free-standing lamp.

Stage: the platform the slide rests on while being viewed. The stage always has a hole in it to permit light to pass through both it and the specimen. Some microscopes have a stage equipped with *spring clips;* others have a clamp-type *mechanical stage* as shown in Figure 3.1. Both hold the slide in position for viewing; in addition, the mechanical stage permits precise movement of the specimen.

Condenser: concentrates the light on the specimen. The condenser may have a height-adjustment knob that raises and lowers the condenser to vary light delivery. Generally, the best position for the condenser is close to the inferior surface of the stage.

Iris diaphragm lever: arm attached to the condenser that regulates the amount of light passing through the condenser. The iris diaphragm permits the best possible contrast when viewing the specimen.

Coarse adjustment knob: used to focus the specimen.

Fine adjustment knob: used for precise focusing once coarse focusing has been completed.

Head or **body tube:** supports the objective lens system (which is mounted on a movable nosepiece), and the ocular lens or lenses.

Arm: vertical portion of the microscope connecting the base and head.

Ocular (or *eyepiece*): depending on the microscope, there will be one or two lenses at the superior end of the head or body tube. Observations are made through the ocular(s). An ocular lens has a magnification of 10× (it increases the apparent size of the object by ten times or ten diameters). If your microscope has a **pointer** (used to indicate a specific area of the viewed specimen), it is attached to the ocular and can be positioned by rotating the ocular lens.

Nosepiece: generally carries three objective lenses and permits sequential positioning of these lenses over the light beam passing through the hole in the stage.

Objective lenses: adjustable lens system that permits the use of a **low-power lens,** a **high-power lens,** or an **oil immersion lens.** The objective lenses have different magnifying and resolving powers.

2. Examine the objectives carefully, noting their relative lengths and the numbers inscribed on their sides. On most microscopes, the low-power (l.p.) objective is the shortest and generally has a magnification of 10×. The high-power (h.p.) objective is of intermediate length and has a magnification range from 40× to 50×, depending on the microscope. The oil immersion objective is usually the longest of the objectives and has a magnifying power of 95× to 100×. Note that some microscopes lack the oil immersion lens but have a very low magnification lens called the **scanning lens.** A scanning lens is a very short objective with a magnification of 4× to 5×. Record the magnification of each objective lens of your microscope in the first row of the chart that follows.

3. Rotate the low-power objective until it clicks into position, and turn the coarse adjustment knob about 180 degrees. Notice how far the stage (or objective) travels during this adjustment. Move the fine adjustment knob 180 degrees, noting again the distance that the stage (or the objective) moves.

Magnification and Resolution

The microscope is an instrument of magnification. In the compound microscope, magnification is achieved through the interplay of two lenses—the ocular lens and the objective lens. The objective lens magnifies the specimen to produce a **real image** that is projected to the ocular. This real image is magnified by the ocular lens to produce the **virtual image** seen by your eye (Figure 3.2).

The **total magnification** of any specimen being viewed is equal to the power of the ocular lens multiplied by the power of the objective lens used. For example, if the ocular lens magnifies 10× and the objective lens being used magnifies 45×, the total magnification is 450×.

Determine the total magnification you may achieve with each of the objectives on your microscope and record the figures on the second row of the chart. At this time, also cross out the column relating to a lens that your microscope does not have, and record the number of your microscope at the top of the chart.

The compound light microscope has certain limitations. Although the level of magnification is almost limitless, the **resolution** (or resolving power), the ability to discriminate two close objects as separate, is not. The human eye can resolve objects about 100 μm apart, but the compound microscope has a resolution of 0.2 μm under ideal conditions. Objects closer than 0.2 μm are seen as a single fused image.

Resolving power (RP) is determined by the amount and physical properties of the visible light that enters the microscope. In general, the greater the amount of light delivered to the objective lens, the greater the resolution. The size of the objective lens aperture (opening) decreases with increasing magnification, allowing less light to enter the objective. Thus, you will probably find it necessary to increase the light intensity at the higher magnifications.

Summary Chart for Microscope # _____

	Scanning		Low power		High power		Oil immersion	
Magnification of objective lens	×		×		×		×	
Total magnification	×		×		×		×	
Detail observed								
Field size (diameter)	mm	μm	mm	μm	mm	μm	mm	μm
Working distance			mm		mm		mm	

VIEWING OBJECTS THROUGH THE MICROSCOPE

1. Obtain a millimeter ruler, a prepared slide of the letter *e* or newsprint, a dropper bottle of immersion oil, and some lens paper. Adjust the condenser to its highest position and switch on the light source of your microscope. (If the light source is not built into the base, use the curved surface of the mirror to reflect the light up into the microscope.)

2. Secure the slide on the stage so that the letter *e* is centered over the light beam passing through the stage. If you are using a microscope with spring clips, make sure the slide is secured at both ends. If your microscope has a mechanical stage, open the jaws of its slide retainer (holder) by using the control lever (typically) located at the rear left corner of the mechanical stage. Insert the slide squarely within the confines of the slide retainer. Check to see that the slide is resting on the stage (and not on the mechanical stage frame) before releasing the control lever.

3. With your lowest power (scanning or low-power) objective in position over the stage, use the coarse adjustment knob to bring the objective and stage as close together as possible.

4. Look through the ocular and adjust the light for comfort. Now use the coarse adjustment knob to focus slowly away from the *e* until it is as clearly focused as possible. Complete the focusing with the fine adjustment knob.

5. Sketch the letter in the circle just as it appears in the **field** (the area you see through the microscope).

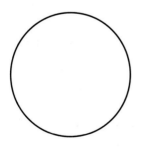

What is the total magnification? _____ ×

How far is the bottom of the objective from the specimen? In other words, what is the **working distance?**

_____ mm

Use a millimeter ruler to make this measurement and record it in the chart on page 18. How has the apparent orientation of the *e* changed top to bottom, right to left, and so on?

_____

_____

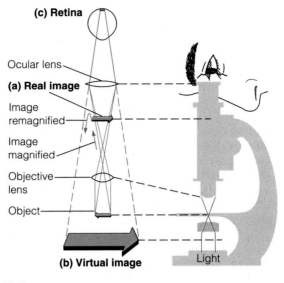

(c) Retina

Ocular lens

(a) Real image

Image remagnified

Image magnified

Objective lens

Object

(b) Virtual image

Light

F3.2

Image formation in light microscopy. Light passing through the objective lens forms a real image **(a)**. The real image serves as the object for the ocular lens, which remagnifies the image and forms the virtual image **(b)**. The virtual image passes through the lens of the eye and is focused on the retina **(c)**.

6. Move the slide slowly away from you on the stage as you view it through the ocular. In what direction does the image move?

_____

Move the slide to the left. In what direction does the image move?

_____

At first this change in orientation will confuse you, but with practice you will learn to move the slide in the desired direction with no problem.

7. Without touching the focusing knobs, increase the magnification by rotating the next higher magnification lens (low-power or high-power) into position over the stage. Using the fine adjustment only, sharpen the focus.* What new details become clear?

_____

_____

What is the total magnification now? _____ ×

* Today most good laboratory microscopes are parfocal; that is, the slide should be in focus (or nearly so) at the higher magnifications once you have properly focused in l.p. If you are unable to swing the objective into position without raising the objective, your microscope is not parfocal. Consult your instructor.

As best you can, measure the distance between the objective and the slide (the working distance) and record it on the chart.

Why should the coarse focusing knob *not* be used when focusing with the higher-powered objective lenses?

_____

_____

Is the image larger or smaller? _____

Approximately how much of the letter *e* is visible now?

_____

Is the field larger or smaller? _____

Why is it necessary to center your object (or the portion of the slide you wish to view) before changing to a higher power?

_____

Move the iris diaphragm lever while observing the field. What happens?

_____

Is it more desirable to increase *or* decrease the light when changing to a higher magnification?

_____ Why? _____

_____

8. If you have just been using the low-power objective, repeat the steps given in direction 7 using the high-power objective lens. Record the total magnification, approximate working distance, and information on detail observed on the chart on page 18.

9. Without touching the focusing knob, rotate the high-power lens out of position so that the area of the slide over the opening in the stage is unobstructed. Place a drop of immersion oil over the *e* on the slide and rotate the oil immersion lens into position. Set the condenser at its highest point (closest to the stage), and open the diaphragm fully. Adjust the fine focus and fine-tune the light for the best possible resolution.

Is the field again decreased in size? _____

What is the total magnification with the immersion lens?

_____ ×

Is the working distance less *or* greater than it was when the high-power lens was focused?

_____

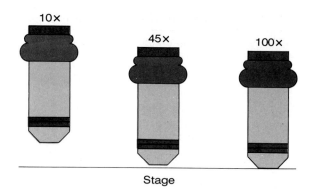

F3.3

Relative working distances of the 10×, 45×, and 100× objectives.

Compare your observations on the relative working distances of the objective lenses with the illustration in Figure 3.3. Explain why it is desirable to begin the focusing process in low power.

_____

_____

_____

10. Rotate the oil immersion lens slightly to the side and remove the slide. Clean the oil immersion lens carefully with lens paper and then clean the slide in the same manner with a fresh piece of lens paper.

DETERMINING THE SIZE OF THE MICROSCOPE FIELD

By this time you should know that the size of the microscope field decreases with increasing magnification. For future microscope work, it will be useful to determine the diameter of each of the microscope fields. This information will allow you to make a fairly accurate estimate of the size of the objects you view in any field. For example, if you have calculated the field diameter to be 4 mm and the object being observed extends across half this diameter, you can estimate the length of the object to be approximately 2 mm.

Microscopic specimens are usually measured in micrometers and millimeters, both units of the metric system. You can get an idea of the relationship and meaning of these units from Table 3.1. A more detailed treatment appears in Appendix A.

1. Return the letter *e* slide and obtain a grid slide, a slide prepared with graph paper ruled in millimeters. Each of the squares in the grid is 1 mm on each side. Use your lowest power objective to bring the grid lines into focus.

TABLE 3.1　Comparison of Metric Units of Length

Metric unit	Abbreviation	Equivalent
Meter	m	(about 39.3 in.)
Centimeter	cm	10^{-2} m
Millimeter	mm	10^{-3} m
Micrometer (or micron)	μm (μ)	10^{-6} m
Nanometer (or millimicrometer, or millimicron)	nm (mμ)	10^{-9} m
Angstrom	Å	10^{-10} m

2.   Move the slide so that one grid line touches the edge of the field on one side, and then count the number of squares you can see across the diameter of the field. If you can see only part of a square, as in the accompanying diagram, estimate the part of a millimeter that the partial square represents.

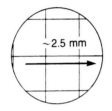

~2.5 mm

For future reference, record this figure in the appropriate space marked "field size" on the summary chart on page 18. (If you have been using the scanning lens, repeat the procedure with the low-power objective lens.) Complete the chart by computing the approximate diameter of the high-power and immersion fields. Say the diameter of the low-power field (total magnification of 50×) is 2 mm. You would compute the diameter of a high-power field with a total magnification of 100× as follows:

$$2 \text{ mm} \times 50 = Y \text{ (diameter of h.p. field)} \times 100$$
$$100 \text{ mm} = 100Y$$
$$1 \text{ mm} = Y \text{ (diameter of the h.p. field)}$$

The formula is:
Diameter of the l.p. field (mm) × Total magnification of the l.p. field = Diameter of field Y × Total magnification of field Y

3.   Estimate the length (longest dimension) of the following microscopic objects. *Base your calculations on the field sizes you have determined for your microscope.*

a. Object seen in low-power field:

approximate length:

————— mm.

b. Object seen in high-power field:

approximate length:

————— mm,

or ————— μm.

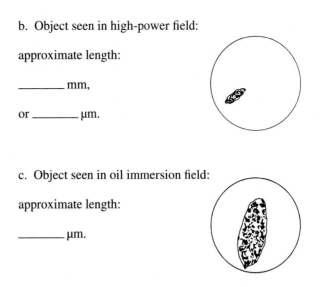

c. Object seen in oil immersion field:

approximate length:

————— μm.

4.   If an object viewed with the oil immersion lens looked like the field depicted just below, could you determine its approximate size from this view?

If not, then how could you determine it? —————

————————————————————————

————————————————————————

PERCEIVING DEPTH

Any microscopic specimen has depth as well as length and width; it is rare indeed to view a tissue slide with just one layer of cells. Normally you can see two or three cell thicknesses. Therefore, it is important to learn how to determine relative depth with your microscope.*

* In microscope work the **depth of field** (the depth of the specimen clearly in focus) is greater at lower magnifications.

1.   Return the grid slide and obtain a slide with colored crossed threads. Focusing at low magnification, locate the point where the three threads cross each other.

2.   Use the iris diaphragm lever to greatly reduce the light, thus increasing the contrast. Focus down with the coarse adjustment until the threads are out of focus, then slowly focus upward again, noting which thread comes into clear focus first. This one is the lowest or most inferior thread. (You will see two or even all three threads, so you must be very careful in determining which one comes into clear focus first.) Record your observations:

_____ thread over _____

Continue to focus upward until the uppermost thread is clearly focused. Again record your observation.

_____ thread over _____

Which thread is uppermost? _____

Lowest? _____

PREPARING AND OBSERVING A WET MOUNT

1.   Obtain the following: a clean microscope slide and coverslip, a flat-tipped toothpick, a dropper bottle of physiologic saline, a dropper bottle of methylene blue stain, forceps, and filter paper.

2.   Place a drop of physiologic saline in the center of the slide. Using the flat end of the toothpick, *gently* scrape the inner lining of your cheek. Agitate the end of the toothpick containing the cheek scrapings in the drop of saline (Figure 3.4a).

Immediately discard the used toothpick in the disposable autoclave bag provided at the supplies area.

3.   Add a tiny drop of the methylene blue stain to the preparation. (These epithelial cells are nearly transparent and thus difficult to see without the stain, which colors the nuclei of the cells and makes them look much darker than the cytoplasm.) Stir again and then dispose of the toothpick as described above.

4.   Hold the coverslip with the forceps so that its bottom edge touches one side of the fluid drop (Figure 3.4b), then *carefully* lower the coverslip onto the preparation (Figure 3.4c). *Do not just drop the coverslip,* or you will trap large air bubbles under it, which will obscure the cells. *A coverslip should always be used with a wet mount* to prevent soiling the lens if you should misfocus.

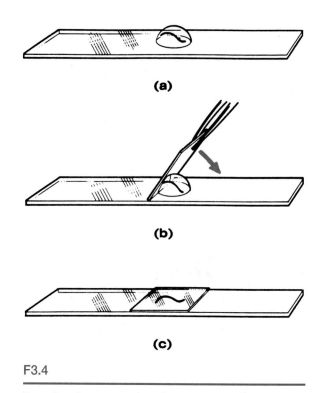

(a)

(b)

(c)

F3.4

Procedure for preparation of a wet mount. (a) The object is placed in a drop of water (or saline) on a clean slide, **(b)** a coverslip is held at a 45° angle with forceps, and **(c)** it is lowered carefully over the water and the object.

5.   Examine your preparation carefully. The coverslip should be closely apposed to the slide. If there is excess fluid around its edges, you will need to remove it. Obtain a piece of filter paper, fold it in half, and use the folded edge to absorb the excess fluid.

Before continuing, discard the filter paper in the disposable autoclave bag.

6.   Place the slide on the stage and locate the cells in low power. You will probably want to dim the light with the iris diaphragm to provide more contrast for viewing the lightly stained cells. Furthermore, a wet mount will dry out quickly in bright light, because a bright light source is hot.

7.   Cheek epithelial cells are very thin, six-sided cells. In the cheek, they provide a smooth, tilelike lining, as shown in Figure 3.5.

8.   Make a sketch of the epithelial cells that you observe.

Approximately how wide are the cheek epithelial cells?

_____ mm

Why do *your* cheek cells look different than those illustrated in Figure 3.5? (Hint: what did you have to *do* to your cheek to obtain them?)

_____

_____

9.   When you have completed your observations, dispose of your wet mount preparation in the beaker of bleach solution.

10.   Before leaving the laboratory, make sure all other materials are properly discarded or returned to the appropriate laboratory station. Clean the microscope lenses and put the dust cover on the microscope before you return it to the storage cabinet.

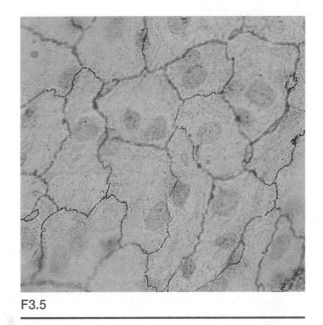

F3.5

Epithelial cells of the cheek cavity (surface view, 488×).

The Cell—Anatomy and Division

OBJECTIVES

1. To define *cell, organelle,* and *inclusion.*

2. To identify on a cell model or diagram the following cellular regions and to list the major function of each: nucleus, cytoplasm, and plasma membrane.

3. To identify and list the major functions of the various organelles studied.

4. To compare and contrast specialized cells with the concept of the "generalized cell."

5. To define *interphase, mitosis,* and *cytokinesis.*

6. To list the stages of mitosis and describe the events of each stage.

7. To identify the mitotic phases on projected slides or appropriate diagrams.

8. To explain the importance of mitotic cell division and its product.

MATERIALS

Three-dimensional model of the "composite" animal cell or laboratory chart of cell anatomy

Prepared slides of simple squamous epithelium (AgNO₃ stain), teased smooth muscle, human blood cell smear, and sperm

Compound microscope

Prepared slides of whitefish blastulae

Three-dimensional models of mitotic states

 See Appendix C, Exercise 4 for links to *Anatomy and PhysioShow: The Videodisc.*

Note to the Instructor: See directions for handling of toothpicks and wet mount preparations on p. 16.

The **cell,** defined as the structural and functional unit of all living things, is a very complex entity. The cells of the human body are highly diverse; and their differences in size, shape, and internal composition reflect their specific roles in the body. Nonetheless cells do have many common anatomical features, and there are some functions that all must perform to sustain life. For example, all cells have the ability to maintain their boundaries, to metabolize, to digest nutrients and dispose of wastes, to grow and reproduce, to move, and to respond to a stimulus. Most of these functions are considered in detail in later exercises. This exercise focuses on structural similarities that typify the "composite," or "generalized," cell and considers only the function of cell reproduction (cell division).

ANATOMY OF THE COMPOSITE CELL

In general, all cells have three major regions, or parts, that can readily be identified with a light microscope: the **nucleus,** the **plasma membrane,** and the **cytoplasm.** The nucleus is usually seen as a round or oval structure near the center of the cell. It is surrounded by cytoplasm, which in turn is enclosed by the plasma membrane. Since the advent of the electron microscope, even smaller cell structures—organelles—have been

identified. Figure 4.1a is a diagrammatic representation of the fine structure of the composite cell; Figure 4.1b depicts cellular structure as revealed by the electron microscope.

Nucleus

The nucleus is often described as the control center of the cell and is necessary for cell reproduction. A cell that has lost or ejected its nucleus (for whatever reason) is literally programmed to die because the nucleus is the site of the "genes," or genetic material—DNA.

When the cell is not dividing, the genetic material is loosely dispersed throughout the nucleus in a thread-like form called **chromatin.** When the cell is in the process of dividing to form daughter cells, the chromatin coils and condenses to form dense, darkly staining rodlike bodies called **chromosomes**—much in the way a stretched spring becomes shorter and thicker when it is released. (Cell division is discussed later in this exercise.) Notice the appearance of the nucleus carefully—it is somewhat nondescript when a cell is healthy. When the nucleus appears dark and the chromatin becomes clumped, this is an indication that the cell is dying and undergoing degeneration.

The nucleus also contains one or more small round bodies, called **nucleoli,** composed primarily of proteins and ribonucleic acid (RNA). The nucleoli are assembly sites for ribosomal particles (particularly abundant in

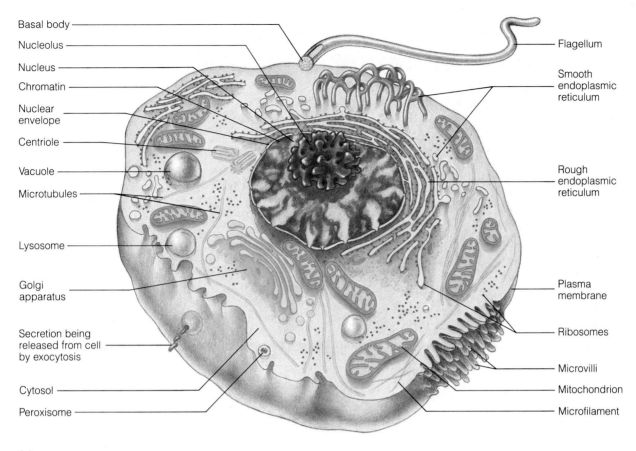

Basal body
Nucleolus
Nucleus
Chromatin
Nuclear envelope
Centriole
Vacuole
Microtubules
Lysosome
Golgi apparatus
Secretion being released from cell by exocytosis
Cytosol
Peroxisome

Flagellum
Smooth endoplasmic reticulum
Rough endoplasmic reticulum
Plasma membrane
Ribosomes
Microvilli
Mitochondrion
Microfilament

(a)

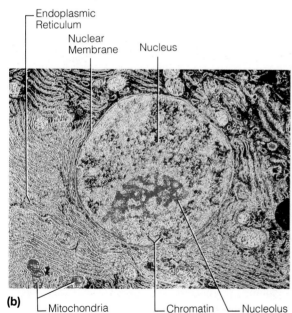

Endoplasmic Reticulum
Nuclear Membrane
Nucleus

(b) Mitochondria   Chromatin   Nucleolus

F4.1

Anatomy of the composite animal cell. (a) Diagrammatic view. **(b)** Transmission electron micrograph (10,000×).

Identify the nuclear membrane, chromatin, nucleoli, and the nuclear pores in Figure 4.1a and b.

Plasma Membrane

The **plasma membrane** separates cell contents from the surrounding environment. Its main structural building blocks are phospholipids (fats) and globular protein molecules, but some of the externally facing proteins and lipids have sugar (carbohydrate) side chains attached to them which are important in cellular interactions (Figure 4.2). Described by the fluid-mosaic model, the membrane appears to have a bimolecular lipid core that the protein molecules float in. Occasional cholesterol molecules dispersed in the fluid phospholipid bilayer help stabilize it.

Besides providing a protective barrier for the cell, the plasma membrane plays an active role in determining which substances may enter or leave the cell and in what quantity. Because of its molecular composition, the plasma membrane is selective about what passes

the cytoplasm), which are the actual protein synthesizing "factories."

The nucleus is bound by a double-layered porous membrane, the **nuclear membran**e (or nuclear envelope). The nuclear membrane is similar in composition to other cellular membranes, but it is distinguished by its large *nuclear pores*, which permit easy passage of protein and RNA molecules.

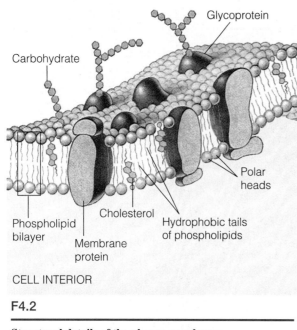

CELL INTERIOR

F4.2

Structural details of the plasma membrane.

through it. It allows nutrients to enter the cell but keeps out undesirable substances. By the same token, valuable cell proteins and other substances are kept within the cell, and excreta or wastes pass to the exterior. This property is known as **selective permeability.** Transport through the plasma membrane occurs in two basic ways. In *active transport,* the cell must provide energy (ATP) to power the transport process. In *passive transport,* the transport process is driven by concentration or pressure differences. Additionally, the plasma membrane maintains a resting potential that is essential to normal functioning of excitable cells, and plays a vital role in cell signaling and cell-to-cell interactions. In some cells the membrane is thrown into minute finger-like projections or folds called **microvilli,** which greatly increase the surface area of the cell available for absorption or passage of materials, and binding of signaling molecules.

 Identify the phospholipid and protein portions of the plasma membrane in Figure 4.2. Also locate the sugar side chains and cholesterol molecules. Identify the microvilli in Figure 4.1.

Cytoplasm and Organelles

The cytoplasm consists of the cell contents outside the nucleus. It is the major site of most activities carried out by the cell. Suspended in the **cytosol,** the fluid cytoplasmic material, are many small structures called **organelles** (literally, small organs). The organelles are the metabolic machinery of the cell, and they are highly organized to carry out specific functions for the cell as a whole. The organelles include the ribosomes, endoplas-

mic reticulum, Golgi apparatus, lysosomes, peroxisomes, mitochondria, cytoskeletal elements, and centrioles.

 Each organelle type is summarized in Table 4.1 and described briefly next. Read through this material and then, as best you can, locate the organelles in both Figure 4.1a and b.

• The **ribosomes** are densely staining spherical bodies composed of RNA and protein. They are the actual sites of protein synthesis. They are seen floating free in the cytoplasm or attached to a membranous structure. When they are attached, the whole ribosome-membrane complex is called the *rough endoplasmic reticulum.*

• The **endoplasmic reticulum** (**ER**) is a highly folded system of membranous tubules and cisternae (sacs) that extends throughout the cytoplasm. The ER is continuous with the Golgi apparatus, nuclear membrane, and plasma membrane. Thus it is assumed that the ER provides a system of channels for the transport of cellular substances (primarily proteins) from one part of the cell to another or to the cell exterior. The ER exists in two forms; a particular cell may have both or only one, depending on its specific functions. The **rough ER,** as noted earlier, is studded with ribosomes. Its cisternae modify and store the newly formed proteins and dispatch them to other areas of the cell. The external face of the rough ER is involved in phospholipid and cholesterol synthesis. The amount of rough ER is closely correlated with the amount of protein a cell manufactures and is especially abundant in cells that make protein products for export—for example, pancreas cells, which produce digestive enzymes destined for the small intestine. The **smooth ER** has no protein synthesis–related function but is present in conspicuous amounts in cells that produce steroid-based hormones—for example, the interstitial cells of the testes, which produce testosterone, and in cells that are highly active in lipid metabolism and drug detoxification activities—liver cells, for instance.

• The **Golgi apparatus** is a stack of flattened sacs with bulbous ends that is generally found close to the nucleus. Within its cisterns, the proteins delivered to it from the rough ER are modified (by attachment of sugar groups), segregated, and packaged into membranous vesicles that are ultimately incorporated into the plasma membrane, or become secretory vesicles that release their contents from the cell, or become lysosomes.

• The **lysosomes,** which appear in various sizes, are membrane-bound sacs containing an array of powerful digestive enzymes. A product of the packaging activities of the Golgi apparatus, the lysosomes contain *acid hydrolase* enzymes capable of digesting worn-out cell structures and foreign substances that enter the cell through engulfment processes (phagocytosis or pinocytosis). Lysosomes also bring about some of the changes

TABLE 4.1   Cytoplasmic Organelles

Organelle	Location and function
Ribosomes	Tiny spherical bodies composed of RNA and protein; actual sites of protein synthesis; floating free or attached to a membranous structure (the rough ER) in the cytoplasm
Endoplasmic reticulum (ER)	Membranous system of tubules that extends throughout the cytoplasm; two varieties: rough or granular ER—studded with ribosomes (tubules of the rough ER provide an area for storage and transport of the proteins made on the ribosomes to other cell areas; external face synthesizes phospholipids and cholesterol); smooth or agranular ER—no protein synthesis–related function (a site of steroid and lipid synthesis, lipid metabolism, and drug detoxification)
Golgi apparatus	Stack of flattened sacs with bulbous ends and associated small vesicles; found close to the nucleus; role in packaging proteins or other substances for export from the cell or incorporation into the plasma membrane and in packaging lysosomal enzymes
Lysosomes	Various-sized membranous sacs containing powerful digestive enzymes; function to digest worn-out cell organelles and foreign substances that enter the cell; since they have the capacity of total cell destruction if ruptured, referred to as "suicide sacs of the cell"
Peroxisomes	Small lysosome-like membranous sacs containing oxidase enzymes that detoxify alcohol, hydrogen peroxide, and other harmful chemicals
Mitochondria	Generally rod-shaped bodies with a double membrane wall; inner membrane is thrown into folds, or cristae; contain enzymes that oxidize foodstuffs to produce cellular energy (ATP); often referred to as "powerhouses of the cell"
Centrioles	Paired, cylindrical bodies lie at right angles to each other, close to the nucleus; direct the formation of the mitotic spindle during cell division; form the bases of cilia and flagella
Cytoskeletal elements: microtubules, intermediate filaments, and microfilaments	Provide cellular support; function in intracellular transport; microtubules form the internal structure of the centrioles and help determine cell shape; intermediate filaments, stable elements composed of a variety of proteins, resist mechanical forces acting on cells; microfilaments are formed largely of actin, a contractile protein, and thus are important in cell mobility (particularly in muscle cells)

that occur during menstruation, when the uterine lining is sloughed off. Since they have the capacity of total cell destruction, the lysosomes are often referred to as the "suicide sacs" of the cell.

● **Peroxisomes,** like lysosomes, are enzyme-containing sacs. However, their *oxidase* enzymes have a different task. Using oxygen, they detoxify a number of harmful substances, most importantly free radicals. Peroxisomes are particularly abundant in kidney and liver cells, cells that are actively involved in detoxification.

● The **mitochondria** are generally rod-shaped bodies with a double-membrane wall; the inner membrane is thrown into folds, or *cristae*. Oxidative enzymes on or within the mitochondria catalyze the reactions of the Krebs cycle and the electron transport chain (collectively called oxidative respiration), in which foods are broken down to produce energy. The released energy is captured in the bonds of ATP (adenosine triphosphate) molecules, which are then transported out of the mitochondria to provide a ready energy supply to power the cell. Every living cell requires a constant supply of ATP for its many activities. Since the mitochondria provide the bulk of this ATP, they are referred to as the powerhouses of the cell.

● The **cytoskeletal elements** ramify throughout the cytoplasm forming an internal scaffolding called the *cytoskeleton* that supports and moves substances within the cell. The **microtubules** are basically slender tubules formed of proteins called *tubulins* which have the ability to aggregate and then disaggregate spontaneously. Microtubules organize the cytoskeleton and direct formation of the spindle formed by the centrioles during cell division. They also act in the transport of substances down the length of elongated cells (such as neurons), suspend organelles, and help maintain cell shape by providing rigidity to the soft cellular substance. **Intermediate filaments** are *stable* proteinaceous cytoskeletal elements that act as internal guy wires to resist mechanical (pulling) forces acting on cells. **Microfilaments,** ribbon or cordlike elements, are formed of contractile proteins. Because of their ability to shorten and then relax to assume a more elongated form, these are important in cell mobility and are very conspicuous in cells that are highly specialized to contract (such as muscle cells). A cross-linked network of microfilaments braces and strengthens the internal face of the plasma membrane.

The cytoskeletal structures are labile and minute. With the exception of the microtubules of the spindle, which are very obvious during cell division (see pp. 30–31), and the microfilaments (myofilaments) of skeletal muscle cells (see p. 99), they are rarely seen, even in electron micrographs, and are not depicted in Figure 4.1b. However, special stains can reveal the plentiful supply of these very important organelles (see Plate 5 in the Histology Atlas).

- The paired **centrioles** lie close to the nucleus in all animal cells capable of reproducing themselves. They are rod-shaped bodies that lie at right angles to each other. Internally each centriole is composed of nine triplets of microtubules. During cell division, the centrioles direct the formation of the mitotic spindle. Centrioles also form the basis for cell projections called cilia and flagella (described below).

In addition to these cell structures, some cells have projections called **flagella,** which propel the cells, or **cilia,** which allow cells to sweep substances along a tract. Identify the cilium in Figure 4.1a.

The cell cytoplasm contains various other substances and structures, including stored foods (glycogen granules and lipid droplets), pigment granules, crystals of various types, water vacuoles, and ingested foreign materials. But these are not part of the active metabolic machinery of the cell and are therefore called **inclusions.**

Once you have located all of these structures in Figure 4.1a, examine the cell model (or cell chart) to repeat and reinforce your identifications.

OBSERVING DIFFERENCES AND SIMILARITIES IN CELL STRUCTURE

1. Obtain a compound microscope and prepared slides of simple squamous epithelium, sperm, smooth muscle cells (teased), and human blood.

2. Observe each slide under the microscope, carefully noting similarities and differences in the cells. (The oil immersion lens will be needed to observe blood and sperm.) Distinguish the limits of the individual cells, and notice the shape and position of the nucleus in each case. When you look at the human blood smear, direct your attention to the red blood cells, the pink-stained cells that are most numerous. The color photomicrographs illustrating a blood smear (Plate 55) and sperm (Plate 50) that appear in the Histology Atlas may be helpful in this cell structure study. Sketch your observations in the circles provided.

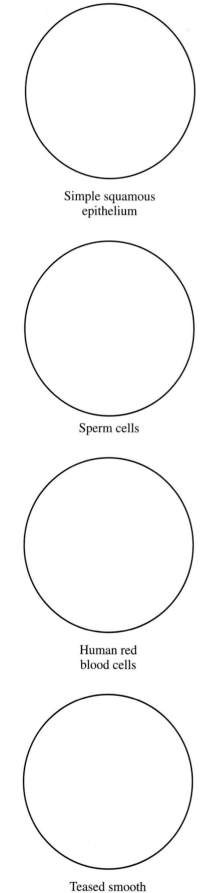

Simple squamous
epithelium

Sperm cells

Human red
blood cells

Teased smooth
muscle cells

3. How do these four cell types differ in shape and size?

_____

_____

_____

_____

_____

How might cell shape affect cell function?

_____

_____

_____

_____

_____

Which cells have visible projections?

_____

How do these projections relate to the function of this cell?

_____

_____

_____

Do any of these cells lack a cell membrane? _____

A nucleus? _____

In the cells with a nucleus, can you discern nucleoli?

_____

Were you able to observe any of the organelles in these

cells? _____Why or why not?

_____

_____

_____

CELL DIVISION: MITOSIS AND CYTOKINESIS

A cell's _life cycle_ is the series of changes it goes through from the time it is formed until it reproduces itself. It encompasses two stages—**interphase,** the longer period during which the cell grows and carries out its usual activities, and **cell division,** when the cell reproduces itself by dividing. In an interphase cell about to divide, the genetic material (DNA) is replicated (duplicated exactly). Once this important event has occurred, cell division ensues.

Cell division in all cells other than bacteria consists of a series of events collectively called mitosis and cytokinesis. **Mitosis** is nuclear division; **cytokinesis** is the division of the cytoplasm, which begins after mitosis is nearly complete. Although mitosis is usually accompanied by cytokinesis, in some instances cytoplasmic division does not occur, leading to the formation of binucleate (or multinucleate) cells. This is relatively common in the human liver and during embryonic development of skeletal muscle cells.

The process of **mitosis** results in the formation of two daughter nuclei that are genetically identical to the mother nucleus. This distinguishes mitosis from **meiosis,** a specialized type of nuclear division that occurs only in the reproductive organs (testes or ovaries). Meiosis, which yields four daughter nuclei that differ genetically and in composition from the mother nucleus, is used only for the production of eggs and sperm (gametes) for sexual reproduction. The function of cell division, including mitosis and cytokinesis in the body, is to increase the number of cells for growth and repair while maintaining their genetic heritage.

The stages of mitosis illustrated in Figure 4.3 include the following events:

Prophase (Figure 4.3b and c): At the onset of cell division, the chromatin threads coil and shorten to form densely staining, short, barlike **chromosomes.** By the middle of prophase the chromosomes appear as double-stranded structures (each strand is a **chromatid**) connected by a small median body called a **centromere.** The centrioles separate from one another and act as focal points for the assembly of two systems of microtubules, the **mitotic spindle** which forms between the centrioles and the **asters** which radiate outward from the ends of the spindle and anchor it to the plasma membrane. Some of the spindle fibers, the _kinetochore fibers_, attach to special protein complexes on each chromosome's centromere. Spindle fibers that do not attach to the chromosomes are called _polar fibers_. The spindle acts as a scaffolding for the attachment and movement of the chromosomes during later mitotic stages. Meanwhile, the nuclear membrane and the nucleolus break down and disappear.

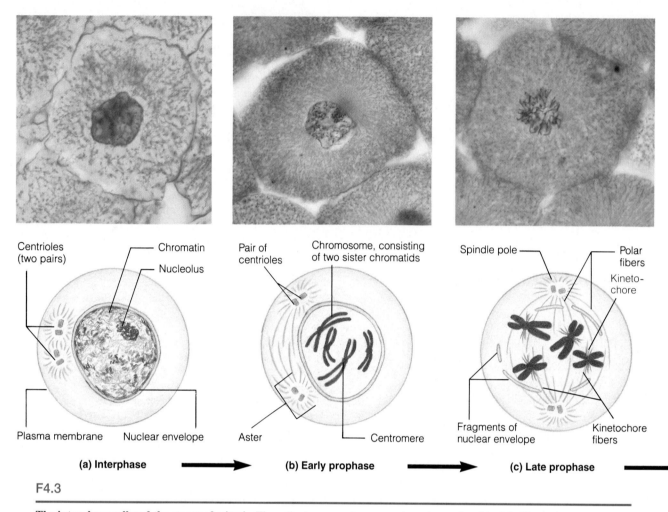

F4.3

The interphase cell and the stages of mitosis. The cells shown are from an early embryo of a whitefish. Photomicrographs are above; corresponding diagrams are below. (Micrographs approximately 600×.)

Metaphase (Figure 4.3d): A brief stage, during which the chromosomes migrate to the central plane or equator of the spindle and align along that plane in a straight line (the so-called *metaphase plate*) from the superior to the inferior region of the spindle (lateral view). Viewed from the poles of the cell (end view), the chromosomes appear to be arranged in a "rosette," or circle, around the widest dimension of the spindle.

Anaphase (Figure 4.3e): During anaphase, the centromeres split, and the chromatids (now called chromosomes again) separate from one another and then progress slowly toward opposite ends of the cell. The chromosomes are pulled by the kinetochore fibers attached to their centromeres, their "arms" dangling behind them. Anaphase is complete when poleward movement ceases.

Telophase (Figure 4.3f): During telophase, the events of prophase are essentially reversed. The chromosomes clustered at the poles begin to uncoil and resume the chromatin form, the spindle breaks down and disappears, a nuclear membrane forms around each chromatin mass, and nucleoli appear in each of the daughter nuclei.

Mitosis is essentially the same in all animal cells, but depending on the type of tissue, it takes from 5 minutes to several hours to complete. In most cells, centriole replication is deferred until interphase of the next cell cycle.

Cytokinesis, or the division of the cytoplasmic mass, begins during telophase (Figure 4.3f) and provides a good guideline for where to look for the mitotic figures of telophase. In animal cells, a *cleavage furrow* begins to form approximately over the equator of the spindle, and eventually splits or pinches the original cytoplasmic mass into two portions. Thus at the end of cell division two daughter cells exist, each smaller in cyto-

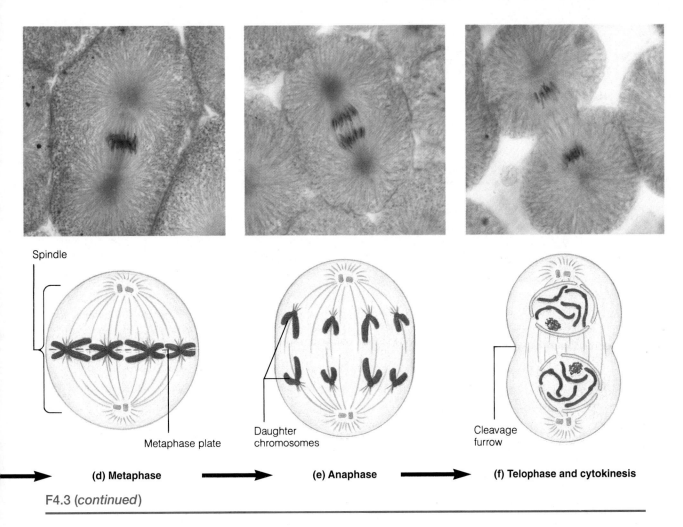

Spindle

Metaphase plate

(d) Metaphase

Daughter
chromosomes

(e) Anaphase

Cleavage
furrow

(f) Telophase and cytokinesis

F4.3 *(continued)*

plasmic mass than the mother cell but genetically identical to it. The daughter cells grow and carry out the normal spectrum of metabolic processes until it is their turn to divide.

Cell division is extremely important during the body's growth period. Most cells (excluding nerve cells) undergo mitosis until puberty, when normal body size is achieved and overall body growth ceases. After this time in life, only certain cells routinely carry out cell division—for example, cells subjected to abrasion (epithelium of the skin and lining of the gut). Other cell populations—such as liver cells—stop dividing but retain this ability should some of them be removed or damaged. Skeletal muscle, cardiac muscle, and nervous tissue completely lose this ability to divide and thus are severely handicapped by injury. Throughout life, the body retains its ability to repair cuts and wounds and to replace some of its aged cells.

Obtain a prepared slide of whitefish blastulae to study the stages of mitosis. The cells of each *blastula* (a stage of embryonic development consisting of a hollow ball of cells) are at approximately the same mitotic stage, so it may be necessary to observe more than one blastula to view all the mitotic stages. The exceptionally high rate of mitosis observed in this tissue is typical of embryos, but if occurring in specialized tissues, it can be an indication of cancerous cells, which also have an extraordinarily high mitotic rate. Examine the slide carefully, identifying the four mitotic stages and the process of cytokinesis. Compare your observations with Figure 4.3, and verify your identifications with your instructor.

Classification of Tissues

OBJECTIVES

1. To name the four major types of tissues in the human body and the major subcategories of each.
2. To identify the tissue subcategories through microscopic inspection or inspection of an appropriate diagram or projected slide.
3. To state the location of the various tissue types in the body.
4. To state the general functions and structural characteristics of each of the four major tissue types.

MATERIALS

Compound microscope
Prepared slides of simple squamous, simple cuboidal, simple columnar, stratified squamous (nonkeratinized), stratified cuboidal, stratified columnar, pseudostratified ciliated columnar, and transitional epithelium
Prepared slides of mesenchyme; of adipose, areolar, reticular, and dense (both regular [tendon] and irregular [dermis]) connective tissues; of hyaline and elastic cartilage; of fibrocartilage; of bone (cross section); and of blood
Prepared slides of skeletal, cardiac, and smooth muscle (longitudinal sections)
Prepared slide of nervous tissue (spinal cord smear)

See Appendix C, Exercise 5 for links to *Anatomy and PhysioShow: The Videodisc.*

Exercise 4 describes cells as the building blocks of life and the all-inclusive functional units of unicellular organisms. But in higher organisms cells do not usually operate as isolated, independent entities. In humans and other multicellular organisms, cells depend on one another and cooperate to maintain homeostasis in the body.

With a few exceptions (parthenogenic organisms), even the most complex animal starts out as a single cell, the fertilized egg, which divides almost endlessly. The trillions of cells that result become specialized for a particular function; some become supportive bone, others the transparent lens of the eye, still others skin cells, and so on. Thus a division of labor exists, with certain groups of cells highly specialized to perform functions that benefit the organism as a whole. Cell specialization brings about great sophistication of achievement but carries with it certain hazards, because when a small specific group of cells is indispensable, any inability to function on its part can paralyze or destroy the entire body.

Groups of cells that are similar in structure and function are called **tissues.** The four primary tissue types—epithelium, connective tissue, nervous tissue, and muscle—have distinctive structures, patterns, and functions. The four primary tissues are further divided into subcategories, as described shortly.

To perform specific body functions, the tissues are organized into such **organs** as the heart, kidneys, and lungs. Most organs contain several representatives of the primary tissues, and the arrangement of these tissues determines the organ's structure and function. Thus **histology,** the study of tissues, complements a study of gross anatomy and provides the structural basis for a study of organ physiology.

The main objective of this exercise is to familiarize you with the major similarities and dissimilarities of the primary tissues, so that when the tissue makeup of an organ is described, you will be able to more easily understand (and perhaps even predict) the organ's major function. Because epithelium and some types of connective tissue will not be considered again, they are emphasized more than muscle, nervous tissue, and bone (a connective tissue), which are covered in more depth in later exercises.

EPITHELIAL TISSUE

Epithelial tissue, or **epithelium,** covers surfaces. For example, epithelium covers the external body surface (as the epidermis), lines its cavities and tubules, and generally marks off our "insides" from our outsides.

(a)

(b)

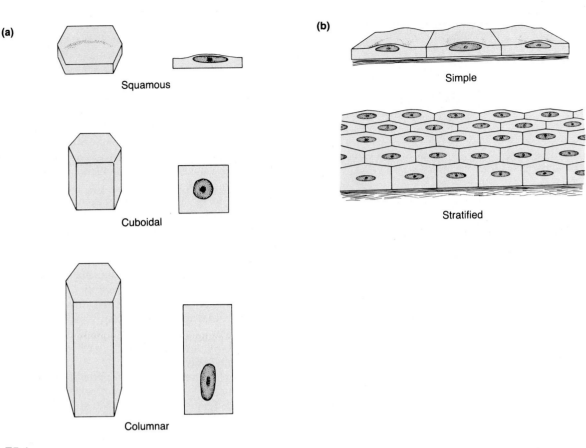

Squamous

Cuboidal

Columnar

Simple

Stratified

F5.1

Classification of epithelia. (a) Classification on the basis of cell shape. For each category, a whole cell is shown on the left and a longitudinal section is shown on the right. **(b)** Classification on the basis of arrangement (relative number of layers).

Since the various endocrine (hormone-producing) and exocrine glands of the body almost invariably develop from epithelial membranes, glands too are logically classed as epithelium.

Epithelial functions include protection, absorption, filtration, excretion, secretion, and sensory reception. For example, the epithelium covering the body protects against bacterial invasion and chemical damage; that lining the respiratory tract is ciliated to sweep dust and other foreign particles away from the lungs. Epithelium specialized to absorb substances lines the stomach and small intestine. In the kidney tubules, the epithelium absorbs, secretes, and filters. Secretion is a specialty of the glands.

Epithelium generally exhibits the following characteristics:

- Cells fit closely together to form membranes, or sheets of cells, and are bound together by specialized junctions.
- The membranes always have one free surface, called the *apical surface.*
- The cells are attached to an adhesive **basement membrane,** an amorphous material secreted partly by the epithelial cells (*basal lamina*) and connective tissue cells (*reticular lamina*) that lie adjacent to each other.

- Epithelial tissues have no blood supply of their own (are avascular), but depend on diffusion of nutrients from the underlying connective tissue.
- If well nourished, epithelial cells can easily regenerate themselves. This is an important characteristic because many epithelia are subjected to a good deal of friction.

The covering and lining epithelia are classified according to two criteria—cell shape and arrangement or relative number of layers (Figure 5.1). **Squamous** (scalelike), **cuboidal** (cubelike), and **columnar** (column-shaped) epithelial cells are the general types based on shape. On the basis of arrangement, there are **simple** epithelia, consisting of one layer of cells attached to the basement membrane, and **stratified** epithelia, consisting of more than one layer of cells. The terms denoting shape and arrangement of the epithelial cells are combined to describe the epithelium fully. *Stratified epithelia are named according to the cells at the apical surface of the epithelial membrane,* not those resting on the basement membrane.

There are, in addition, two less easily categorized types of epithelia. **Pseudostratified epithelium** is actually a simple columnar epithelium (one layer of cells), but because its cells extend varied distances from the basement membrane, it gives the false appearance of

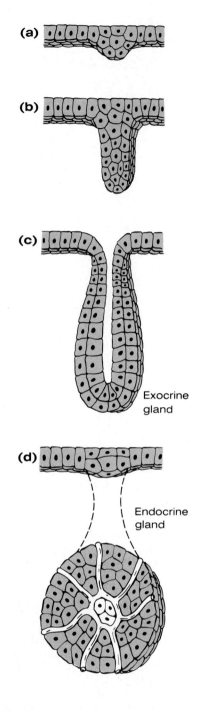

(a)

(b)

(c)

Exocrine
gland

(d)

Endocrine
gland

F5.2

Formation of endocrine and exocrine glands from epithelial sheets. (a) Epithelial cells grow and push into the underlying tissue. **(b)** A cord of epithelial cells forms. **(c)** In an exocrine gland, a lumen (cavity) forms. The inner cells form the duct, the outer cells produce the secretion. **(d)** In a forming endocrine gland, the connecting duct cells atrophy, leaving the secretory cells with no connection to the epithelial surface. However, they do become heavily invested with blood and lymphatic vessels that receive the secretions.

being stratified. This epithelium is often ciliated. **Transitional epithelium** is a rather peculiar stratified squamous epithelium formed of rounded, or "plump," cells with the ability to slide over one another to allow the organ to be stretched. Transitional epithelium is found only in urinary system organs subjected to periodic distension, such as the bladder. The superficial cells are flattened (like true squamous cells) when the organ is distended and rounded when the organ is empty.

Epithelial cells forming glands are highly specialized to remove materials from the blood and to manufacture them into new materials, which they then secrete. There are two types of glands, as shown in Figure 5.2. **Endocrine glands** lose their surface connection (duct) as they develop; thus they are referred to as ductless glands. Their secretions (all hormones) are extruded directly into the blood or the lymphatic vessels that weave through the glands. **Exocrine glands** retain their ducts, and their secretions empty through these ducts to an epithelial surface. The exocrine glands—including the sweat and oil glands, liver, and pancreas—are both external and internal. They will be discussed in conjunction with the organ systems to which their products are functionally related.

The most common types of epithelia, their most common locations in the body, and their functions are described in Figure 5.3.

 Obtain slides of simple squamous, simple cuboidal, simple columnar, stratified squamous (nonkeratinized), pseudostratified ciliated columnar, stratified cuboidal, stratified columnar, and transitional epithelia. Examine each carefully, and notice how the epithelial cells fit closely together to form intact sheets of cells, a necessity for a tissue that forms linings or covering membranes. Scan each epithelial type for modifications for specific functions, such as cilia (motile cell projections that help to move substances along the cell surface), and microvilli, which increase the surface area for absorption. Also be alert for goblet cells, which secrete lubricating mucus (see Plate 1 of the Histology Atlas). Compare your observations with the photomicrographs in Figure 5.3.

While working, check the questions in the laboratory review section for this exercise. A number of the questions there refer to some of the observations you are asked to make during your microscopic study.

CONNECTIVE TISSUE

Connective tissue is found in all parts of the body as discrete structures or as part of various body organs. It is the most abundant and widely distributed of the tissue types.

The connective tissues perform a variety of functions, but they primarily protect, support, and bind together other tissues of the body. For example, bones are composed of connective tissue (**bone** or **osseous tissue**),

(*Text continues on p. 39*)

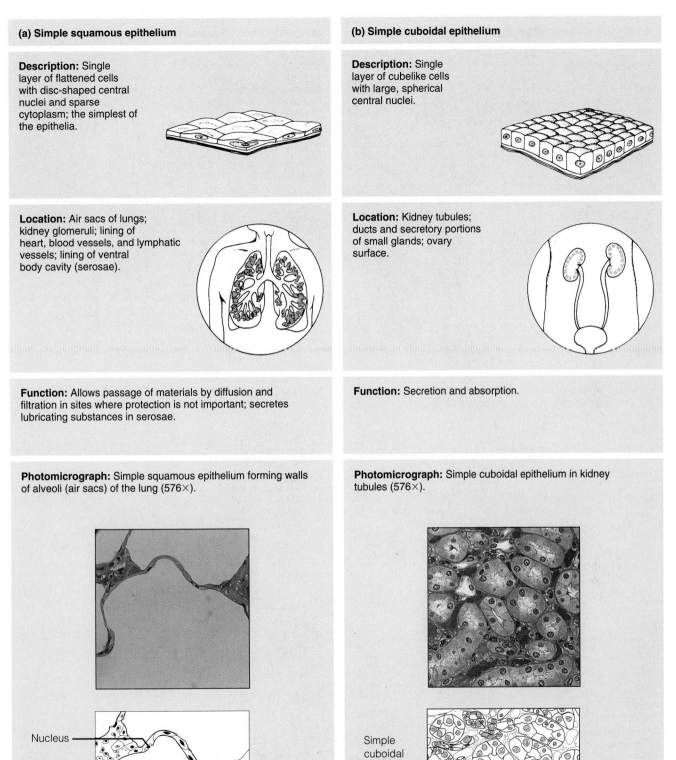

(a) Simple squamous epithelium

Description: Single layer of flattened cells with disc-shaped central nuclei and sparse cytoplasm; the simplest of the epithelia.

Location: Air sacs of lungs; kidney glomeruli; lining of heart, blood vessels, and lymphatic vessels; lining of ventral body cavity (serosae).

Function: Allows passage of materials by diffusion and filtration in sites where protection is not important; secretes lubricating substances in serosae.

Photomicrograph: Simple squamous epithelium forming walls of alveoli (air sacs) of the lung (576×).

Nucleus

Simple squamous epithelial cell

(b) Simple cuboidal epithelium

Description: Single layer of cubelike cells with large, spherical central nuclei.

Location: Kidney tubules; ducts and secretory portions of small glands; ovary surface.

Function: Secretion and absorption.

Photomicrograph: Simple cuboidal epithelium in kidney tubules (576×).

Simple cuboidal epithelial cells

Basement membrane

Connective tissue

F5.3

Epithelial tissues. Simple epithelia (a–b).

(c) Simple columnar epithelium

Description: Single layer of tall cells with *oval* nuclei; some cells bear cilia; layer may contain mucus-secreting glands (goblet cells).

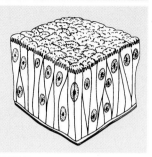

Location: Nonciliated type lines most of the digestive tract (stomach to anal canal), gallbladder and excretory ducts of some glands; ciliated variety lines small bronchi, uterine tubes, and some regions of the uterus.

Function: Absorption; secretion of mucus, enzymes, and other substances; ciliated type propels mucus (or reproductive cells) by ciliary action.

Photomicrograph: Simple columnar epithelium of the gallbladder mucosa (576×).

(d) Pseudostratified columnar epithelium

Description: Single layer of cells of differing heights, some not reaching the free surface; nuclei seen at different levels; may contain goblet cells and bear cilia.

Location: Nonciliated type in ducts of large glands, parts of male urethra; ciliated variety lines the trachea, most of the upper respiratory tract.

Function: Secretion, particularly of mucus; propulsion of mucus by ciliary action.

Photomicrograph: Pseudostratified ciliated columnar epithelium lining the human trachea (612×).

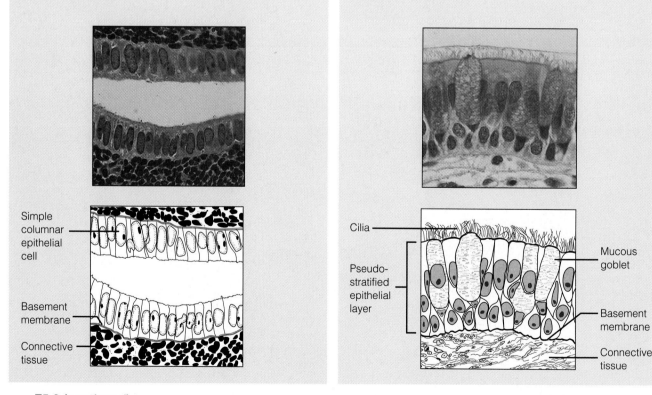

F5.3 (*continued*)

Simple epithelia (c and d).

(e) Stratified squamous epithelium

Description: Thick membrane composed of several cell layers; basal cells are cuboidal or columnar and metabolically active; surface cells are flattened (squamous); in the keratinized type, the surface cells are full of keratin and dead; basal cells are active in mitosis and produce the cells of the more superficial layers.

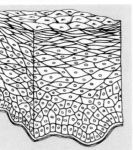

Location: Nonkeratinized type forms the moist linings of the esophagus, mouth, and vagina; keratinized variety forms the epidermis of the skin, a dry membrane.

Function: Protects underlying tissues in areas subjected to abrasion.

Photomicrograph: Stratified squamous epithelium lining of the esophagus (144×).

(f) Stratified cuboidal epithelium

Description: Generally two layers of cubelike cells.

Location: Largest ducts of sweat glands, mammary glands, and salivary glands.

Function: Protection.

Photomicrograph: Stratified cuboidal epithelium forming a salivary gland duct (86×).

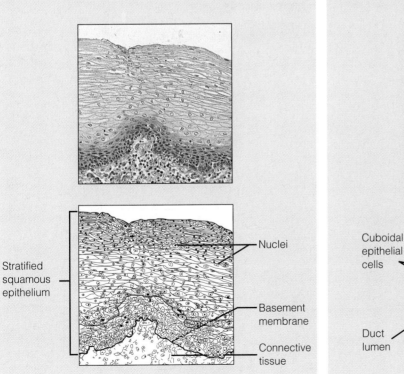

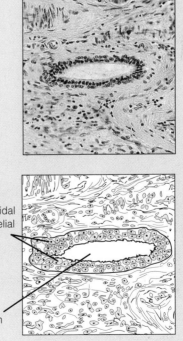

F5.3 (*continued*)

Stratified epithelia (e and f).

(g) Stratified columnar epithelium

Description: Several cell layers; basal cells usually cuboidal; superficial cells elongated and columnar.

Location: Rare in the body; small amounts in male urethra and in large ducts of some glands.

Function: Protection; secretion.

Photomicrograph: Stratified columnar epithelium lining of the male urethra (429×).

(h) Transitional epithelium

Description: Resembles both stratified squamous and stratified cuboidal; basal cells cuboidal or columnar; surface cells dome-shaped or squamous-like, depending on degree of organ stretch.

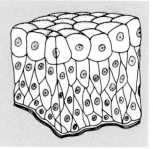

Location: Lines the ureters, bladder, and part of the urethra.

Function: Stretches readily and permits distension of urinary organ by contained urine.

Photomicrograph: Transitional epithelium lining of the bladder, relaxed state (252×); note the bulbous, or rounded, appearance of the cells at the surface; these cells flatten and become elongated when the bladder is filled with urine.

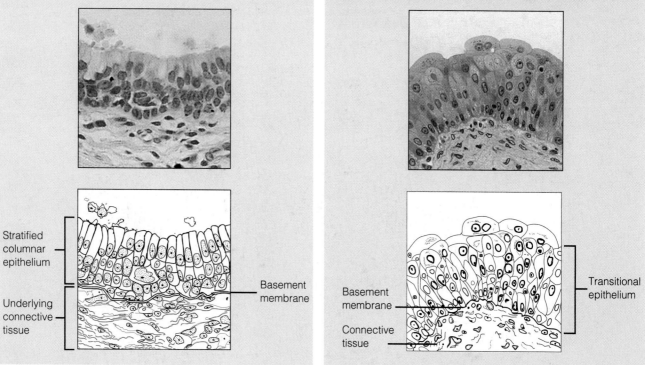

Stratified columnar epithelium

Underlying connective tissue

Basement membrane

Basement membrane

Connective tissue

Transitional epithelium

F5.3 (*continued*)

Stratified epithelia (g and h).

and they protect and support other body tissues and organs. The ligaments and tendons (**dense connective tissue**) bind the bones together or bind skeletal muscles to bones.

Areolar connective tissue is a soft packaging material that cushions and protects body organs. Fat (**adipose**) tissue provides insulation for the body tissues and a source of stored food. Blood-forming (**hematopoietic**) tissue replenishes the body's supply of red blood cells. In addition, connective tissue serves a vital function in the repair of all body tissues since many wounds are repaired by connective tissue in the form of scar tissue.

The characteristics of connective tissue include the following:

- With a few exceptions (cartilages, tendons, and ligaments), connective tissues are well vascularized.
- Connective tissues are composed of many types of cells.
- There is a great deal of noncellular, nonliving material (matrix) between the cells of connective tissue.

The nonliving material between the cells—the **extracellular matrix**—deserves a bit more explanation because it distinguishes connective tissue from all other tissues. It is produced by the cells and then extruded. The matrix is primarily responsible for the strength associated with connective tissue, but there is variation. At one extreme, adipose tissue is composed mostly of cells. At the opposite extreme, bone and cartilage have very few cells and large amounts of matrix.

The matrix has two components—ground substance and fibers. The **ground substance** is composed chiefly of glycoproteins and large charged polysaccharide molecules. Depending on its specific composition, the ground substance may be liquid, semisolid, gel-like, or very hard. When the matrix is firm, as in cartilage and bone, the connective tissue cells reside in cavities in the matrix called *lacunae*. The fibers, which provide support, include **collagenic** (white) **fibers, elastic** (yellow) **fibers,** and **reticular** (fine collagenic) **fibers.** Of these, the collagen fibers are most abundant.

Generally speaking, the ground substance functions as a molecular sieve, or medium through which nutrients and other dissolved substances can diffuse between the blood capillaries and the cells. The fibers in the matrix hinder diffusion somewhat and make the ground substance less pliable.

The properties of the connective tissue cells and the makeup and arrangement of their matrix elements vary tremendously, accounting for the amazing diversity of this tissue type. Nonetheless, the connective tissues have a common structural plan seen best in *areolar connective tissue* (Figure 5.4), a soft packing tissue that occurs throughout the body. Since all other connective tissues are variations of areolar, it is considered the model or prototype of the connective tissues. Notice in Figure 5.4 that areolar tissue has all three varieties of fibers, but they are sparsely arranged in its transparent gel-like

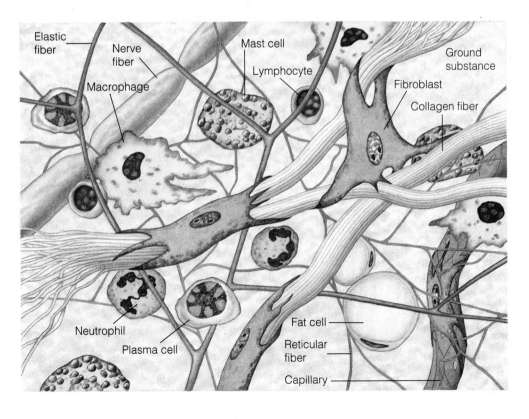

F5.4

Areolar connective tissue: A prototype (model) connective tissue. Note the various cell types and the three classes of fibers (collagen, reticular, elastic) embedded in the ground substance.

ground substance. The cell type that secretes its matrix is the *fibroblast,* but a wide variety of other cells including phagocytic cells like macrophages and certain white blood cells and mast cells which act in the inflammatory response are present as well. The more durable connective tissues, such as bone, cartilage, and the dense fibrous varieties, characteristically have a firm ground substance and many more fibers.

There are four main types of adult connective tissue, all of which typically have large amounts of matrix. These are connective tissue proper (which includes areolar, adipose, reticular, and dense [fibrous] connective tissues), cartilage, bone, and blood. All of these derive from an embryonic tissue called *mesenchyme.* Figure 5.5 lists the general characteristics, location, and function of some of the connective tissues found in the body.

 Obtain prepared slides of mesenchyme; of adipose, areolar, reticular, dense regular and irregular connective tissue; of hyaline and elastic cartilage and fibrocartilage; and of osseous connective tissue (bone). Compare your observations with the views in Figure 5.5.

Distinguish between the living cells and the matrix and pay particular attention to the denseness and arrangement of the matrix. For example, notice how the matrix of the dense fibrous connective tissues, making up tendons and the dermis of the skin, is chock-full of collagenic fibers, and that in the regular variety (tendon), the fibers are all running in the same direction, whereas in the dermis they appear to be running in many directions.

While examining the areolar connective tissue, a soft "packing tissue," notice how much empty space there appears to be, and distinguish between the collagen fibers and the coiled elastic fibers. Also, try to locate a **mast cell,** which has large darkly staining granules in its cytoplasm. This cell type releases histamine that makes capillaries quite permeable during inflammatory reactions and allergies and thus is partially responsible for that "runny nose" of allergies.

In adipose tissue, locate a cell (signet ring cell) in which the nucleus can be seen pushed to one side by the large fat-filled vacuole which appears to be a large empty space. Also notice how little matrix there is in fat or adipose tissue.

Distinguish between the living cells and the matrix in the dense fibrous, bone, and hyaline cartilage preparations.

Embryonic connective tissue

(a) Mesenchyme

Description: Embryonic connective tissue; gel-like ground substance containing fine fibers; star-shaped mesenchymal cells.

Location: Primarily in embryo.

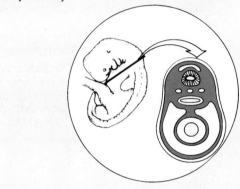

Function: Gives rise to all other connective tissue types.

Photomicrograph: Mesenchymal tissue, an embryonic connective tissue (1072×); the clear-appearing background is the fluid ground substance of the matrix; notice the fine, sparse fibers.

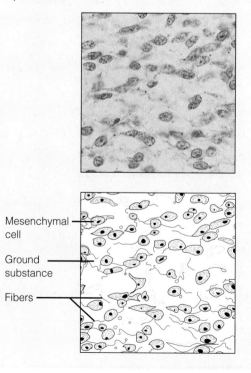

Mesenchymal cell

Ground substance

Fibers

F5.5

Connective tissues.

Connective tissue proper: Loose connective tissue (b to d)

(b) Areolar connective tissue

Description: Gel-like matrix with all three fiber types; cells: fibroblasts, macrophages, mast cells, and some white blood cells.

Location: Widely distributed under epithelia of body, e.g., forms lamina propria of mucous membranes; packages organs, surrounds capillaries.

Epithelium

Lamina propria

Function: Wraps and cushions organs; its macrophages phagocytize bacteria; plays important role in inflammation; holds and conveys tissue fluid.

Photomicrograph: Areolar connective tissue, a soft packaging tissue of the body (170×).

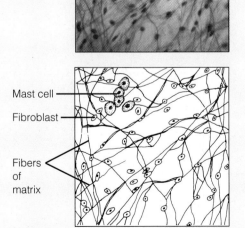

Mast cell

Fibroblast

Fibers of matrix

(c) Adipose tissue

Description: Matrix as in areolar, but very sparse; closely packed adipocytes, or fat cells, have nucleus pushed to the side by large fat droplet.

Location: Under skin; around kidneys and eyeballs; in bones and within abdomen; in breasts.

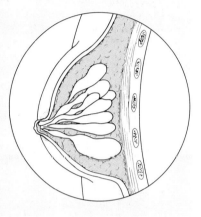

Function: Provides reserve food fuel; insulates against heat loss; supports and protects organs.

Photomicrograph: Adipose tissue from the subcutaneous layer under the skin (864×).

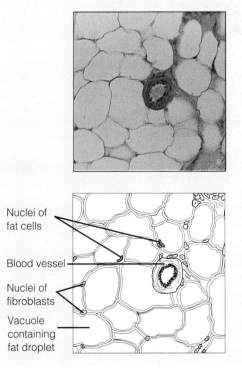

Nuclei of fat cells

Blood vessel

Nuclei of fibroblasts

Vacuole containing fat droplet

F5.5 (continued)

| | **Connective tissue proper: Dense connective tissue (e and f)** |

(d) Reticular connective tissue

Description: Network of reticular fibers in a typical loose ground substance; reticular cells predominate.

Location: Lymphoid organs (lymph nodes, bone marrow, and spleen).

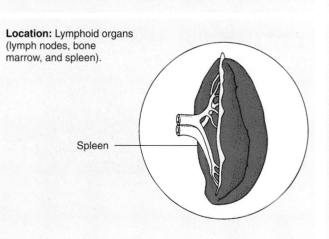

Spleen

Function: Fibers form a soft internal skeleton that supports other cell types.

Photomicrograph: Dark-staining network of reticular connective tissue fibers forming the internal skeleton of the spleen (1125×).

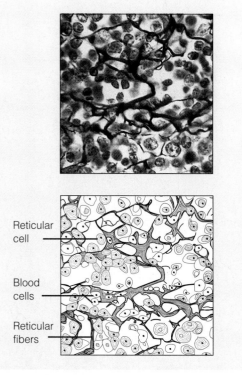

Reticular cell

Blood cells

Reticular fibers

(e) Dense regular connective tissue

Description: Primarily parallel collagen fibers; a few elastin fibers; major cell type is the fibroblast.

Location: Tendons, most ligaments, aponeuroses.

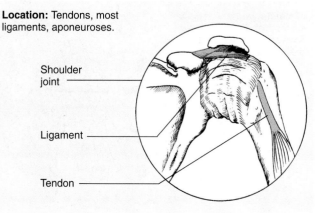

Shoulder joint

Ligament

Tendon

Function: Attaches muscles to bones or to muscles; attaches bones to bones; withstands great tensile stress when pulling force is applied in one direction.

Photomicrograph: Dense regular connective tissue from a tendon (576×).

Collagen fibers

Nuclei of fibroblasts (fiber forming cells)

F5.5 (*continued*)

(f) Dense irregular connective tissue

Description: Primarily irregularly arranged collagen fibers; some elastic fibers; major cell type is the fibroblast.

Location: Dermis of the skin; submucosa of digestive tract; fibrous capsules of organs and of joints.

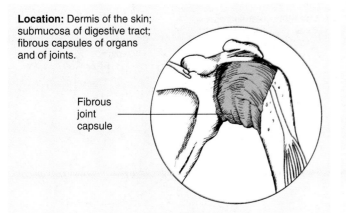

Fibrous joint capsule

Function: Able to withstand tension exerted in many directions; provides structural strength.

Photomicrograph: Dense irregular connective tissue from the dermis of the skin (268×).

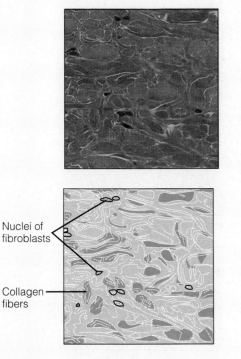

Nuclei of fibroblasts

Collagen fibers

Cartilage: (g to i)

(g) Hyaline cartilage

Description: Amorphous but firm matrix; collagen fibers form an imperceptible network; chondroblasts produce the matrix and when mature (chondrocytes) lie in lacunae.

Location: Forms most of the embryonic skeleton; covers the ends of long bones in joint cavities; forms costal cartilages of the ribs; cartilages of the nose, trachea, and larynx.

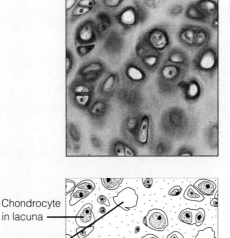

Costal cartilages

Function: Supports and reinforces; has resilient cushioning properties; resists compressive stress.

Photomicrograph: Hyaline cartilage from the trachea (469×).

Chondrocyte in lacuna

Empty lacunae

Matrix

F5.5 *(continued)*

(h) Elastic cartilage

Description: Similar to hyaline cartilage, but more elastic fibers in matrix.

Location: Supports the external ear (pinna); epiglottis.

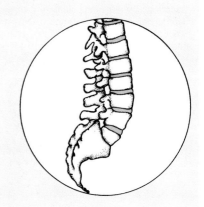

Function: Maintains the shape of a structure while allowing great flexibility.

Photomicrograph: Elastic cartilage from the human ear pinna; forms the flexible skeleton of the ear (144×).

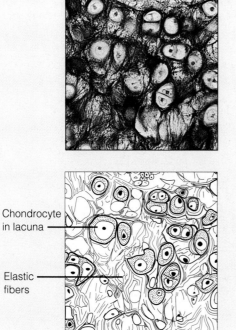

Chondrocyte in lacuna

Elastic fibers

(i) Fibrocartilage

Description: Matrix similar but less firm than in hyaline cartilage; thick collagen fibers predominate.

Location: Intervertebral discs; pubic symphysis; discs of knee joint.

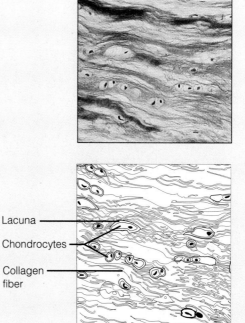

Function: Tensile strength with the ability to absorb compressive shock.

Photomicrograph: Fibrocartilage of an intervertebral disc (823×).

Lacuna

Chondrocytes

Collagen fiber

F5.5 (*continued*)

Others: (j and k)

(j) Bone (osseous tissue)

Description: Hard, calcified matrix containing many collagen fibers; osteocytes lie in lacunae. Very well vascularized.

Location: Bones

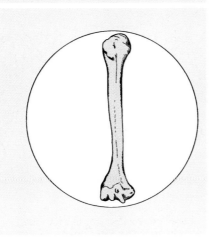

Function: Bone supports and protects (by enclosing); provides levers for the muscles to act on; stores calcium and other minerals and fat; marrow inside bones is the site for blood cell formation (hematopoiesis).

Photomicrograph: Cross-sectional view of bone (144×).

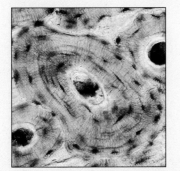

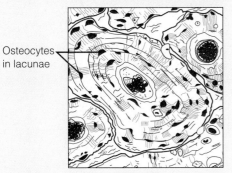

Osteocytes in lacunae

(k) Blood

Description: Red and white blood cells in a fluid matrix (plasma).

Location: Contained within blood vessels.

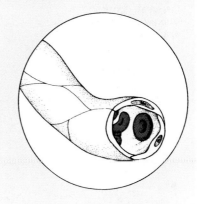

Function: Transport of respiratory gases, nutrients, wastes, and other substances.

Photomicrograph: Smear of human blood (1076×); two white blood cells (neutrophils) are seen surrounded by red blood cells.

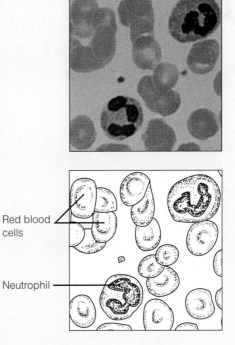

Red blood cells

Neutrophil

F5.5 (*continued*)

MUSCLE TISSUE

Muscle tissue is highly specialized to contract (shorten) in order to produce movement of some body parts. As you might expect, muscle cells tend to be quite elongated, providing a long axis for contraction. The three basic types of muscle tissue are described briefly here; cardiac muscle and skeletal muscle are treated more completely in later exercises.

Skeletal muscle, the "meat" or flesh of the body, is attached to the skeleton. It is under voluntary control (consciously controlled), and its contraction moves the limbs and other external body parts. The cells of skeletal muscles are long, cylindrical, and multinucleate (several nuclei per cell); they have obvious *striations* (stripes).

Cardiac muscle is found only in the heart. As it contracts, the heart acts as a pump, propelling the blood through the blood vessels. Cardiac muscle, like skeletal muscle, has striations. But cardiac cells are branching uninucleate (or occasionally binucleate) cells that interdigitate (fit together) at junctions called **intercalated discs.** These structural modifications allow the cardiac muscle to act as a unit. Cardiac muscle is under involuntary control, which means that we cannot voluntarily or consciously control the operation of the heart.

Smooth muscle, or *visceral muscle,* is found mainly in the walls of hollow organs (digestive and urinary tract organs, uterus, blood vessels). Typically there are two layers that run at right angles to each other; consequently its contraction can constrict or dilate the lumen (cavity) of an organ and propel substances along predetermined pathways. Smooth muscle cells are quite different in appearance from those of skeletal or cardiac muscle. No striations are visible, and the uninucleate smooth muscle cells are spindle-shaped.

 Obtain and examine prepared slides of skeletal, cardiac, and smooth muscle. Notice their similarities and dissimilarities in both your observations and in the illustrations in Figure 5.6.

(a) Skeletal muscle

Description: Long, cylindrical, multinucleate cells; obvious striations.

Location: In skeletal muscles attached to bones or occasionally to skin.

Function: Voluntary movement; locomotion; manipulation of the environment; facial expression. Voluntary control.

Photomicrograph: Skeletal muscle (approx. 576×). Notice the obvious banding pattern and the fact that these large cells are multinucleate.

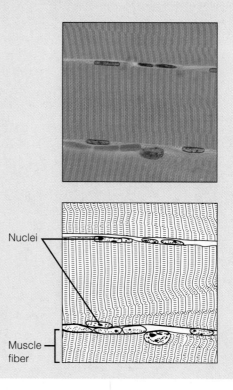

Nuclei

Muscle fiber

F5.6

Muscle tissues.

(b) Cardiac muscle

Description: Branching, striated, generally uninucleate cells that interdigitate at specialized junctions (intercalated discs).

Location: The walls of the heart.

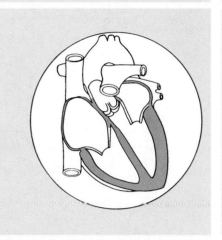

Function: As it contracts, it propels blood into the circulation; involuntary control.

Photomicrograph: Cardiac muscle (576×); notice the striations, branching of fibers, and the intercalated discs.

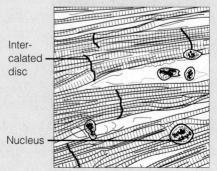

Inter-
calated
disc

Nucleus

(c) Smooth muscle

Description: Spindle-shaped cells with central nuclei; cells arranged closely to form sheets; no striations.

Location: Mostly in the walls of hollow organs.

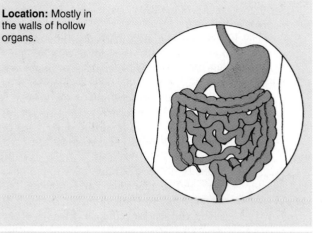

Function: Propels substances or objects (foodstuffs, urine, a baby) along internal passageways; involuntary control.

Photomicrograph: Sheet of smooth muscle (324×).

Smooth
muscle
cell

Nuclei

F5.6 (*continued*)

NERVOUS TISSUE

Nervous tissue is composed of two major cell populations. The **neuroglia** are special supporting cells that protect, support, and insulate the more delicate neurons. The **neurons** are highly specialized to receive stimuli (irritability) and to conduct waves of excitation, or impulses, to all parts of the body (conductivity). They are the cells that are most often associated with nervous system functioning.

The structure of neurons is markedly different from that of all other body cells. They all have a nucleus-containing cell body, and their cytoplasm is drawn out into long extensions (cell processes)—sometimes as long as 3 feet (about 1 m), which allows a single neuron to conduct an impulse over relatively long distances. More detail about the anatomy of the different classes of neurons and neuroglia appears in Exercise 15.

Obtain a prepared slide of a spinal cord smear. Locate a neuron and compare it to Figure 5.7. Keep the light dim—this will help you see the cellular extensions of the neurons. Also see Plates 4 and 5 in the Histology Atlas.

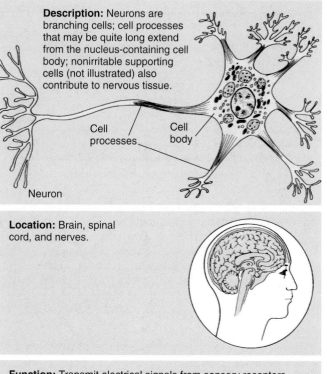

Description: Neurons are branching cells; cell processes that may be quite long extend from the nucleus-containing cell body; nonirritable supporting cells (not illustrated) also contribute to nervous tissue.

Cell processes

Cell body

Neuron

Location: Brain, spinal cord, and nerves.

Function: Transmit electrical signals from sensory receptors and to effectors (muscles and glands) which control their activity.

Photomicrograph: Neuron (170×).

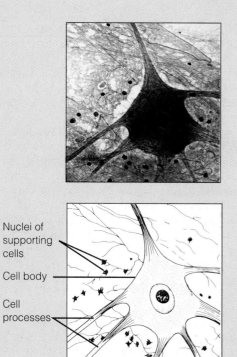

Nuclei of supporting cells

Cell body

Cell processes

F5.7

Nervous tissue.

The Integumentary System

OBJECTIVES

1. To recount several important functions of the skin, or integumentary system.

2. To recognize and name during observation of an appropriate model, diagram, projected slide, or microscopic specimen the following skin structures: epidermis (and note relative positioning of its strata), dermis (papillary and reticular layers), hair follicles and hair, sebaceous glands, and sweat glands.

3. To name the layers of the epidermis and describe the characteristics of each.

4. To compare the properties of the epidermis to those of the dermis.

5. To describe the distribution and function of the skin derivatives—sebaceous glands, sweat glands, and hairs.

6. To differentiate between eccrine and apocrine sweat glands.

7. To enumerate the factors determining skin color.

8. To describe the function of melanin.

9. To identify the major regions of nails.

MATERIALS

Skin model (three-dimensional, if available)
Compound microscope
Prepared slide of human skin with hair follicles
Sheet of #20 bond paper ruled to mark off cm^2 areas
Scissors
Betadine swabs, or Lugol's iodine and cotton swabs
Adhesive tape

See Appendix C, Exercise 6 for links to
Anatomy and PhysioShow: The Videodisc.

The **skin,** or **integument,** is often considered an organ system because of its extent and complexity. It is much more than an external body covering; architecturally, the skin is a marvel. It is tough yet pliable, a characteristic that enables it to withstand constant insult from outside agents.

The skin has many functions, most (but not all) concerned with protection. It insulates and cushions the underlying body tissues and protects the entire body from mechanical damage (bumps and cuts), chemical damage (acids, alkalis, and the like), thermal damage (heat), and bacterial invasion (by virtue of its acid mantle and continuous surface). The hardened uppermost layer of the skin (the cornified layer) prevents water loss from the body surface. The skin's abundant capillary network (under the control of the nervous system) plays an important role in regulating heat loss from the body surface.

The skin has other functions as well. For example, it acts as a mini-excretory system; urea, salts, and water are lost through the skin pores in sweat. The skin also has important metabolic duties. For example, like liver cells, it carries out some chemical conversions that activate or inactivate certain drugs and hormones, and it is the site of vitamin D synthesis for the body. Finally, the cutaneous sense organs are located in the dermis.

BASIC STRUCTURE OF THE SKIN

The skin has two distinct regions—the superficial *epidermis* composed of epithelium and an underlying connective tissue *dermis.* These layers are firmly "cemented" together along an undulating border. But friction, such as the rubbing of a poorly fitting shoe, may cause them to separate, resulting in a blister. Immediately deep to the dermis is the **hypodermis** or **superficial fascia** (primarily adipose tissue), which is not considered part of the skin. The main skin areas and structures are described below.

 As you read, locate the following structures on Figure 6.1 and on a skin model.

Epidermis

Structurally, the avascular epidermis is a keratinized stratified squamous epithelium consisting of four distinct cell types and four or five distinct layers.

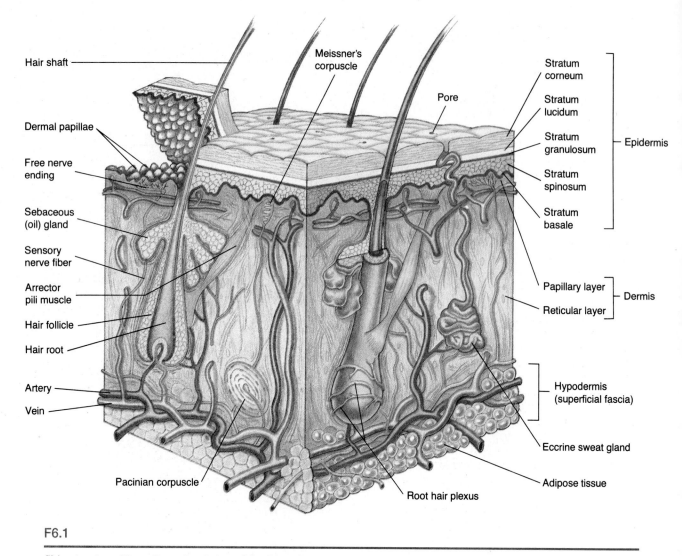

Hair shaft

Dermal papillae

Free nerve
ending

Sebaceous
(oil) gland

Sensory
nerve fiber

Arrector
pili muscle

Hair follicle

Hair root

Artery

Vein

Pacinian corpuscle

Meissner's
corpuscle

Pore

Root hair plexus

Stratum
corneum

Stratum
lucidum

Stratum
granulosum

Stratum
spinosum

Stratum
basale

Epidermis

Papillary layer

Reticular layer

Dermis

Hypodermis
(superficial fascia)

Eccrine sweat gland

Adipose tissue

F6.1

Skin structure. Three-dimensional view of the skin and the underlying hypodermis. The epidermis and dermis have been pulled apart at the left corner to reveal the dermal papillae.

CELLS OF THE EPIDERMIS   Most epidermal cells are **keratinocytes** (literally, keratin cells), epithelial cells that function mainly to produce keratin fibrils. **Keratin** is a fibrous protein that gives the epidermis its durability and protective capabilities.

Far less abundant are the following types of epidermal cells (Figure 6.2):

- **Melanocytes**—spidery black cells that produce the brown-to-black pigment called **melanin.** The skin tans because melanin production increases when the skin is exposed to sunlight. The melanin provides a protective pigment umbrella over the nuclei of the cells in the deeper epidermal layers, thus shielding their genetic material (DNA) from the damaging effects of ultraviolet radiation. A concentration of melanin in one spot is called a *freckle*.

- **Langerhans' cells**—phagocytic cells (macrophages) that play a role in immunity.
- **Merkel cells**—in conjunction with sensory nerve endings, Merkel cells form sensitive touch receptors called *Merkel discs* located at the epidermal-dermal junction.

LAYERS OF THE EPIDERMIS   From deep to superficial, the layers of the epidermis are the stratum basale, stratum spinosum, stratum granulosum, stratum lucidum, and stratum corneum (Figure 6.1).

The **stratum basale** (basal layer) is a single row of cells immediately adjacent to the dermis. Its cells are constantly undergoing mitotic cell division to produce millions of new cells daily, hence its alternate name *stratum germinativum*. About a quarter of the cells in this stratum are melanocytes, whose processes thread their way through this and the adjacent layers of keratinocytes (see Figure 6.2).

The **stratum spinosum** (spiny layer) is a stratum consisting of several cell layers immediately superficial to the basal layer. Its cells contain *tonofilaments*, thick bundles of intermediate filaments made of a prekeratin protein. The stratum spinosum cells appear spiky (hence their name) because, as the skin tissue is prepared for histological examination, they shrink but their desmosomes hold tight. Cells divide fairly rapidly in this stratum, but less so than in the stratum basale. Cells in the basal and spiny layers are the only ones to receive adequate nourishment (via diffusion of nutrients from the dermis). So as their daughter cells are pushed upward and away from the source of nutrition, they gradually die.

The **stratum granulosum** (granular layer) is a thin layer named for the abundant granules its cells contain. These granules are of two types: (1) *laminated granules,* which contain a waterproofing lipid that is secreted into the extracellular space; and (2) *keratohyalin granules,* which combine with the tonofilaments in the more superficial layers to form the keratin fibrils. At the upper border of this layer, the cells are beginning to die.

The **stratum lucidum** (clear layer) is a very thin translucent band of flattened dead keratinocytes with indistinct boundaries. It is not present in thin skin.

The outermost epidermal layer, the **stratum corneum** (horny layer), consists of some 20 to 30 cell layers, and accounts for the bulk of the epidermal thickness. Cells in this layer, like those in the stratum lucidum (where it exists), are dead and their flattened scalelike remnants are fully keratinized. They are constantly rubbing off and being replaced by division of the deeper cells.

Dermis

The dense irregular connective tissue making up the dermis consists of two principal regions—the papillary and reticular areas. Like the epidermis, the dermis varies in thickness. For example, the skin is particularly thick on the palms of the hands and soles of the feet and is quite thin on the eyelids.

The **papillary layer** is the more superficial dermal region. It is very uneven and has fingerlike projections from its superior surface, the **dermal papillae,** which attach it to the epidermis above. These projections produce unique patterns of ridges that remain unchanged throughout life, which are reflected in fingerprints. Abundant capillary networks in the papillary layer furnish nutrients for the epidermal layers and allow heat to radiate to the skin surface. The pain and touch receptors (Meissner's corpuscles) are also found here.

The **reticular layer** is the deepest skin layer. It contains many arteries and veins, sweat and sebaceous glands, and pressure receptors.

Both the papillary and reticular layers are heavily invested with collagenic and elastic fibers. The elastic fibers give skin its exceptional elasticity in youth. In old age, the number of elastic fibers decreases and the subcutaneous layer loses fat, which leads to wrinkling and

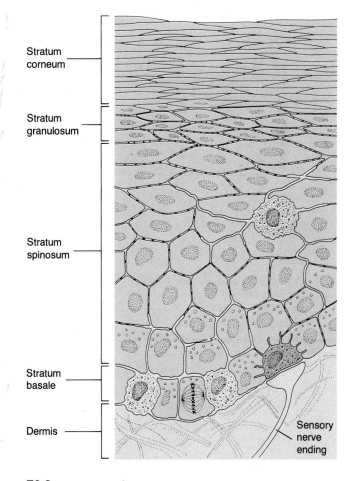

F6.2

Diagram showing the main features—layers and relative numbers of the different cell types—in epidermis of thin skin. The keratinocytes (pink) form the bulk of the epidermis. Less numerous are the melanocytes (gray), which produce the pigment melanin; Langerhans' cells (blue), which function as macrophages; and Merkel cells (purple). A sensory nerve ending (yellow), extending from the dermis, is depicted in association with the Merkel cell forming a Merkel disc (touch receptor). Notice that the keratinocytes, but not the other cell types, are joined by numerous desmosomes.

inelasticity of the skin. Fibroblasts, adipose cells, various types of macrophages (which are important in the body's defense), and other cell types are found throughout the dermis.

The abundant dermal blood supply allows the skin to play a role in the regulation of body temperature. When body temperature is high, the arterioles serving the skin dilate, and the capillary network of the dermis becomes engorged with the heated blood. Thus body heat is allowed to radiate from the skin surface. If the environment is cool and body heat must be conserved, the arterioles constrict so that blood bypasses the dermal capillary networks.

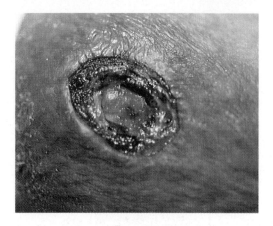

F6.3

Photograph of a deep (stage III) decubitus ulcer.

Any restriction of the normal blood supply to the skin results in cell death and, if severe enough, skin ulcers (Figure 6.3). **Bedsores (decubitus ulcers)** occur in bedridden patients who are not turned regularly enough. The weight of the body exerts pressure on the skin, especially over bony projections (hips, heels, etc.), which leads to restriction of the blood supply and death of tissue. ■

The dermis is also richly provided with lymphatic vessels and a nerve supply. Many of the nerve endings bear highly specialized receptor organs that, when stimulated by environmental changes, transmit messages to the central nervous system for interpretation. Some of these receptors—bare nerve endings (pain receptors), a Meissner's corpuscle, a Pacinian corpuscle, and a root hair plexus—are shown in Figure 6.1. (These receptors are discussed in depth in Exercise 15.)

Skin Color

Skin color is a result of three factors—the relative amount of two pigments (melanin and carotene) in skin and the degree of oxygenation of the blood. People who produce large amounts of melanin have brown-toned skin. In light-skinned people, who have less melanin, the dermal blood supply flushes through the rather transparent cell layers above, giving the skin a rosy glow. *Carotene* is a yellow-orange pigment present primarily in the stratum corneum and in the adipose tissue of the hypodermis. Its presence is most noticeable when large amounts of carotene-rich foods (carrots, for instance) are eaten.

Skin color may be an important diagnostic tool. For example, flushed skin may indicate hypertension, fever, or embarrassment, whereas pale skin is typically seen in anemic individuals. When the blood is inadequately oxygenated, as during asphyxiation and serious lung disease, both the blood and the skin take on a bluish or cyanotic cast. **Jaundice,** in which the tissues become yellowed, is almost always diagnostic for liver disease, whereas a bronzing of the skin hints that a person's adrenal cortex is hypoactive (**Addison's disease**). ■

APPENDAGES OF THE SKIN

The appendages of the skin—hair, nails, and cutaneous glands—are all derivatives of the epidermis, but they reside in the dermis. They originate from the stratum basale and grow downward into the deeper skin regions.

Cutaneous Glands

The cutaneous glands fall primarily into two categories: the sebaceous glands and the sweat glands (Figure 6.1). The **sebaceous glands** are found nearly all over the skin, except for the palms of the hands and the soles of the feet. Their ducts usually empty into a hair follicle, but some open directly onto the skin surface.

The product of the sebaceous glands, called **sebum,** is a mixture of oily substances and fragmented cells. The sebum is a lubricant that keeps the skin soft and moist (a natural skin cream) and keeps the hair from becoming brittle. The sebaceous glands become particularly active during puberty when more male hormones (androgens) begin to be produced; thus, the skin tends to become oilier during this period of life. *Blackheads* are accumulations of dried sebum and bacteria; *acne* is due to active infection of the sebaceous glands.

Epithelial openings, called *pores*, are the outlets for the **sweat (sudoriferous) glands.** These exocrine glands are widely distributed in the skin. Sweat glands are subcategorized by the composition of their secretions. The **eccrine glands,** which are distributed all over the body, produce clear perspiration, consisting primarily of water, salts (NaCl), and urea. The **apocrine glands,** found predominantly in the axillary and genital areas, secrete a milky protein- and fat-rich substance (also containing water, salts, and urea) that is an excellent nutrient medium for the microorganisms typically found on the skin.

The sweat glands, under the control of the nervous system, are an important part of the body's heat-regulating apparatus. They secrete perspiration when the external temperature or body temperature is high. When this water-based substance evaporates, it carries excess body heat with it. Thus evaporation of greater amounts of perspiration provides an efficient means of dissipating body heat when the capillary cooling system is not sufficient or is unable to maintain body temperature homeostasis.

Hair

Hairs are found over the entire body surface, except for thick-skinned areas (the palms of the hands, the soles of the feet), parts of the external genitalia, the nipples, and the lips. A hair, enclosed in a hair **follicle,** is also an epithelial structure (Figure 6.4). The portion of the hair enclosed within the follicle is called the **root;** the portion projecting from the scalp surface is called the **shaft.** The hair is formed by mitosis of the well-nourished germinal epithelial cells at the basal end of the follicle (the **hair bulb**). As the daughter cells are pushed farther away from the growing region, they die and become keratinized; thus the bulk of the hair shaft, like the bulk of the epidermis, is dead material.

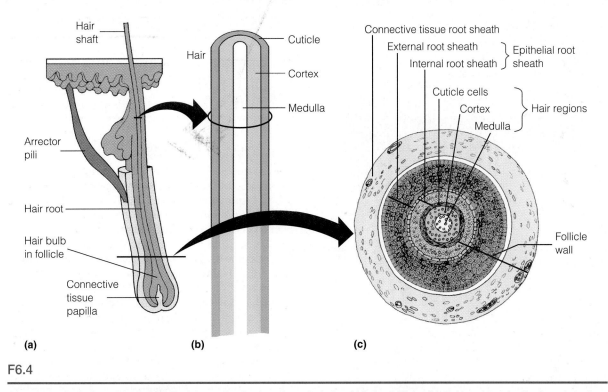

F6.4

Structure of a hair and hair follicle. (a) Longitudinal section of a hair within its follicle. **(b)** Enlarged longitudinal section of a hair. **(c)** Cross section of a hair and hair follicle.

A hair (Figure 6.4b) consists of a central region (medulla) surrounded first by the cortex and then by a protective cuticle. Abrasion of the cuticle results in "split ends." Hair color is a manifestation of the amount and kind of melanin pigment within the hair cortex.

The hair follicle is structured from both epidermal and dermal cells (Figure 6.4c). Its inner *epithelial root sheath*, with two parts (internal and external), is enclosed by the *connective tissue root sheath*, which is essentially dermal tissue. A small nipple of dermal tissue, the *connective tissue papilla*, protrudes into the hair bulb from the connective tissue sheath and provides nutrition to the growing hair. If you look carefully at the structure of the hair follicle (see Figure 6.1), you will see that it generally is in a slanted position. Small bands of smooth muscle cells—**arrector pili**—connect each hair follicle to the papillary layer of the dermis. When these muscles contract (during cold or fright), the hair follicle is pulled upright, dimpling the skin surface with "goose bumps." This phenomenon is especially dramatic in a scared cat, whose fur actually stands on end to increase its apparent size. The activity of the arrector pili muscles also exerts pressure on the sebaceous glands surrounding the follicle, causing a small amount of sebum to be released.

Nails

Nails, the hornlike derivatives of the epidermis, consist of a *free edge*, a *body* (visible attached portion), and a *root* (embedded in the skin and adhering to an epithelial **nail bed**). The borders of the nail are overlapped by skin folds called **nail folds;** the thick proximal nail fold is the **eponychium,** commonly called the cuticle (Figure 6.5).

The germinal cells in the **nail matrix,** the thickened proximal part of the nail bed, are responsible for nail growth. As the nail cells are produced by the matrix, they become heavily keratinized and die. Thus, nails, like hairs, are mostly nonliving material.

Nails are transparent and nearly colorless, but they appear pink because of the blood supply in the underlying dermis. The exception to this is the proximal region of the thickened nail matrix which appears as a white crescent called the *lunula.* When someone is cyanotic due to a lack of oxygen in the blood, the nail beds take on a blue cast.

Identify the nail structures shown in Figure 6.5 on yourself or your lab partner.

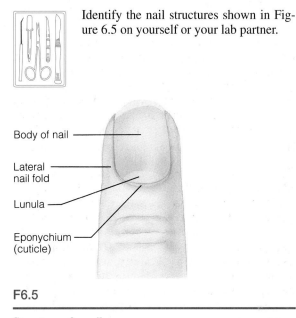

F6.5

Structure of a nail.

53

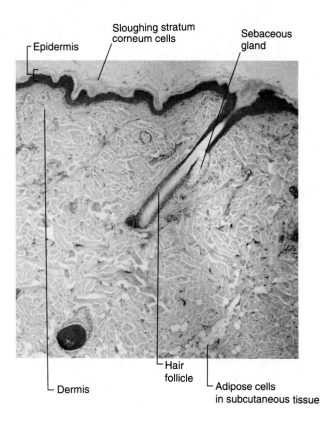

Labels on image: Epidermis · Sloughing stratum corneum cells · Sebaceous gland · Dermis · Hair follicle · Adipose cells in subcutaneous tissue

F6.6

Photomicrograph of skin (20×).

EXAMINATION OF THE MICROSCOPIC STRUCTURE OF THE SKIN

Obtain a prepared slide of human skin, and study it carefully under the microscope. Compare your tissue slide to the view shown in Figure 6.6, and identify as many of the structures diagrammed in Figure 6.1 as possible.

How is this stratified squamous epithelium different from that observed in Exercise 5?

_____

_____

How do these differences relate to the functions of these two similar epithelia?

_____

_____

PLOTTING THE DISTRIBUTION OF SWEAT GLANDS

1.   For this simple experiment you will need two squares of bond paper (each 1 cm × 1 cm), adhesive tape, and a Betadine (iodine) swab *or* Lugol's iodine and a cotton-tipped swab. (The bond paper has been preruled in cm^2—just cut along the lines to obtain the required squares.)

2.   Paint the medial aspect of your left palm (avoid the crease lines) and a region of your left forearm with the iodine solution, and allow it to dry thoroughly. The painted area in each case should be slightly larger than the paper squares to be used.

3.   Have your lab partner *securely* tape a square of bond paper over each iodine-painted area, and leave them in place for 20 minutes. (If it is very warm in the laboratory while this test is being conducted, good results may be obtained within 10 to 15 minutes.)

4.   After 20 minutes, remove the paper squares, and count the number of blue-black dots on each square. The presence of a blue-black dot on the paper indicates an active sweat gland. (The iodine in the pore is dissolved in the sweat and reacts chemically with the starch in the bond paper to produce the blue-black color.) Thus "sweat maps" have been produced for the two skin areas.

5.   Which skin area tested has the greater density of sweat glands?

_____

_____

_____

Classification of Body Membranes

OBJECTIVES

1. To compare the structure and function of the major membrane types.
2. To list the general functions of each membrane type and note its location in the body.
3. To recognize by microscopic examination cutaneous, mucous, and serous membranes.

MATERIALS

Compound microscope
Prepared slides of trachea (cross section) and small intestine (cross section)
Prepared slide of serous membrane (e.g., mesentery)
Longitudinally cut fresh beef joint (if available)

 See Appendix B, Exercise 7 for links to A.D.A.M. Standard.

See Appendix C, Exercise 7 for links to *Anatomy and PhysioShow: The Videodisc.*

The body membranes, which cover surfaces, line body cavities, and form protective (and often lubricating) sheets around organs, fall into two major categories. These are the so-called *epithelial membranes* and the *synovial membranes.*

EPITHELIAL MEMBRANES

The term "epithelial membrane" is used in various ways. Here we will define an **epithelial membrane** as a simple organ consisting of an epithelial sheet bound to an underlying layer of connective tissue. Most of the covering and lining epithelia take part in forming one of the three common varieties of epithelial membranes: cutaneous, mucous, or serous.

The **cutaneous membrane** (Figure 7.1a) is the skin, a dry membrane with a keratinizing epithelium (the epidermis). Since the skin is discussed in some detail in Exercise 6, the mucous and serous membranes will receive our attention here.

Mucous Membranes

The **mucous membranes** (**mucosae**) are composed of epithelial cells resting on a layer of loose connective tissue called the **lamina propria.** They line all body cavities that open to the body exterior—the respiratory (Figure 7.1b), digestive, and urinary tracts. All mucosae are "wet" membranes because they are continuously bathed by secretions (or, in the case of urinary tract mucosae, urine). Although mucous membranes often secrete mucus, this is not a requirement. The mucous membranes of both the digestive and respiratory tracts secrete mucus, but that of the urinary tract does not.

 Using Plates 31 and 36 of the Histology Atlas as guides, examine a slide made from a cross section of the trachea and another of the small intestine. Draw the mucosa of each in the appropriate circle, and fully identify each epithelial type. Re-

member to look for the epithelial cells at the free surface. Also search the epithelial sheets for **goblet cells**—columnar epithelial cells with a large mucus-containing vacuole (goblet) in their apical cytoplasm. Which mucosa type contains goblet cells?

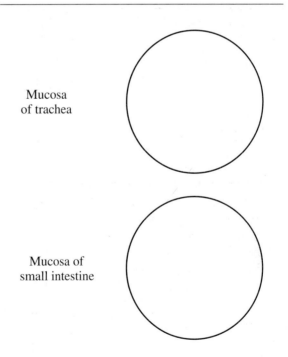

Mucosa of trachea

Mucosa of small intestine

Compare and contrast the roles of these two mucous membranes.

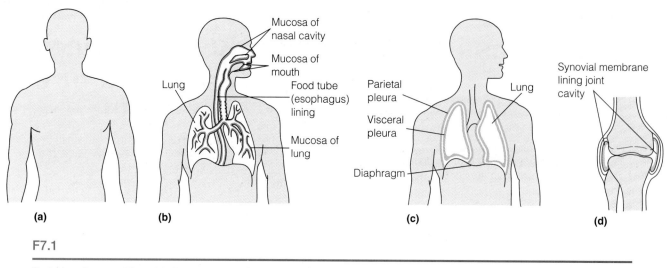

F7.1

Body membranes. The epithelial membranes (a–c) are composite membranes with epithelial and connective tissue elements. **(a)** The cutaneous membrane, or skin, covers and protects the body surface. **(b)** Mucous membranes line body cavities (hollow organs) that open to the exterior. **(c)** Serous membranes line the closed ventral cavity of the body. One example, the pleura, is illustrated here. **(d)** Synovial membranes, derived solely from connective tissue, form smooth linings in joint cavities.

Serous Membranes

The **serous membranes (serosae)** are also epithelial membranes (Figure 7.1c). They are composed of a layer of simple squamous epithelium on a scant amount of loose connective tissue. The serous membranes generally occur in twos. The parietal layer lines a body cavity, and the visceral layer covers the outside of the organs in that cavity (see also Figure 1.8, p. 8). In contrast to the mucous membranes, which line open body cavities, the serous membranes line body cavities that are closed to the exterior (with the exception of the female peritoneal cavity and the dorsal body cavity). The serosae secrete a thin fluid (serous fluid) that lubricates the organs and body walls and thus reduces friction as the organs slide across one another and against the body cavity walls. A serous membrane also lines the interior of blood vessels (endothelium) and the heart (endocardium). In capillaries, the entire wall is composed of serosa that serves as a selectively permeable membrane between the blood and the tissue fluid of the body.

 Examine a prepared slide of a serous membrane and diagram it in the circle provided here.

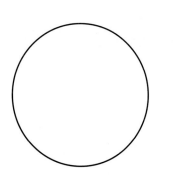

What are the specific names of the serous membranes covering the heart and lining the cavity in which it resides (respectively)?

_____

The abdominal viscera and visceral cavity (respectively)?

_____

SYNOVIAL MEMBRANES

Synovial membranes, unlike the mucous and serous membranes, are composed entirely of connective tissue; they contain no epithelial cells. These membranes line the cavities surrounding the joints, providing a smooth surface and secreting a lubricating fluid. They also line smaller sacs of connective tissue (bursae and tendon sheaths), which cushion structures moving against each other, as during muscle activity. Figure 7.1d illustrates the positioning of a synovial membrane in the joint cavity.

 If a freshly sawed beef joint is available, visually examine the interior surface of the joint capsule to observe the smooth texture of the synovial membrane.

Overview of the Skeleton: Classification and Structure of Bones and Cartilages

OBJECTIVES

1. To list at least three functions of the skeletal system.
2. To identify the four main kinds of bones.
3. To identify surface bone markings and their function.
4. To identify the major anatomical areas on a longitudinally cut long bone (or diagram of one).
5. To identify the major regions and structures of an osteon in a histologic specimen of compact bone (or diagram of one).
6. To explain the role of the inorganic salts and organic matrix in providing flexibility and hardness to bone.
7. To locate and identify the three major types of skeletal cartilages.

MATERIALS

Disarticulated bones (identified by name or number) that demonstrate classic examples of the four bone classifications (long, short, flat, and irregular)

Long bone sawed longitudinally (beef bone from a slaughterhouse, if possible, or prepared laboratory specimen)

Compound microscope

Prepared slides of ground bone (cross section), hyaline cartilage, elastic cartilage, and fibrocartilage

3-D model of microscopic structure of compact bone

Long bone soaked in 10% nitric acid (or vinegar) until flexible

Long bone baked at 250°F for more than 2 hours

Disposable plastic gloves

Articulated skeleton

 See Appendix B, Exercise 8 for links to A.D.A.M. Standard.

See Appendix C, Exercise 8 for links to *Anatomy and PhysioShow: The Videodisc.*

The **skeleton** is constructed of two of the most supportive tissues found in the human body—cartilage and bone. In embryos, the skeleton is predominantly composed of hyaline cartilage, but in the adult, most of the cartilage is replaced by more rigid bone. Cartilage persists only in such isolated areas as the bridge of the nose, the larynx, the trachea, joints, and parts of the rib cage.

Besides supporting and protecting the body as an internal framework, the skeleton provides a system of levers with which the skeletal muscles work to move the body. In addition, the bones store lipids and many minerals (most importantly calcium). Finally, the red marrow cavities of bones provide a site for hematopoiesis (blood cell formation).

The skeleton is made up of bones that are connected at joints, or articulations. The skeleton is subdivided into two divisions: the **axial skeleton** (those bones that lie around the body's center of gravity) and the **appendicular skeleton** (bones of the limbs, or appendages) (Figure 8.1).

Before beginning your study of the skeleton, imagine for a moment that your bones have turned to putty. What if you were running when this metamorphosis took place? Now imagine your bones forming a continuous metal framework within your body, somewhat like a network of plumbing pipes. What problems could you envision with this arrangement? These images should help you understand how well the skeletal system provides support and protection, as well as facilitating movement.

BONE MARKINGS

Even a casual observation of the bones will reveal that bone surfaces are not featureless smooth areas but are scarred with an array of bumps, holes, and ridges. These **bone markings** reveal where bones form joints with other bones, where muscles, tendons, and ligaments were attached, and where blood vessels and nerves passed. Bone markings fall into two categories: projections, or processes which grow out from the bone and serve as sites of muscle attachment or help form joints; and depressions or cavities, indentations or openings in the bone that often serve as conduits for nerves and blood vessels. The bone markings are summarized in Table 8.1.

CLASSIFICATION OF BONES

The 206 bones of the adult skeleton are composed of two basic kinds of osseous tissue that differ in their texture. **Compact** bone looks smooth and homogeneous; **spongy** (or *cancellous*) bone is composed of small trabeculae (bars) of bone and lots of open space.

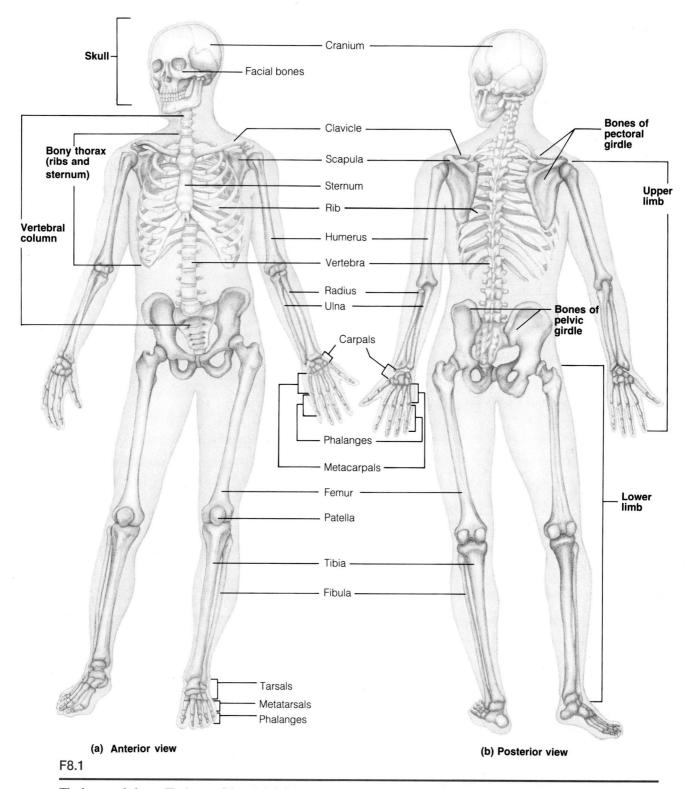

Skull

Cranium

Facial bones

Bony thorax (ribs and sternum)

Clavicle

Scapula

Sternum

Rib

Humerus

Vertebral column

Vertebra

Radius

Ulna

Carpals

Phalanges

Metacarpals

Femur

Patella

Tibia

Fibula

Tarsals

Metatarsals

Phalanges

Bones of pectoral girdle

Upper limb

Bones of pelvic girdle

Lower limb

(a) Anterior view

(b) Posterior view

F8.1

The human skeleton. The bones of the axial skeleton are colored green to distinguish them from the bones of the appendicular skeleton.

Bones may be classified further on the basis of their relative gross anatomy into four groups: long, short, flat, and irregular bones.

Long bones, such as the femur (Figure 8.1), are much longer than they are wide, generally consisting of a shaft with heads at either end. Long bones are composed predominantly of compact bone. **Short bones** are typically cube-shaped, and they contain more spongy

bone than compact bone. See the tarsals and carpals in Figure 8.1.

Flat bones are generally thin, with two thin layers of compact bone sandwiching a layer of spongy bone between them. Although the name "flat bone" implies a structure that is level or horizontal, many flat bones are curved (for example, the bones of the skull). Bones that do not fall into one of the preceding categories are clas-

TABLE 8.1   Bone Markings

Name of bone marking	Description	Illustration
Projections that are sites of muscle and ligament attachment		
Tuberosity	Large rounded projection; may be roughened	
Crest	Narrow ridge of bone; usually prominent	
Trochanter	Very large, blunt, irregularly shaped process. (The only examples are on the femur.)	
Line	Narrow ridge of bone; less prominent than a crest	
Tubercle	Small rounded projection or process	
Epicondyle	Raised area on or above a condyle	
Spine	Sharp, slender, often pointed projection	
Projections that help to form joints		
Head	Bony expansion carried on a narrow neck	
Facet	Smooth, nearly flat articular surface	
Condyle	Rounded articular projection	
Ramus	Armlike bar of bone	
Depressions and openings allowing blood vessels and nerves to pass		
Meatus	Canal-like passageway	
Sinus	Cavity within a bone, filled with air and lined with mucous membrane	
Fossa	Shallow, basinlike depression in a bone, often serving as an articular surface	
Groove	Furrow	
Fissure	Narrow, slitlike opening	
Foramen	Round or oval opening through a bone	

sified as **irregular bones.** The vertebrae are irregular bones (see Figure 8.1).

Some anatomists also recognize two other subcategories of bones. **Sesamoid bones** are small bones formed in tendons. The patellas (kneecaps) are sesamoid bones. **Wormian bones** are tiny bones between cranial bones. Except for the patellas, the sesamoid and Wormian bones are not included in the bone count given above because they vary in number and location in different individuals.

 Examine the isolated (disarticulated) bones (numbered) on display. See if you can find specific examples of the bone markings described in Table 8.1. Then classify each of the bones into one of the four anatomical groups by recording its name or number in the chart at right. Verify your identifications with your instructor before leaving the laboratory.

Long	Short	Flat	Irregular

GROSS ANATOMY OF THE TYPICAL LONG BONE

1. Obtain a long bone that has been sawed along its longitudinal axis. If a cleaned dry bone is provided, no special preparations need be made.

⚠️ If the bone supplied is a fresh beef bone, don plastic gloves before beginning your observations.

With the help of Figure 8.2, identify the shaft, or **diaphysis.** Observe its smooth surface, which is composed of compact bone. If you are using a fresh specimen, carefully pull away the **periosteum,** or fibrous membrane covering, to view the bone surface. Notice that many fibers of the periosteum penetrate into the bone. These fibers are called **Sharpey's fibers.** The periosteum is the source of the blood vessels and nerves that invade the bone. **Osteoblasts** (bone-forming cells) on its inner face secrete the bony matrix that increases the girth of the long bone.

2. Now inspect the **epiphysis,** the end of the long bone. Notice that it is composed of a thin layer of compact bone that encloses spongy bone.

3. Identify the **articular cartilage,** which covers the epiphyseal surface in place of the periosteum. Since it is composed of glassy hyaline cartilage, it provides a smooth surface to prevent friction at joint surfaces.

4. If the animal was still young and growing, you will be able to see the **epiphyseal plate,** a thin area of hyaline cartilage that provides for longitudinal growth of the bone during youth. Once the long bone has stopped growing, these areas are replaced with bone and appear as thin, barely discernible remnants—the **epiphyseal lines.**

5. In an adult animal, the central cavity of the shaft (*medullary cavity*) is essentially a storage region for adipose tissue, or **yellow marrow.** In the infant, this area is involved in forming blood cells, and so **red marrow** is found in the marrow cavities. In adult bones, the red marrow is confined to the interior of the epiphyses, where it occupies the spaces between the trabeculae of spongy bone.

6. If you are examining a fresh bone, look carefully to see if you can distinguish the delicate **endosteum** lining the shaft. In a living bone, **osteoclasts** (bone-destroying cells) are found on the inner surface of the endosteum, against the compact bone of the diaphysis. As the bone grows in diameter on its external surface, it is constantly being broken down on its inner surface. Thus the thickness of the compact bone layer composing the shaft remains relatively constant.

⚠️ 7. If you have been working with a fresh bone specimen, return it to the appropriate area and properly dispose of your gloves, as designated by your instructor. Wash your hands before continuing on to the microscope study.

▶ Longitudinal bone growth at epiphyseal discs follows a predictable sequence and provides a reliable indicator of the age of children exhibiting normal growth. In cases in which problems of long-bone growth are suspected (for example, pituitary dwarfism), X rays are taken to view the width of the growth plates. An abnormally thin epiphyseal plate indicates growth retardation. ■

CHEMICAL COMPOSITION OF BONE

Bone is one of the hardest materials in the body. Although relatively light, bone has a remarkable ability to resist tension and shear forces that continually act on it. An engineer would tell you that a cylinder (like a long bone) is one of the strongest structures for its mass. Thus nature has given us an extremely strong, exceptionally simple (almost crude), and flexible supporting system without sacrificing mobility.

The hardness of bone is due to the inorganic calcium salts deposited in its ground substance. Its flexibility comes from the organic elements of the matrix, particularly the collagenic fibers.

Obtain a bone sample that has been soaked in nitric acid (or vinegar) and one that has been baked. Heating removes the organic part of bone, while acid dissolves out the minerals. Do the treated bones retain the structure of untreated specimens?

_____

Gently apply pressure to each bone sample. What happens to the heated bone?

_____

_____

The bone treated with acid?

_____

_____

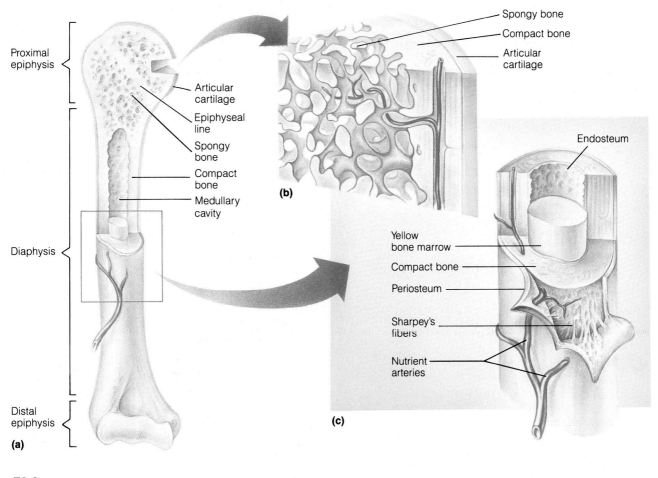

F8.2

The structure of a long bone (humerus of the arm). (a) Anterior view with longitudinal section cut away at the proximal end. **(b)** Pie-shaped, three-dimensional view of spongy bone and compact bone of the epiphysis. **(c)** Cross section of shaft (diaphysis). Note that the external surface of the diaphysis is covered by a periosteum, but the articular surface of the epiphysis is covered with hyaline cartilage.

What does the acid appear to remove from the bone?

_____

_____

What does baking appear to do to the bone?

_____

_____

In rickets, the bones are not properly calcified. Which of the demonstration specimens would more closely resemble the bones of a child with rickets?

_____

_____

MICROSCOPIC STRUCTURE OF COMPACT BONE

As you have seen, spongy bone has a spiky, open-work appearance, resulting from the arrangement of the **trabeculae** that compose it, while compact bone appears to be dense and homogeneous. Microscopic examination of compact bone, however, reveals that it is riddled with passageways carrying blood vessels, nerves, and lymphatic vessels that provide the living bone cells with needed substances and a way to eliminate wastes. Indeed, bone histology is much easier to understand when you recognize that bone tissue is organized around its blood supply.

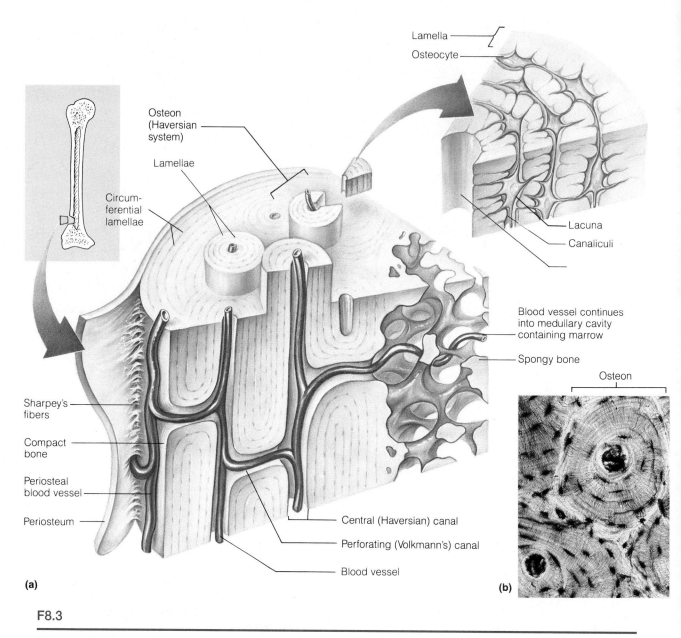

F8.3

Microscopic structure of compact bone. (a) Diagrammatic view of a pie-shaped segment of compact bone, illustrating its structural units (osteons). The inset shows a more highly magnified view of a portion of one osteon. Notice the position of osteocytes in lacunae (cavities of the matrix). **(b)** Photomicrograph of a cross-sectional view of one osteon (90×).

1. Obtain a prepared slide of ground bone and examine it under low power. Using Figure 8.3 as a guide, focus on a **central (Haversian) canal.** The central canal runs parallel to the long axis of the bone and carries blood vessels, nerves, and lymph vessels through the bony matrix. Identify the **osteocytes** (mature bone cells) in **lacunae** (chambers), which are arranged in concentric circles (concentric **lamellae**) around the central canal. A central canal and all the concentric lamellae surrounding it are referred to as an **osteon** or **Haversian system.** Also identify **canaliculi,** tiny canals radiating outward from a central canal to the lacunae of the first lamella and then from lamella to lamella. The canaliculi form a dense transportation network through the hard bone matrix, connecting all the living cells of the osteon to the nutrient supply. The canaliculi allow each cell to take what it needs for nourishment and to pass along the excess to the next osteocyte. You may need a higher-power magnification to see the fine canaliculi.

2. Also note the **perforating (Volkmann's) canals** in Figure 8.3. These canals run into the compact bone and marrow cavity from the periosteum, at right angles to the shaft. With the central canals, the perforating canals complete the communication pathway between the bone interior and its external surface.

3. If a model of bone histology is available, identify the same structures on the model.

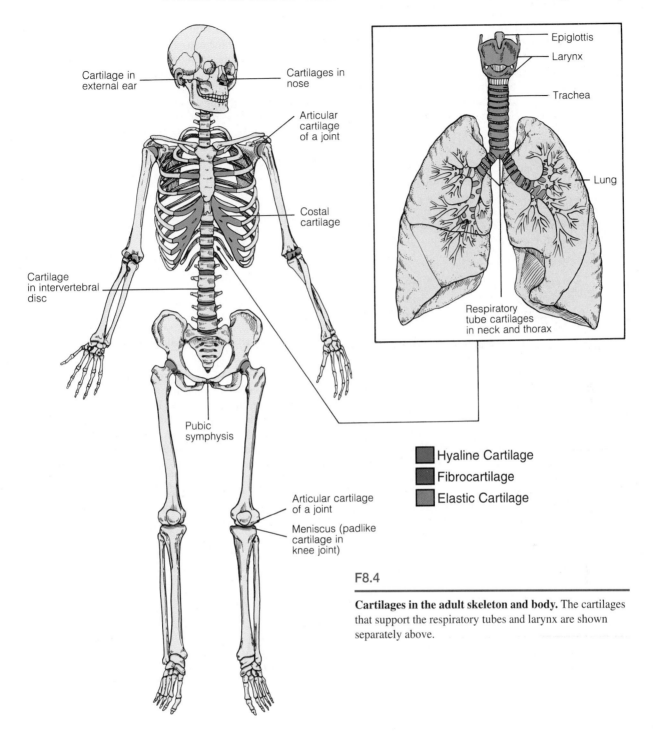

Cartilage in external ear

Cartilages in nose

Articular cartilage of a joint

Costal cartilage

Cartilage in intervertebral disc

Pubic symphysis

Articular cartilage of a joint

Meniscus (padlike cartilage in knee joint)

Epiglottis

Larynx

Trachea

Lung

Respiratory tube cartilages in neck and thorax

Hyaline Cartilage

Fibrocartilage

Elastic Cartilage

F8.4

Cartilages in the adult skeleton and body. The cartilages that support the respiratory tubes and larynx are shown separately above.

CARTILAGES OF THE SKELETON

Location and Basic Structure

As mentioned earlier, cartilaginous regions of the skeleton have a fairly limited distribution in adults (Figure 8.4). The most important of these skeletal cartilages are (1) **articular cartilages,** which cover the bone ends at movable joints; (2) **costal cartilages,** found connecting the ribs to the sternum (breastbone); (3) **laryngeal cartilages,** which largely construct the larynx (voicebox); (4) **tracheal** and **bronchial cartilages,** which reinforce other passageways of the respiratory system; (5) **nasal cartilages,** which support the external nose; (6) **intervertebral discs,** which separate and cushion bones of the spine (vertebrae); and (7) the cartilage supporting the external ear.

The skeletal cartilages consist of some variety of *cartilage tissue,* which typically consists primarily of water and is fairly resilient. Additionally, cartilage tissues are distinguished by the fact that they contain no nerves or blood vessels. Like bones, each cartilage is surrounded by a covering of dense connective tissue, called a *perichondrium* (rather than a periosteum). The perichondrium acts like a girdle to resist distortion of the cartilage when the cartilage is subjected to pressure. It also plays a role in cartilage growth and repair.

Classification of Cartilage

The skeletal cartilages have representatives from each of the three cartilage tissue types—hyaline, elastic, and fibrocartilage. Although you have already studied cartilage tissues (Exercise 5), some of that information will be recapped briefly here and you will have a chance to review the microscope structure unique to each cartilage type.

Obtain prepared slides of hyaline cartilage, elastic cartilage, and fibrocartilage and bring them to your laboratory bench for viewing.

As you read through the descriptions of these cartilage types, keep in mind that the bulk of cartilage tissue consists of a nonliving *matrix* (containing a jellylike ground substance and fibers) secreted by chondrocytes.

HYALINE CARTILAGE **Hyaline cartilage** looks like frosted glass when viewed by the unaided eye. As easily seen in Figure 8.4, most skeletal cartilages are composed of hyaline cartilage. Its chondrocytes, snugly housed in lacunae, appear spherical and collagen fibers are the only fiber type in its matrix. Hyaline cartilage provides sturdy support with some resilience or "give." Draw a small section of hyaline cartilage in the circle below. Label the chrondrocytes, lacunae, and the cartilage matrix. Compare your drawing to Figure 5.5g, p. 43.

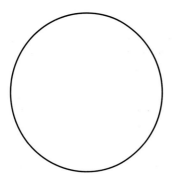

ELASTIC CARTILAGE **Elastic cartilage** can be envisioned as "hyaline cartilage with more elastic fibers." Consequently, it is much more flexible than hyaline cartilage and it tolerates repeated bending better. Essentially, only the cartilages of the external ear and the epiglottis (which flops over and covers the larynx when we swallow) are made of elastic cartilage. Focus on how this cartilage differs from hyaline cartilage as you diagram it in the circle below. Compare your illustration to Figure 5.5h, p. 44.

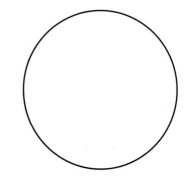

FIBROCARTILAGE **Fibrocartilage** consists of rows of chondrocytes alternating with rows of thick collagen fibers. This tissue looks like a cartilage-dense regular connective tissue hybrid, and it is always found where hyaline cartilage joins a tendon or ligament. Fibrocartilage has great tensile strength and can withstand heavy compression. Hence, its use to construct the intervertebral discs and the cartilages within the knee joint makes a lot of sense (see Figure 8.4). Sketch a section of this tissue in the circle provided and then compare your sketch to the view seen in Figure 5.5i, p. 44.

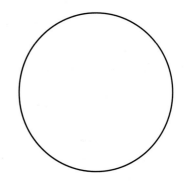

The Axial Skeleton

OBJECTIVES

1. To identify and name the three bone groups composing the axial skeleton (skull, bony thorax, and vertebral column).

2. To identify the bones composing the axial skeleton, either by examining the isolated bones or by pointing them out on an articulated skeleton or a skull, and to name the important bone markings on each.

3. To distinguish by examination the different types of vertebrae.

4. To discuss the importance of the intervertebral fibrous discs and spinal curvatures.

5. To distinguish the three abnormal spinal curvatures (lordosis, kyphosis, and scoliosis).

MATERIALS

Intact skull and Beauchene skull
X rays of individuals with scoliosis, lordosis, and kyphosis (if available)
Articulated skeleton, articulated vertebral column
Isolated cervical, thoracic, and lumbar vertebrae, sacrum, and coccyx

See Appendix B, Exercise 9 for links to A.D.A.M. Standard.

See Appendix C, Exercise 9 for links to *Anatomy and PhysioShow: The Videodisc.*

The **axial skeleton** (the green portion of Figure 8.1 on p. 58) can be divided into three parts: the skull, the vertebral column, and the bony thorax.

THE SKULL

The **skull** is composed of two sets of bones. Those of the **cranium** enclose and protect the fragile brain tissue. The **facial bones** present the eyes in an anterior position and form the base for the facial muscles, which make it possible for us to present our feelings to the world. All but one of the bones of the skull are joined by interlocking joints called *sutures;* the mandible, or lower jawbone, is attached to the rest of the skull by a freely movable joint.

The bones of the skull, shown in Figures 9.1 through 9.4, are described below. As you read through this material, identify each bone on an intact (and/or Beauchene) skull. Note that important bone markings are listed beneath the bones on which they appear and that a color-coding dot before each bone name indicates its color in the figures.

The Cranium

The cranium may be divided into two major areas for study—the **cranial vault** or **calvaria,** forming the superior, lateral, and posterior walls of the skull, and the **cranial floor** or **base,** forming the skull bottom. Internally, the cranial floor has three distinct concavities, the **anterior, middle,** and **posterior cranial fossae** (see Figure 9.4). The brain sits in these fossae, completely enclosed by the cranial vault.

Eight large flat bones construct the cranium. *With the exception of two paired bones (the parietals and the temporals), all are single bones.* Sometimes the six ossicles of the middle ear are also considered part of the cranium. Because the ossicles are functionally part of the hearing apparatus, their consideration is deferred to Exercise 19, Special Senses: Hearing and Equilibrium.

○ FRONTAL   See Figures 9.1, 9.2, and 9.4. Anterior portion of cranium; forms the forehead, superior part of the orbit, and floor of anterior cranial fossa.

Supraorbital foramen (notch): opening above each orbit allowing blood vessels and nerves to pass.
Glabella: smooth area between the eyes.

○ PARIETAL   See Figures 9.1 and 9.2. Posterolateral to the frontal bone, forming sides of cranium.

Sagittal suture: midline articulation point of the two parietal bones.
Coronal suture: point of articulation of parietals with frontal bone.

○ TEMPORAL   See Figures 9.1 through 9.4. Inferior to parietal bone on lateral skull. The temporals can be divided into four major parts: the **squamous region** abuts the parietals; the **tympanic region** surrounds the external ear opening; the **mastoid region** is the area posterior to the ear; and the **petrous region** forms the lateral region of the skull base.

Important markings associated with the flaring squamous region (Figures 9.2 and 9.3) include:

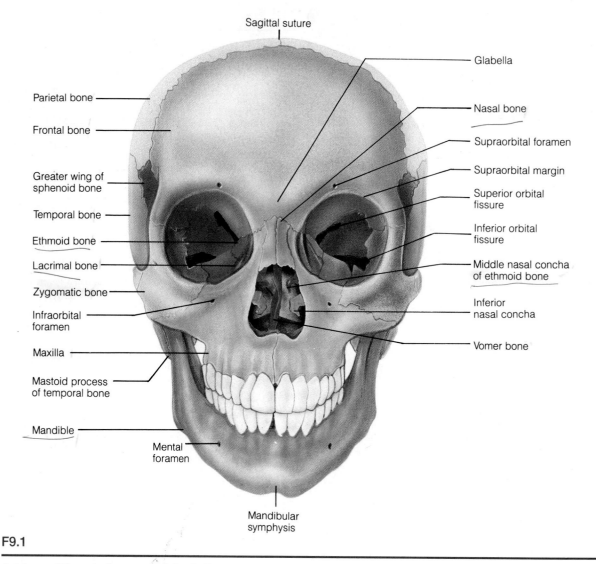

F9.1

Anatomy of the anterior aspect of the skull.

Squamosal suture: point of articulation of the temporal bone with parietal bone.

Zygomatic process: a bridgelike projection joining the zygomatic bone (cheekbone) anteriorly. Together these two bones form the *zygomatic arch.*

Mandibular fossa: rounded depression on the inferior surface of the zygomatic process (anterior to the ear); forms the socket for the mandibular condyle, the point where the mandible (lower jaw) joins the cranium.

The markings of the tympanic region (Figures 9.2 and 9.3) include:

External auditory meatus: canal leading to eardrum and middle ear.

Styloid (*stylo*=stake, pointed object) **process:** needle-like projection inferior to external auditory meatus; attachment point for muscles and ligaments of the neck. This process is often missing from (broken off) demonstration skulls.

Prominent structures in the mastoid region (Figures 9.2 and 9.3) are:

Mastoid process: rough projection inferior and posterior to external auditory meatus; attachment site for muscles.

The mastoid process is full of air cavities and is so close to the middle ear, a trouble spot for infections, that it often becomes infected too, a condition referred to as **mastoiditis.** Because the mastoid area is separated from the brain by only a very thin layer of bone, an ear infection that has spread to the mastoid process can inflame the brain coverings or the meninges. The latter condition is known as **meningitis.** ■

Stylomastoid foramen: tiny opening between the mastoid and styloid processes through which cranial nerve VII leaves the cranium.

The petrous region (Figure 9.3), which helps form the middle and posterior cranial fossae, exhibits several obvious foramina with important functions:

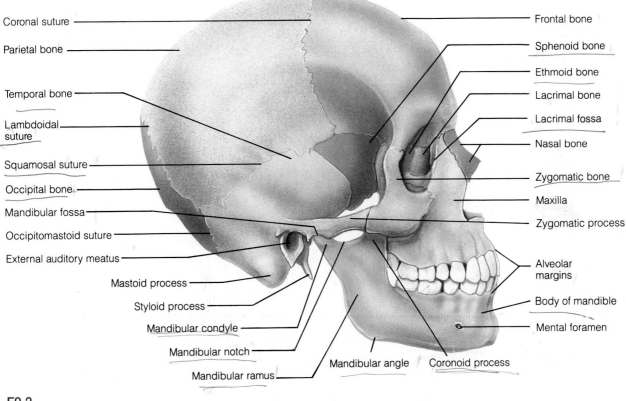

Coronal suture

Parietal bone

Temporal bone

Lambdoidal suture

Squamosal suture

Occipital bone

Mandibular fossa

Occipitomastoid suture

External auditory meatus

Mastoid process

Styloid process

Mandibular condyle

Mandibular notch

Mandibular ramus

Frontal bone

Sphenoid bone

Ethmoid bone

Lacrimal bone

Lacrimal fossa

Nasal bone

Zygomatic bone

Maxilla

Zygomatic process

Alveolar margins

Body of mandible

Mental foramen

Mandibular angle   Coronoid process

F9.2

External anatomy of the right lateral aspect of the skull.

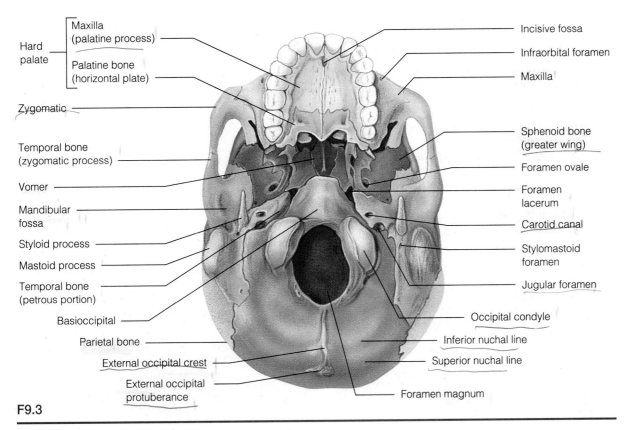

Hard palate

Maxilla (palatine process)

Palatine bone (horizontal plate)

Zygomatic

Temporal bone (zygomatic process)

Vomer

Mandibular fossa

Styloid process

Mastoid process

Temporal bone (petrous portion)

Basioccipital

Parietal bone

External occipital crest

External occipital protuberance

Incisive fossa

Infraorbital foramen

Maxilla

Sphenoid bone (greater wing)

Foramen ovale

Foramen lacerum

Carotid canal

Stylomastoid foramen

Jugular foramen

Occipital condyle

Inferior nuchal line

Superior nuchal line

Foramen magnum

F9.3

Inferior superficial view of the skull, mandible removed.

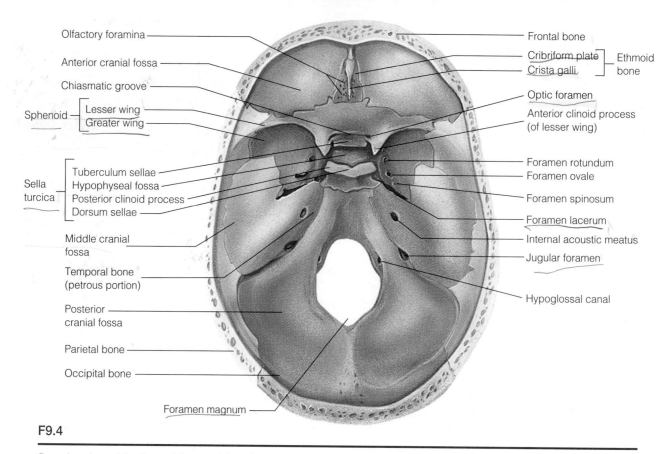

Olfactory foramina

Anterior cranial fossa

Chiasmatic groove

Sphenoid — Lesser wing / Greater wing

Sella turcica — Tuberculum sellae / Hypophyseal fossa / Posterior clinoid process / Dorsum sellae

Middle cranial fossa

Temporal bone (petrous portion)

Posterior cranial fossa

Parietal bone

Occipital bone

Foramen magnum

Frontal bone

Cribriform plate — Ethmoid bone
Crista galli

Optic foramen

Anterior clinoid process (of lesser wing)

Foramen rotundum
Foramen ovale

Foramen spinosum

Foramen lacerum

Internal acoustic meatus

Jugular foramen

Hypoglossal canal

F9.4

Superior view of the floor of the cranial cavity, calvaria removed.

Jugular foramen: opening medial to styloid process through which the internal jugular vein and cranial nerves IX, X, and XI pass.

Carotid canal: opening medial to the styloid process, through which the internal carotid artery passes into the cranial cavity.

Internal acoustic meatus: opening on posterior aspect (petrous portion) of temporal bone allowing passage of cranial nerves VII and VIII (Figure 9.4).

Foramen lacerum: a jagged opening between the petrous temporal bone and the sphenoid providing passage for a number of small nerves, and for the internal carotid artery to enter the middle cranial fossa (after it passes through part of the temporal bone).

● OCCIPITAL   See Figures 9.2, 9.3, 9.4. Most posterior bone of cranium—forms floor and back wall. Joins sphenoid bone anteriorly via its narrow basioccipital region.

Lambdoidal suture: site of articulation of occipital bone and parietal bones.

Foramen magnum: large opening in base of occipital, which allows the spinal cord to join with the brain.

Occipital condyles: rounded projections lateral to the foramen magnum that articulate with the first cervical vertebra (atlas).

Hypoglossal canal: opening medial and superior to the occipital condyle through which the hypoglossal nerve (cranial nerve XII) passes.

External occipital crest and protuberance: midline prominences posterior to the foramen magnum.
Superior and inferior nuchal lines: inconspicuous ridges that serve as sites of muscle attachment.

● SPHENOID   See Figures 9.1 through 9.4. Bat-shaped bone forming the anterior plateau of the middle cranial fossa across the width of the skull.

Greater wings: portions of the sphenoid seen exteriorly anterior to the temporal and forming a portion of the orbits of the eyes.
Superior orbital fissures: jagged openings in orbits providing passage for cranial nerves III, IV, V, and VI to enter the orbit where they serve the eye.

The sphenoid bone can be seen in its entire width if the top of the cranium (calvaria) is removed (Figure 9.4).

Sella turcica (Turk's saddle): a saddle-shaped region in the sphenoid midline which nearly encloses the pituitary gland in a living person. The pituitary gland sits in the **hypophyseal fossa** portion of the sella turcica. This fossa is abutted before and aft respectively by the **tuberculum sellae** and the **dorsum sellae.** The dorsum sellae terminates laterally in the **posterior clinoid processes.**
Lesser wings: bat-shaped portions of the sphenoid anterior to the sella turcica. Posteromedially these terminate in the pointed **anterior clinoid processes,** which provide an anchoring site for securing the brain within the skull.

68

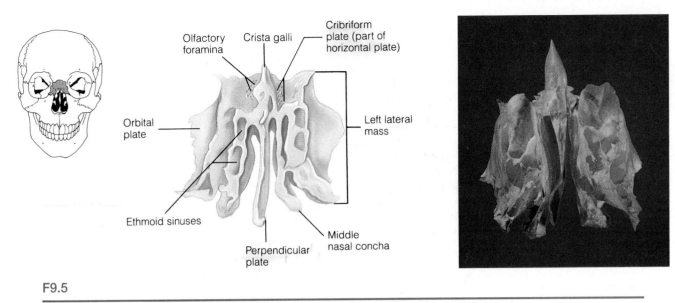

F9.5

The ethmoid bone. Anterior view.

Optic foramina: openings in the bases of the lesser wings through which the optic nerves enter the orbits to serve the eyes; these foramina are connected by the *chiasmatic groove.*

Foramen rotundum: opening lateral to sella turcica providing passage for a branch of the fifth cranial nerve. (This foramen is not visible on an inferior view of the skull.)

Foramen ovale: opening posterior to the sella turcica that allows passage of a branch of the fifth cranial nerve.

○ **ETHMOID** See Figures 9.1, 9.2, 9.4, and 9.5. Irregularly shaped bone anterior to the sphenoid. Forms the roof of the nasal cavity, upper nasal septum, and part of the medial orbit walls.

Crista galli (cock's comb): vertical projection providing a point of attachment for the dura mater (outermost membrane covering of the brain).

Cribriform plates: bony plates lateral to the crista galli through which olfactory fibers pass to the brain from the nasal mucosa. Together the cribriform plates and the midline crista galli form the *horizontal plate* of the ethmoid bone.

Perpendicular plate: Inferior projection of the ethmoid that forms the superior part of the nasal septum.

Lateral masses: Irregularly shaped thin-walled bony regions flanking the perpendicular plate laterally. Their lateral surfaces (*orbital plates*) shape part of the medial orbit wall.

Superior and middle nasal conchae (turbinates): thin, delicately coiled plates of bone extending medially from the lateral masses of the ethmoid into the nasal cavity. The conchae make air flow through the nasal cavity more efficient and greatly increase the surface area of the mucosa that covers them, thus increasing the mucosa's ability to warm and humidify incoming air.

Facial Bones

Of the 14 bones composing the face, 12 are paired. *Only the mandible and vomer are single bones.* An additional bone, the hyoid bone, although not a facial bone, is considered here because of its location. Refer to Figures 9.1 through 9.4 to find the structures described below.

○ **MANDIBLE** See Figures 9.1, 9.2, and 9.6. The lower jawbone, which articulates with the temporal bones to provide the only freely movable joints of the skull.

Body: horizontal portion; forms the chin.
Ramus: vertical extension of the body on either side.
Mandibular condyle: articulation point of the mandible with the mandibular fossa of the temporal bone.
Coronoid process: jutting anterior portion of the ramus; site of muscle attachment.
Angle: posterior point at which ramus meets the body.
Mental foramen: prominent opening on the body (lateral to the midline) that transmits the mental blood vessels and nerve to the lower jaw.
Mandibular foramen: open the lower jaw of the skull to identify this prominent foramen on the medial aspect of the mandibular ramus. This foramen permits passage of the nerve involved with tooth sensation (mandibular branch of cranial nerve V) and is the site where the dentist injects Novocain to prevent pain while working on the lower teeth.
Alveolar margin: superior margin of mandible, contains sockets in which the teeth lie.
Mandibular symphysis: anterior median depression indicating point of mandibular fusion.

○ **MAXILLAE** See Figures 9.1, 9.2, and 9.3. Two bones fused in a median suture; form the upper jawbone and part of the orbits. All facial bones, except the mandible, join the maxillae. Thus they are the main, or keystone, bones of the face.

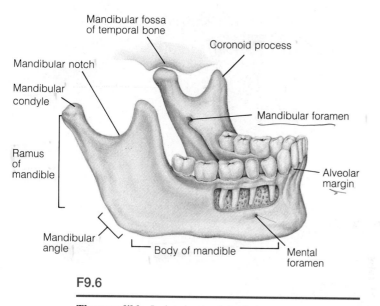

F9.6

The mandible. Isolated anterolateral view.

Alveolar margin: inferior margin containing sockets (alveoli) in which teeth lie.

Palatine processes: form the anterior hard palate.

Infraorbital foramen: opening under the orbit carrying the infraorbital nerves and blood vessels to the nasal region.

Incisive fossa: large bilateral opening located posterior to the central incisor tooth of the maxilla and piercing the hard palate; transmits the nasopalatine arteries and blood vessels.

○ PALATINE   See Figure 9.3. Paired bones posterior to the palatine processes; form posterior hard palate and part of the orbit.

○ ZYGOMATIC   See Figures 9.1, 9.2, and 9.3. Lateral to the maxilla; forms the portion of the face commonly called the cheekbone, and forms part of the lateral orbit. Its three processes are named for the bones with which they articulate.

○ LACRIMAL   See Figures 9.1 and 9.2. Fingernail-sized bones forming a part of the medial orbit walls between the maxilla and the ethmoid. Each lacrimal bone is pierced by an opening, the **lacrimal fossa,** which serves as a passageway for tears (*lacrima* means "tear").

● NASAL   See Figures 9.1 and 9.2. Small rectangular bones forming the bridge of the nose.

● VOMER (*vomer* = plow)   See Figures 9.1 and 9.3. Blade-shaped bone in median plane of nasal cavity that forms the posterior and inferior nasal septum.

○ INFERIOR NASAL CONCHAE (turbinates) See Figure 9.1. Thin curved bones protruding medially from the lateral walls of the nasal cavity;

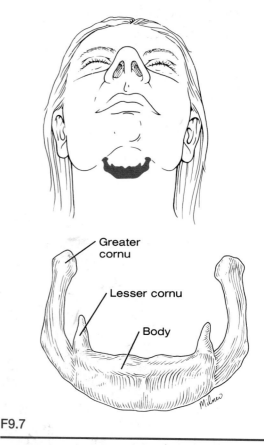

F9.7

Hyoid bone.

serve the same purpose as the turbinate portions of the ethmoid bone (described earlier).

Hyoid Bone

Not really considered or counted as a skull bone. Located in the throat above the larynx (Figure 9.7); serves as a point of attachment for many tongue and neck muscles. Does not articulate with any other bone, and is thus unique. Horseshoe-shaped with a body and two pairs of horns, or **cornua.**

Paranasal Sinuses

Four skull bones—maxillary, sphenoid, ethmoid, and frontal—contain sinuses (mucosa-lined air cavities), which lead into the nasal passages (see Figure 9.8). These paranasal sinuses lighten the facial bones and may act as resonance chambers for speech. The maxillary sinus is the largest of the sinuses found in the skull.

▲ **Sinusitis,** or inflammation of the sinuses, sometimes occurs as a result of an allergy or bacterial invasion of the sinus cavities. In such cases, some of the connecting passageways between the sinuses and nasal passages may become blocked with thick mucus or infectious material. Then, as the air in the sinus cavities is absorbed, a partial vacuum forms. The result is a sinus headache localized over the inflamed sinus area. Severe sinus infections may require surgical drainage to relieve this painful condition. ■

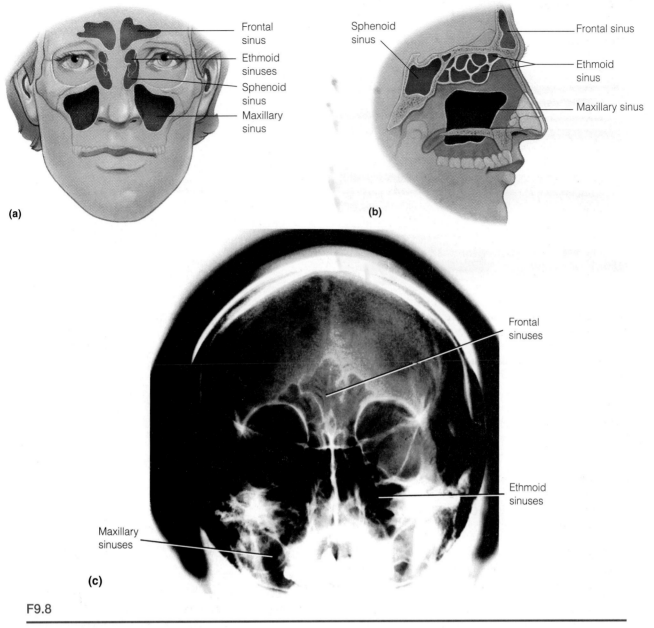

(a)

Frontal sinus

Ethmoid sinuses

Sphenoid sinus

Maxillary sinus

(b)

Sphenoid sinus

Frontal sinus

Ethmoid sinus

Maxillary sinus

(c)

Frontal sinuses

Ethmoid sinuses

Maxillary sinuses

F9.8

Paranasal sinuses. (a) Anterior "see-through" view. **(b)** As seen in a sagittal section of the head. **(c)** Skull X ray showing three of the paranasal sinuses, anterior view.

Palpation of Selected Skull Markings

Palpate the following areas on yourself:

- Zygomatic bone and arch. (The most prominent part of your cheek is your zygomatic bone. Follow the posterior course of the zygomatic arch to its junction with your temporal bone.)
- Mastoid process (the rough area behind your ear).
- Temporomandibular joints. (Open and close your jaws to locate these.)
- Greater wing of sphenoid. (Find the indentation posterior to the orbit and superior to the zygomatic arch on your lateral skull.)

- Superior orbital foramen. (Apply firm pressure along the superior orbital margin to find the indentation resulting from this foramen.)
- Inferior orbital foramen. (Apply firm pressure along the inferomedial border of the orbit to locate this large foramen.)
- Mandibular angle (most inferior and posterior aspect of the mandible).
- Mandibular symphysis (midline of chin).
- Nasal bones. (Run your index finger and thumb along opposite sides of the bridge of your nose until they "slip" medially at the inferior end of the nasal bones.)
- External occipital protuberance. (This midline projection is easily felt by running your fingers up the furrow at the back of your neck to the skull.)
- Hyoid bone. (Place a thumb high behind the lateral edge of the mandible and squeeze medially.)

THE VERTEBRAL COLUMN

The **vertebral column,** extending from the skull to the pelvis, forms the body's major axial support. Additionally, it surrounds and protects the delicate spinal cord while allowing the spinal nerves to issue from the cord via openings between adjacent vertebrae. The term *vertebral column* might suggest a rather rigid supporting rod, but this is far from the truth. The vertebral column consists of 24 single bones called **vertebrae** and two composite, or fused, bones (the sacrum and coccyx) that are connected in such a way as to provide a flexible curved structure (Figure 9.9). Of the 24 single vertebrae, the seven bones of the neck are called *cervical vertebrae;* the next 12 are *thoracic vertebrae;* and the 5 supporting the lower back are *lumbar vertebrae.* Remembering common mealtimes for breakfast, lunch, and dinner (7 A.M., 12 noon, and 5 P.M.) may help you to remember the number of bones in each region.

The vertebrae are separated by pads of fibrocartilage, **intervertebral discs,** that cushion the vertebrae and absorb shocks. Each disc is composed of two major regions, a central gelatinous *nucleus pulposus* that behaves like a fluid, and an outer ring of encircling collagen fibers called the *annulus fibrosus* that stabilizes the disc and contains the pulposus.

As a person ages, the water content of the discs decreases (as it does in other tissues throughout the body), and the discs become thinner and less compressible. This situation, along with other degenerative changes such as weakening of the ligaments and tendons of the vertebral column, predisposes older people to **ruptured discs.** A ruptured disc is a situation in which the nucleus pulposus herniates through the annulus portion and typically compresses adjacent nerves. ■

The presence of the discs and the S-shaped or springlike construction of the vertebral column prevent shock to the head in walking and running and provide flexibility to the body trunk. The thoracic and sacral curvatures of the spine are referred to as *primary curvatures,* since they are present and well developed at birth. Later the *secondary curvatures* are formed. The cervical curvature becomes prominent when the baby begins to hold its head up independently, and the lumbar curvature develops when the baby begins to walk.

 1.   Observe the normal curvature of the vertebral column in your laboratory specimen, and compare it to Figure 9.9. Then examine Figure 9.10, which depicts three abnormal spinal curvatures—*scoliosis, kyphosis,* and *lordosis.* These abnormalities may result from disease or poor posture. Also examine X rays, if they are available, showing these same conditions in a living patient.

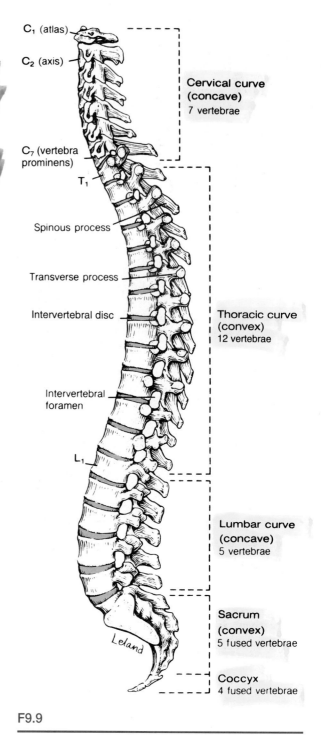

F9.9

The vertebral column. Notice the curvatures in the lateral view. (The terms *convex* and *concave* refer to the curvature of the posterior aspect of the vertebral column.)

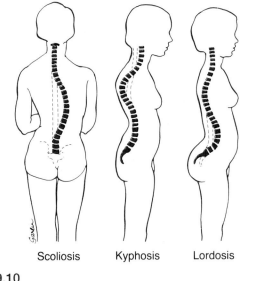

Scoliosis   Kyphosis   Lordosis

F9.10

Abnormal spinal curvatures.

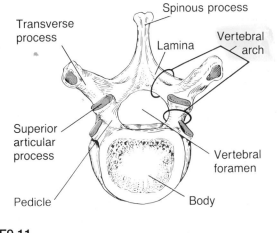

Spinous process
Transverse process
Lamina
Vertebral arch
Superior articular process
Vertebral foramen
Pedicle
Body

F9.11

A typical vertebra, superior view. Inferior articulating surfaces not shown.

2.   Using an articulated vertebral column (or an articulated skeleton), examine the freedom of movement between two lumbar vertebrae separated by an intervertebral disc.

When the fibrous disc is properly positioned, are the spinal cord or peripheral nerves impaired in any way?

_____

Remove the disc and put the two vertebrae back together. What happens to the nerve?

_____

_____

What would happen to the spinal nerves in areas of malpositioned or "slipped" discs?

_____

_____

_____

Structure of a Typical Vertebra

Although they differ in size and specific features, all vertebrae have some features in common (Figure 9.11).

Body (or centrum): rounded central portion of the vertebra, which faces anteriorly in the human vertebral column.

Vertebral arch: composed of pedicles, laminae, and a spinous process, it represents the junction of all posterior extensions from the vertebral body.

Vertebral foramen: opening enclosed by the body and vertebral arch; a conduit for the spinal cord.

Transverse processes: two lateral projections from the vertebral arch.

Spinous process: single medial and posterior projection from the vertebral arch.

Superior and inferior articular processes: paired projections lateral to the vertebral foramen that enable articulation with adjacent vertebrae. The superior articular processes typically face toward the spinous process, whereas the inferior articular processes face away from the spinous process.

Intervertebral foramina: the right and left pedicles have notches on their inferior and superior surfaces that create openings, the intervertebral foramina, for spinal nerves to leave the spinal cord between adjacent vertebrae.

Figures 9.12 and 9.13 show how specific vertebrae differ; refer to them as you read the following sections.

Cervical Vertebrae

The seven cervical vertebrae (referred to as C_1 through C_7) form the neck portion of the vertebral column. The first two cervical vertebrae (atlas and axis) are highly modified to perform special functions (see Figure 9.12). The **atlas** (C_1) lacks a body, and its lateral processes contain large concave depressions on their superior surfaces that receive the occipital condyles of the skull. This joint enables you to nod "yes." The **axis** (C_2) acts as a pivot for the rotation of the atlas (and skull) above. It bears a large vertical process, the **odontoid process,** or **dens,** which serves as the pivot point. The articulation between C_1 and C_2 allows you to rotate your head from side to side to indicate "no."

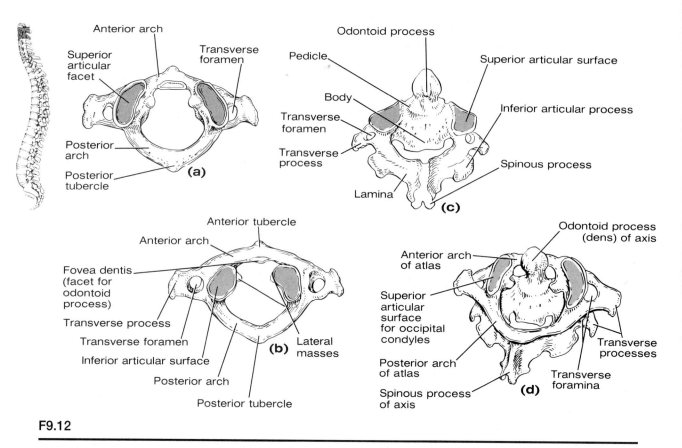

F9.12

Cervical vertebrae C_1 and C_2. (a) Superior view of the atlas (C_1). (b) Inferior view of the atlas (C_1). (c) Superior view of the axis (C_2). (d) Superior view of the articulated atlas and axis.

The more typical cervical vertebrae (C_3 through C_7) are distinguished from the thoracic and lumbar vertebrae by several features (see Figure 9.13a). They are the smallest, lightest vertebrae and the vertebral foramen is triangular. The spinous process is short and often bifurcated, or divided into two branches. The spinous process of C_7 is not branched, however, and is substantially longer than that of the other cervical vertebrae. Because the spinous process of C_7 is visible through the skin, it is called the *vertebra prominens* and is used as a landmark for counting the vertebrae. Transverse processes of the cervical vertebrae are wide, and they contain foramina through which the vertebral arteries pass superiorly on their way to the brain. Any time you see these foramina in a vertebra, you should know immediately that it is a cervical vertebra.

• Palpate your vertebra prominens.

Thoracic Vertebrae

The 12 thoracic vertebrae (referred to as T_1 through T_{12}) may be recognized by the following structural characteristics. As shown in Figure 9.13b, they have a larger body than the cervical vertebrae. The body is somewhat heart shaped, with two small articulating surfaces, or *costal demifacets,* on each side (one superior, the other inferior) close to the origin of the vertebral arch. These demifacets articulate with the heads of the correspond-

ing ribs. The vertebral foramen is oval or round, and the spinous process is long, with a sharp downward hook. The closer the thoracic vertebra is to the lumbar region, the less sharp and shorter the spinous process. Articular facets on the transverse processes articulate with the tubercles of the ribs. Besides forming the thoracic part of the spine, these vertebrae form the posterior aspect of the bony thoracic cage (rib cage). Indeed, they are the only vertebrae that articulate with the ribs.

Lumbar Vertebrae

The five lumbar vertebrae (L_1 through L_5) have massive blocklike bodies and short, thick, hatchet-shaped spinous processes extending backward horizontally (see Figure 9.13c). The superior articular facets are directed posteromedially; the inferior ones are directed anterolaterally. These structural features reduce the mobility of the lumbar region of the spine. Since most stress on the vertebral column occurs in the lumbar region, these are also the sturdiest of the vertebrae.

The spinal cord ends at the superior edge of L_2, but the outer covering of the cord, filled with cerebrospinal fluid, extends an appreciable distance beyond. Thus a *lumbar puncture* (for examination of the cerebrospinal fluid) or the administration of "saddle block" anesthesia for childbirth is normally done between L_3 and L_4 or L_4 and L_5, where there is little or no chance of injuring the delicate spinal cord.

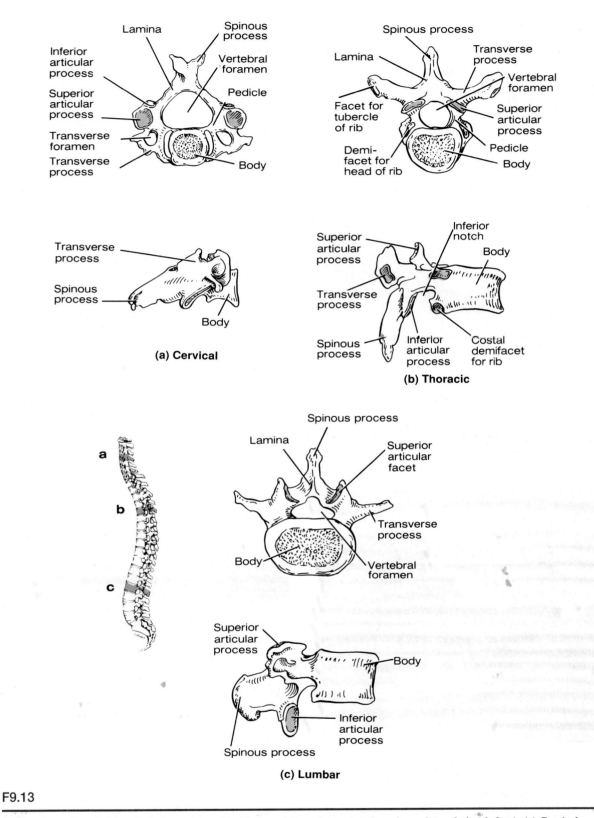

(a) Cervical

(b) Thoracic

(c) Lumbar

F9.13

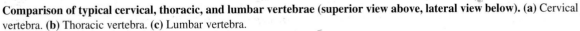

Comparison of typical cervical, thoracic, and lumbar vertebrae (superior view above, lateral view below). (a) Cervical vertebra. **(b)** Thoracic vertebra. **(c)** Lumbar vertebra.

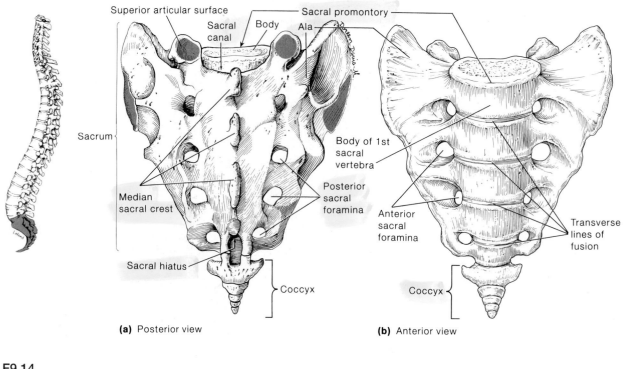

(a) Posterior view

(b) Anterior view

F9.14

Sacrum and coccyx. (a) Posterior view. **(b)** Anterior view.

The Sacrum

The **sacrum** (Figure 9.14) is a composite bone formed from the fusion of five vertebrae. Superiorly it articulates with L_5, and inferiorly it connects with the coccyx. The **median sacral crest** is a remnant of the spinous processes of the fused vertebrae. The winglike **alae,** formed by fusion of the transverse processes, articulate laterally with the hip bones. The sacrum is slightly concave anteriorly and forms the posterior border of the pelvis. Four ridges (lines of fusion) cross the anterior part of the sacrum, and **sacral foramina** are located at either end of these ridges. These foramina allow blood vessels and nerves to pass. The vertebral canal continues inside the sacrum as the **sacral canal** and terminates near the coccyx via an enlarged opening called the **sacral hiatus.** The **sacral promontory** (anterior border of the body of S_1) is an important anatomical landmark for obstetricians.

- Attempt to palpate the median sacral crest of your sacrum. (This is more easily done by thin people and [obviously] in privacy.)

The Coccyx

The **coccyx** (see Figure 9.14) is formed from the fusion of three to five small irregularly shaped vertebrae. It is literally the human tailbone, a vestige of the tail that other vertebrates have. The coccyx is attached to the sacrum by ligaments.

Obtain examples of each type of vertebra and examine them carefully, comparing them to Figures 9.12, 9.13, and 9.14 and to each other.

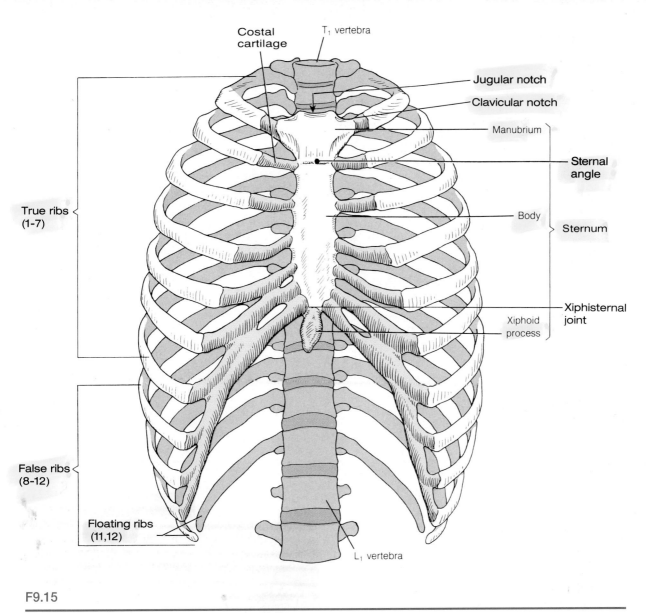

Costal cartilage · **T₁ vertebra**

Jugular notch

Clavicular notch

Manubrium

Sternal angle

True ribs (1–7)

Body

Sternum

Xiphisternal joint

Xiphoid process

False ribs (8–12)

Floating ribs (11,12)

L₁ vertebra

F9.15

Bony thorax, anterior view. (The pink areas are cartilages.)

THE BONY THORAX

The **bony thorax** is composed of the sternum, ribs, and thoracic vertebrae (Figure 9.15). It is also referred to as the **thoracic cage** because of its appearance and because it forms a protective cone-shaped enclosure around the organs of the thoracic cavity (heart and lungs, for example).

The Sternum

The **sternum** (breastbone), a typical flat bone, is a result of the fusion of three bones—the manubrium, body, and xiphoid process. It is attached to the first seven pairs of ribs. The superiormost **manubrium** looks like the knot of a tie; it articulates with the clavicle (collarbone) laterally. The **body (gladiolus)** forms the bulk of the sternum. The **xiphoid process** constructs the inferior end of

the sternum and lies at the level of the fifth intercostal space. Although it is made of hyaline cartilage in children, it is usually ossified in adults.

In some people, the xiphoid process projects dorsally. This may present a problem because physical trauma to the chest can push such a xiphoid into the heart or liver (both immediately deep to the process), causing massive hemorrhage. ■

The sternum has three important bony landmarks—the jugular notch and the sternal angle. The **jugular notch** (concave upper border of the manubrium) can be palpated easily; generally it is at the level of the third thoracic vertebra. The **sternal angle** is a result of the manubrium and body meeting at a slight angle to each other, so that a transverse ridge is formed at the level of the second ribs. It provides a handy reference point for counting ribs to locate the second intercostal space for

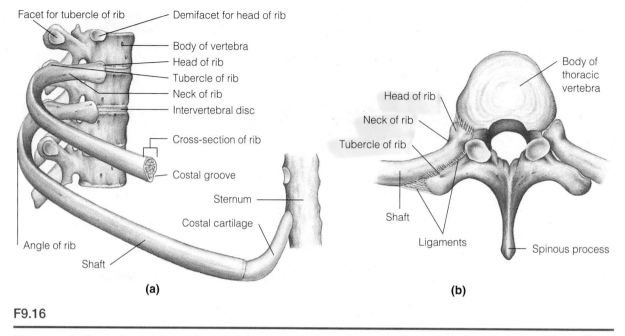

Facet for tubercle of rib
Demifacet for head of rib
Body of vertebra
Head of rib
Tubercle of rib
Neck of rib
Intervertebral disc
Cross-section of rib
Costal groove
Sternum
Costal cartilage
Angle of rib
Shaft

(a)

Head of rib
Neck of rib
Tubercle of rib
Shaft
Ligaments
Body of thoracic vertebra
Spinous process

(b)

F9.16

Structure of a "typical" true rib and its articulations. (a) Vertebral and sternal articulations of a typical true rib. **(b)** Superior view of the articulation between a rib and a thoracic vertebra, with costovertebral ligaments shown on left side only.

listening to certain heart valves, and is an important anatomical landmark for thoracic surgery. The **xiphisternal joint,** the point where the sternal body and xiphoid process fuse, lies at the level of the ninth thoracic vertebra.

- Palpate your sternal angle and jugular notch.

Because of its accessibility, the sternum is a favored site for obtaining samples of blood-forming (hematopoietic) tissue for the diagnosis of suspected blood diseases. A needle is inserted into the marrow of the sternum and the sample withdrawn (sternal puncture).

The Ribs

The 12 pairs of **ribs** form the walls of the thoracic cage (see Figures 9.15 and 9.16). All of the ribs articulate

posteriorly with the vertebral column via their heads and tubercles and then curve downward and toward the anterior body surface. The first seven pairs, called the *true,* or *vertebrosternal, ribs,* attach directly to the sternum by their "own" costal cartilages. The next three pairs, called *false,* or *vertebrochondral, ribs,* have indirect cartilage attachments to the sternum. The last two pairs are referred to as *floating,* or *vertebral,* ribs. Since these ribs have no sternal attachment at all, they are also called *false ribs.*

First take a deep breath to expand your chest. Notice how your ribs seem to move outward and how your sternum rises. Then examine an articulated skeleton to observe the relationship between the ribs and the vertebrae.

The Appendicular Skeleton

OBJECTIVES

1. To identify on an articulated skeleton the bones of the pectoral and pelvic girdles and their attached limbs.

2. To arrange unmarked, disarticulated bones in their proper relative position to form the entire skeleton.

3. To differentiate between a male and a female pelvis.

4. To discuss the common features of the human appendicular girdles (pectoral and pelvic), and to note how their structure relates to their specialized functions.

5. To identify specific bone markings in the appendicular skeleton.

MATERIALS

Articulated skeletons
Disarticulated skeletons (complete)
Articulated pelves (male and female for comparative study)
X rays of bones of the appendicular skeleton

See Appendix B, Exercise 10 for links to A.D.A.M. Standard.

See Appendix C, Exercise 10 for links to *Anatomy and PhysioShow: The Videodisc.*

The **appendicular skeleton** (the gold-colored portion of Figure 8.1) is composed of the 126 bones of the appendages and the pectoral and pelvic girdles, which attach the limbs to the axial skeleton. Although the bones of the upper and lower limbs are quite different in their functions and mobility, they have the same fundamental plan, with each limb composed of three major segments connected together by freely movable joints.

Carefully examine each of the bones described and identify the characteristic bone markings of each. The markings aid in determining whether a bone is the right or left member of its pair. *This is a very important instruction because, before* *completing this laboratory exercise, you will be constructing your own skeleton.* Additionally, when corresponding X rays are available, compare the actual bone specimen to its X-ray image.

BONES OF THE PECTORAL GIRDLE AND UPPER EXTREMITY

The Pectoral Girdle

The paired **pectoral,** or **shoulder, girdles** (Figure 10.1) each consist of two bones—the anterior clavicle and the posterior scapula. The shoulder girdles function to at-

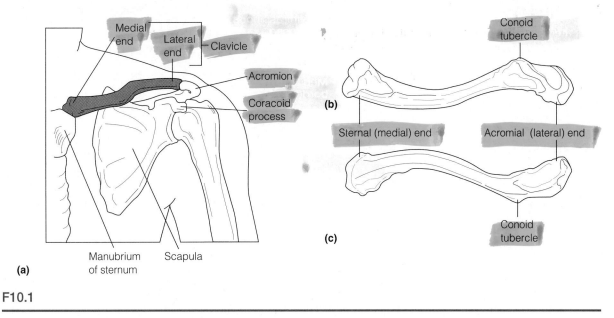

(a)

F10.1

Bones of the pectoral (shoulder) girdle. (a) Left pectoral girdle articulated to show the relationship of the girdle to the bones of the thorax and arm. **(b)** Left clavicle, superior view. **(c)** Left clavicle, inferior view.

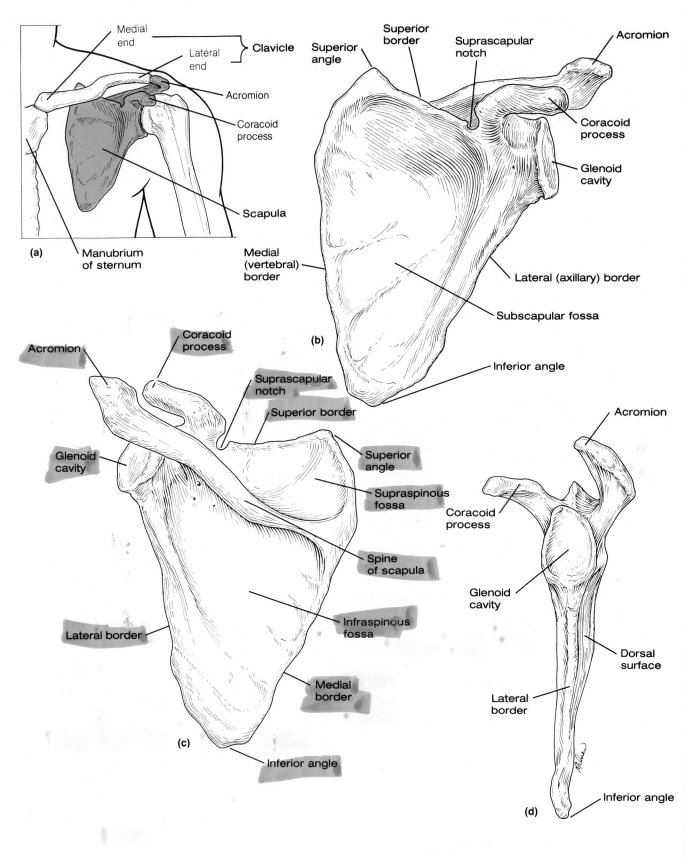

F10.2

Bones of the pectoral (shoulder) girdle. (a) Left pectoral girdle articulated to show the relationship of the girdle to the bones of the thorax and arm. (b) Left scapula, anterior view. (c) Left scapula, posterior view. (d) Left scapula, lateral view.

tach the upper limbs to the axial skeleton. In addition, the bones of the shoulder girdles serve as attachment points for many trunk and neck muscles.

The **clavicle,** or collarbone, is a slender doubly curved bone—convex forward on its medial two-thirds and concave laterally. Its *sternal* (medial) *end,* which attaches to the sternal manubrium, is rounded or triangular in cross section. The sternal end projects above the manubrium and can be felt and (usually) seen forming the lateral walls of the *jugular notch* (see Figure 9.15, p. 77). The *acromial* (lateral) *end* of the clavicle is flattened where it articulates with the scapula to form part of the shoulder joint. On its posteroinferior surface is the prominent **conoid tubercle.** This projection serves to attach a ligament and provides a handy landmark for determining whether a given clavicle is from the right or left side of the body. The clavicle serves as an anterior brace, or strut, to hold the arm away from the top of the thorax.

The **scapulae** (Figure 10.2), or shoulder blades, are generally triangular and are commonly called the "wings" of humans. Each scapula has a flattened body and two important processes—the **acromion** (the enlarged end of the spine of the scapula) and the beaklike **coracoid process** (*corac* = crow, raven). The acromion connects with the clavicle; the coracoid process points anteriorly over the tip of the shoulder joint and serves as a point of attachment for some of the muscles of the upper limb. The **suprascapular notch** at the base of the coracoid process allows nerves to pass. The scapula has no direct attachment to the axial skeleton but is loosely held in place by trunk muscles.

The scapula has three angles: superior, inferior, and lateral. The inferior angle provides a landmark for auscultating (listening to) lung sounds. The scapula also has three named borders: superior, medial (vertebral), and lateral (axillary). Several shallow depressions (fossae) appear on both sides of the scapula and are named according to location; i.e., there are the *anterior subscapular fossa* and the *posterior infraspinous* and *supraspinous fossae.* The **glenoid cavity,** a shallow socket that receives the head of the arm bone, is located in the lateral angle.

The shoulder girdle is exceptionally light and allows the upper limb a degree of mobility not seen anywhere else in the body. This is due to the following factors:

- The sternoclavicular joints are the *only* site of attachment of the shoulder girdles to the axial skeleton.
- The relative looseness of the scapular attachment allows it to slide back and forth against the thorax with muscular activity.
- The glenoid cavity is shallow, and does little to stabilize the shoulder joint.

However, this exceptional flexibility exacts a price: the arm bone (humerus) is very susceptible to dislocation, and fracture of the clavicle disables the entire upper limb.

The Arm

The arm (Figure 10.3) consists of a single bone—the **humerus,** a typical long bone. At its proximal end is the rounded *head,* which fits into the shallow glenoid cavity of the scapula. The head is separated from the shaft by the *anatomical neck* and the more constricted *surgical neck,* which is a common site of fracture. Opposite the head are two prominences, the **greater** and **lesser tubercles** (from lateral to medial aspect), separated by a groove (the **intertubercular** or **bicipital groove**) that guides the tendon of the biceps muscle to its point of attachment (the superior rim of the glenoid cavity). In the midpoint of the shaft is a roughened area, the **deltoid tuberosity,** where the large fleshy shoulder muscle, the deltoid, attaches. Just inferior to the deltoid tuberosity is the **radial groove,** which indicates the pathway of the radial nerve.

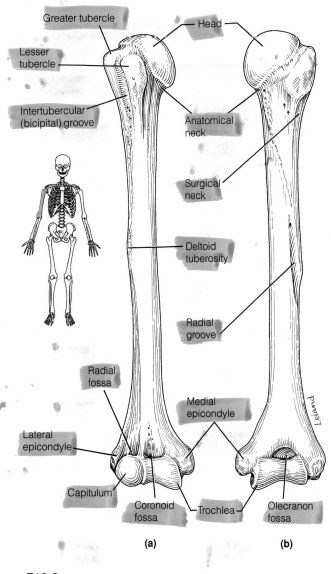

Bones of the right arm. (a) Humerus, anterior view. **(b)** Humerus, posterior view.

At the distal end of the humerus are two condyles—the medial **trochlea** (looking rather like a spool), which articulates with the ulna, and the lateral **capitulum,** which articulates with the radius of the forearm. This condyle pair is flanked medially by the **medial epicondyle** and laterally by the **lateral epicondyle.**

The medial epicondyle is commonly referred to as the "funny bone." The large ulnar nerve runs in a groove beneath the medial epicondyle, and when this region is sharply bumped, we are quite likely to experience a temporary, but excruciatingly painful, tingling sensation. This event is called "hitting the funny bone," a strange expression, because it is certainly *not* funny!

Above the trochlea on the anterior surface is a depression, the **coronoid fossa;** on the posterior surface is the **olecranon fossa.** These two depressions allow the corresponding processes of the ulna to move freely when the elbow is flexed and extended. A small **radial fossa,** lateral to the coronoid fossa, receives the head of the radius when the elbow is flexed.

The Forearm

Two bones, the radius and the ulna, compose the skeleton of the forearm, or antebrachium (see Figure 10.4). When the body is in the anatomical position, the **radius** is in the lateral position in the forearm and the radius and ulna are parallel. Proximally, the disc-shaped head of the radius articulates with the capitulum of the humerus. Just below the head, on the medial aspect of the shaft, is a prominence called the **radial tuberosity,** the point of attachment for the tendon of the biceps muscle of the arm. Distally, the small **ulnar notch** reveals where it articulates with the end of the ulna.

The **ulna** is the medial bone of the forearm. Its proximal head bears the anterior **coronoid process** and the posterior **olecranon process,** which are separated by the **trochlear notch.** Together these processes grip the trochlea of the humerus in a plierslike joint. The small **radial notch** on the lateral side of the coronoid process articulates with the head of the radius. The slimmer distal end of the ulna bears a small medial **styloid process,** which serves as a point of attachment for the ligaments of the wrist.

The Wrist

The wrist is referred to anatomically as the **carpus,** and the eight bones composing it are the **carpals.** The carpals are arranged in two irregular rows of four bones each, which are illustrated in Figure 10.5. In the proximal row (lateral to medial) are the scaphoid, lunate, triangular, and pisiform bones; the scaphoid and lunate articulate with the distal end of the radius. In the distal row are the trapezium, trapezoid, capitate, and hamate. The carpals are bound closely together by ligaments, which restrict movements between them.

The Hand

The hand, or manus (see Figure 10.5), consists of two groups of bones: the **metacarpals** (bones of the palm) and the **phalanges** (bones of the fingers). The metacarpals are numbered 1 to 5 from the thumb side of the hand toward the little finger. When the fist is clenched, the heads of the metacarpals become prominent as the knuckles. Each hand contains 14 phalanges. There are 3 phalanges in each finger except the thumb, which has only proximal and distal phalanges.

Surface Anatomy of the Pectoral Girdle and the Upper Limb

Before continuing on to study the bones of the pelvic girdle, take the time to identify the following bone markings related to the upper appendage on the skin surface. It is usually preferable to observe and palpate the bone markings on your lab partner, particularly since many of these markings can only be seen from the dorsal aspect.

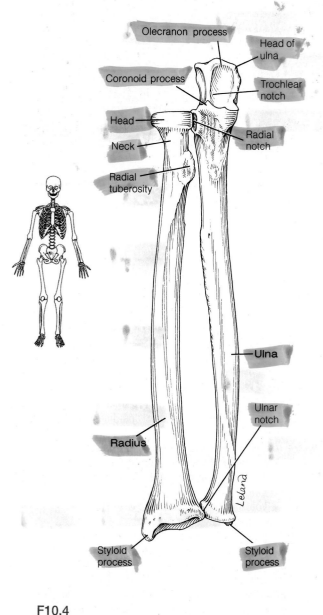

Olecranon process
Head of ulna
Coronoid process
Trochlear notch
Head
Radial notch
Neck
Radial tuberosity
Ulna
Ulnar notch
Radius
Leland
Ulna
Styloid process
Styloid process

F10.4

Bones of the right forearm. Radius and ulna, anterior view.

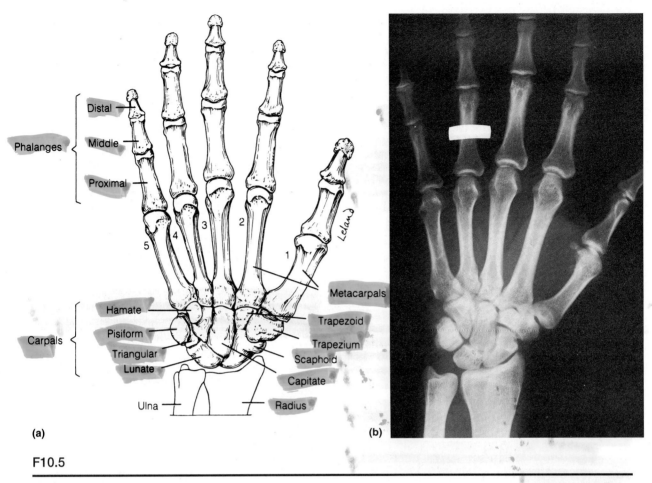

(a)

(b)

F10.5

Bones of the right wrist and hand. (a) Anterior view showing the relationships of the carpals, metacarpals, and phalanges. (b) X ray. White bar on phalanx 1 of the ring finger shows the position at which a ring would be worn.

- Clavicle: Palpate the clavicle along its entire length from sternum to shoulder.
- Acromioclavicular joint: The high point of the shoulder, which represents the junction point between the clavicle and the acromion of the scapular spine.
- Spine of the scapula: Extend your arm at the shoulder so that your scapula is moved posteriorly. As you do this, your scapular spine will be seen as a winglike protrusion on your dorsal thorax and can be easily palpated by your lab partner.
- Lateral epicondyle of the humerus: The inferior-most projection at the lateral aspect of the distal humerus. After you have located the epicondyle, run your finger posteriorly into the hollow immediately dorsal to the epicondyle. This is the site where the extensor muscles of the hand are attached and is a common site of the often excruciating pain of tennis elbow, a condition in which those muscles and their tendons are abused physically.
- Medial epicondyle of the humerus: Feel this medial projection at the distal end of the humerus.

- Olecranon process of the ulna: Work your elbow—flexing and extending—as you palpate its dorsal aspect to feel the olecranon process of the ulna moving into and out of the olecranon fossa on the dorsal aspect of the humerus.
- Styloid process of the ulna: With the hand in the anatomical position, feel out this small inferior projection on the medial aspect of the distal end of the ulna.
- Styloid process of the radius: Find this projection at the distal end of the radius (lateral aspect). It is most easily located by moving the hand medially at the wrist. Once you have palpated the styloid process, move your fingers just medially to the process (onto the anterior wrist). Press firmly and then let up slightly on the pressure. You should be able to feel your pulse at this pressure point, which lies over the radial artery (radial pulse).
- Metacarpophalangeal joints (knuckles): Clench your fist and find the first set of flexed-joint protrusions beyond the wrist—these are your metacarpophalangeal joints.

BONES OF THE PELVIC GIRDLE AND LOWER LIMB

The Pelvic Girdle

The **pelvic girdle,** or **hip girdle,** (Figure 10.6) is formed by the two **coxal bones** (**ossa coxae** or hip bones). The two coxal bones together with the sacrum and coccyx form the **bony pelvis.** In contrast to the bones of the shoulder girdle, those of the pelvic girdle are heavy and massive, and they are attached securely to the axial skeleton. The sockets for the heads of the femurs (thigh bones) are deep and heavily reinforced by ligaments to ensure a stable, strong limb attachment. The ability to bear weight is more important here than exceptional mobility and flexibility. The combined weight of the upper body rests on the pelvis (specifically, where the hip bones meet the sacrum).

Each coxal bone is a result of the fusion of three bones—the ilium, ischium, and pubis—which are distinguishable in the young child. The **ilium** is a large flaring bone forming the major portion of the coxal bone. It connects posteriorly, via its **auricular surface,** with the sacrum at the **sacroiliac joint.** The superior margin of the iliac bone, the **iliac crest,** is rough; when you rest your hands on your hips, you are palpating your iliac crests. The iliac crest terminates anteriorly in the **anterior superior spine** and posteriorly in the **posterior superior spine.** Two inferior spines are located below these. The shallow **iliac fossa** marks its internal surface, and a shallow ridge, the **arcuate line,** outlines the pelvic inlet, or pelvic brim.

The **ischium** is the "sit-down" bone, forming the most inferior and posterior portion of the coxal bone. The most outstanding marking on the ischium is the **ischial tuberosity,** which receives the weight of the body when sitting. The **ischial spine,** superior to the ischial tuberosity, is an important anatomical landmark of the pelvic cavity. (See Comparison of the Male and Female Pelves, below.) The obvious **lesser** and **greater sciatic notches** allow nerves and blood vessels to pass to and from the thigh. The sciatic nerve passes through the latter.

The **pubis** is the most anterior portion of the coxal bone. Fusion of the **rami** of the pubic bone anteriorly and the ischium posteriorly forms a bar of bone enclosing the **obturator foramen,** through which blood vessels and nerves run from the pelvic cavity into the thigh. The pubic bones of each hip bone meet anteriorly at the **pubic crest** to form a cartilaginous joint called the **pubic symphysis.** At the lateral end of the pubic crest is the *pubic tubercle* (not obvious in Figure 10.6) to which the important *inguinal ligament* attaches.

The ilium, ischium, and pubis fuse at the deep hemispherical socket called the **acetabulum** (literally, "vinegar cup"), which receives the head of the thigh bone.

Before continuing with the bones of the lower limbs, take the time to examine an articulated pelvis. Note how each coxal bone articulates with the sacrum posteriorly and how the two coxal bones join at the pubic symphysis. The sacroiliac joint, because of the pressure it must bear, is a common site of lower back problems.

COMPARISON OF THE MALE AND FEMALE PELVES Although bones of males are usually larger, heavier, and have more prominent bone markings, the male and female skeletons are very similar. The outstanding exception to this generalization is pelvic structure (Figure 10.7).

The female pelvis reflects modifications for childbearing. Generally speaking, the female pelvis is wider, shallower, lighter, and rounder than that of the male. Not only must her pelvis support the increasing size of a fetus, but it must also be large enough to allow the infant's head (its largest dimension) to descend through the birth canal at birth.

To describe pelvic sex differences, a few more terms must be introduced. Anatomically, the pelvis can be described in terms of a false pelvis and a true pelvis. The **false pelvis** is that portion superior to the arcuate line; it is bounded by the alae of the ilia laterally and the sacral promontory and lumbar vertebrae posteriorly. Although the false pelvis supports the abdominal viscera, it does not restrict childbirth in any way. The **true pelvis** is the region inferior to the arcuate line that is almost entirely surrounded by bone. Its posterior boundary is formed by the sacrum. The ilia, ischia, and pubic bones define its limits laterally and anteriorly.

The dimensions of the true pelvis, particularly its inlet and outlet, are critical if delivery of a baby is to be uncomplicated; and they are carefully measured by the obstetrician. The **pelvic inlet,** or **pelvic brim,** is the opening delineated by the sacral promontory posteriorly and the arcuate lines of the ilia anterolaterally. It is the superiormost margin of the true pelvis. Its widest dimension is from left to right, that is, along the frontal plane. The **pelvic outlet** is the inferior margin of the true pelvis. It is bounded anteriorly by the pelvic arch, laterally by the ischia, and posteriorly by the sacrum and coccyx. Since both the coccyx and the ischial spines protrude into the outlet opening, a sharply angled coccyx or large, sharp ischial spines can dramatically narrow the outlet. The largest dimension of the outlet is the anterior-posterior diameter.

The major differences between the male and female pelves are summarized below (see Figure 10.7).

Examine male and female pelves for the following differences:

- The female inlet is larger and more circular.
- The female pelvis as a whole is shallower, and the bones are lighter and thinner.
- The female sacrum is broader and less curved, and the pubic arch is more rounded.
- The female acetabula are smaller and farther apart, and the ilia flare more laterally.
- The female ischial spines are shorter, farther apart, and everted, thus enlarging the pelvic outlet.

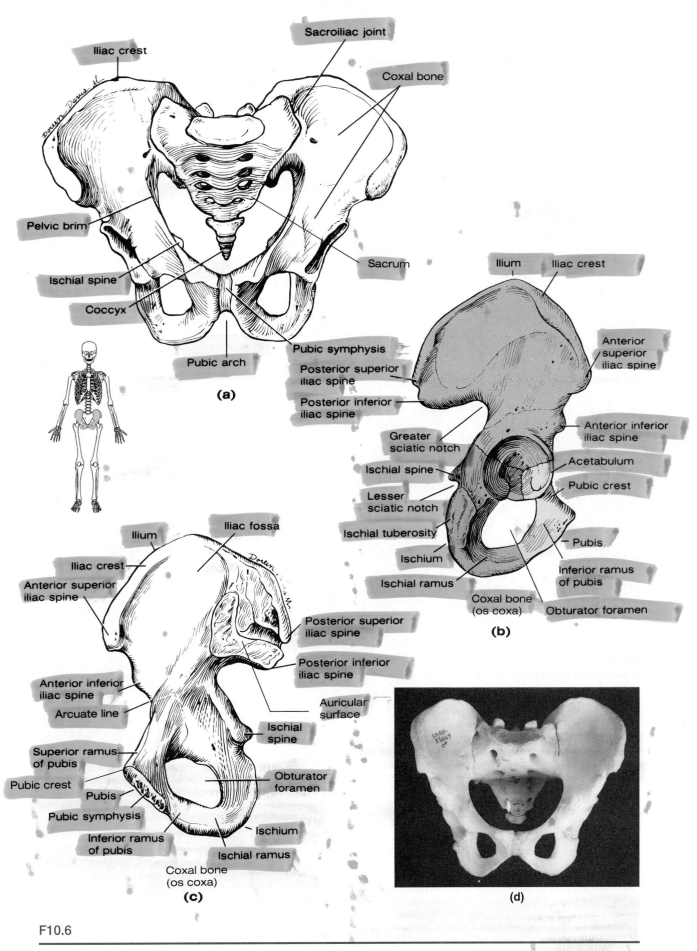

Iliac crest

Sacroiliac joint

Coxal bone

Pelvic brim

Ischial spine

Coccyx

Pubic arch

Sacrum

Pubic symphysis

(a)

Ilium   Iliac crest

Anterior superior iliac spine

Posterior superior iliac spine

Posterior inferior iliac spine

Greater sciatic notch

Ischial spine

Lesser sciatic notch

Ischial tuberosity

Ischium

Ischial ramus

Coxal bone (os coxa)

Anterior inferior iliac spine

Acetabulum

Pubic crest

Pubis

Inferior ramus of pubis

Obturator foramen

(b)

Ilium   Iliac fossa

Iliac crest

Anterior superior iliac spine

Anterior inferior iliac spine

Arcuate line

Superior ramus of pubis

Pubic crest

Pubis

Pubic symphysis

Inferior ramus of pubis

Posterior superior iliac spine

Posterior inferior iliac spine

Auricular surface

Ischial spine

Obturator foramen

Ischium

Ischial ramus

Coxal bone (os coxa)

(c)

(d)

F10.6

Bones of the pelvic girdle. (a) Articulated bony pelvis, showing the two coxal bones, which together comprise the pelvic girdle, and the sacrum. **(b)** Right coxal bone, lateral view, showing the point of fusion of the ilium, ischium, and pubic bones. **(c)** Right coxal bone, medial view. **(d)** Photograph of male pelvis.

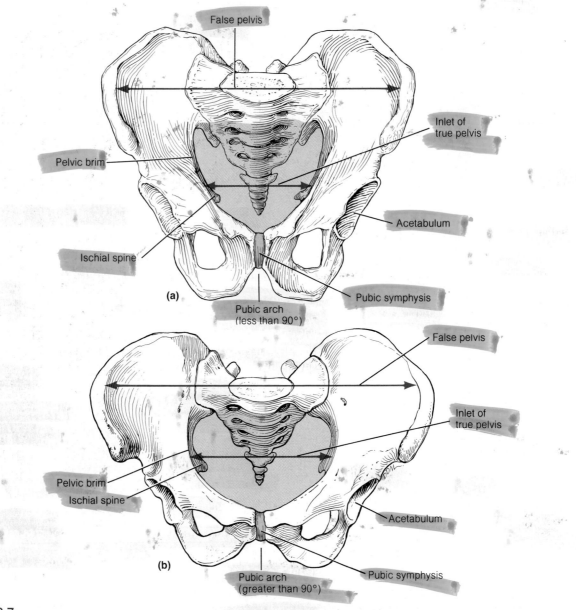

(a)

(b)

F10.7

Comparison of male and female pelves, anterior views. (a) Male. **(b)** Female.

The Thigh

The **femur,** or thigh bone (Figure 10.8), is the sole bone of the thigh and is the heaviest, strongest bone in the body. The ball-like head of the femur articulates with the hip bone via the deep, secure socket of the acetabulum. Obvious in the femur's head is a small central pit called the **fovea capitis** ("pit of the head") from which a small ligament runs to the acetabulum. The head of the femur is carried on a short, constricted *neck,* which angles laterally to join the shaft. The neck is the weakest part of the femur and is a common fracture site (an injury called a broken hip), particularly in the elderly. At the junction of the shaft and neck are the **greater** and **lesser trochanters** (separated posteriorly by the **intertrochanteric crest** and anteriorly by the **intertrochanteric line**).

The femur inclines medially as it runs downward to the leg bones; this brings the knees in line with the body's center of gravity, or maximum weight. The medial course of the femur is even more noticeable in females because of the wider female pelvis.

Distally, the femur terminates in the **lateral** and **medial condyles,** which articulate with the tibia below, and the **patellar surface,** which forms a joint with the patella anteriorly (see Figure 8.1, p. 58). The **lateral** and **medial epicondyles,** just superior to the condyles, are separated by the **intercondylar notch.**

The trochanters and trochanteric crest, as well as the **gluteal tuberosity** and the **linea aspera** located on the shaft, are sites of muscle attachment.

The Leg

Two bones, the tibia and the fibula, form the skeleton of the leg (see Figure 10.9). The **tibia,** or *shinbone,* is the larger and more medial of the two leg bones. At the

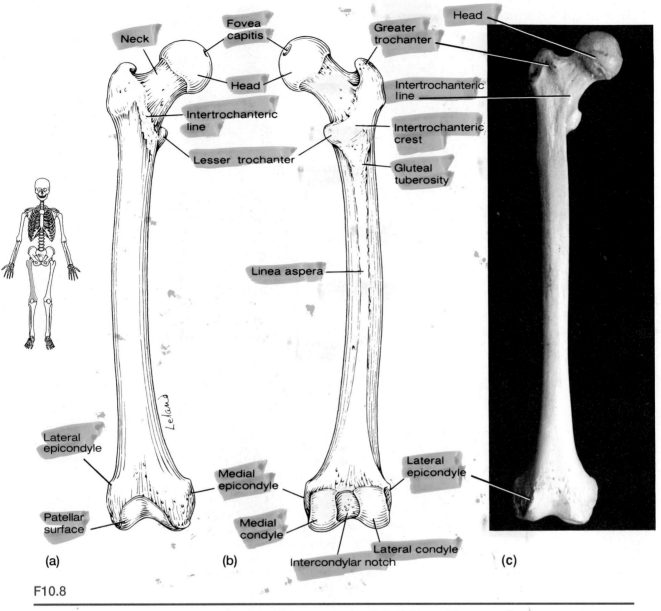

F10.8

Bone of the right thigh. (a) Femur, anterior view. **(b)** Femur, posterior view. **(c)** Photo of femur, anterior view.

proximal end, the **medial** and **lateral condyles** (separated by the **intercondylar eminence**) receive the distal end of the femur to form the knee joint. The **tibial tuberosity,** a roughened protrusion on the anterior tibial surface (just below the condyles), is the site of attachment of the patellar (kneecap) ligament. Small facets on its superior and inferior lateral surface articulate with the fibula. Distally, a process called the **medial malleolus** forms the inner (medial) bulge of the ankle, and the smaller distal end articulates with the talus bone of the foot. The anterior surface of the tibia is a sharpened ridge (anterior crest) that is relatively unprotected by muscles. It is easily felt beneath the skin.

The **fibula,** which lies parallel to the tibia, takes no part in forming the knee joint. Its proximal head articulates with the lateral condyle of the tibia. The fibula is thin and sticklike with a sharp anterior crest. It terminates distally in the **lateral malleolus,** which forms the outer part, or lateral bulge, of the ankle.

The Foot

The bones of the foot include the 7 **tarsal** bones, 5 **metatarsals,** which form the instep, and 14 **phalanges,** which form the toes (see Figure 10.10). Body weight is concentrated on the two largest tarsals which form the posterior aspect of the foot, the *calcaneus* (heel bone) and the *talus,* which lies between the tibia and the calcaneus. The other tarsals are named and identified in Figure 10.10. Like the fingers of the hand, each toe has 3 phalanges except the great toe, which has 2.

The bones in the foot are arranged to produce three strong arches—two longitudinal arches (medial and lateral) and one transverse arch (Figure 10.11). Ligaments, binding the foot bones together, and tendons of the foot muscles hold the bones firmly in the arched position but still allow a certain degree of give. Weakened arches are referred to as fallen arches or flat feet.

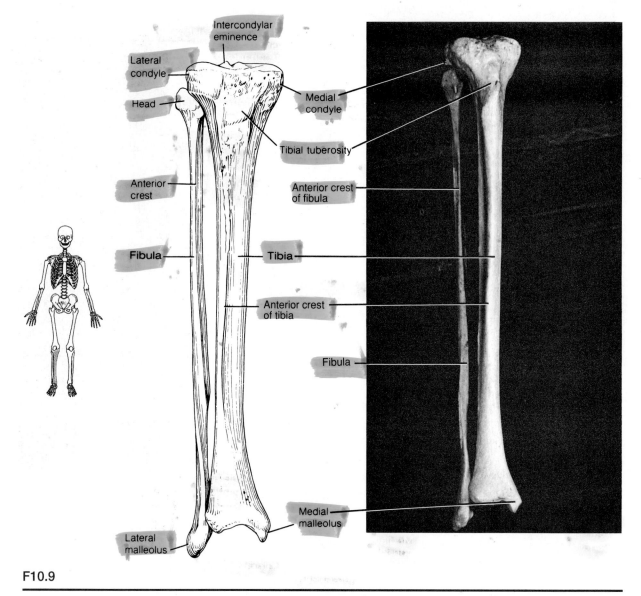

F10.9

Bones of the right leg. Tibia and fibula, anterior view.

☯ Surface Anatomy of the Pelvic Girdle and Lower Limb

Locate and palpate the following bone markings on yourself and/or your lab partner.

- Iliac crest and anterior superior iliac spine: Rest your hands on your hips—they will be overlying the iliac crests. Trace the crest as far posteriorly as you can and then follow it anteriorly to the anterior superior iliac spine. This latter bone marking is easily felt in almost everyone, and is clearly visible through the skin (and perhaps the clothing) of very slim people. (The posterior superior iliac spine is much less obvious and is usually indicated only by a dimple in the overlying skin. Check it out in the mirror tonight.)

- Greater trochanter of the femur: This is easier to locate in females than in males because of the wider female pelvis; also it is more likely to be clothed by bulky muscles in males. Try to locate it on yourself as the most lateral point of the proximal femur. It typically lies about 6–8 inches below the iliac crest.
- Patella and tibial tuberosity: Feel your kneecap and palpate the ligaments attached to its borders. Follow the inferior patellar ligament to the tibial tuberosity.
- Medial and lateral condyles of the femur and tibia: As you move from the patella inferiorly on the medial (and then the lateral) knee surface, you will feel first the femoral and then the tibial condyle.
- Medial malleolus: Feel the medial protrusion of your ankle, the medial malleolus of the distal tibia.
- Lateral malleolus: Feel the bulge of the lateral aspect of your ankle, the lateral malleolus of the fibula.
- Calcaneus: Attempt to follow the extent of your calcaneus or heel bone.

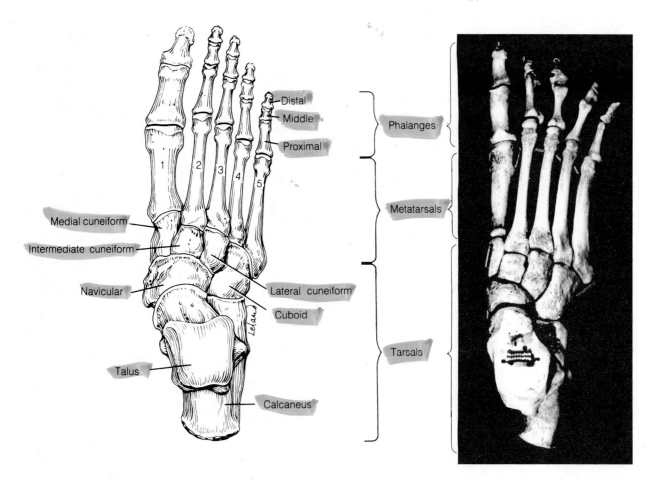

F10.10

Bones of the right ankle and foot, superior view.

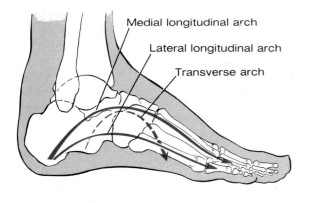

F10.11

Arches of the foot.

APPLYING KNOWLEDGE: CONSTRUCTING A SKELETON

1.   When you finish examining the disarticulated bones of the appendicular skeleton and yourself, work with your lab partner to arrange the disarticulated bones on the laboratory bench in their proper relative positions to form an entire skeleton. Careful observations of the bone markings should help you distinguish between right and left members of bone pairs.

2.   When you believe that you have accomplished this task correctly, ask the instructor to check your arrangement to ensure that it is correct. If it is not, go to the articulated skeleton and check your bone arrangements. Also review the descriptions of the bone markings as necessary to correct your bone arrangement.

The Fetal Skeleton

OBJECTIVES	MATERIALS
1. To define *fontanel* and discuss the function and fate of fontanels in the fetus. **2.** To demonstrate important differences between the fetal and adult skeletons.	Isolated fetal skull Fetal skeleton Adult skeleton See Appendix C, Exercise 11 for links to *Anatomy and PhysioShow: The Videodisc.*

A human fetus about to be born has 275 bones, many more than the 206 bones found in the adult skeleton. This is because many of the bones described as single bones in the adult skeleton (for example, the coxal bone, sternum, and sacrum) have not yet fully ossified and fused in the fetus.

1. Obtain a fetal skull and study it carefully. Make observations as needed to answer the following questions. Does it have the same bones as the adult skull? How does the size of the fetal face relate to the cranium? How does this compare to what is seen in the adult?

2. Indentations between the bones of the fetal skull, called **fontanels,** are fibrous membranes. These areas will become bony (ossify) as the fetus ages, completing the process by the age of 20 to 22 months. The fontanels allow the fetal skull to be compressed slightly during birth and also allow for brain growth during late fetal life. Locate the following fontanels on the fetal skull with the aid of Figure 11.1: anterior (or frontal) fontanel, mastoid fontanel, sphenoidal fontanel, and posterior (or occipital) fontanel.

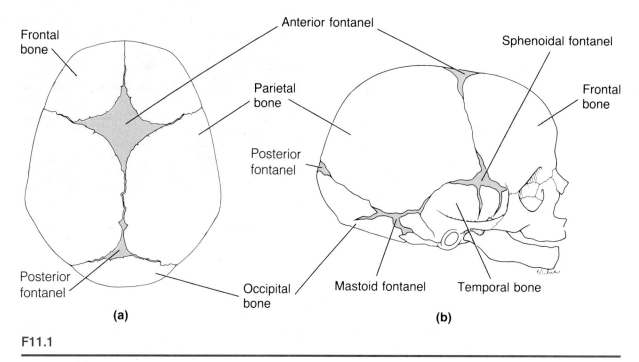

F11.1

The fetal skull. (a) Superior view. **(b)** Lateral view.

3.   Notice that some of the cranial bones have conical protrusions. These are growth centers. Notice also that the frontal bone is still bipartite, and the temporal bone is incompletely ossified, little more than a ring of bone in the fetus.

4.   Obtain a fetal skeleton (Figure 11.2) and examine it carefully, noting differences between it and an adult skeleton. Pay particular attention to the vertebrae, sternum, frontal bone of the cranium, patellae (kneecaps), coxal bones, carpals and tarsals, and rib cage.

5.   Check the questions in the review section before completing this study to ensure that you have made all of the necessary observations.

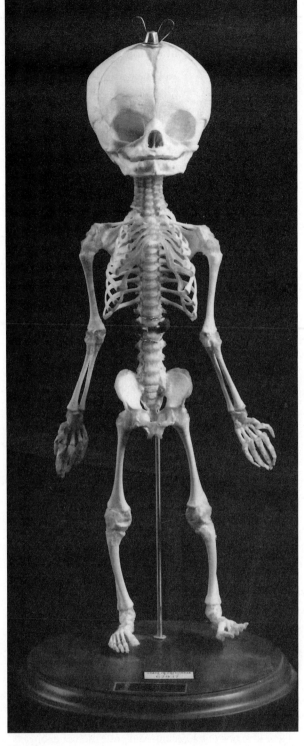

F11.2

The fetal skeleton.

Articulations and Body Movements

OBJECTIVES

1. To name the three structural categories of joints, and to compare their structure and mobility.
2. To identify the types of synovial joints.
3. To define *origin* and *insertion* of muscles.
4. To demonstrate or identify the various body movements.

MATERIALS

Articulated skeleton
Skull
Diarthrotic beef joint (fresh or preserved), preferably a knee joint
Disposable gloves
Anatomical chart of joint types (if available)
X rays of normal and arthritic joints (if available)

See Appendix B, Exercise 12 for links to A.D.A.M. Standard.

See Appendix C, Exercise 12 for links to *Anatomy and PhysioShow: The Videodisc.*

With rare exceptions, every bone in the body is connected to, or forms a joint with, at least one other bone. **Articulations,** or joints, perform two functions for the body. They (1) hold the bones together and (2) allow the rigid skeletal system some flexibility so that gross body movements can occur.

TYPES OF JOINTS

Joints may be classified structurally or functionally. The *structural classification* is based on whether there is connective tissue fiber, cartilage, or a joint cavity between the articulating bones. Structurally, there are *fibrous, cartilaginous,* and *synovial joints.*

The *functional classification* focuses on the amount of movement allowed at the joint. On this basis, there are **synarthroses,** or immovable joints; **amphiarthroses,** or slightly movable joints; and **diarthroses,** or freely movable joints. Freely movable joints predominate in the limbs, whereas immovable and slightly movable joints are largely restricted to the axial skeleton, where firm bony attachments and protection of enclosed organs are a priority.

As a general rule, fibrous joints are immovable, and synovial joints are freely movable. Cartilaginous joints offer both rigid and slightly movable examples. Since the structural categories are more clear-cut, we will use the structural classification here and indicate functional properties as appropriate.

Fibrous Joints

In **fibrous joints,** the bones are joined by fibrous tissue. No joint cavity is present. The amount of movement allowed depends on the length of the fibers uniting the

bones. Although some fibrous joints are slightly movable, most are synarthrotic and permit virtually no movement.

The two major types of fibrous joints are sutures and syndesmoses. In **sutures** (Figure 12.1d) the irregular edges of the bones interlock and are united by very short connective tissue fibers, as in most joints of the skull. In **syndesmoses** the articulating bones are connected by short ligaments of dense fibrous tissue; the bones do not interlock. The joint at the distal end of the tibia and fibula is an example of a syndesmosis (Figure 12.1e). Although this syndesmosis allows some give, it is classed functionally as a synarthrosis.

Examine a human skull again. Notice that adjacent bone surfaces do not actually touch but are separated by fibrous connective tissue. Also examine a skeleton and anatomical chart of joint types for examples of fibrous joints.

Cartilaginous Joints

In **cartilaginous joints,** the articulating bone ends are connected by a plate or pad of cartilage. No joint cavity is present. The two major types of cartilaginous joints are synchondroses and symphyses. Although there is variation, most cartilaginous joints are *slightly movable* (amphiarthroses) functionally. In **symphyses** (*symphysis* means "a growth together") the bones are connected by a broad, flat disc of **fibrocartilage.** The intervertebral joints and the pubic symphysis of the pelvis are symphyses (see Figure 12.1b and c). In **synchondroses** the bony portions are united by hyaline cartilage. The articulation of the costal cartilage of the first rib with the sternum (Figure 12.1a) is a synchondrosis, but perhaps the best examples of synchondroses are the epiphyseal

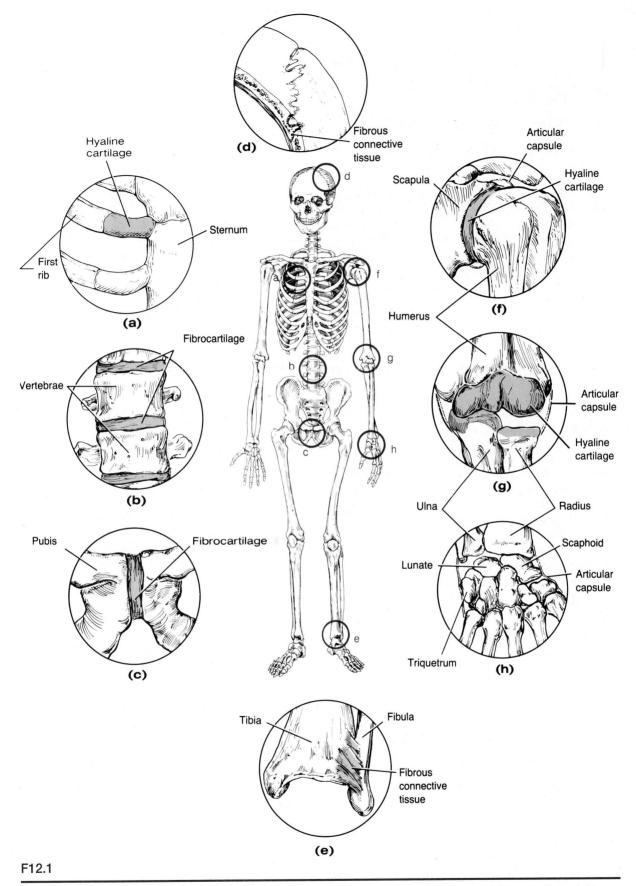

F12.1

Types of joints. Joints to the left of the skeleton are cartilaginous joints; joints above and below the skeleton are fibrous joints; joints to the right of the skeleton are synovial joints. **(a)** Synchondrosis (joint between costal cartilage of rib 1 and the sternum). **(b)** Symphyses (intervertebral discs of fibrocartilage connecting adjacent vertebrae). **(c)** Symphysis (fibrocartilaginous pubic symphysis connecting the pubic bones anteriorly). **(d)** Suture (fibrous connective tissue connecting interlocking skull bones). **(e)** Syndesmosis (fibrous connective tissue connecting the distal ends of the tibia and fibula). **(f)** Synovial joint (multiaxial shoulder joint). **(g)** Synovial joint (uniaxial elbow joint). **(h)** Synovial joints (biaxial intercarpal joints of the hand).

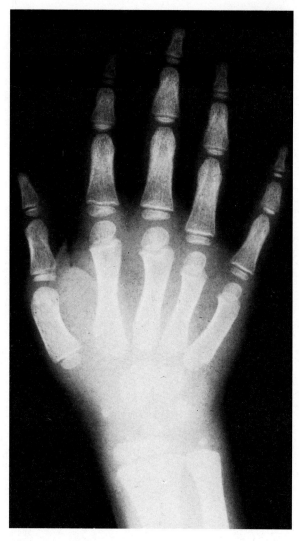

F12.2

X ray of the hand of a child. Notice the cartilaginous epiphyseal plates, examples of temporary synchondroses.

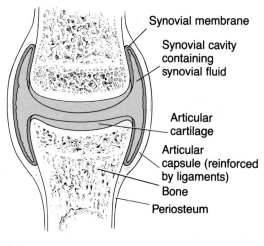

F12.3

Major structural features of a synovial joint.

Labels on Figure 12.3:
- Synovial membrane
- Synovial cavity containing synovial fluid
- Articular cartilage
- Articular capsule (reinforced by ligaments)
- Bone
- Periosteum

plates seen in the long bones of growing children (Figure 12.2). The epiphyseal plates are flexible during childhood but eventually they are totally ossified.

 Identify the cartilaginous joints on a human skeleton and on an anatomical chart of joint types.

Synovial Joints

Synovial joints are those in which the articulating bone ends are separated by a joint cavity containing synovial fluid (see Figure 12.1f–h). All synovial joints are diarthroses, or freely movable joints. Their mobility varies, however; some synovial joints can move in only one plane, and others can move in several directions (multiaxial movement). Most joints in the body are synovial joints.

All synovial joints are characterized by the following structural characteristics (Figure 12.3):

- The joint surfaces are enclosed by an *articular capsule* (a sleeve of fibrous connective tissue).
- The interior of this capsule is lined with a smooth connective tissue membrane, called *synovial membrane,* which produces a lubricating fluid (synovial fluid) that reduces friction.
- Articulating surfaces of the bones forming the joint are covered with hyaline (*articular*) cartilage.
- The articular capsule is typically reinforced with ligaments and may contain bursae (fluid-filled sacs that reduce friction where tendons cross bone).
- Fibrocartilage pads may be present within the capsule.

 1.   Examine a beef joint to identify the general structural features of diarthrotic joints.

⚠ If the joint is freshly obtained from the slaughterhouse and you will be handling it, don plastic gloves before beginning your observations.

2.   Compare and contrast the structure of the hip and knee joints (Figure 12.4). Both of these joints are large weight-bearing joints of the lower limb but they differ substantially in their security. Read through the questions in the Exercise 12 Review Sheet before beginning your comparison.

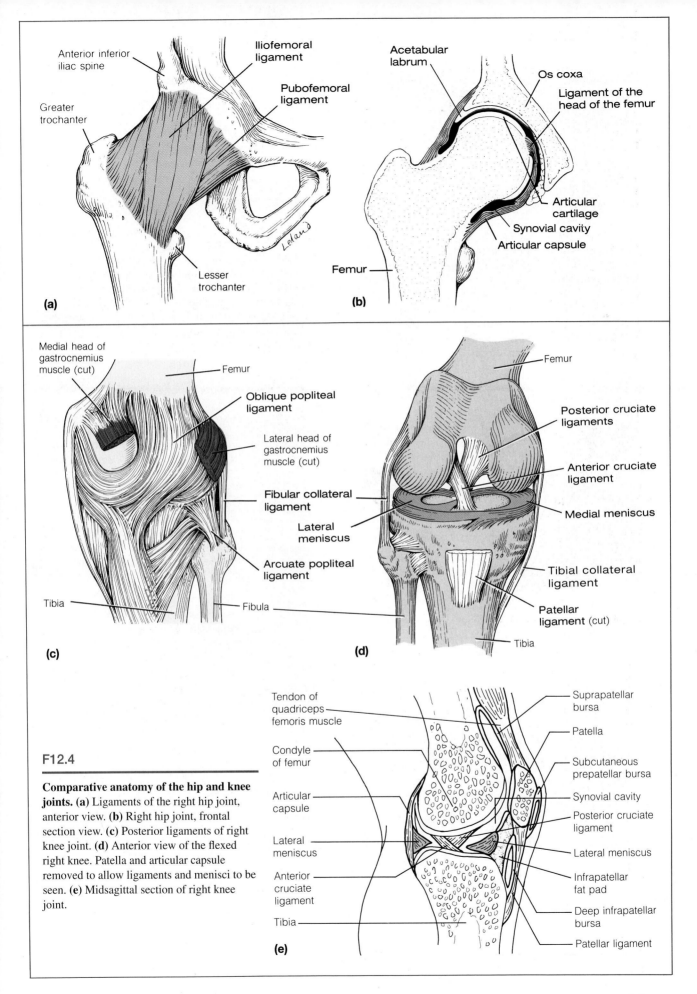

(a)

Anterior inferior iliac spine

Iliofemoral ligament

Pubofemoral ligament

Greater trochanter

Lesser trochanter

(b)

Acetabular labrum

Os coxa

Ligament of the head of the femur

Articular cartilage

Synovial cavity

Articular capsule

Femur

(c)

Medial head of gastrocnemius muscle (cut)

Femur

Oblique popliteal ligament

Lateral head of gastrocnemius muscle (cut)

Fibular collateral ligament

Lateral meniscus

Arcuate popliteal ligament

Tibia

Fibula

(d)

Femur

Posterior cruciate ligaments

Anterior cruciate ligament

Medial meniscus

Tibial collateral ligament

Patellar ligament (cut)

Tibia

F12.4

Comparative anatomy of the hip and knee joints. (a) Ligaments of the right hip joint, anterior view. (b) Right hip joint, frontal section view. (c) Posterior ligaments of right knee joint. (d) Anterior view of the flexed right knee. Patella and articular capsule removed to allow ligaments and menisci to be seen. (e) Midsagittal section of right knee joint.

(e)

Tendon of quadriceps femoris muscle

Condyle of femur

Articular capsule

Lateral meniscus

Anterior cruciate ligament

Tibia

Suprapatellar bursa

Patella

Subcutaneous prepatellar bursa

Synovial cavity

Posterior cruciate ligament

Lateral meniscus

Infrapatellar fat pad

Deep infrapatellar bursa

Patellar ligament

TYPES OF SYNOVIAL JOINTS   Because there are so many types of synovial joints, they have been divided into the following subcategories on the basis of movements allowed:

- Gliding: Articulating surfaces are flat or slightly curved, allowing sliding movements in one or two planes. Examples are the intercarpal and intertarsal joints and the vertebrocostal joints.
- Hinge: The rounded process of one bone fits into the concave surface of another to allow movement in one plane (uniaxial), usually flexion and extension. Examples are the elbow and interphalangeal joints.
- Pivot: The rounded or conical surface of one bone articulates with a shallow depression or foramen in another bone to allow uniaxial rotation, as in the joint between the atlas and axis (C_1 and C_2).
- Condyloid: The oval condyle of one bone fits into an ellipsoidal depression in another bone, allowing biaxial (two-way) movement. The wrist joint and the metacarpal-phalangeal joints (knuckles) are examples.
- Saddle: Articulating surfaces are saddle shaped; the articulating surface of one bone is convex, and the reciprocal surface is concave. Saddle joints, which are biaxial, include the joint between the thumb metacarpal and the trapezium of the wrist.
- Ball and socket: The ball-shaped head of one bone fits into a cuplike depression of another. These are multiaxial joints, allowing movement in all directions and pivotal rotation. Examples are the shoulder and hip joints.

 Examine the articulated skeleton, anatomical charts, and yourself to identify the subcategories of synovial joints. Make sure you understand the terms *uniaxial*, *biaxial*, and *multiaxial*.

BODY MOVEMENTS

Every muscle of the body is attached to bone (or other connective tissue structures) at two points—the **origin** (the stationary, immovable, or less movable attachment) and the **insertion** (the movable attachment). Body movement occurs when muscles contract across the diarthrotic synovial joints. When the muscle contracts and its fibers shorten, the insertion moves toward the origin. The type of movement depends on the construction of the joint (uniaxial, biaxial, or multiaxial) and on the placement of the muscle relative to the joint. The most common types of body movements are described below and illustrated in Figure 12.5.

 Attempt to demonstrate each movement as you read through the following material:

Flexion: a movement, generally in the sagittal plane, that decreases the angle of the joint and lessens the distance between the two bones. Flexion is typical of hinge joints (bending the knee or elbow), but is also common at ball-and-socket joints (bending forward at the hip).

Extension: a movement that increases the angle of a joint and the distance between two bones or parts of the body (straightening the knee or elbow). Extension is the opposite of flexion. If extension is greater than 180 degrees (bending the trunk backward), it is termed *hyperextension*.

Abduction: movement of a limb away from the midline or median plane of the body, generally on the frontal plane, or the fanning movement of fingers or toes when they are spread apart.

Adduction: movement of a limb toward the midline of the body. Adduction is the opposite of abduction.

Rotation: movement of a bone around its longitudinal axis without lateral or medial displacement. Rotation, a common movement of ball-and-socket joints, also describes the movement of the atlas around the odontoid process of the axis.

Circumduction: a combination of flexion, extension, abduction, and adduction commonly observed in ball-and-socket joints like the shoulder. The proximal end of the limb remains stationary, and the distal end moves in a circle. The limb as a whole outlines a cone.

Pronation: movement of the palm of the hand from an anterior or upward-facing position to a posterior or downward-facing position. This action moves the distal end of the radius across the ulna.

Supination: movement of the palm from a posterior position to an anterior position (the anatomical position). Supination is the opposite of pronation. During supination, the radius and ulna are parallel.

The last four terms refer to movements of the foot:

Inversion: a movement that results in the medial turning of the sole of the foot.

Eversion: a movement that results in the lateral turning of the sole of the foot; the opposite of inversion.

Dorsiflexion: a movement of the ankle joint in a dorsal direction (standing on one's heels).

Plantar flexion: a movement of the ankle joint in which the foot is flexed downward (standing on one's toes or pointing the toes).

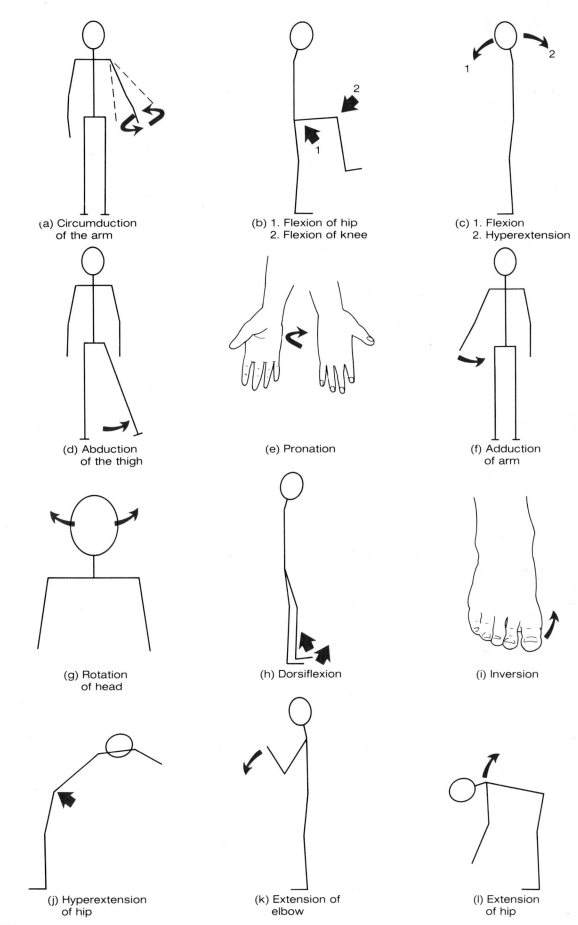

(a) Circumduction of the arm

(b) 1. Flexion of hip
    2. Flexion of knee

(c) 1. Flexion
    2. Hyperextension

(d) Abduction of the thigh

(e) Pronation

(f) Adduction of arm

(g) Rotation of head

(h) Dorsiflexion

(i) Inversion

(j) Hyperextension of hip

(k) Extension of elbow

(l) Extension of hip

F12.5

Movements occurring at synovial joints of the body.

JOINT DISORDERS

Most of us don't think about our joints until something goes wrong with them. Joint pains and malfunctions may be caused by a variety of things. For example, a hard blow to the knee can cause a painful bursitis, known as "water on the knee," due to damage to, or inflammation of, the patellar bursa. Slippage of a fibrocartilage pad or the tearing of a ligament may result in a painful condition that persists over a long period, since these poorly vascularized structures heal so slowly.

Sprains and dislocations are other types of joint problems. In a **sprain,** the ligaments reinforcing a joint are damaged by excessive stretching or are torn away from the bony attachment. Since both ligaments and tendons are cords of dense connective tissue with a poor blood supply, sprains heal slowly and are quite painful.

Dislocations occur when bones are forced out of their normal position in the joint cavity. They are normally accompanied by torn or stressed ligaments and considerable inflammation. The process of returning the bone to its proper position, called *reduction,* should be done only by a physician. Attempts by the untrained person to "snap the bone back into its socket" are often more harmful than helpful.

Advancing years also take their toll on joints. Weight-bearing joints in particular eventually begin to degenerate. *Adhesions* (fibrous bands) may form between the surfaces where bones join, and extraneous bone tissue (*spurs*) may grow along the joint edges. Such degenerative changes lead to the complaint so often heard from the elderly: "My joints are getting so stiff. . . ."

• If possible compare an X ray of an arthritic joint to one of a normal joint. ■

Microscopic Anatomy, Organization, and Classification of Skeletal Muscle

OBJECTIVES

1. To describe the structure of skeletal muscle from gross to microscopic levels.

2. To define and explain the role of the following:
   actin   myofilament   tendon
   myosin   perimysium   endomysium
   fiber   aponeurosis   epimysium
   myofibril

3. To describe the structure of a neuromuscular junction and to explain its role in muscle function.

4. To define: *agonist* (prime mover), *antagonist, synergist, fixator, origin,* and *insertion.*

5. To cite criteria used in naming skeletal muscles.

MATERIALS

Three-dimensional model of skeletal muscle cells (if available)
Forceps
Dissecting needles
Microscope slides and coverslips
0.9% saline solution in dropper bottles
Chicken breast or thigh muscle (freshly obtained from the meat market)
Compound microscope
Histologic slides of skeletal muscle (longitudinal and cross-sectional) and skeletal muscle showing neuromuscular junctions
Three-dimensional model of skeletal muscle showing neuromuscular junction (if available)

See Appendix C, Exercise 13 for links to *Anatomy and PhysioShow: The Videodisc.*

The bulk of the body's muscle is called **skeletal muscle** because it is attached to the skeleton (or associated connective tissue structures). Skeletal muscle influences body contours and shape, allows you to grin and frown, provides a means of locomotion, and enables you to manipulate the environment. The balance of the body's muscle—smooth and cardiac muscle—as the major component of the walls of hollow organs and the heart is involved with the transport of materials within the body.

Each of the three muscle types has a structure and function uniquely suited to its task in the body. However, because the term *muscular system* applies specifically to skeletal muscle, the primary objective of this exercise is to investigate the functional anatomy of skeletal muscle.

Skeletal muscle is also known as *voluntary muscle* (because it can be consciously controlled) and as *striated muscle* (because it appears to be striped). As you might guess from both of these alternative names, skeletal muscle has some very special characteristics. Thus an investigation of skeletal muscle should begin at the cellular level.

THE CELLS OF SKELETAL MUSCLE

Skeletal muscle is composed of relatively large, long cylindrical cells ranging from 10 to 100 μm in diameter and up to 6 cm in length. However, the cells of large, hard-working muscles like the antigravity muscles of the hip are extremely coarse, ranging up to 25 cm in length, and can be seen with the naked eye.

Skeletal muscle cells (Figure 13.1a) are multinucleate: Multiple oval nuclei can be seen just beneath the plasma membrane (called the *sarcolemma* in these cells). The nuclei are pushed peripherally by the longitudinally arranged **myofibrils,** which nearly fill the sarcoplasm (Figure 13.1b). Alternating light (I) and dark (A) bands along the length of the perfectly aligned myofibrils give the muscle fiber as a whole its striped appearance.

Electron microscope studies have revealed that the myofibrils are made up of even smaller threadlike structures called **myofilaments** (Figure 13.1b and d). The myofilaments are composed largely of two varieties of contractile proteins—**actin** and **myosin**—which slide past each other during muscle activity to bring about shortening or contraction of the muscle cells. It is the highly specific arrangement of the myofilaments within the myofibrils that is responsible for the banding pattern in skeletal muscle. The actual contractile units of muscle, called **sarcomeres,** extend from the middle of one I band (its Z line) to the middle of the next along the length of the myofibrils. (See Figure 13.1c and d.)

1. Look at the three-dimensional model of skeletal muscle cells, noting the relative shape and size of the cells. Identify the nuclei, myofibrils, and light and dark bands.

2. Obtain forceps, two dissecting needles, slide and coverslip, and a dropper bottle of saline solution. With forceps, remove a very small piece of muscle from the chicken breast (or thigh). Place the tis-

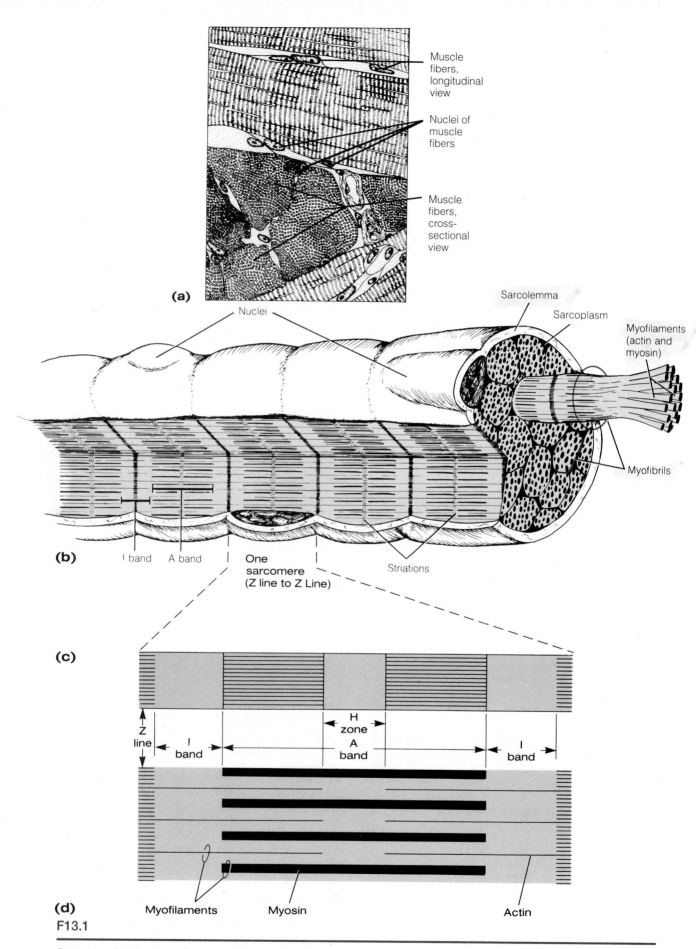

(a) Muscle fibers, longitudinal view

Nuclei of muscle fibers

Muscle fibers, cross-sectional view

Nuclei

Sarcolemma

Sarcoplasm

Myofilaments (actin and myosin)

Myofibrils

(b) I band · A band · One sarcomere (Z line to Z Line) · Striations

(c)

(d) F13.1

Z line · I band · H zone · A band · I band

Myofilaments · Myosin · Actin

Structure of skeletal muscle cells. (a) Muscle fibers, longitudinal and transverse views. (See corresponding photomicrograph in Plate 2 of the Histology Atlas.) **(b)** A portion of a skeletal muscle cell; one myofibril has been extended and disrupted to indicate its myofilament composition. **(c)** One sarcomere of the myofibril. **(d)** Banding pattern in the sarcomere. (From H. E. Huxley, "The Contraction of Muscle." © November 1958 by Scientific American, Inc. All rights reserved.)

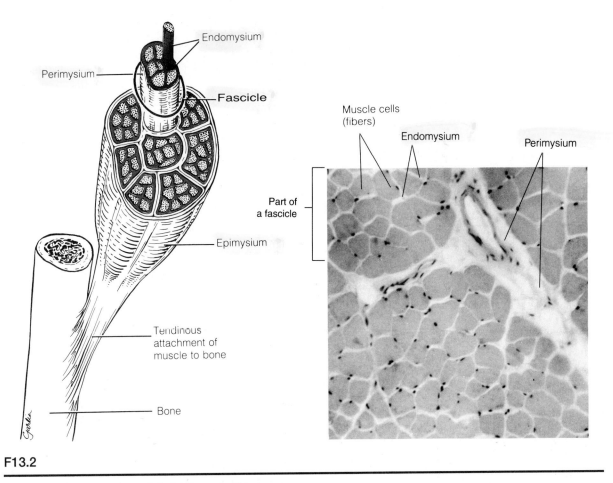

F13.2

Connective tissue coverings of skeletal muscle (64×).

sue on a clean microscope slide, and add a drop of the saline solution.

3.   Pull the muscle fibers apart with the dissecting needles (tease them) until you have a fluffy-looking mass of tissue. Cover the teased tissue with a coverslip, and observe under the high-power lens of a microscope. Look for the banding pattern. Regulate the light carefully to obtain the highest possible contrast.

4.   Now compare your observations with Figure 13.1a and with what can be seen with professionally prepared muscle tissue. Obtain a slide of skeletal muscle (longitudinal section), and view it under high power. From your observations, draw a small section of a muscle fiber in the space provided here. Label the nuclei, sarcolemma, and A and I bands.

What structural details become apparent with the prepared slide?

_____

_____

ORGANIZATION OF SKELETAL MUSCLE CELLS INTO MUSCLES

Muscle fibers are soft and surprisingly fragile. Thus thousands of muscle fibers are bundled together with connective tissue to form the organs we refer to as skeletal muscles (Figure 13.2). Each muscle fiber is enclosed in a delicate, areolar connective tissue sheath called **endomysium.** Several sheathed muscle fibers are wrapped by a collagenic membrane called **perimysium,** forming a bundle of fibers called a **fascicle,** or **fasciculus.** A large number of fascicles are bound together by a substantially coarser "overcoat" of dense connective tissue called an **epimysium,** which sheathes the entire muscle. These epimysia blend into the **deep fascia,** still coarser sheets of dense connective tissue that bind muscles into functional groups, and into strong cordlike **ten-**

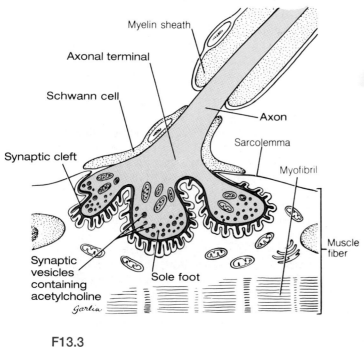

F13.3

The neuromuscular junction.

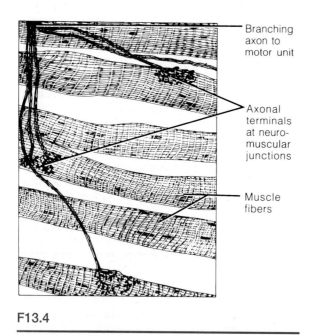

F13.4

A portion of a motor unit. (Corresponding photomicrograph is Plate 3 in the Histology Atlas.)

dons or sheetlike **aponeuroses,** which attach muscles to each other or indirectly to bones. As noted in Exercise 12, a muscle's more movable attachment is called its *insertion* whereas its fixed (or immovable) attachment is the *origin*.

Tendons perform several functions, two of the most important being to provide durability and to conserve space. Because tendons are tough collagenic connective tissue, they can span rough bony prominences that would destroy the more delicate muscle tissues. Because of their relatively small size, more tendons than fleshy muscles can pass over a joint.

In addition to supporting and binding the muscle fibers, and providing strength to the muscle as a whole, the connective tissue wrappings provide a route for the entry and exit of nerves and blood vessels that serve the muscle fibers. The larger, more powerful muscles have relatively more connective tissue than muscles involved in fine or delicate movements.

As we age, the mass of the muscle fibers decreases, and the amount of connective tissue increases; thus the skeletal muscles gradually become more sinewy, or "stringier." ■

Obtain a slide showing a cross section of skeletal muscle tissue. Using Figure 13.2 as a reference, identify the muscle fibers, endomysium, perimysium, and epimysium (if visible).

THE NEUROMUSCULAR JUNCTION

Voluntary muscle cells are always stimulated by motor neurons via nerve impulses. The junction between a nerve fiber (axon) and a muscle cell is called a **neuromuscular,** or **myoneural, junction** (Figure 13.3).

Each motor axon breaks up into many branches called *axonal terminals* as it approaches the muscle, and each of these branches participates in forming a neuromuscular junction with a single muscle cell. Thus a single neuron may stimulate many muscle fibers. Together, a neuron and all the muscle cells it stimulates make up the functional structure called the **motor unit.** Part of a motor unit is shown in Figure 13.4 and in Plate 3 of the Histology Atlas.

Each axonal terminal has numerous projections called **sole feet.** The neuron and muscle fiber membranes, close as they are, do not actually touch. They are separated by a small fluid-filled gap called the **synaptic cleft** (see Figure 13.3).

Within the sole foot are many mitochondria and vesicles containing a neurotransmitter chemical called acetylcholine. When a nerve impulse reaches the axonal endings, some of these vesicles liberate their contents into the synaptic cleft. The acetylcholine rapidly diffuses across the junction and combines with the receptors on the sarcolemma. If sufficient acetylcholine has been released, a transient change in the permeability of the sarcolemma briefly allows more sodium ions to diffuse into the muscle fiber. The result is depolarization of the sarcolemma and subsequent contraction of the muscle fiber.

1. If possible, examine a three-dimensional model of skeletal muscle cells that illustrates the neuromuscular junction. Identify the structures just described.

2. Obtain a slide of skeletal muscle stained to show a portion of a motor unit. Examine the slide under high power to identify the axonal fibers extending leashlike to the muscle cells. Follow one of the axonal fibers to its terminus to identify the oval-shaped axonal terminal. Compare your observations to Figure 13.4. Sketch a small section in the space provided, labeling the motor axon, its terminal branches, sole feet, and muscle fibers.

CLASSIFICATION OF SKELETAL MUSCLES

Naming Skeletal Muscles

Remembering the names of the skeletal muscles is a monumental task, but certain clues help. Muscles are named on the basis of the following criteria:

- **Direction of muscle fibers:** Some muscles are named in reference to some imaginary line, usually the midline of the body or the longitudinal axis of a limb bone. A muscle with fibers (and fascicles) running parallel to that imaginary line will have the term *rectus* (straight) in its name. For example, the rectus abdominis is the straight muscle of the abdomen. Likewise, the terms *transverse* and *oblique* indicate that the muscle fibers run at right angles and obliquely (respectively) to the imaginary line.
- **Relative size of the muscle:** Terms such as *maximus* (largest), *minimus* (smallest), *longus* (long),

and *brevis* (short) are often used in naming muscles—as in gluteus maximus and gluteus minimus.
- **Location of the muscle:** Some muscles are named according to the bone with which they are associated. For example, the frontalis muscle overlies the frontal bone.
- **Number of origins:** When the term *biceps, triceps,* or *quadriceps* forms part of a muscle name, you can generally assume that the muscle has two, three, or four origins (respectively). For example, the biceps muscle of the arm has two heads, or origins.
- **Location of the muscle's origin and insertion:** For example, the sternocleidomastoid muscle has its origin on the sternum (*sterno*) and clavicle (*cleido*), and inserts on the mastoid process of the temporal bone.
- **Shape of the muscle:** For example, the deltoid muscle is roughly triangular (*deltoid* = "triangle"), and the trapezius muscle resembles a trapezoid.
- **Action of the muscle:** For example, all the adductor muscles of the anterior thigh bring about its adduction, and all the extensor muscles of the wrist extend the wrist.

Types of Muscles

Most often, body movements are not a result of the contraction of a single muscle but instead reflect the coordinated action of several muscles acting together. Muscles that are primarily responsible for producing a particular movement are called **prime movers,** or **agonists.**

Muscles that oppose or reverse a movement are called **antagonists.** When a prime mover is active, the fibers of the antagonist are stretched and in the relaxed state. The antagonist can also regulate the prime mover by providing some resistance, to prevent overshoot or to stop its action.

It should be noted that antagonists can be prime movers in their own right. For example, the biceps muscle of the arm (a prime mover of elbow flexion) is antagonized by the triceps (a prime mover of elbow extension).

Synergists contribute substantially to the action of agonists by reducing undesirable or unnecessary movement. Contraction of a muscle crossing two or more joints would cause movement at all joints spanned if the synergists were not there to stabilize them. For example, you can make a fist without bending your wrist only because synergist muscles stabilize the wrist joint and allow the prime mover to exert its force at the finger joints.

Fixators, or fixation muscles, are specialized synergists. They immobilize the origin of a prime mover so that all the tension is exerted at the insertion. Muscles that help maintain posture are fixators; so too are muscles of the back that stabilize or "fix" the scapula during arm movements.

Gross Anatomy of the Muscular System

OBJECTIVES

1. To name and locate the major muscles of the human body (on a torso model, a human cadaver, laboratory chart, or diagram) and state the action of each.
2. To explain how muscle actions are related to their location.
3. To name muscle origins and insertions as required by the instructor.
4. To identify antagonists of the major prime movers.
5. To name and locate muscles on a dissection animal.
6. To recognize similarities and differences between human and cat musculature.

MATERIALS

Disposable gloves or protective skin cream
Preserved and injected cat (one for every two to four students)
Dissecting trays and instruments
Name tag and large plastic bag
Paper towels
Embalming fluid
Human torso model or large anatomical chart showing human musculature
Human cadaver for demonstration (if available)
Human Musculature videotape*

 See Appendix B, Exercise 14 for links to A.D.A.M. Standard.

 See Appendix C, Exercise 14 for links to *Anatomy and PhysioShow: The Videodisc.*

*Available to qualified adopters from Benjamin/Cummings.

IDENTIFICATION OF HUMAN MUSCLES

Muscles of the Head and Neck

The muscles of the head serve many specific functions. For instance, the muscles of facial expression differ from most skeletal muscles because they insert into the skin (or other muscles) rather than into bone. As a result, they move the facial skin, allowing a wide range of emotions to be shown on the face. Other muscles of the head are the muscles of mastication, which manipulate the mandible during chewing, and the six extrinsic eye muscles located within the orbit, which aim the eye. (Orbital muscles are studied in conjunction with the anatomy of the eye in Exercise 18.) Neck muscles are primarily concerned with the movement of the head and shoulder girdle. Figures 14.1 and 14.2 are summary figures illustrating the superficial musculature of the body as a whole. Head and neck muscles are discussed in Tables 14.1 and 14.2 and shown in Figures 14.3 and 14.4.

Carefully read the description of each muscle and visualize what happens when the muscle contracts. After reading the tables and identifying the head and neck muscles in Figures 14.3 and 14.4, use a torso model or

an anatomical chart to again identify as many of these muscles as possible. (If a human cadaver is available for observation, specific instructions for muscle examination will be provided by your instructor.) Then carry out the following palpations on yourself:

- To demonstrate how the temporalis works, clench your teeth. The masseter can also be palpated at this time at the angle of the jaw.

Muscles of the Trunk

The trunk musculature includes muscles that move the vertebral column; anterior thorax muscles that act to move ribs, head, and arms; and muscles of the abdominal wall that play a role in the movement of the vertebral column but more importantly form the "natural girdle," or the major portion of the abdominal body wall.

 The trunk muscles are described in Tables 14.3 and 14.4 and shown in Figures 14.5 and 14.6. As before, identify the muscles in the figure as you read the tabular descriptions and then identify them on the torso or laboratory chart.

(*Text continues on p. 117*)

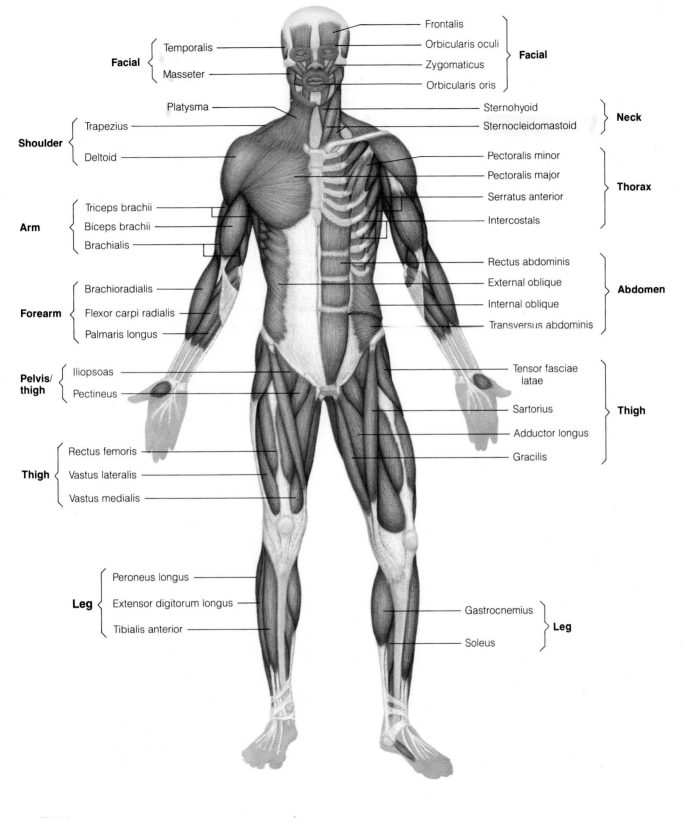

F14.1

Anterior view of superficial muscles of the body. The abdominal surface has been partially dissected on the left side of the body to show somewhat deeper muscles.

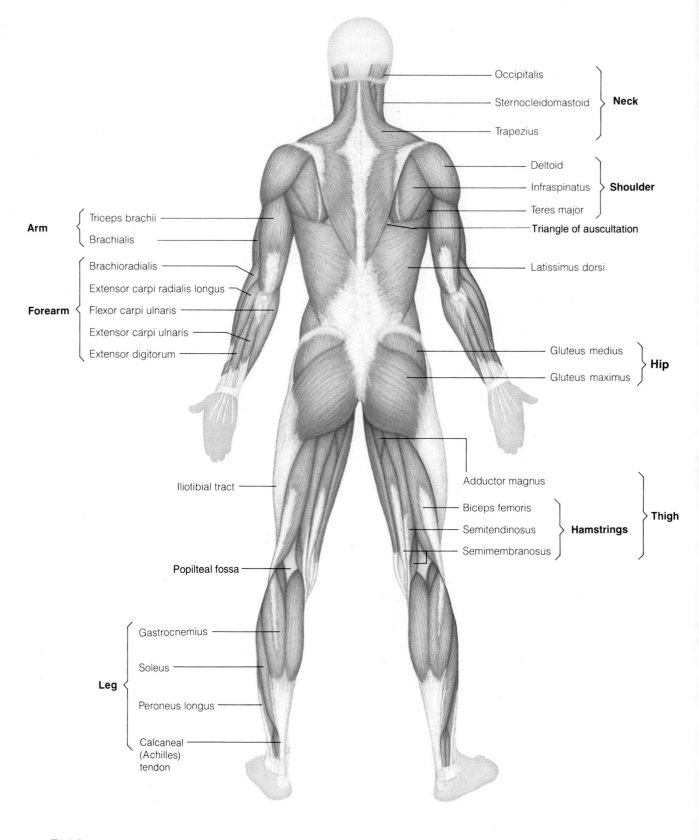

Neck
- Occipitalis
- Sternocleidomastoid
- Trapezius

Shoulder
- Deltoid
- Infraspinatus
- Teres major
- **Triangle of auscultation**

- Latissimus dorsi

Arm
- Triceps brachii
- Brachialis

Forearm
- Brachioradialis
- Extensor carpi radialis longus
- Flexor carpi ulnaris
- Extensor carpi ulnaris
- Extensor digitorum

Hip
- Gluteus medius
- Gluteus maximus

Thigh

Hamstrings
- Adductor magnus
- Biceps femoris
- Semitendinosus
- Semimembranosus

- Iliotibial tract
- **Popilteal fossa**

Leg
- Gastrocnemius
- Soleus
- Peroneus longus
- Calcaneal (Achilles) tendon

F14.2

Posterior view of superficial muscles of the body.

TABLE 14.1   Major Muscles of Human Head (see Figure 14.3)

Muscle	Comments	Origin	Insertion	Action
Facial Expression (Figure 14.3a)				
Epicranius— frontalis and occipitalis	Bipartite muscle consisting of frontalis and occipitalis, which covers dome of skull	Frontalis: cranial aponeurosis (galea aponeurotica); occipitalis: occipital bone	Frontalis: skin of eyebrows and root of nose; occipitalis: cranial aponeurosis	With aponeurosis fixed, frontalis raises eyebrows; occipitalis fixes aponeurosis and pulls scalp posteriorly
Orbicularis oculi	Sphincter muscle of eyelids	Frontal and maxillary bones and ligaments around orbit	Encircles orbit and inserts in tissue of eyelid	Various parts can be activated individually; closes eyes, produces blinking, squinting, and draws eyebrows downward
Corrugator supercilii	Small muscle; activity associated with that of orbicularis oculi	Arch of frontal bone above nasal bone	Skin of eyebrow	Draws eyebrows medially; wrinkles skin of forehead vertically
Levator labii superioris	Thin muscle between orbicularis oris and inferior eye margin	Zygomatic bone and infraorbital margin of maxilla	Skin and muscle of upper lip and border of nostril	Raises and furrows upper lip; flares nostril (as in disgust)
Zygomaticus— major and minor	Extends diagonally from corner of mouth to cheekbone	Zygomatic bone	Skin and muscle at corner of mouth	Raises lateral corners of mouth upward (smiling muscle)
Risorius	Slender muscle; runs laterally to zygomaticus	Fascia of masseter muscle	Skin at corner of mouth	Draws corner of lip laterally; tenses lip; zygomaticus synergist
Depressor labii inferioris	Small muscle from lower lip to jawbone	Body of mandible lateral to its midline	Skin and muscle of lower lip	Draws lower lip downward
Depressor anguli oris	Small muscle lateral to depressor labii inferioris	Body of mandible below incisors	Skin and muscle at angle of mouth below insertion of zygomaticus	Zygomaticus antagonist; draws corners of mouth downward and laterally
Orbicularis oris	Multilayered sphincter muscle of lips with fibers that run in many different directions	Arises indirectly from maxilla and mandible; fibers blended with fibers of other muscles associated with lips	Encircles mouth; inserts into muscle and skin at angles of mouth	Closes mouth; purses and protrudes lips (kissing muscle)
Mentalis	One of muscle pair forming V-shaped muscle mass on chin	Mandible below incisors	Skin of chin	Protrudes lower lip; wrinkles chin
Buccinator	Principal muscle of cheek; runs horizontally, deep to the masseter	Molar region of maxilla and mandible	Orbicularis oris	Draws corner of mouth laterally; compresses cheek (as in whistling); holds food between teeth during chewing

(*continued*)

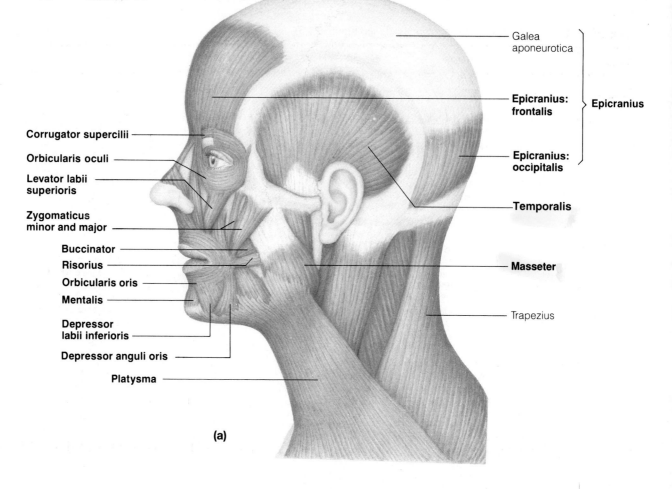

Galea
aponeurotica

Epicranius:
frontalis

Epicranius

Epicranius:
occipitalis

Corrugator supercilii

Orbicularis oculi

Levator labii
superioris

Zygomaticus
minor and major

Buccinator

Risorius

Orbicularis oris

Mentalis

Depressor
labii inferioris

Depressor anguli oris

Platysma

Temporalis

Masseter

Trapezius

(a)

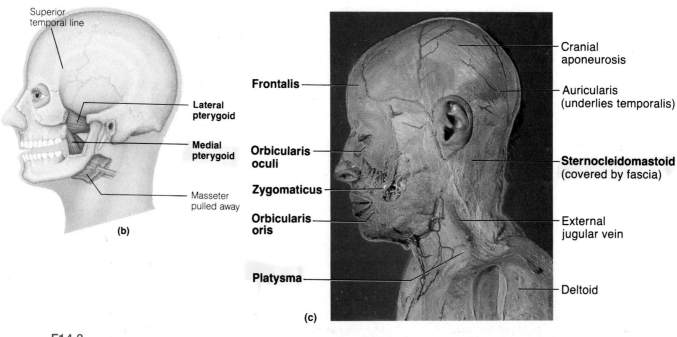

Superior
temporal line

**Lateral
pterygoid**

**Medial
pterygoid**

Masseter
pulled away

(b)

Frontalis

**Orbicularis
oculi**

Zygomaticus

**Orbicularis
oris**

Platysma

Cranial
aponeurosis

Auricularis
(underlies temporalis)

Sternocleidomastoid
(covered by fascia)

External
jugular vein

Deltoid

(c)

F14.3

Muscles of the scalp, face, and neck; left lateral view. (**a**) Superficial muscles. (**b**) The deep chewing muscles, the medial and lateral pterygoid muscles. (**c**) Photo of superficial structures of head and neck.

TABLE 14.1   (*Continued*)

Muscle	Comments	Origin	Insertion	Action
Mastication (Figure 14.3a,b)				
Masseter	Extends across jawbone; can be palpated on forcible closure of jaws	Zygomatic process and arch	Angle and ramus of mandible	Closes jaw and elevates mandible
Temporalis	Fan-shaped muscle over temporal bone	Temporal fossa	Coronoid process of mandible	Closes jaw; elevates and retracts mandible
Buccinator	(See muscles of facial expression.)			
Pterygoid—medial	Runs along internal (medial) surface of mandible (thus largely concealed by that bone)	Sphenoid, palatine, and maxillary bones	Medial surface of mandibular ramus and angle	Synergist of temporalis and masseter; closes and elevates mandible; in conjunction with lateral pterygoid, aids in grinding movements of teeth
Pterygoid—lateral	Superior to medial pterygoid	Greater wing of sphenoid bone	Mandibular condyle	Protracts jaw (moves it anteriorly); in conjunction with medial pterygoid, aids in grinding movements of teeth

TABLE 14.2   Anterolateral Muscles of Human Neck (see Figure 14.4)

Muscle	Comments	Origin	Insertion	Action
Superficial				
Platysma	Unpaired muscle: thin, sheetlike superficial neck muscle, not strictly a head muscle but plays role in facial expression (see Fig. 14.3a)	Fascia of chest (over pectoral muscles) and deltoid	Lower margin of mandible, skin, and muscle at corner of mouth	Depresses mandible; pulls lower lip back and down; i.e., produces downward sag of the mouth
Sternocleidomastoid	Two-headed muscle located deep to platysma on anterolateral surface of neck; fleshy parts on either side indicate limits of anterior and posterior triangles of neck	Manubrium of sternum and medial portion of clavicle	Mastoid process of temporal bone	Simultaneous contraction of both muscles of pair causes flexion of neck forward, generally against resistance (as when lying on the back); acting independently, rotate head toward shoulder on opposite side
Scalenes—anterior, middle, and posterior	Located more on lateral than anterior neck; deep to platysma (see Fig. 14.4b)	Transverse processes of cervical vertebrae	Anterolaterally on first two ribs	Flex and slightly rotate neck; elevate first two ribs (aid in inspiration)

(*continued*)

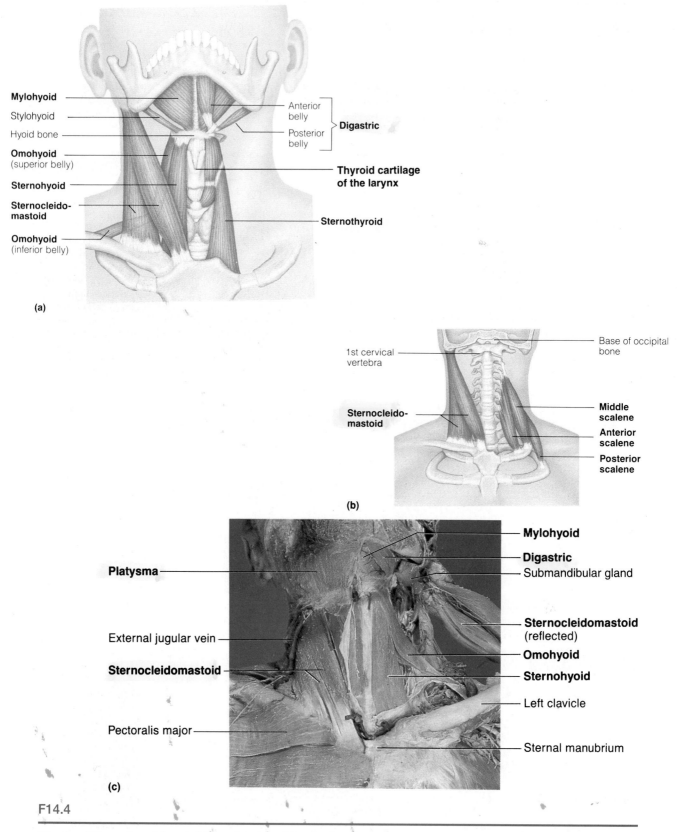

(a)

(b)

(c)

F14.4

Muscles of the neck and throat. (a) Anterior view of deep neck (suprahyoid and infrahyoid) muscles. **(b)** Muscles of the anterolateral neck. The superficial platysma muscle and deeper neck muscles have been removed to show the origins and insertions of
the sternocleidomastoid and scalene muscles clearly. **(c)** Photo of the anterior and lateral regions of the neck. The fascia has been
partially removed (left side of photo) to expose the sternocleidomastoid muscle. On the right side of the photo, the sternocleidomastoid muscle is reflected to expose the sternohyoid and omohyoid muscles.

TABLE 14.2   (*Continued*)

Muscle	Comments	Origin	Insertion	Action
Deep (Figure 14.4a,c)				
Digastric	Consists of two bellies united by an intermediate tendon; assumes a V-shaped configuration under chin	Lower margin of mandible (anterior belly) and mastoid process (posterior belly)	By a connective tissue loop to hyoid bone	Acting in concert, elevate hyoid bone; open mouth and depress mandible
Mylohyoid	Just deep to digastric; forms floor of mouth	Medial surface of mandible	Hyoid bone	Elevates hyoid bone and base of tongue during swallowing
Sternohyoid	Runs most medially along neck; straplike	Posterior surface of manubrium	Lower margin of body of hyoid bone	Acting with sternothyroid and omohyoid (all inferior to hyoid bone), depresses larynx and hyoid bone if mandible is fixed; may also flex skull
Sternothyroid	Lateral to sternohyoid; straplike	Manubrium and medial end of clavicle	Thyroid cartilage of larynx	(See Sternohyoid, above)
Omohyoid	Straplike with two bellies; lateral to sternohyoid	Superior surface of scapula	Hyoid bone	(See Sternohyoid, above)

TABLE 14.3   Anterior Muscles of Human Thorax, Shoulder, and Abdominal Wall (see Figure 14.5)

Muscle	Comments	Origin	Insertion	Action
Thorax and Shoulder (Figure 14.5a)				
Pectoralis major	Large fan-shaped muscle covering upper portion of chest	Clavicle, sternum, cartilage of first six ribs, and aponeurosis of external oblique muscle	Fibers converge to insert by short tendon into greater tubercle of humerus	Prime mover of arm flexion; adducts, medially rotates arm; with arm fixed, pulls chest upward (thus also acts in forced inspiration)
Serratus anterior	Deep and superficial portions; beneath and inferior to pectoral muscles on lateral rib cage	Lateral aspect of first to eighth (or ninth) ribs	Vertebral border of anterior surface of scapula	Moves scapula forward toward chest wall; rotates scapula causing inferior angle to move laterally and upward

(*continued*)

TABLE 14.3 (*Continued*)

Muscle	Comments	Origin	Insertion	Action
Deltoid	Fleshy triangular muscle forming shoulder muscle mass	Lateral third of clavicle; acromion and spine of scapula	Deltoid tuberosity of humerus	Acting as a whole, prime mover of arm abduction; when only specific fibers are active, can aid in flexion, extension, and rotation of humerus
Pectoralis minor	Flat, thin muscle directly beneath and obscured by pectoralis major	Anterior surface of third, fourth, and fifth ribs, near their costal cartilages	Coracoid process of scapula	With ribs fixed, draws scapula forward and inferiorly; with scapula fixed, draws rib cage superiorly
Intercostals—external	11 pairs lie between ribs; fibers run obliquely downward and forward toward sternum	Inferior border of rib above (not shown in figure)	Superior border of rib below	Pulls ribs toward one another to elevate rib cage; aids in inspiration
Intercostals—internal	11 pairs lie between ribs; fibers run deep and at right angles to those of external intercostals	Superior border of rib below	Inferior border of rib above (not shown in figure)	Draws ribs together to depress rib cage; aids in forced expiration; antagonistic to external intercostals

Abdominal Wall (Figure 14.5b and c)

Muscle	Comments	Origin	Insertion	Action
Rectus abdominis	Medial superficial muscle, extends from pubis to rib cage; ensheathed by aponeuroses of oblique muscles; segmented	Pubic crest and symphysis	Xiphoid process and costal cartilages of fifth through seventh ribs	Flexes vertebral column; increases abdominal pressure; fixes and depresses ribs; stabilizes pelvis during walking
External oblique	Most superficial lateral muscle; fibers run downward and medially; ensheathed by an aponeurosis	Anterior surface of last eight ribs	Linea alba,* pubic tubercles, and iliac crest	See Rectus abdominis, above; also aids muscles of back in trunk rotation and lateral flexion
Internal oblique	Fibers run at right angles to those of external oblique, which it underlies	Lumbodorsal fascia, iliac crest, and inguinal ligament	Linea alba, pubic crest, and costal cartilages of last three ribs	As for External oblique
Transversus abdominis	Deepest muscle of abdominal wall; fibers run horizontally	Inguinal ligament, iliac crest, and cartilages of last five or six ribs	Linea alba and pubic crest	Compresses abdominal contents

*The linea alba ("white line") is a narrow, tendinous sheath that runs along the middle of the abdomen from the sternum to the pubic symphysis. It is formed by the fusion of the aponeurosis of the external oblique and transversus muscles.

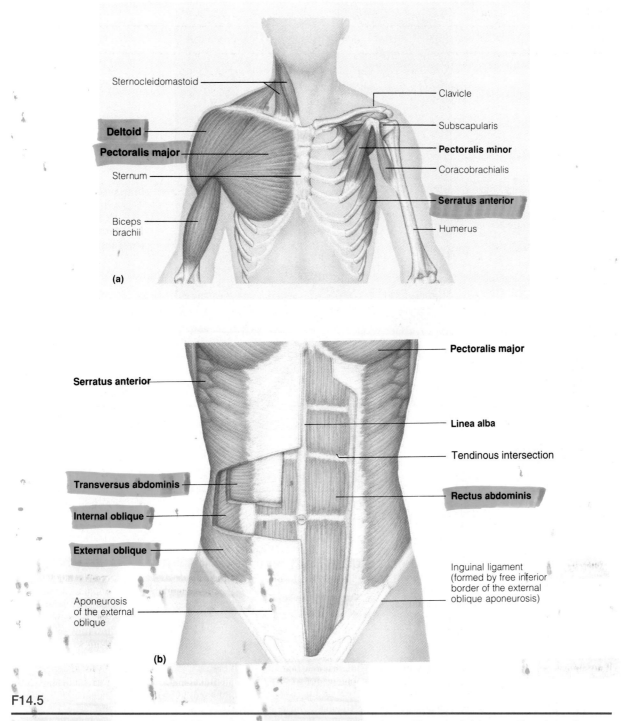

Sternocleidomastoid

Clavicle

Deltoid

Subscapularis

Pectoralis major

Pectoralis minor

Sternum

Coracobrachialis

Serratus anterior

Biceps brachii

Humerus

(a)

Pectoralis major

Serratus anterior

Linea alba

Tendinous intersection

Transversus abdominis

Rectus abdominis

Internal oblique

External oblique

Inguinal ligament (formed by free inferior border of the external oblique aponeurosis)

Aponeurosis of the external oblique

(b)

F14.5

Anterior muscles of the thorax, shoulder, and abdominal wall. (a) Anterior thorax. The superficial pectoralis major and deltoid muscles that effect arm movements are illustrated on the left. These muscles have been removed on the right side of the figure to illustrate the pectoralis minor, serratus anterior, and subscapularis muscles. **(b)** Anterior view of the muscles forming the anterolateral abdominal wall. The superficial muscles have been partially cut away on the left side of the diagram to reveal the deeper internal oblique and transversus abdominis muscles.

(continued)

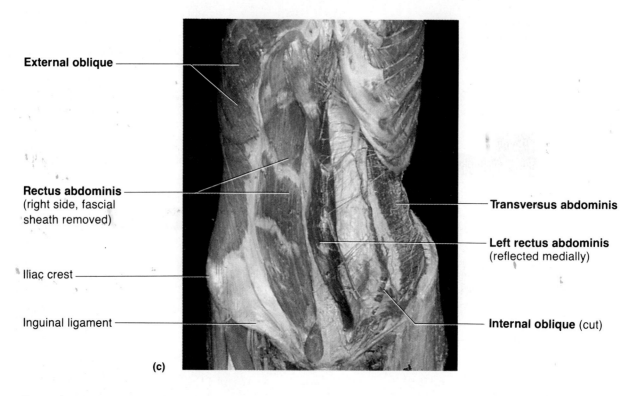

External oblique

Rectus abdominis
(right side, fascial
sheath removed)

Iliac crest

Inguinal ligament

Transversus abdominis

Left rectus abdominis
(reflected medially)

Internal oblique (cut)

(c)

F14.5 (*continued*)

Anterior muscles of the thorax, shoulder, and abdominal wall. (c) Photo of the anterolateral abdominal wall.

TABLE 14.4   Posterior Muscles of Human Trunk (see Figure 14.6)

Muscle	Comments	Origin	Insertion	Action
Muscles of the Neck, Shoulder, and Thorax (Figure 14.6a)				
Trapezius	Most superficial muscle of posterior neck and thorax; very broad origin and insertion	Occipital bone; ligamentum nuchae; spines of C_7 and all thoracic vertebrae	Acromion and spinous process of scapula; lateral third of clavicle	Extends head; retracts (adducts) scapula and stabilizes it; upper fibers elevate scapula; lower fibers depress it
Latissimus dorsi	Broad flat muscle of lower back (lumbar region); extensive superficial origins	Indirect attachment to spinous processes of lower six thoracic vertebrae, lumbar vertebrae, lower 3 to 4 ribs, and iliac crest	Floor of intertubercular groove of humerus	Prime mover of arm extension; adducts and medially rotates arm; depresses scapula; brings arm down in power stroke, as in striking a blow
Infraspinatus	Partially covered by deltoid and trapezius; a rotator cuff muscle	Infraspinous fossa of scapula	Greater tubercle of humerus	Lateral rotation of humerus; helps hold head of humerus in glenoid cavity
Teres minor	Small muscle inferior to infraspinatus; a rotator cuff muscle	Lateral margin of scapula	Greater tuberosity of humerus	As for infraspinatus

(continued)

TABLE 14.4   (*Continued*)

Muscle	Comments	Origin	Insertion	Action
Teres major	Located inferiorly to teres minor	Posterior surface at inferior angle of scapula	Crest of lesser tubercle of humerus	Extends, medially rotates, and adducts humerus; synergist of latissimus dorsi
Supraspinatus	Obscured by trapezius and deltoid; a rotator cuff muscle	Supraspinous fossa of scapula	Greater tubercle of humerus	Assists abduction of humerus; stabilizes shoulder joint
Levator scapulae	Located at back and side of neck, deep to trapezius	Transverse processes of C_1 through C_4	Superior vertebral border of scapula	Raises and adducts scapula; with fixed scapula, flexes neck to the same side
Rhomboids—major and minor	Beneath trapezius and inferior to levator scapulae; run from vertebral column to scapula	Spinous processes of C_7 and T_1 through T_5	Vertebral border of scapula	Pull scapula medially (retraction) and elevate it

Muscles Associated with the Vertebral Column (Figure 14.6b)

Muscle	Comments	Origin	Insertion	Action
Semispinalis	Deep composite muscle of the back— thoracis, cervicis, and capitis portions	Transverse processes of C_7–T_{12}	Occipital bone and spinous processes of cervical vertebrae and T_1–T_4	Acting together, extend head and vertebral column; acting independently (right vs. left) causes rotation toward the opposite side

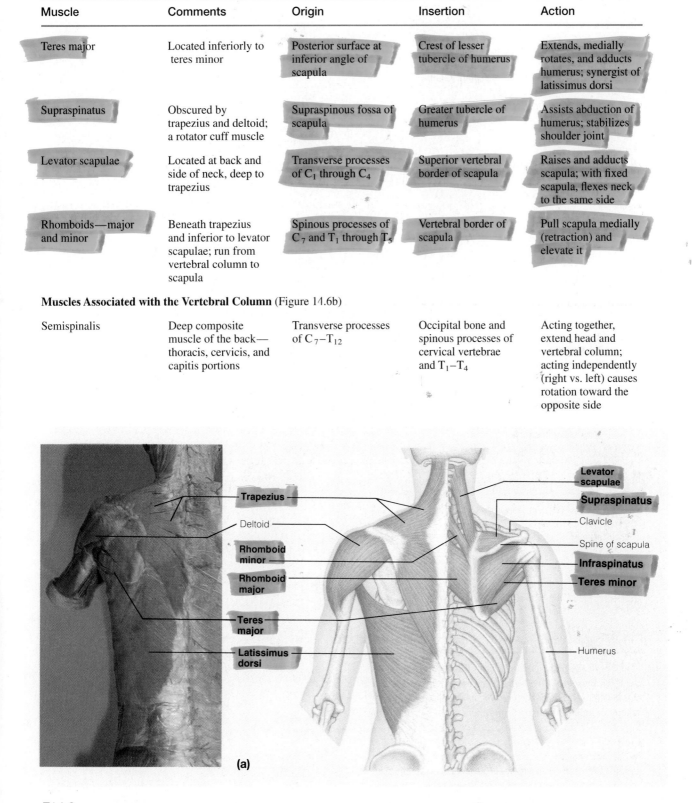

(a)

F14.6

Muscles of the neck, shoulder, and thorax, posterior view. **(a)** The superficial muscles of the back are shown for the left side of the body, with a corresponding photograph. The superficial muscles are removed on the right side of the illustration to reveal the deeper muscles acting on the scapula and the rotator cuff muscles that help to stabilize the shoulder joint.

(continued)

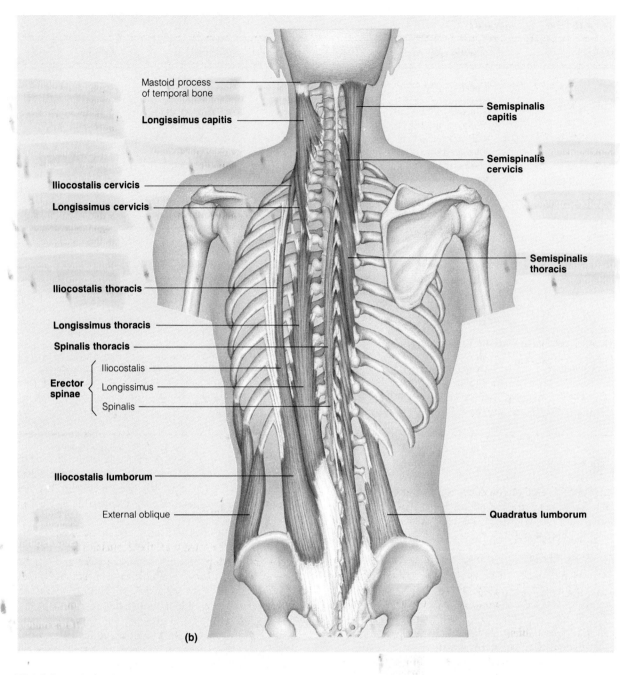

Mastoid process
of temporal bone

Longissimus capitis

Iliocostalis cervicis

Longissimus cervicis

Iliocostalis thoracis

Longissimus thoracis

Spinalis thoracis

**Erector
spinae**
Iliocostalis
Longissimus
Spinalis

Iliocostalis lumborum

External oblique

**Semispinalis
capitis**

**Semispinalis
cervicis**

**Semispinalis
thoracis**

Quadratus lumborum

(b)

F14.6 (*continued*)

Muscles of the neck, shoulder, and thorax, posterior view.
(b) Deep muscles of the back. The superficial muscles and
splenius muscles have been removed. The three muscle
columns (iliocostalis, longissimus, and spinalis) composing
the erector spinae are shown on the left. (Note: The
iliocostalis has cervicis, thoracis, and lumborum parts. The
longissimus has capitis, cervicis, and thoracis parts. The
spinalis has cervicis and thoracis parts.) The semispinalis
muscles are shown on the right. **(c)** Deep (splenius) muscles
of the posterior neck. Superficial muscles have been
removed.

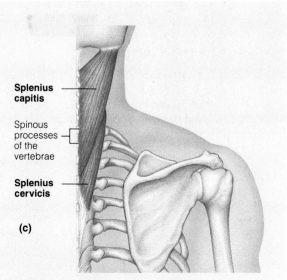

**Splenius
capitis**

Spinous
processes
of the
vertebrae

**Splenius
cervicis**

(c)

TABLE 14.4   *(Continued)*

Muscle	Comments	Origin	Insertion	Action
Erector spinae	A long tripartite muscle composed of lateral iliocostalis (cervicis, thoracis, and lumborum parts), longissimus (capitis, cervicis, and thoracis parts) and medial spinalis (cervicis and thoracis parts); superficial to semispinalis muscles; extends from pelvis to head	Sacrum, iliac crest, transverse processes of lumbar, thoracic, and cervical vertebrae, and/or ribs 3–12 depending on specific part	Ribs and transverse processes of vertebrae about six segments above origin. Longissimus also inserts into mastoid process	All act to extend and abduct the vertebral column; fibers of the longissimus also extend head
Splenius (see Figure 14.6c)	Superficial muscle (capitis and cervicis parts) just deep to levator scapulae and superficial to erector spinae	Ligamentum nuchae and spinous processes of C_7–T_6	Mastoid process, occipital bone, and transverse processes of C_2–C_4	As a group, extend or hyperextend head; when only one side is active, head is rotated and bent toward the same side
Quadratus lumborum	Forms greater portion of posterior abdominal wall	Iliac crest and iliolumbar fascia	Inferior border twelfth rib; transverse processes of lumbar vertebrae	Each flexes vertebral column laterally; together extend the lumbar spine and fix the twelfth rib

When you have completed this study, work with a partner to demonstrate the operation of the following muscles. One of you can demonstrate the movement (the following steps are addressed to this partner). The other can supply the necessary resistance and palpate the muscle being tested.

1. Start by fully abducting the arm and extending the elbow. Now try to adduct the arm against resistance. You are exercising the *latissimus dorsi.*
2. To observe the *deltoid,* attempt to abduct your arm against resistance. Now attempt to elevate your shoulder against resistance; you are contracting the upper portion of the *trapezius.*
3. The *pectoralis major* comes into play when you press your hands together at chest level with your elbows widely abducted.

Muscles of the Upper Limb

The muscles that act on the upper limb fall into four groups: those that move the arm, those causing movement at the elbow, those effecting movements of the wrist and hand, and those producing fine movements of the fingers.

The muscles that cross the shoulder joint to insert on the humerus and move the arm (subscapularis, supraspinatus and infraspinatus, deltoid, and so on) are primarily trunk muscles that originate on the axial skeleton or shoulder girdle. These muscles are included with the trunk muscles.

The second group of muscles, which cross the elbow joint and move the forearm, consists of muscles forming the musculature of the humerus. These muscles

arise primarily from the humerus and insert in forearm bones. They are responsible for flexion, extension, pronation, and supination. The origins, insertions, and actions of these muscles are summarized in Table 14.5 and the muscles are shown in Figure 14.7.

The third group composes the musculature of the forearm. For the most part, these muscles insert on the digits and produce movements at the wrist and fingers. These forearm muscles are more easily identified if their insertion tendons are located first. These muscles are described in Table 14.6 and illustrated in Figure 14.8.

The last group is the muscles that lie entirely in the hand. All are in the palm and all move the metacarpals and fingers. Since these muscles are small, they are weak. Hence, they mostly control precise finger movements including thumb opposition, leaving the powerful movements of the fingers to the forearm muscles. These muscles are described in Table 14.7 and illustrated in Figure 14.9.

 First study the tables and figures, then see if you can identify these muscles on a torso model, anatomical chart, or cadaver. Complete this portion of the exercise with palpation demonstrations as outlined next.

*   To observe the *biceps brachii,* attempt to flex your forearm (hand supinated) against resistance. The insertion tendon of this biceps muscle can also be felt in the lateral aspect of the antecubital fossa (where it runs toward the radius to attach).
*   If you acutely flex your elbow and then try to extend it against resistance, you can demonstrate the action of your *triceps brachii.*

(Text continues on p. 123)

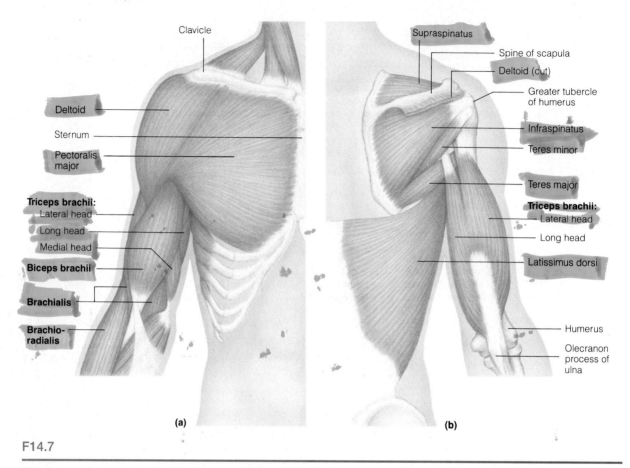

F14.7

Muscles causing movements of the forearm. (a) Superficial muscles of the anterior thorax, shoulder, and arm, anterior view.
(b) Posterior aspect of the arm showing the lateral and long heads of the triceps brachii muscle.

TABLE 14.5   Muscles of Human Humerus That Act on the Forearm (see Figure 14.7)

Muscle	Comments	Origin	Insertion	Action
Triceps brachii	Sole, large fleshy muscle of posterior humerus; three-headed origin	Long head: inferior margin of glenoid cavity; lateral head: posterior humerus; medial head: distal radial groove on posterior humerus	Olecranon process of ulna	Powerful forearm extensor; antagonist of forearm flexors (brachialis and biceps brachii)
Biceps brachii	Most familiar muscle of anterior humerus because this two-headed muscle bulges when forearm is flexed	Short head: coracoid process; tendon of long head runs in intertubercular groove and within capsule of shoulder joint	Radial tuberosity	Flexion (powerful) of elbow and supination of forearm; "it turns the corkscrew and pulls the cork"; weak arm flexor
Brachioradialis	Superficial muscle of lateral forearm; forms lateral boundary of antecubital fossa	Lateral ridge at distal end of humerus	Base of styloid process of radius	Forearm flexor (weak)
Brachialis	Immediately deep to biceps brachii	Distal portion of anterior humerus	Coronoid process of ulna	A major flexor of forearm

TABLE 14.6   Muscles of Human Forearm That Act on Hand and Fingers (see Figure 14.8)

Muscle	Comments	Origin	Insertion	Action
Anterior Compartment (Figure 14.8a,b,c)				
Superficial				
Pronator teres	Seen in a superficial view between proximal margins of brachioradialis and flexor carpi ulnaris	Medial epicondyle of humerus and coronoid process of ulna	Midshaft of radius	Acts synergistically with pronator quadratus to pronate forearm; weak forearm flexor
Flexor carpi radialis	Superficial; runs diagonally across forearm	Medial epicondyle of humerus	Base of second and third metacarpals	Powerful flexor of wrist; abducts hand
Palmaris longus	Small fleshy muscle with a long tendon; medial to flexor carpi radialis	Medial epicondyle of humerus	Palmar aponeurosis	Flexes wrist (weak)
Flexor carpi ulnaris	Superficial; medial to palmaris longus	Medial epicondyle of humerus and olecranon process of ulna	Base of fifth metacarpal	Powerful flexor of wrist; adducts hand

(continued)

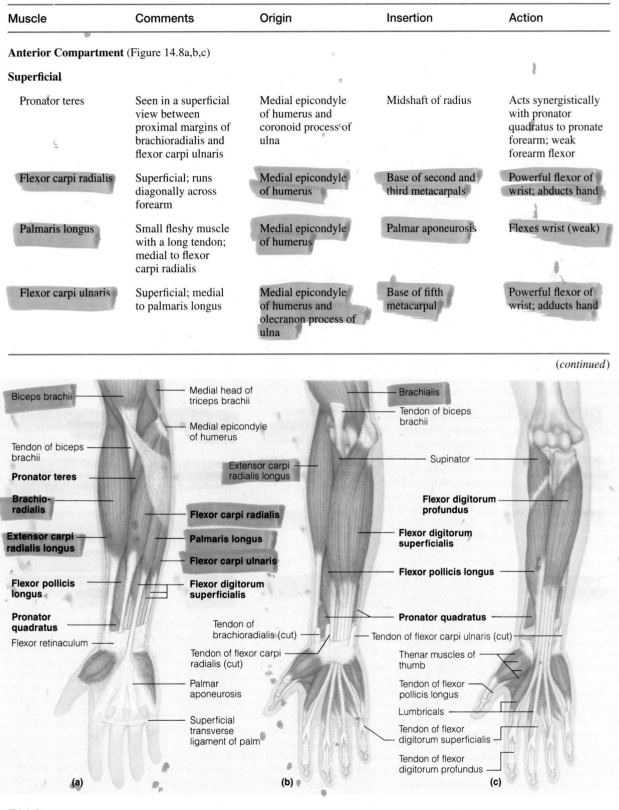

F14.8

Muscles of the forearm and wrist. (a) Superficial anterior view of right forearm and hand. **(b)** The brachioradialis, flexors carpi radialis and ulnaris, and palmaris longus muscles have been removed to reveal the position of the somewhat deeper flexor digitorum superficialis. **(c)** Deep muscles of the anterior compartment. Superficial muscles have been removed.

(continued)

TABLE 14.6 (*Continued*)

Muscle	Comments	Origin	Insertion	Action
Flexor digitorum superficialis	Deeper muscle; overlain by muscles named above; visible at distal end of forearm	Medial epicondyle of humerus, medial surface of ulna, and anterior border of radius	Middle phalanges of second through fifth fingers	Flexes wrist and middle phalanges of second through fifth fingers
Deep				
Flexor pollicis longus	Deep muscle of anterior forearm; distal to and paralleling lower margin of flexor digitorum superficialis	Anterior surface of radius, and interosseous membrane	Distal phalanx of thumb	Flexes thumb (*pollix* is Latin for "thumb"); weak flexor of wrist
Flexor digitorum profundus	Deep muscle; overlain entirely by flexor digitorum superficialis	Anteromedial surface of ulna and interosseous membrane	Distal phalanges of second through fifth fingers	Sole muscle that flexes distal phalanges; assists in wrist flexion
Pronator quadratus	Deepest muscle of distal forearm	Distal portion of anterior ulnar surface	Anterior surface of radius, distal end	Pronates forearm

Posterior Compartment (Figure 14.8d,e,f)

Superficial

Muscle	Comments	Origin	Insertion	Action
Extensor carpi radialis longus	Superficial; parallels brachioradialis on lateral forearm	Lateral supracondylar ridge of humerus	Base of second metacarpal	Extends and abducts wrist
Extensor carpi radialis brevis	Posterior to extensor carpi radialis longus	Lateral epicondyle of humerus	Base of third metacarpal	Extends and abducts wrist; steadies wrist during finger flexion
Extensor digitorum	Superficial; between extensor carpi ulnaris and extensor carpi radialis brevis	Lateral epicondyle of humerus	By four tendons into distal phalanges of second through fifth fingers	Prime mover of finger extension; extends wrist; can flare (abduct) fingers
Extensor carpi ulnaris	Superficial; medial posterior forearm	Lateral epicondyle of humerus	Base of fifth metacarpal	Extends and adducts wrist

Deep

Muscle	Comments	Origin	Insertion	Action
Extensor pollicis longus and brevis	Deep muscle pair with a common origin and action; overlain by extensor carpi ulnaris	Dorsal shaft of ulna and radius, interosseous membrane	Base of distal phalanx of thumb (longus) and proximal phalanx of thumb (brevis)	Extends thumb
Abductor pollicis longus	Deep muscle; lateral and parallel to extensor pollicis longus	Posterior surface of radius and ulna; interosseous membrane	First metacarpal	Abducts and extends thumb
Supinator	Deep muscle at posterior aspect of elbow	Lateral epicondyle of humerus	Proximal end of radius	Acts with biceps brachii to supinate forearm; antagonist of pronator muscles

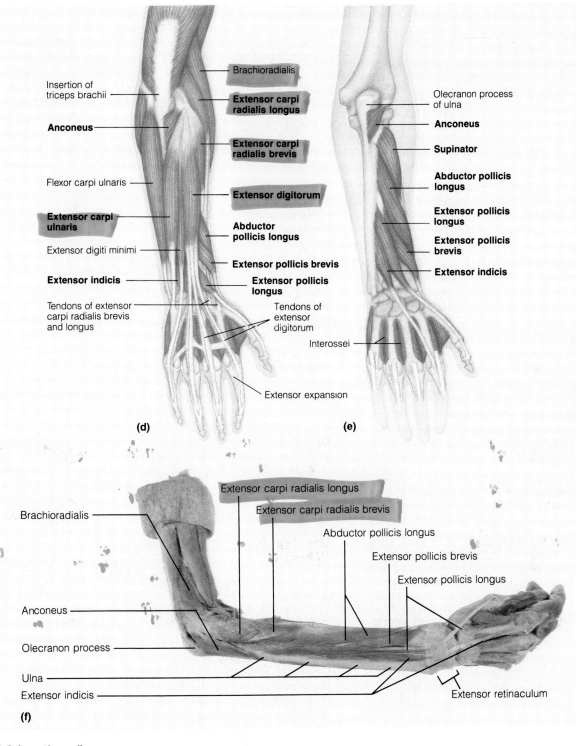

Insertion of
triceps brachii

Anconeus

Flexor carpi ulnaris

**Extensor carpi
ulnaris**

Extensor digiti minimi

Extensor indicis

Tendons of extensor
carpi radialis brevis
and longus

Brachioradialis

**Extensor carpi
radialis longus**

**Extensor carpi
radialis brevis**

Extensor digitorum

**Abductor
pollicis longus**

Extensor pollicis brevis

**Extensor pollicis
longus**

Tendons of
extensor
digitorum

Extensor expansion

(d)

Olecranon process
of ulna

Anconeus

Supinator

**Abductor pollicis
longus**

**Extensor pollicis
longus**

**Extensor pollicis
brevis**

Extensor indicis

Interossei

(e)

Brachioradialis

Anconeus

Olecranon process

Ulna

Extensor indicis

Extensor carpi radialis longus

Extensor carpi radialis brevis

Abductor pollicis longus

Extensor pollicis brevis

Extensor pollicis longus

Extensor retinaculum

(f)

F14.8 *(continued)*

(d) Superficial muscles, posterior view. **(e)** Deep posterior muscles; superficial muscles have been removed. The interossei, the deepest layer of intrinsic hand muscles, are also illustrated. **(f)** Photo of deep posterior muscles of the right forearm. The superficial muscles have been removed.

TABLE 14.7 Intrinsic Muscles of the Hand: Fine Movements of the Fingers (Figure 14.9)

Muscle	Comments	Origin	Insertion	Action
Thenar Muscles in Ball of Thumb (Figure 14.9a and b)				
Abductor pollicis brevis	Lateral muscle of thenar group; superficial	Flexor retinaculum	Lateral base of thumb's proximal phalanx	Abducts thumb (at carpometacarpal joint)
Flexor pollicis brevis	Medial and deep muscle of thenar group	Flexor retinaculum and nearby carpals	Lateral side of base of proximal phalanx of thumb	Flexes thumb (at carpometacarpal and metacarpophalangeal joints)
Opponens pollicis	Deep to abductor pollicis brevis, on metacarpal 1	Flexor retinaculum and nearby carpals	Whole anterior side of metacarpal 1	Opposition: moves thumb to touch tip of little finger
Adductor pollicis	Fan-shaped with horizontal fibers; distal to other thenar muscles; oblique and transverse heads	Bases of metacarpals 2–4 (oblique head); whole front of metacarpal 3 (transverse head)	Medial side of base of proximal phalanx of thumb	Adducts thumb
Hypothenar Muscles in Ball of Little Finger (Figure 14.9a and b)				
Abductor digiti minimi	Medial muscle of hypothenar group; superficial	Pisiform bone	Medial side of proximal phalanx of little finger	Abducts (and flexes) little finger at metacarpophalangeal joint
Flexor digiti minimi brevis	Lateral deep muscle of hypothenar group	Hamate bone and flexor retinaculum	Same as abductor digiti minimi	Flexes little finger at metacarpophalangeal joint
Opponens digiti minimi	Deep to abductor digiti minimi	Same as flexor digiti minimi brevis	Most of length of lateral side of metacarpal 5	Helps in opposition: brings metacarpal 5 toward thumb to cup the hand
Midpalmar Muscles (Figure 14.9a,b,c, and d)				
Lumbricals	Four worm-shaped muscles in palm, one to each finger (except thumb); odd because they originate from the tendons of another muscle	Lateral side of each tendon of flexor digitorum profundus in palm	Lateral edge of extensor expansion on back of first phalanx of fingers 2–5	By pulling on the extensor expansion over the first phalanx, they flex fingers at metacarpophalangeal joints but extend fingers at interphalangeal joints
Palmar interossei	Four long, cone-shaped muscles; lie ventral to the dorsal interossei and between the metacarpals	The side of each metacarpal that faces the midaxis of the hand (metacarpal 3); but absent from metacarpal 3	Extensor expansion on first phalanx of each finger (except finger 3), on side facing midaxis of hand	Adduct fingers; pull fingers in toward third digit; act with lumbricals to extend fingers at interphalangeal joints and flex them at metacarpophalangeal joints
Dorsal interossei	Four bipennate muscles filling spaces between the metacarpals; deepest palm muscles, also visible on dorsal side of hand	Sides of metacarpals	Extensor expansion over first phalanx of fingers 2–4 on side opposite midaxis of hand (finger 3), but on *both* sides of finger 3	Abduct fingers; extend fingers at interphalangeal joints and flex them at metacarpophalangeal joints

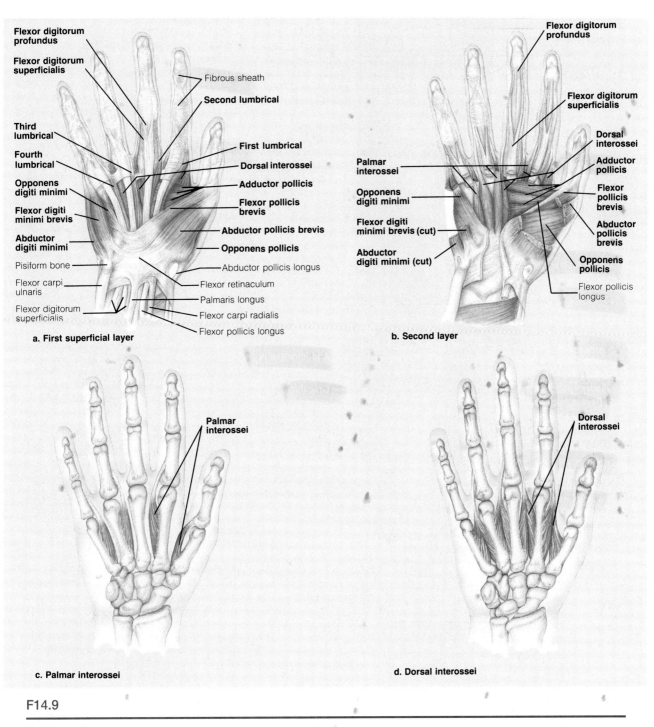

a. First superficial layer

Flexor digitorum profundus
Flexor digitorum superficialis
Third lumbrical
Fourth lumbrical
Opponens digiti minimi
Flexor digiti minimi brevis
Abductor digiti minimi
Pisiform bone
Flexor carpi ulnaris
Flexor digitorum superficialis
Fibrous sheath
Second lumbrical
First lumbrical
Dorsal interossei
Adductor pollicis
Flexor pollicis brevis
Abductor pollicis brevis
Opponens pollicis
Abductor pollicis longus
Flexor retinaculum
Palmaris longus
Flexor carpi radialis
Flexor pollicis longus

b. Second layer

Flexor digitorum profundus
Flexor digitorum superficialis
Dorsal interossei
Adductor pollicis
Flexor pollicis brevis
Abductor pollicis brevis
Opponens pollicis
Flexor pollicis longus
Palmar interossei
Opponens digiti minimi
Flexor digiti minimi brevis (cut)
Abductor digiti minimi (cut)

c. Palmar interossei

Palmar interossei

d. Dorsal interossei

Dorsal interossei

F14.9

Hand muscles.

- Strongly flex your wrist and make a fist. Palpate your contracting wrist flexor muscles (which originate from the medial epicondyle of the humerus) and their insertion tendons, which can be easily felt at the anterior aspect of the wrist.
- Flare your fingers to identify the tendons of the *extensor digitorum* muscle on the dorsum of your hand.

Muscles of the Lower Limb

Muscles that act on the lower limb cause movement at the hip, knee, and foot joints. Since the human pelvic girdle is composed of heavy fused bones that allow very little movement, no special group of muscles is necessary to stabilize it. This is unlike the shoulder girdle, where several muscles (mainly trunk muscles) are needed to stabilize the scapulae.

Muscles acting on the thigh (femur) cause various movements at the multiaxial hip joint (flexion, extension, rotation, abduction, and adduction). These include the iliopsoas, the adductor group, and other muscles summarized in Tables 14.8 and 14.9 and illustrated in Figures 14.10 and 14.11.

(*Text continues on p. 124*)

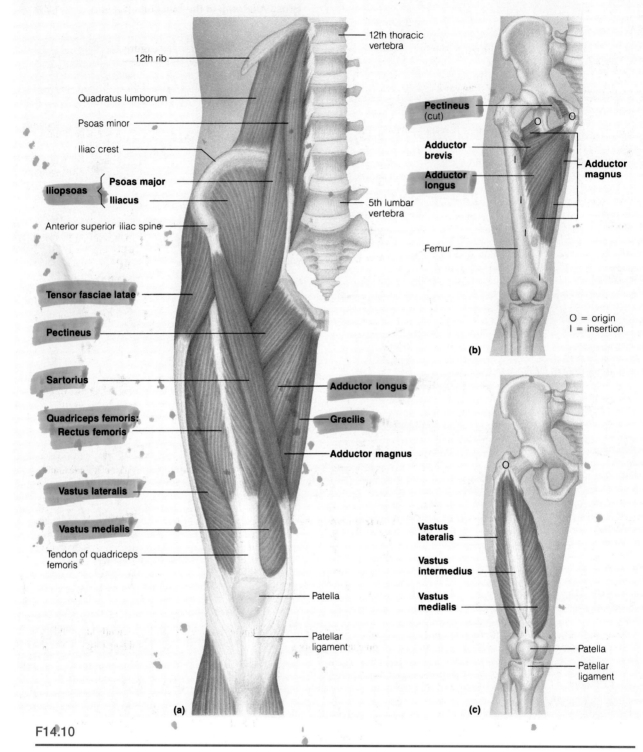

12th thoracic vertebra

12th rib

Quadratus lumborum

Psoas minor

Iliac crest

Iliopsoas { **Psoas major** **Iliacus** }

5th lumbar vertebra

Anterior superior iliac spine

Tensor fasciae latae

Pectineus

Sartorius

Quadriceps femoris: **Rectus femoris**

Vastus lateralis

Vastus medialis

Tendon of quadriceps femoris

Patella

Patellar ligament

Adductor longus

Gracilis

Adductor magnus

(a)

Pectineus (cut)

Adductor brevis

Adductor longus

O   O

I

I

I

Adductor magnus

Femur

I

O = origin
I = insertion

(b)

Vastus lateralis

Vastus intermedius

Vastus medialis

O

I

Patella

Patellar ligament

(c)

F14.10

Anterior and medial muscles promoting movements of the thigh and leg. (**a**) Anterior view of the deep muscles of the pelvis and superficial muscles of the right thigh. (**b**) Adductor muscles of the medial compartment of the thigh. (**c**) The vastus muscles (isolated) of the quadriceps group.

Muscles acting on the leg form the major musculature of the thigh. (Anatomically the term *leg* refers only to that portion between the knee and the ankle.) The thigh muscles cross the knee to allow its flexion and extension. They include the hamstrings and the quadriceps and, along with the muscles acting on the thigh, are described in Tables 14.8 and 14.9 and illustrated in Figures 14.10 and 14.11. Since some of these muscles also have attachments on the pelvic girdle, they can cause movement at the hip joint.

The muscles originating on the leg and acting on the foot and toes are described in Table 14.10 and shown in Figures 14.12 and 14.13. The final muscle group of the lower limb is the intrinsic muscles of the foot, which move the toes and help support the arches of the foot. Of these, all are on the plantar surface (sole) of the foot except one, the extensor digitorum brevis of the foot dorsum. These muscles are described in Table 14.11 and illustrated in Figure 14.14. Identify the muscles as instructed previously.

(*Text continues on p. 135*)

TABLE 14.8   Muscles Acting on Human Thigh and Leg, Anterior and Medial Aspects (see Figure 14.10)

Muscle	Comments	Origin	Insertion	Action
Origin on the Pelvis				
Iliopsoas—iliacus and psoas major	Two closely related muscles; fibers pass under inguinal ligament to insert into femur via a common tendon	Iliacus: iliac fossa; psoas major: transverse processes, bodies, and discs of T_{12} and lumbar vertebrae	Lesser trochanter of femur	Flex trunk on thigh; major flexor of hip (or thigh on pelvis when pelvis is fixed)
Sartorius	Straplike superficial muscle running obliquely across anterior surface of thigh to knee	Anterior superior iliac spine	By an aponeurosis into medial aspect of proximal tibia	Flexes and laterally rotates thigh; flexes knee; known as "tailor's muscle" because it helps bring about cross-legged position in which tailors are often depicted
Medial Compartment				
Adductors—magnus, longus, and brevis	Large muscle mass forming medial aspect of thigh; arise from front of pelvis and insert at various levels on femur	Magnus: ischial and pubic rami; longus: pubis near pubic symphysis; brevis: body and inferior ramus of pubis	Magnus: linea aspera and adductor tubercle of femur; longus and brevis: linea aspera	Adduct and laterally rotate and flex thigh; posterior part of magnus is also a synergist in thigh extension
Pectineus	Overlies adductor brevis on proximal thigh	Pectineal line of pubis	Inferior to lesser trochanter of femur	Adducts, flexes, and laterally rotates thigh
Gracilis	Straplike superficial muscle of medial thigh	Inferior ramus and body of pubis	Medial surface of head of tibia	Adducts thigh; flexes and medially rotates leg, especially during walking
Anterior Compartment				
Quadriceps*				
Rectus femoris	Superficial muscle of thigh; runs straight down thigh; only muscle of group to cross hip joint; arises from two heads	Anterior inferior iliac spine and superior margin of acetabulum	Tibial tuberosity	Extends knee and flexes thigh at hip
Vastus lateralis	Forms lateral aspect of thigh	Greater trochanter and linea aspera	Tibial tuberosity	Extends knee
Vastus medialis	Forms medial aspect of thigh	Linea aspera	Tibial tuberosity	Extends knee
Vastus intermedius	Obscured by rectus femoris; lies between vastus lateralis and vastus medialis on anterior thigh	Anterior and lateral surface of femur (not shown in figure)	Tibial tuberosity	Extends knee
Tensor fasciae latae	Enclosed between fascia layers of thigh	Anterior aspect of iliac crest and anterior superior iliac spine	Iliotibial band of fascia lata	Flexes, abducts, and medially rotates thigh

*The quadriceps form the flesh of the anterior thigh and have a common insertion in the tibial tuberosity via the patellar tendon. They are powerful leg extensors, enabling humans to kick a football, for example.

TABLE 14.9 Muscles Acting on Human Thigh and Leg, Posterior Aspect (see Figure 14.11)

Muscle	Comments	Origin	Insertion	Action
Origin on Pelvis				
Gluteus maximus	Largest and most superficial of gluteal muscles (which form buttock mass)	Dorsal ilium, sacrum, and coccyx	Gluteal tuberosity of femur and iliotibial tract*	Complex, powerful hip extensor (most effective when hip is flexed, as in climbing stairs—but not as in walking); antagonist of iliopsoas; laterally rotates thigh
Gluteus medius	Partially covered by gluteus maximus	Upper lateral surface of ilium	Greater trochanter of femur	Abducts and medially rotates thigh; steadies pelvis during walking
Gluteus minimus	Smallest and deepest gluteal muscle	Inferior surface of ilium (not shown in figure)	Greater trochanter of femur	Abducts and medially rotates thigh
Posterior Compartment				
Hamstrings†				
Biceps femoris	Most lateral muscle of group; arises from two heads	Ischial tuberosity (long head); linea aspera and distal femur (short head)	Tendon passes laterally to insert into head of fibula and lateral condyle of tibia	Extends thigh; laterally rotates leg on thigh; flexes knee
Semitendinosus	Medial to biceps femoris	Ischial tuberosity	Medial aspect of upper tibial shaft	Extends thigh; flexes knee; medially rotates leg
Semimembranosus	Deep to semitendinosus	Ischial tuberosity	Medial condyle of tibia	Extends thigh; flexes knee; medially rotates leg

*The iliotibial tract, a thickened lateral portion of the fascia lata, ensheathes all the muscles of the thigh. It extends as a tendinous band from the iliac crest to the knee.

†The hamstrings are the fleshy muscles of the posterior thigh. The name comes from the butchers' practice of using the tendons of these muscles to hang hams for smoking. As a group, they are strong extensors of the hip; they counteract the powerful quadriceps by stabilizing the knee joint when standing.

TABLE 14.10 Muscles Acting on Human Foot and Ankle (see Figures 14.12 and 14.13)

Muscle	Comments	Origin	Insertion	Action
Posterior Compartment				
Superficial (Figure 14.12a,b)				
Triceps surae	Muscle pair that shapes posterior calf		Via common tendon (calcaneal or Achilles) into heel	Plantar flex foot
Gastrocnemius	Superficial muscle of pair; two prominent bellies	By two heads from medial and lateral condyles of femur	Calcaneus via calcaneal tendon	Crosses knee joint; thus also can flex knee (when foot is dorsiflexed)
Soleus	Deep to gastrocnemius	Proximal portion of tibia and fibula	Calcaneus via calcaneal tendon	Plantar flexion; is an important muscle for locomotion

(*continued on p. 128*)

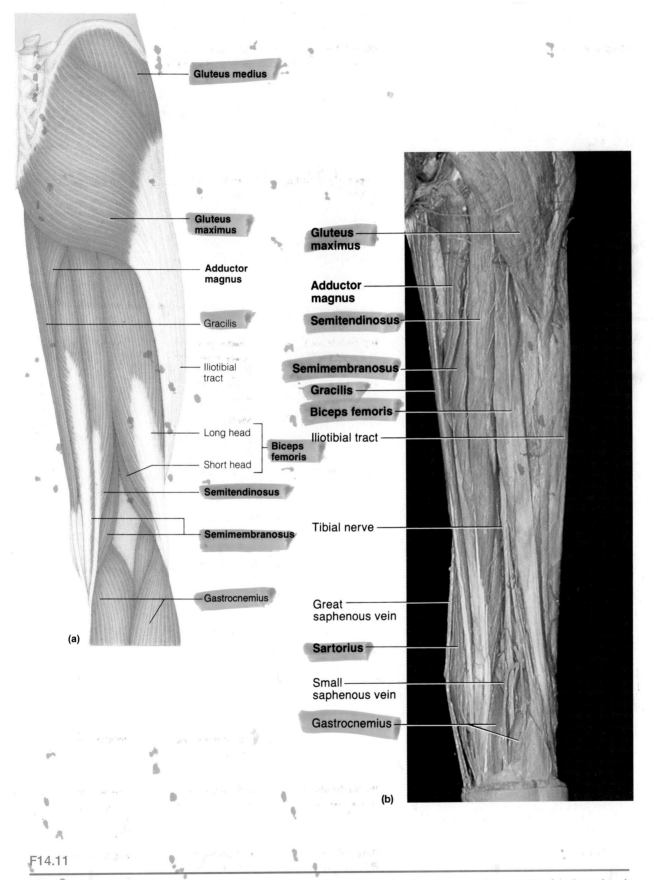

Gluteus medius

Gluteus maximus

Adductor magnus

Gracilis

Iliotibial tract

Long head

Short head

Biceps femoris

Semitendinosus

Semimembranosus

Gastrocnemius

(a)

Gluteus maximus

Adductor magnus

Semitendinosus

Semimembranosus

Gracilis

Biceps femoris

Iliotibial tract

Tibial nerve

Great saphenous vein

Sartorius

Small saphenous vein

Gastrocnemius

(b)

F14.11

Muscles of the posterior aspect of the right hip and thigh. (a) Superficial view showing the gluteus muscles of the buttock and hamstring muscles of the thigh. **(b)** Photo of muscles of the posterior thigh.

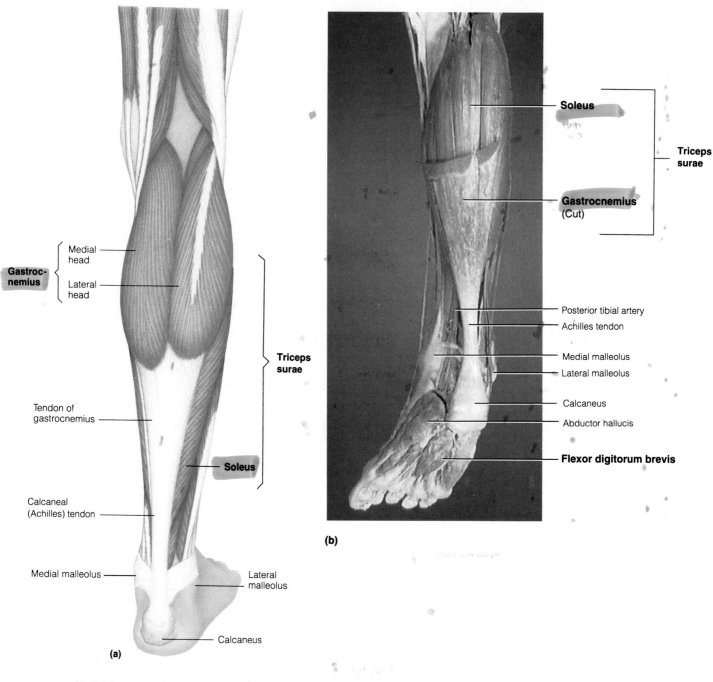

(a)

Gastroc-nemius
- Medial head
- Lateral head

Triceps surae

Tendon of gastrocnemius

Soleus

Calcaneal (Achilles) tendon

Medial malleolus

Lateral malleolus

Calcaneus

(b)

Soleus

Triceps surae

Gastrocnemius (Cut)

Posterior tibial artery

Achilles tendon

Medial malleolus

Lateral malleolus

Calcaneus

Abductor hallucis

Flexor digitorum brevis

F14.12

Muscles of the posterior aspect of the right leg. (a) Superficial view of the posterior leg. (b) Photo of posterior aspect of right leg. The gastrocnemius has been transected and its superior part removed.

(continued)

TABLE 14.10 *(Continued)*

Muscle	Comments	Origin	Insertion	Action
Deep (Figure 14.12c,d)				
Popliteus	Thin muscle at posterior aspect of knee	Lateral condyle of femur	Proximal tibia	Flexes and rotates leg medially to "unlock" extended knee when knee flexion begins

(continued)

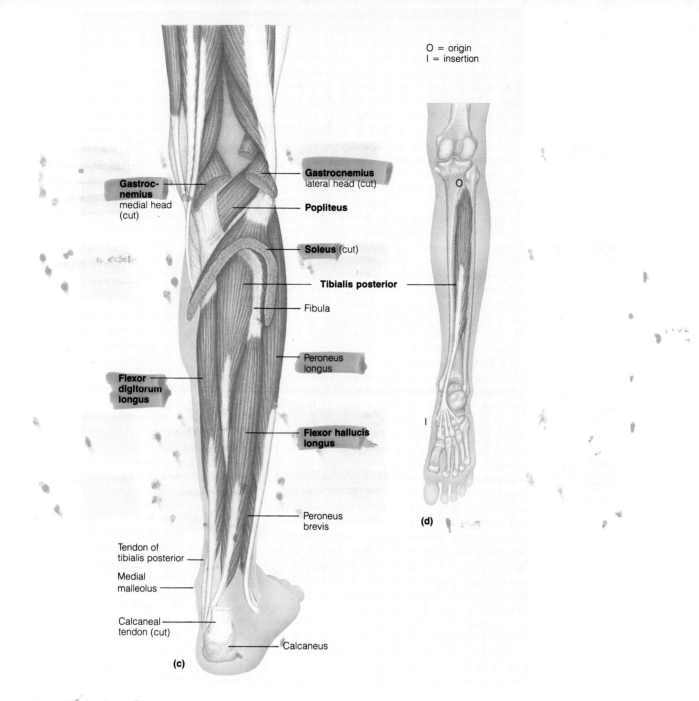

O = origin
I = insertion

Gastrocnemius lateral head (cut)

Popliteus

Soleus (cut)

Tibialis posterior

Fibula

Peroneus longus

Flexor hallucis longus

Peroneus brevis

(d)

Gastroc-nemius medial head (cut)

Flexor digltorum longus

Tendon of tibialis posterior

Medial malleolus

Calcaneal tendon (cut)

Calcaneus

(c)

F14.12 (continued)

Muscles of the posterior aspect of the right leg. (c) The triceps surae has been removed to show the deep muscles of the posterior compartment. **(d)** Tibialis posterior shown in isolation so that its origin and insertion may be visualized.

TABLE 14.10   (Continued)

Muscle	Comments	Origin	Insertion	Action
Tibialis posterior	Thick muscle deep to soleus	Superior portion of tibia and fibula and interosseous membrane	Tendon passes obliquely behind medial malleolus and under arch of foot; inserts into several tarsals and metatarsals 2–4	Prime mover of foot inversion; plantar flexes foot
Flexor digitorum longus	Runs medial to and partially overlies tibialis posterior	Posterior surface of tibia	Distal phalanges of second through fifth toes	Flexes toes; plantar flexes and inverts foot

(continued)

TABLE 14.10   (*Continued*)

Muscle	Comments	Origin	Insertion	Action
Flexor hallucis longus (see also Figure 14.13a)	Lies lateral to inferior aspect of tibialis posterior	Middle portion of fibula shaft	Tendon runs under foot to insert on distal phalanx of great toe	Flexes great toe; plantar flexes and inverts foot; the "push-off muscle" during walking

Lateral Compartment (Figure 14.12c and Figure 14.13a,b,c)

Muscle	Comments	Origin	Insertion	Action
Peroneus longus	Superficial lateral muscle; overlies fibula	Head and upper portion of fibula	By long tendon under foot to first metatarsal and medial cuneiform	Plantar flexes and everts foot; helps keep foot flat on ground
Peroneus brevis	Smaller muscle; deep to peroneus longus	Distal portion of fibula shaft	By tendon running behind lateral malleolus to insert on proximal end of fifth metatarsal	Plantar flexes and everts foot, as part of peronei group

Anterior Compartment (Figure 14.13a,b,c)

Muscle	Comments	Origin	Insertion	Action
Tibialis anterior	Superficial muscle of anterior leg; parallels sharp anterior margin of tibia	Lateral condyle and upper 2/3 of tibia; interosseous membrane	By tendon into inferior surface of first cuneiform and metatarsal 1	Prime mover of dorsiflexion; inverts foot
Extensor digitorum longus	Anterolateral surface of leg; lateral to tibialis anterior	Lateral condyle of tibia; proximal 3/4 of fibula; interosseous membrane	Tendon divides into four parts; insert into middle and distal phalanges of toes 2–5	Prime mover of toe extension; dorsiflexes foot
Peroneus tertius	Small muscle; often fused to distal part of extensor digitorum longus	Distal anterior surface of fibula	Tendon passes anterior to lateral malleolus and inserts on dorsum of fifth metatarsal	Dorsiflexes and everts foot
Extensor hallucis longus	Deep to extensor digitorum longus and tibialis anterior	Anteromedial shaft of fibula and interosseous membrane	Tendon inserts on distal phalanx of great toe	Extends great toe; dorsiflexes foot

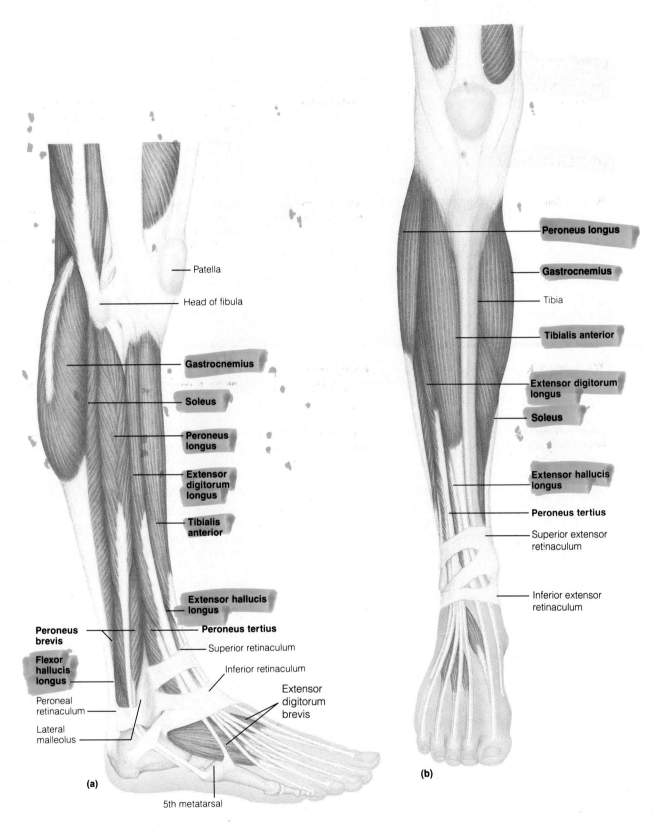

Patella

Head of fibula

Gastrocnemius

Soleus

**Peroneus
longus**

**Extensor
digitorum
longus**

**Tibialis
anterior**

**Extensor hallucis
longus**

Peroneus tertius

Superior retinaculum

Inferior retinaculum

**Peroneus
brevis**

**Flexor
hallucis
longus**

Peroneal
retinaculum

Lateral
malleolus

Extensor
digitorum
brevis

5th metatarsal

(a)

Peroneus longus

Gastrocnemius

Tibia

Tibialis anterior

**Extensor digitorum
longus**

Soleus

**Extensor hallucis
longus**

Peroneus tertius

Superior extensor
retinaculum

Inferior extensor
retinaculum

(b)

F14.13

Muscles of the anterolateral aspect of the right leg. (a) Superficial view of lateral aspect of the leg, illustrating the positioning of the lateral compartment muscles (peroneus longus and brevis) relative tos anterior and posterior leg muscles. **(b)** Superficial view of anterior leg muscles.

(continued)

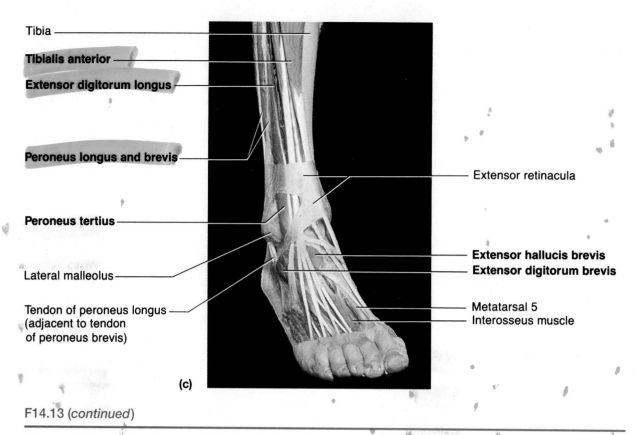

Tibia

Tibialis anterior

Extensor digitorum longus

Peroneus longus and brevis

Peroneus tertius

Lateral malleolus

Tendon of peroneus longus
(adjacent to tendon
of peroneus brevis)

Extensor retinacula

Extensor hallucis brevis
Extensor digitorum brevis

Metatarsal 5
Interosseus muscle

(c)

F14.13 (*continued*)

Muscles of the anterolateral aspect of the right leg. (c) Photo of anteroinferior aspect of right leg and foot, and extensor retinacula, anterolateral view.

TABLE 14.11   Intrinsic Muscles of the Foot: Toe Movement and Foot Support (Figure 14.14)

The intrinsic muscles of the foot help to flex, extend, abduct, and adduct the toes. Furthermore, along with the tendons of some leg muscles that enter the sole, the foot muscles support the arches of the foot. There is a single muscle on the foot's dorsum (superior aspect), and many muscles on the plantar aspect (the sole). The plantar muscles occur in four layers, from superficial to deep. Overall, the foot muscles are remarkably similar to those in the palm of the hand.

Muscle	Comments	Origin	Insertion	Action
Muscle on Dorsum of Foot (Figure 14.13c)				
Extensor digitorum brevis	Small, four-part muscle on dorsum of foot; deep to the tendons of extensor digitorum longus; corresponds to the extensor indicis and extensor pollicis muscles of forearm	Anterior part of calcaneus bone; extensor retinaculum	Base of proximal phalanx of big toe; extensor expansions on toes 2–4	Helps extend toes at metatarsophalangeal joints
Muscles on Sole of Foot (Figure 14.14)				
First layer				
Flexor digitorum brevis	Bandlike muscle in middle of sole; corresponds to flexor digitorum superficialis of forearm and inserts into digits in the same way	Tuber calcanei	Middle phalanx of toes 2–4	Helps flex toes

(Table continues on next page)

TABLE 14.11   (*continued*)

Muscle	Comments	Origin	Insertion	Action
Abductor hallucis	Lies medial to flexor digitorum brevis; recall the similar thumb muscle, abductor pollicis brevis	Tuber calcanei and flexor retinaculum	Proximal phalanx of big toe, medial side, via a sesamoid bone in the tendon of flexor hallucis brevis (see below)	Abducts big toe
Abductor digiti minimi	Most lateral of the three superficial sole muscles; recall the similar abductor muscle in palm	Tuber calcanei	Lateral side of base of little toe's proximal phalanx	Abducts little toe
Second layer Flexor accessorius (quadratus plantae)	Rectangular muscle just deep to flexor digitorum brevis in posterior half of sole; two heads	Medial and lateral sides of calcaneus	Tendon of flexor digitorum longus in midsole	Straightens out the oblique pull of flexor digitorum longus
Lumbricals	Four little "worms," like lumbricals in hand	From each tendon of flexor digitorum longus	Extensor expansion on proximal phalanx of toes 2–5, medial side	Function same as lumbricals of hand

(*Table continues on next page*)

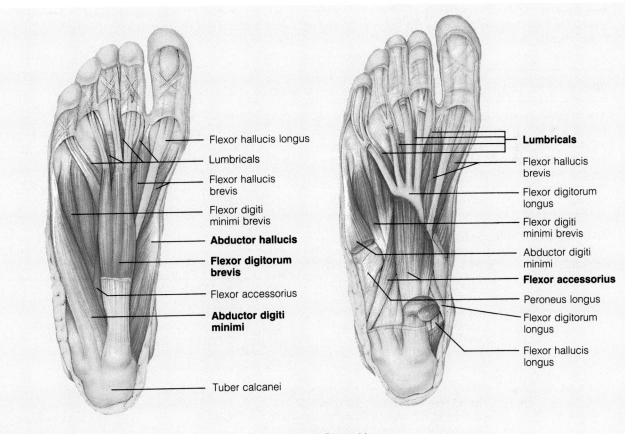

Flexor hallucis longus
Lumbricals
Flexor hallucis brevis
Flexor digiti minimi brevis
Abductor hallucis
Flexor digitorum brevis
Flexor accessorius
Abductor digiti minimi
Tuber calcanei

Lumbricals
Flexor hallucis brevis
Flexor digitorum longus
Flexor digiti minimi brevis
Abductor digiti minimi
Flexor accessorius
Peroneus longus
Flexor digitorum longus
Flexor hallucis longus

a. First layer

b. Second layer

F14.14

Muscles of the right foot, first and second layers.

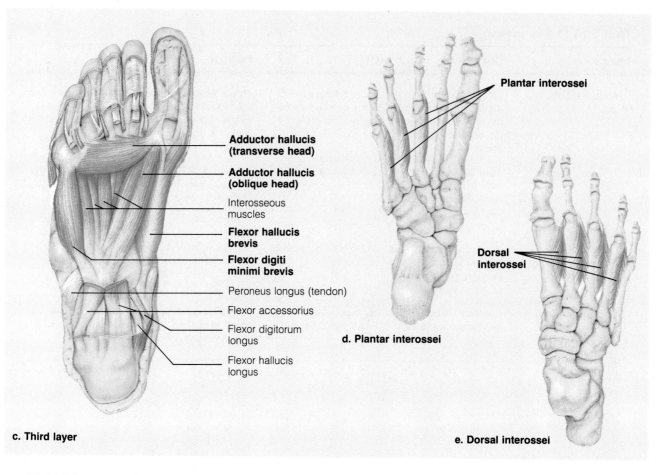

c. Third layer

d. Plantar interossei

e. Dorsal interossei

F14.14 *(continued)*

Muscles of the right foot, third layer and interossei.

TABLE 14.11   *(continued)*

Muscle	Comments	Origin	Insertion	Action
Third layer				
Flexor hallucis brevis	Covers metatarsal 1; splits into two bellies; recall the flexor pollicis brevis of thumb	Mostly from cuboid bone	Via two tendons onto both sides of the base of the proximal phalanx of big toe; each tendon has a sesamoid bone in it	Flexes big toe's metatarsophalangeal joint
Adductor hallucis	Oblique and transverse heads; deep to lumbricals; recall adductor pollicis in thumb	From bases of metatarsals 2–4 and from peroneus longus tendon (oblique head); from a ligament across metatarsophalangeal joints (transverse head)	Base of proximal phalanx of big toe, lateral side	Helps maintain the transverse arch of foot; weak adductor of big toe

(Table continues on next page)

TABLE 14.11   (*continued*)

Muscle	Comments	Origin	Insertion	Action
Flexor digiti minimi brevis	Covers metatarsal 5; recall same muscle in hand	Base of metatarsal 5 and tendon of peroneus longus	Base of proximal phalanx of toe 5	Flexes little toe at metatarsophalangeal joint
Fourth layer Plantar and dorsal interossei	Three plantar and four dorsal interossei; similar to the palmar and dorsal interossei of hand in locations, attachments, and actions; however, the long axis of foot around which these muscles orient is the second digit, not the third	See palmar and dorsal interossei (Table 14.7)	See palmar and dorsal interossei (Table 14.7)	See palmar and dorsal interossei (Table 14.7)

Complete this exercise by performing the following palpation demonstrations with your lab partner.

- Go into a deep knee bend and palpate your own *gluteus maximus* muscle as you extend your hip to resume the upright posture.
- Demonstrate the contraction of the anterior *quadriceps femoris* by trying to extend your knee against resistance. Do this while seated and note how the patellar tendon reacts. The *biceps femoris* of the posterior thigh comes into play when you flex your knee against resistance.
- Now stand on your toes. Have your partner palpate the lateral and medial heads of the *gastrocnemius* and follow it to its insertion in the calcaneal tendon.
- Dorsiflex and invert your foot while palpating your *tibialis anterior* muscle (which parallels the sharp anterior crest of the tibia laterally).

DISSECTION AND IDENTIFICATION OF CAT MUSCLES

The skeletal muscles of all mammals are named in a similar fashion. However, some muscles that are separate in other animals are fused in humans, and some muscles present in lower animals are lacking in humans. This exercise involves dissection of the cat musculature, in conjunction with the study of human muscles, to enhance your knowledge of the human muscular system.

Since the aim is to become familiar with the muscles of the human body, you should pay particular attention to the similarities between cat and human muscles. However, pertinent differences will be pointed out as they are encountered.

Preparing the Cat for Dissection

The preserved laboratory animals purchased for dissection have been embalmed with a solution that prevents deterioration of the tissues. The animals are generally delivered in plastic bags that contain a small amount of the embalming fluid. Do not dispose of this fluid when you remove the cat; the fluid prevents the cat from drying out. It is very important to keep the cat's tissues moist, because you will probably use the same cat from now until the end of the course. The embalming fluid may cause your eyes to smart and may dry your skin, but these small irritants are preferable to working with a cat that has become hard and odoriferous due to bacterial action. You can alleviate the possibility of skin irritation by using disposable gloves or applying skin cream before beginning the dissection.

1. Obtain a cat, dissection tray, dissection instruments, and a name tag. Mark the name tag with the names of the members of your group and set it aside. The name tag will be attached to the plastic bag at the end of the dissection so that you may identify your animal in subsequent laboratories.

2. To begin removing the skin, place the cat ventral side down on the dissecting tray. With a scalpel, make a short and shallow incision in the midline of the neck, just to penetrate the skin. From this point on, use scissors. Continue the cut the length of the back to the sacrolumbar region, stopping at the tail (Figure 14.15).

3. From the dorsal surface of the tail region, make an incision around the tail, encircling the anus and genital organs. The skin will not be removed from this region. Beginning again at the dorsal tail region, make an incision through the skin down each hind limb nearly to the ankle. Continue the cuts completely around the ankles.

4. Return to the neck. Cut the skin around the neck, and then cut down each foreleg to the wrist. Completely cut the skin around the wrists.

5. Now free the skin from the loose connective tissue (superficial fascia) that binds it to the underlying structures. With one hand, grasp the skin on one side of the midline dorsal incision. Then, using your fingers or a blunt probe, break through the "cottony" connective tissue fibers to release the skin from the muscle beneath. Work toward the ventral surface and then toward the neck. As you pull the skin from the body, you should see small, white, cordlike structures extending from the skin to the muscles at fairly regular intervals. These are the cutaneous nerves that serve the skin. You will also see (particularly as you approach the ventral surface) that a thin layer of muscle fibers remains adhered to the skin. This is the **cutaneous maximus** muscle, which enables the cat to move its skin rather like our facial muscles allow us to express emotion. Where the cutaneous maximus fibers cling to those of the deeper muscles, they should be carefully cut free.

6. You will notice as you start to free the skin in the neck region that it is more difficult to remove. Take extra care and time in this region. The large flat **platysma** muscle in the ventral neck region (a skin muscle like the cutaneous maximus) will remain attached to the skin. The skin will not be removed from the head since the cat's muscles are not sufficiently similar to human head muscles to merit study.

7. Complete the skinning process by freeing the skin from the forelimbs, lower torso, and hindlimbs in the same manner.

8. Inspect your skinned cat. Notice that it is difficult to see any cleavage lines between the muscles because of the overlying connective tissue, which is white or yellow. If time allows, carefully remove as much of the fat and fascia from the surface of the muscles as possible, using forceps or your fingers. The muscles, when exposed, look grainy or threadlike and are light brown. If this clearing process is done carefully and thoroughly, you will be ready to begin your identification of the superficial muscles.

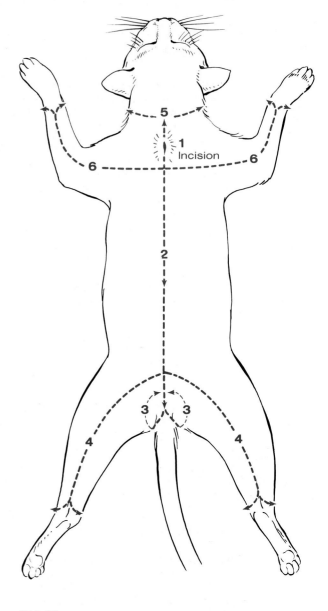

F14.15

Incisions to be made in skinning a cat. Numbers indicate sequence.

9. If the muscle dissection exercises are to be done at a later laboratory session, carefully rewrap the cat's skin around its body. Then dampen several paper towels in embalming fluid and wrap these around the animal. Return the cat to the plastic bag (add more embalming fluid to the bag if necessary), seal the bag, and attach your name tag. Prepare your cat for storage in this way every time the cat is used. Place the cat in the storage container designated by your instructor.

10. Before leaving the laboratory, dispose of any tissue remnants in the organic debris container, wash the dissecting tray and instruments with soapy water, rinse and dry, and wash down the laboratory bench.

Dissection of Cat Trunk and Neck Muscles

The proper dissection of muscles involves careful separation of one muscle from another and transection of superficial muscles in order to study those lying deeper. In general, when directions are given to transect a muscle, it should be completely freed from all adhering connective tissue and then cut through about halfway between its origin and insertion points.

As a rule, all the fibers of one muscle are held together by a connective tissue sheath (epimysium) and run in the same general direction. Before you begin dissection, observe your skinned cat. If you look carefully, you can see changes in the direction of the muscle fibers, which will help you to locate the muscle borders. Pulling in slightly different directions on two adjacent muscles will usually allow you to expose the normal cleavage line between them. Once cleavage lines have been identified, use a blunt probe to break the connective tissue between muscles and to separate them. If the muscles separate as clean, distinct bundles, your procedure is probably correct. If they appear ragged or chewed up, you are probably tearing a muscle apart rather than sep-

arating it from adjacent muscles. Only the muscles that are most easily identified and separated out will be identified in this exercise because of time considerations.

ANTERIOR NECK MUSCLES

1. Using Figure 14.16 as a guide, examine the anterior neck surface of the cat and identify the following superficial neck muscles. (The platysma belongs in this group but was probably removed during the skinning process.) The **sternomastoid** muscle and the more lateral and deeper **cleidomastoid** muscle are joined in humans to form the sternocleidomastoid. The large external jugular vein, which drains the head, should be obvious crossing the anterior aspect of these muscles. The **mylohyoid** muscle parallels the bottom aspect of the chin, and the **digastric** muscles form a V over the mylohyoid muscle. Although it is not one of the neck muscles, you can now identify the fleshy **masseter** muscle, which flanks the digastric muscle laterally. Finally, find the **sternohyoid**, a narrow muscle between the mylohyoid (superiorly) and the inferior sternomastoid.

2. The deeper muscles of the anterior neck of the cat are small and straplike and hardly worth the effort of dissection. However, one of these deeper muscles can be seen with a minimum of extra effort. Transect the sternomastoid and sternohyoid muscles approximately

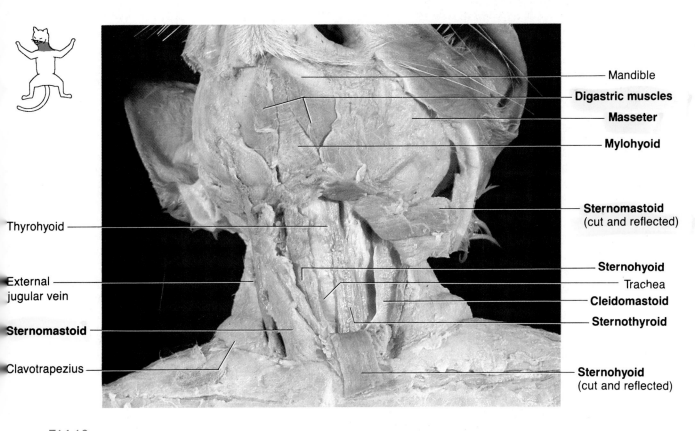

F14.16

Anterior neck muscles of the cat. On the left side of the neck, the sternomastoid and the sternohyoid are transected and reflected to reveal the cleidomastoid and sternothyroid muscles.

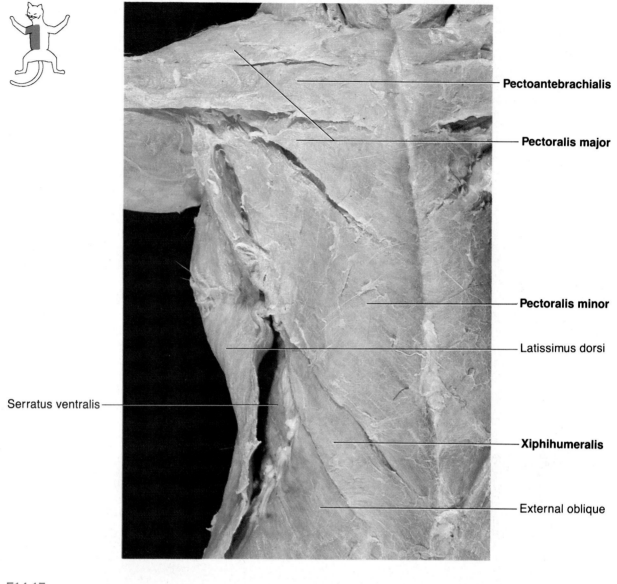

Pectoantebrachialis

Pectoralis major

Pectoralis minor

Latissimus dorsi

Serratus ventralis

Xiphihumeralis

External oblique

F14.17

Superficial thorax muscles, inferior view. Latissimus dorsi is reflected away from the thorax. (Compare to human thorax muscles, anterior view, Figure 14.5a.)

at midbelly. Reflect the cut ends to reveal the **sternothyroid** muscle, which runs along the anterior surface of the throat just deep and lateral to the sternohyoid muscle. The cleidomastoid muscle, which lies deep to the sternomastoid, is also more easily identified now.

SUPERFICIAL CHEST MUSCLES   In the cat, the chest or pectoral muscles adduct the arm, just as they do in humans. However, humans have only two pectoral muscles, and cats have four—the pectoralis major, pectoralis minor, xiphihumeralis, and pectoantebrachialis (see Figure 14.17). Because of their relatively great degree of fusion, the cat's pectoral muscles appear to be a single muscle. The pectoral muscles are rather difficult to dissect and identify, as they do not separate from one another easily.

The **pectoralis major** is 2 to 3 inches wide and can be seen arising on the manubrium, just inferior to the sternomastoid muscle of the neck, and running to the humerus. Its fibers run at right angles to the longitudinal axis of the cat's body.

The **pectoralis minor** lies beneath the pectoralis major and extends posterior to it on the abdominal surface. It originates on the sternum and inserts on the humerus. Its fibers run obliquely to the long axis of the body, which helps to distinguish it from the pectoralis major. Contrary to what its name implies, the pectoralis minor is a larger and thicker muscle than the pectoralis major.

The **xiphihumeralis** can be distinguished from the posterior edge of the pectoralis minor only by virtue of the fact that its origin is lower—on the xiphoid process

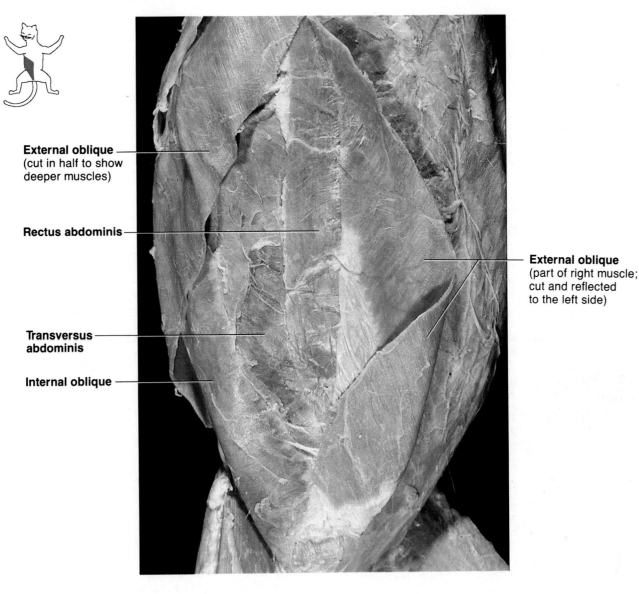

External oblique
(cut in half to show
deeper muscles)

Rectus abdominis

External oblique
(part of right muscle;
cut and reflected
to the left side)

**Transversus
abdominis**

Internal oblique

F14.18

Muscles of the abdominal wall of the cat.

of the sternum. Its fibers run parallel to and are fused with those of the pectoralis minor.

The **pectoantebrachialis** is a thin, straplike muscle, about ½ inch wide, lying over the pectoralis major. It originates from the manubrium, passes laterally over the pectoralis major, and merges with the muscles of the forelimb approximately halfway down the humerus. It has no homologue in humans.

Identify, free, and trace out the origin and insertion of the cat's chest muscles. Refer to Figure 14.17 as you work.

MUSCLES OF THE ABDOMINAL WALL   The superficial anterior trunk muscles include those of the abdominal wall (see Figure 14.18). Cat musculature in this area is quite similar in function to that of humans.

1.   Complete the dissection of the more superficial anterior trunk muscles of the cat by identifying the origins and insertions of the muscles of the abdominal wall.

Work carefully here. These muscles are very thin, and it is easy to miss their boundaries. Begin with the **rectus abdominis,** a long band of muscle approximately 1 inch wide running immediately lateral to the midline of the body on the abdominal surface. Humans have four transverse *tendinous intersections* in the rectus abdominis (see Figure 14.5b), but they are absent or difficult to identify in the cat. Identify the **linea alba,** which separates the rectus abdominis muscles. Note the relationship of the rectus abdominis to the other abdominal muscles and their fascia.

2.   The **external oblique** is a sheet of muscle immediately beside (and running beneath) the rectus abdominis (see Figure 14.18). Carefully free and then transect the external oblique to reveal the anterior attachment of the rectus abdominis and the deeper **internal oblique.** Reflect the external oblique; observe the deeper muscle. Note which way the fibers run.

How does the fiber direction of the internal oblique compare to the external oblique?

_____

_____

3.   Free and then transect the internal oblique muscle to reveal the fibers of the **transversus abdominis,** whose fibers run transversely across the abdomen.

SUPERFICIAL MUSCLES OF THE SHOULDER AND POSTERIOR TRUNK AND NECK   Refer to Figure 14.19 as you dissect the superficial muscles of the posterior surface of the trunk.

1.   Turn your cat on its ventral surface and start your observations with the **trapezius group.** Humans have a single large trapezius muscle, but the cat has three separate muscles—the clavotrapezius, acromiotrapezius, and spinotrapezius. The prefix (clavo-, acromio-, and spino-) in each case reveals the muscle's site of insertion. The **clavotrapezius,** the most superior muscle of the group, is homologous to that part of the human trapezius that inserts into the clavicle. Slip a probe under this muscle and follow it to its apparent origin.

Where does the clavotrapezius appear to originate?

_____

Is this similar to its origin in humans?_____

The fibers of the clavotrapezius are continuous inferiorly with those of the clavicular part of the cat's deltoid muscle (clavodeltoid), and the two muscles work together to bring about the extension of the humerus. Release the clavotrapezius muscle from the adjoining muscles. The **acromiotrapezius** is a large, nearly square muscle easily identified by its aponeurosis, which passes over the vertebral border of the scapula. It originates from the cervical and T_1 vertebrae and inserts into the scapular spine. The triangular **spinotrapezius** runs from the thoracic vertebrae to the scapular spine. This is the most posterior of the trapezius muscles in the cat. Now that you know where they are located, pull on the three trapezius muscles.

Do the trapezius muscles appear to have the same

functions in cats as in humans?_____

2.   The **levator scapulae ventralis,** a flat, straplike muscle, can be located in the triangle created by the division of the fibers of the clavotrapezius and acromiotrapezius. Its anterior fibers run underneath the clavotrapezius from its origin at the base of the skull (occipital bone), and it inserts on the vertebral border of the scapula. In the cat it helps to hold the upper edges of the scapulae together and draws them toward the head.

What is the function of the levator scapulae in humans?

_____

_____

3.   The **deltoid group.** Like the trapezius, the human deltoid muscle is represented by three separate muscles in the cat—the clavodeltoid, acromiodeltoid, and spinodeltoid. The **clavodeltoid,** the most superficial muscle of the shoulder, is a continuation of the clavotrapezius below the clavicle, which is this muscle's point of origin (see Figure 14.19). Follow its course down the forelimb to the point where it merges along a white line with the pectoantebrachialis. Separate it from the pectoantebrachialis, and then transect it and pull it back.

Where does the clavodeltoid insert? _____

What do you think the function of this muscle is?

_____

The **acromiodeltoid** lies posterior to the clavodeltoid and runs over the top of the shoulder. This small triangular muscle originates on the acromion process of the scapula and inserts into the spinodeltoid posterior to it. The **spinodeltoid** is posterior to the acromiodeltoid and approximately the same size. This muscle is covered with fascia near the anterior end of the scapula. Its tendon extends under the acromiodeltoid muscle and inserts on the humerus. Notice that its fibers run obliquely to those of the acromiodeltoid. Like the human deltoid muscle, the acromiodeltoid and clavodeltoid muscles in the cat raise and rotate the humerus.

4.   The **latissimus dorsi** is a large flat muscle covering most of the lateral surface of the posterior trunk. Its upper edge is covered by the spinotrapezius. As in humans, it inserts into the humerus. But before inserting, its fibers merge with the fibers of many other muscles, among them the xiphihumeralis of the pectoralis group.

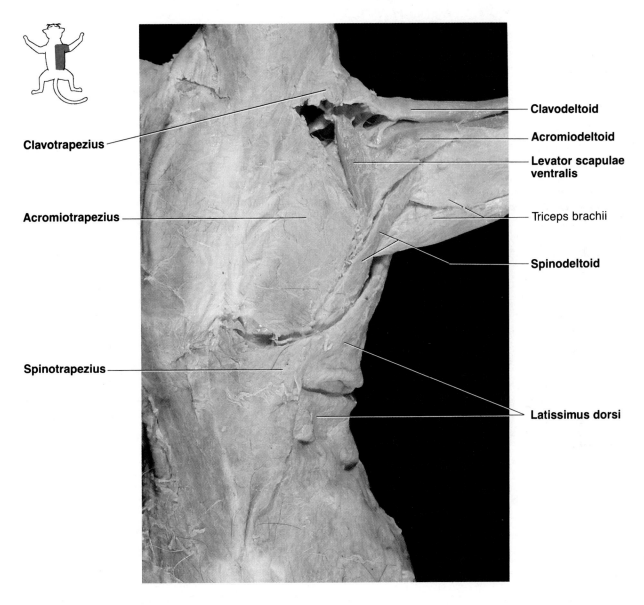

Clavotrapezius

Acromiotrapezius

Spinotrapezius

Clavodeltoid

Acromiodeltoid

Levator scapulae ventralis

Triceps brachii

Spinodeltoid

Latissimus dorsi

F14.19

Superficial muscles of the posterosuperior aspect of the right shoulder, trunk, and neck of the cat. (Compare to human thorax muscles, posterior view, Figure 14.6a.)

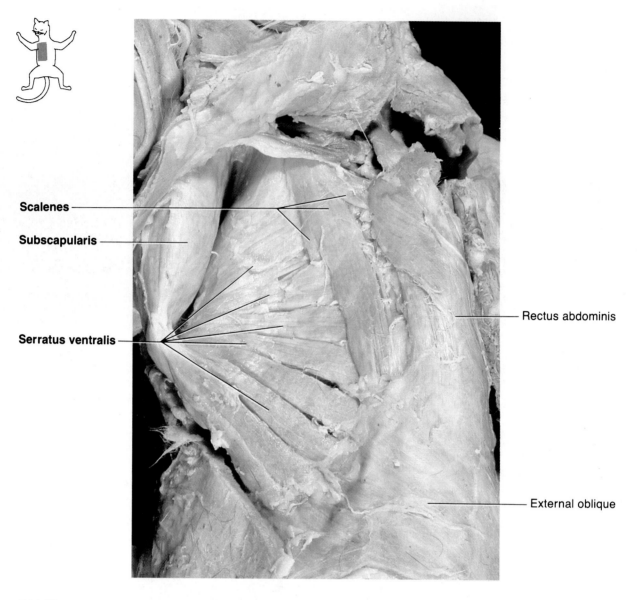

F14.20

Deep muscles of the right inferolateral thorax of the cat.

DEEP MUSCLES OF THE TRUNK AND POSTERIOR NECK

1.   In preparation, transect the latissimus dorsi, the muscles of the pectoralis group, and the spinotrapezius and reflect them back. Be careful not to damage the large brachial nerve plexus, which lies in the axillary space beneath the pectoralis group.

2.   The **serratus ventralis,** homologous to the serratus anterior of humans, arises deep to the pectoral muscles and covers the lateral surface of the rib cage. It is easily identified by its fingerlike muscular origins, which arise on the first 9 or 10 ribs. It inserts into the scapula. The anterior portion of the serratus ventralis, which arises

from the cervical vertebrae, is homologous to the levator scapulae in humans; both pull the scapula toward the sternum. Trace this muscle to its insertion. In general, in the cat, this muscle acts to pull the scapula posteriorly and downward. Refer to Figure 14.20.

3.   Reflect the upper limb to reveal the **subscapularis,** which occupies most of the ventral surface of the scapula. Humans have a homologous muscle.

4.   Locate the anterior, posterior, and middle **scalene** muscles on the lateral surface of the cat's neck and trunk. The most prominent and longest of these muscles is the middle scalene, which lies between the anterior and posterior members. The scalenes originate on the

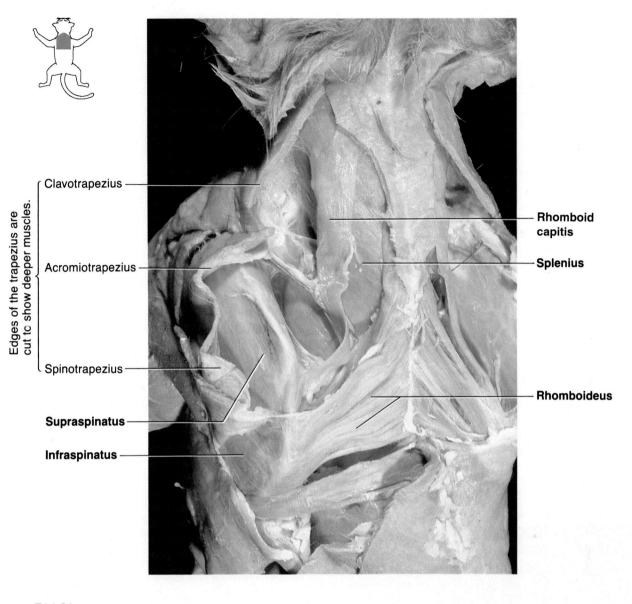

Edges of the trapezius are cut to show deeper muscles.

Clavotrapezius

Acromiotrapezius

Spinotrapezius

Supraspinatus

Infraspinatus

Rhomboid capitis

Splenius

Rhomboideus

F14.21

Deep muscles of the superior aspect of the thorax of the cat.

ribs and run cephalad over the serratus ventralis to insert in common on the cervical vertebrae. These muscles draw the ribs anteriorly and bend the neck downward; thus they are homologous to the human scalene muscles, which elevate the ribs and flex the neck. (Note that the difference is only one of position. Humans walk erect, but cats are quadrupeds.)

5. Transect and reflect the latissimus dorsi, spinodeltoid, acromiodeltoid, and levator scapulae ventralis. The **splenius** is a large flat muscle occupying most of the side of the neck close to the vertebrae. As in humans, it originates on the ligamentum nuchae at the back of the neck and inserts into the occipital bone. It functions to raise the head. Refer to Figure 14.21.

6. To view the rhomboid muscles, lay the cat on its side and hold its forelegs together to spread the scapulae apart. The rhomboid muscles lie between the scapulae and beneath the acromiotrapezius. All the rhomboid muscles originate on the vertebrae and insert on the scapula.

There are three rhomboids in the cat. The **rhomboid capitis,** the most anterolateral muscle of the group, has no counterpart in the human body. The **rhomboid minor,** located posteriorly to the rhomboid capitis, is much larger. The fibers of the rhomboid minor run transversely to those of the rhomboid capitis. The most posterior muscle of the group, the **rhomboid major,** is so closely fused to the rhomboid minor that many consider them to be one muscle—the **rhomboideus,** which is homologous to human rhomboid muscles.

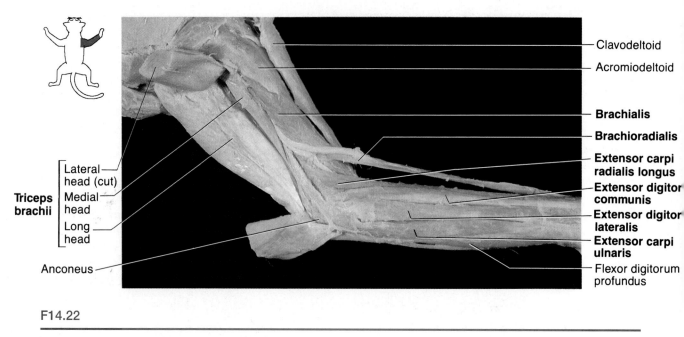

F14.22

Lateral surface of the right forelimb of the cat. The lateral head of the triceps brachii has been transected and reflected.

7. The **supraspinatus** and **infraspinatus** muscles are similar to the same muscles in humans. The supraspinatus can be found under the acromiotrapezius, and the infraspinatus under the spinotrapezius. Both originate on the lateral scapular surface and insert on the humerus.

Dissection of Cat Forelimb Muscles

Cat forelimb muscles fall into the same three categories as human upper extremity muscles, but in this section the muscles of the entire forelimb are considered together. Refer to Figure 14.22 as you study these muscles.

MUSCLES OF THE LATERAL SURFACE

1. The triceps muscle (**triceps brachii**) of the cat can easily be identified if the cat is placed on its side. It is a large fleshy muscle covering the posterior aspect and much of the side of the humerus. As in humans, this muscle arises from three heads, which originate from the humerus and scapula and insert jointly into the olecranon process of the ulna. Remove the fascia from the superior region of the lateral arm surface to identify the lateral and long heads of the triceps. The long head is approximately twice as long as the lateral head and lies medial to it on the posterior arm surface. The medial head can be exposed by transecting the lateral head and pulling it aside. Now pull on the triceps muscle.

How does the function of the triceps muscle compare in cats and in humans?

Anterior and distal to the medial head of the triceps is the tiny anconeus muscle, sometimes called the fourth head of the triceps muscle. Notice its darker color and the way it wraps the tip of the elbow.

2. The **brachialis** can be located anterior to the lateral head of the triceps muscle. Identify its origin on the humerus, and trace its course as it crosses the elbow and inserts on the ulna. It flexes the cat's foreleg.

Identification of the forearm muscles is difficult because of the tough fascia sheath that encases them.

3. Remove as much of the connective tissue as possible and cut through the ligaments that secure the tendons at the wrist (transverse carpal ligaments) so that you will be able to follow the muscles to their insertions. Begin your identification of the forearm muscles by examining the lateral surface of the forearm. The muscles of this region are very much alike in appearance and are difficult to identify accurately unless a definite order is followed. Thus you will begin with the most anterior muscles and proceed to the posterior aspect. Remember to check carefully the tendons of insertion to verify your muscle identifications.

4. The ribbonlike muscle on the lateral surface of the humerus is the **brachioradialis.** Observe how it passes down the forearm to insert on the styloid process of the radius. (If your removal of the fascia was not very careful, this muscle may have been removed.)

5. The **extensor carpi radialis longus** has a broad origin and is larger than the brachioradialis. It extends down the anterior surface of the radius (see Figure

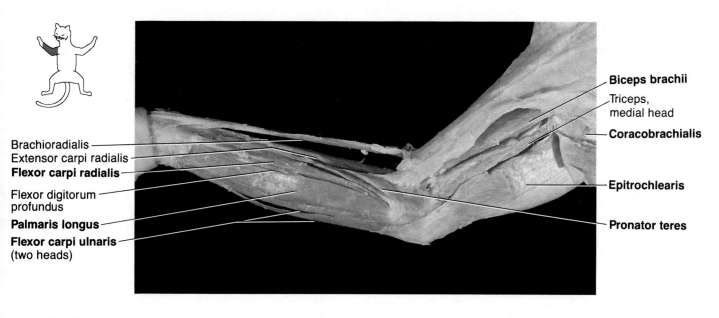

Brachioradialis

Extensor carpi radialis

Flexor carpi radialis

Flexor digitorum profundus

Palmaris longus

Flexor carpi ulnaris (two heads)

Biceps brachii

Triceps, medial head

Coracobrachialis

Epitrochlearis

Pronator teres

F14.23

Medial surface of the right forelimb of the cat.

14.22). Transect this muscle to view the **extensor carpi radialis brevis,** which is partially covered by and sometimes fused with the extensor carpi radialis longus. Both muscles have origins, insertions, and actions similar to their human counterparts.

6. You can see the entire **extensor digitorum communis** along the lateral surface of the forearm. Trace it to its four tendons, which insert on the second to fifth digits. This muscle extends these digits. The **extensor digitorum lateralis** (not present in humans) also extends the digits. This muscle lies immediately posterior to the extensor digitorum communis.

7. Follow the **extensor carpi ulnaris** from the lateral epicondyle of the humerus to the ulnar side of the fifth metacarpal. Very often this muscle has a shiny tendon, which helps in its identification.

MUSCLES OF THE MEDIAL SURFACE

1. The **biceps brachii** (refer to Figure 14.23) is a large spindle-shaped muscle medial to the brachialis on the anterior surface of the humerus. Pull back the cut ends of the pectoral muscles to get a good view of the biceps. Although this muscle is much more prominent in humans, its origin, insertion, and action are very similar in cats and in humans. Follow the muscle to its origin.

Does the biceps have two heads in the cat? _____

2. The broad, flat, exceedingly thin muscle on the posteromedial surface of the arm is the **epitrochlearis.** Its tendon originates from the fascia of the latissimus dorsi, and the muscle inserts into the olecranon process of the

ulna. This muscle extends the forearm of the cat; it is not found in humans.

3. The **coracobrachialis** of the cat is insignificant (approximately ½ in. long) and can be seen as a very small muscle crossing the ventral aspect of the shoulder joint. It runs beneath the biceps brachii to insert on the humerus and has the same function as the human coracobrachialis.

4. Referring again to Figure 14.23, turn the cat so that the ventral forearm muscles (mostly flexors and pronators) can easily be observed. As in humans, most of these muscles arise from the medial epicondyle of the humerus. The **pronator teres** runs from the medial epicondyle of the humerus and declines in size as it approaches its insertion on the radius. Do not bother to trace it to its insertion.

5. The **flexor carpi radialis** runs from the medial epicondyle of the humerus to insert into the second and third metacarpals, as in humans.

6. The large flat muscle in the center of the medial surface is the **palmaris longus.** Its origin on the medial epicondyle of the humerus abuts that of the pronator teres and is shared with the flexor carpi radialis. The palmaris longus extends down the forearm to terminate in four tendons on the digits. Comparatively speaking, this muscle is much larger in cats than in humans. The **flexor carpi ulnaris** arises from a two-headed origin (medial epicondyle of the humerus and olecranon of the ulna). Its two bellies (fleshy parts) pass downward to the wrist, where they are united by a single tendon that inserts into the carpals of the wrist. As in humans, this muscle flexes the wrist.

Dissection of Cat Hindlimb Muscles

Remove the fat and fascia from all the thigh surfaces, but do not cut through or remove the **fascia lata,** which is a tough white aponeurosis covering the anterolateral surface of the thigh from the hip to the leg. If the cat is a male, the cordlike sperm duct will be embedded in the fat near the pubic symphysis. Carefully clear around, but not in, this region.

POSTEROLATERAL HINDLIMB MUSCLES

1.   Turn the cat on its ventral surface and identify the following superficial muscles of the hip and thigh, referring to Figure 14.24. (The deeper hip muscles of the cat will not be identified.) Viewing the lateral aspect of the hindlimb, you will identify these muscles in sequence from the anterior to the posterior aspects of the hip and thigh. Most anterior is the **sartorius.** Approximately 1½ in. wide, it extends around the lateral aspect of the thigh to the anterior surface, where the major portion of it lies. Free it from the adjacent muscles and pass a blunt probe under it to trace its origin and insertion. Homologous to the sartorius muscle in humans, it adducts and rotates the thigh, but in addition, the cat sartorius acts as a knee extensor. Transect this muscle.

2.   The **tensor fasciae latae** is posterior to the sartorius. It is wide at its superior end, where it originates on the iliac crest, and narrows as it approaches its insertion into the fascia lata (or iliotibial band), which runs to the proximal tibial region. Transect its superior end and pull it back to expose the gluteus medius lying beneath it. This is the largest of the gluteus muscles in the cat. It originates on the ilium and inserts on the greater trochanter of the femur. The gluteus medius overlays and obscures the gluteus minimus, pyriformis, and gemellus muscles (which will not be identified here).

3.   The **gluteus maximus** is a small triangular hip muscle posterior to the superior end of the tensor fasciae latae and paralleling it. In humans the gluteus maximus is a large fleshy muscle forming most of the buttock mass. In the cat it is only about ½ in. wide and 2 in. long, and is smaller than the gluteus medius. The gluteus maximus covers part of the gluteus medius as it extends from the sacral region and the end of the femur. It abducts the thigh.

4.   Posterior to the gluteus maximus, identify the triangular **caudofemoralis,** which originates on the caudal vertebrae and inserts into the patella via an aponeurosis. There is no homologue to this muscle in humans; in cats it abducts the thigh and flexes the vertebral column.

5.   The **hamstring muscles,** a group in the hindlimb, include the biceps femoris, the semitendinosus, and the semimembranosus muscles. The **biceps femoris** is a large, powerful muscle that covers most of the posterolateral surface of the thigh. It is 1½ to 2 in. wide throughout its length. Trace it from its origin on the ischial tuberosity to its insertion on the tibia. The **semitendinosus** can be partially seen beneath the posterior border of the biceps femoris. Transect and reflect the biceps muscle to reveal the whole length of the semitendinosus. Contrary to what its name implies ("half-tendon"), this muscle is muscular and fleshy except at its insertion. It is uniformly about ¾ in. wide as it runs down the thigh from the ischial tuberosity to the medial side of the ulna. It acts to bend the knee. The **semimembranosus,** a large muscle lying medial to the semitendinosus and largely obscured by it, is best seen in an anterior view of the thigh (see Figure 14.26b). If desired, however, the semitendinosus can be transected to view it from the posterior aspect. The semimembranosus is larger and broader than the semitendinosus. Like the other hamstrings, it originates on the ischial tuberosity and inserts on the medial epicondyle of the femur and the medial tibial surface.

How does the semimembranosus compare with its human homologue?

_____

_____

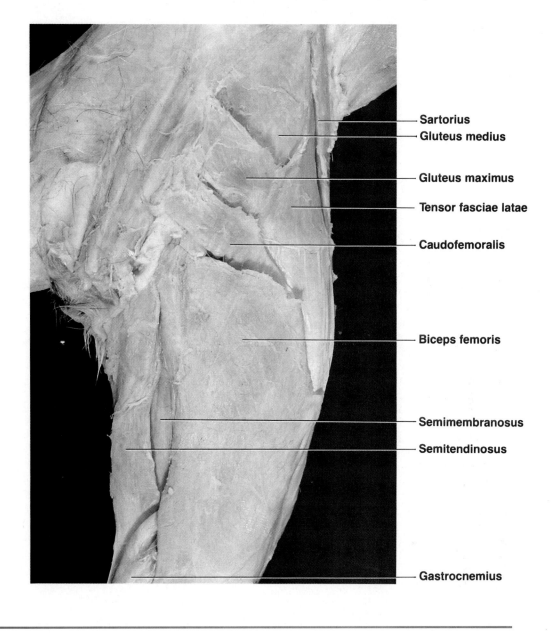

- Sartorius
- Gluteus medius
- Gluteus maximus
- Tensor fasciae latae
- Caudofemoralis
- Biceps femoris
- Semimembranosus
- Semitendinosus
- Gastrocnemius

F14.24

Muscles of the right posterolateral thigh in the cat; superficial view.

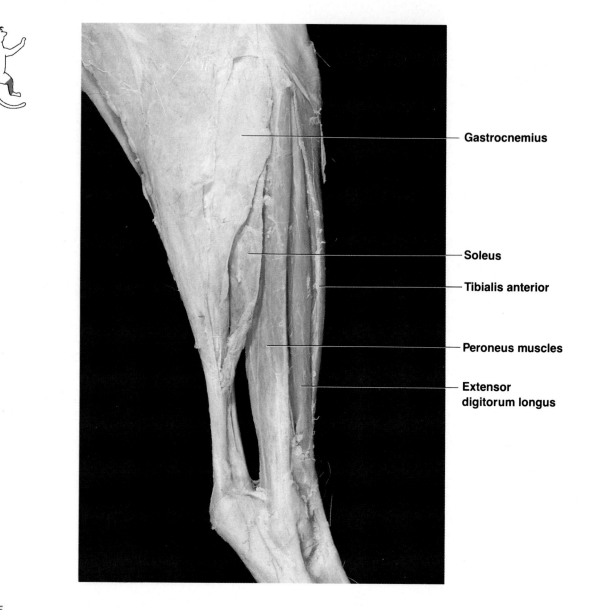

Gastrocnemius

Soleus

Tibialis anterior

Peroneus muscles

Extensor
digitorum longus

F14.25

Superficial muscles of the posterolateral aspect of the right shank (leg).

6.   Remove the heavy fascia covering the lateral surface of the shank. Moving from the posterior to the anterior aspect, identify the following muscles on the posterolateral shank (leg). (Refer to Figure 14.25.) First reflect the lower portion of the biceps femoris to see the origin of the **triceps surae,** the large composite muscle of the calf. Humans also have a triceps surae. The **gastrocnemius,** part of the triceps surae, is the largest muscle on the shank. As in humans, it has two heads and inserts via the calcaneal (Achilles) tendon into the calcaneus. Run a probe beneath this muscle and then transect it to reveal the **soleus,** which is deep to the gastrocnemius.

7.   Another important group of muscles in the leg is the **peroneus muscles** (see Figure 14.25), which collectively appear as a slender, evenly shaped superficial muscle lying anterior to the triceps surae. Originating on the fibula and inserting on the digits and metatarsals, the peroneus muscles flex the foot.

8.   The **extensor digitorum longus** is anterior to the peroneus muscles. Its origin, insertion, and action in cats are similar to the homologous human muscle. The **tibialis anterior** is anterior to the extensor digitorum longus. The tibialis anterior is roughly triangular in cross section and heavier at its proximal end. Locate its origin on the proximal fibula and tibia and its insertion on the first metatarsal. You can see the sharp edge of the tibia at the anterior border of this muscle. As in humans, it is a foot flexor.

ANTEROMEDIAL HINDLIMB MUSCLES

1.   Turn the cat onto its dorsal surface to identify the muscles of the anteromedial hindlimb (refer to Figure 14.26). Note once again the straplike sartorius at the surface of the thigh, which you have already identified and transected. It originates on the ilium and inserts on the medial region of the tibia.

2.   Reflect the cut ends of the sartorius to identify the **quadriceps** muscles. The **vastus medialis** lies just beneath the sartorius. Resting close to the femur, it arises from the ilium and inserts into the patellar ligament. The small cylindrical muscle anterior and lateral to the vastus medialis is the **rectus femoris.** In cats this muscle originates entirely from the femur.

What is the origin of the rectus femoris in humans?

_____

_____

Free the rectus femoris from the most lateral muscle of this group, the large, fleshy **vastus lateralis,** which lies deep to the tensor fasciae latae. The vastus lateralis arises from the lateral femoral surface and inserts, along with the other vasti muscles, into the patellar ligament. Transect this muscle to identify the deep **vastus intermedius,** the smallest of the vasti muscles. It lies medial to the vastus lateralis and merges superiorly with the vastus medialis. (The vastus intermedius is not shown in the figure.)

3.   The **gracilis** is a broad muscle that covers the posterior portion of the medial aspect of the thigh (see Figure 14.26a). It originates on the pubic symphysis and inserts on the medial proximal tibial surface. In cats the gracilis adducts the leg and draws it posteriorly.

How does this compare with the human gracilis?

_____

_____

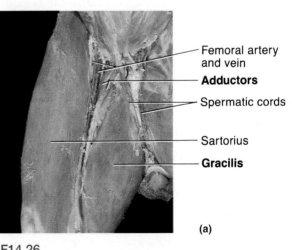

Femoral artery and vein
Adductors
Spermatic cords
Sartorius
Gracilis

(a)

F14.26

Superficial muscles of the anteromedial thigh. (a) Gracilis and sartorius are intact in this superficial view of the right thigh.

4. Free and transect the gracilis to view the adductor muscles deep to it. The **adductor femoris** is a large muscle that lies beneath the gracilis and abuts the semi-membranosus medially. Its origin is the pubic ramus and the ischium, and its fibers pass downward to insert on most of the length of the femoral shaft. The adductor femoris is homologous to the human adductor magnus, brevis, and longus. Its function is to extend the thigh after it has been drawn forward, and to adduct the thigh. A small muscle about an inch long—the **adductor longus**—touches the superior margin of the adductor femoris. It originates on the pubic bone and inserts on the proximal surface of the femur.

5. Before continuing your dissection, locate the **femoral triangle** (Scarpa's triangle), an important area bordered by the proximal edge of the sartorius and the adductor muscles. It is usually possible to identify the femoral artery (injected with red latex) and the femoral vein (injected with blue latex), which span the triangle

(see Figure 14.26b). (You will identify these vessels again in your study of the cardiovascular system.) If your instructor wishes you to identify the pectineus and iliopsoas, remove these vessels and go on to steps 6 and 7.

6. Examine the superolateral margin of the adductor longus to locate the small **pectineus.** It is normally covered by the gracilis (which you have cut and reflected). The pectineus, which originates on the pubis and inserts on the proximal end of the femur, is similar in all ways to its human homologue.

7. Just lateral to the pectineus you can see a small portion of the **iliopsoas,** a long and cylindrical muscle. Its origin is on the transverse processes of T_1 through T_{12} and the lumbar vertebrae, and it passes posteriorly toward the body wall to insert on the medial aspect of the proximal femur. The iliopsoas flexes and laterally rotates the thigh. It corresponds to the human iliopsoas and psoas minor.

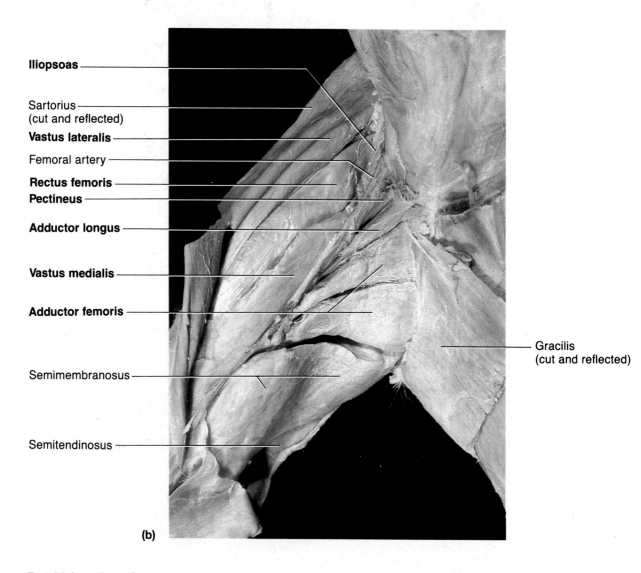

(b)

F14.26 (*continued*)

Superficial muscles of the anteromedial thigh. (b) The gracilis and sartorius are transected and reflected to show deeper muscles.

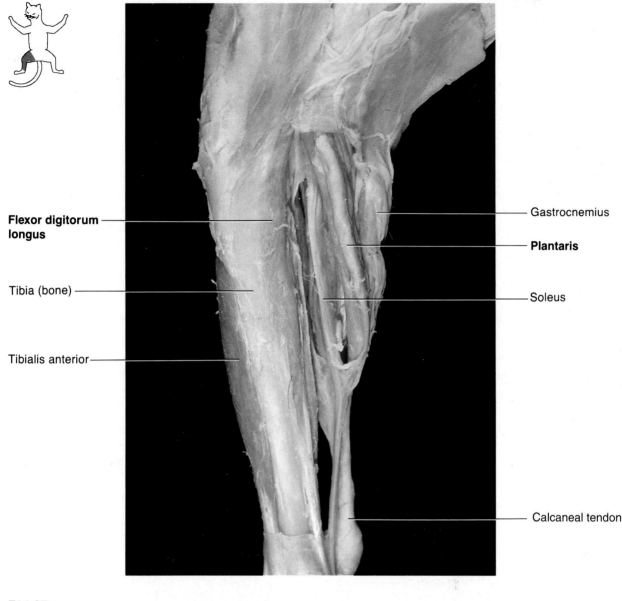

Flexor digitorum longus

Tibia (bone)

Tibialis anterior

Gastrocnemius

Plantaris

Soleus

Calcaneal tendon

F14.27

Superficial muscles of the right anteromedial (leg) of the cat.

8. Reidentify the gastrocnemius of the shank and then the **plantaris,** which is fused with the lateral head of the gastrocnemius. (See Figure 14.27.) It originates from the lateral aspect of the femur and patella, and its tendon passes around the calcaneus to insert on the second phalanx. Working with the triceps surae, it flexes the digits and extends the foot.

9. Anterior to the plantaris is the **flexor digitorum longus,** a long, tapering muscle with two heads. It originates on the lateral surfaces of the proximal fibula and tibia and inserts via four tendons into the terminal phalanges. As in humans, it flexes the toes.

10. The **tibialis posterior** is a long, flat muscle lateral and deep to the flexor digitorum longus. Its origin is on the medial surface of the head of the fibula and the ventral tibia. It merges with a flat, shiny tendon to insert into the tarsals.

11. The **flexor hallucis longus** is a long muscle that lies lateral to the tibialis posterior. It originates from the posterior tibia and passes downward to the ankle. It is a uniformly broad muscle in the cat. As in humans, it is a flexor of the great toe.

15

EXERCISE

Histology of Nervous Tissue

OBJECTIVES

1. To differentiate between the functions of neurons and neuroglia.
2. To list four types of neuroglia cells.
3. To identify the important anatomical characteristics of a neuron on an appropriate diagram or projected slide.
4. To state the functions of axons, dendrites, axonal terminals, neurofibrils, and myelin sheaths.
5. To explain how a nerve impulse is transmitted from one neuron to another.
6. To explain the role of Schwann cells in the formation of the myelin sheath.
7. To classify neurons according to structure and function.
8. To distinguish between a nerve and a tract and between a ganglion and a nucleus.

9. To describe the structure of a nerve, identifying the connective tissue coverings (endoneurium, perineurium, and epineurium) and citing their functions.
10. To recognize various types of general sensory receptors as studied in the laboratory, and to describe the function and location of each type.

MATERIALS

Model of a "typical" neuron (if available)
Compound microscope
Histologic slides of an ox spinal cord smear and teased myelinated nerve fibers
Prepared slides of Purkinje cells (cerebellum), pyramidal cells (cerebrum), and a dorsal root ganglion
Prepared slide of a nerve (cross section)
Histologic slides of Pacinian corpuscles (longitudinal section), Meissner's corpuscles, Golgi tendon organs, and muscle spindles

See Appendix C, Exercise 15 for links to *Anatomy and PhysioShow: The Videodisc.*

The nervous system is the master integrating and coordinating system, continuously monitoring and processing sensory information both from the external environment and from within the body. Every thought, action, and sensation is a reflection of its activity. Like a computer, it processes and integrates new "inputs" with information previously fed into it ("programmed") to produce an appropriate response ("readout"). However, no computer can possibly compare in complexity and scope to the human nervous system.

Despite its complexity, nervous tissue is made up of just two principal cell populations: **neurons** and the **neuroglia** (glial cells). The *neuroglia,* literally "nerve glue," include *astrocytes, oligodendrocytes, microglia, and ependymal cells* (Figure 15.1). These cells serve the needs of the neurons by acting as supportive and protective cells, myelinating cells, and phagocytes. In addition, they probably serve some nutritive function by acting as a selective barrier between the capillary blood supply and the neurons. Although neuroglia resemble neurons in some ways (they have fibrous cellular extensions), they are not capable of generating and transmitting nerve impulses, a capability that is highly developed in neurons. Our focus in this exercise is the highly irritable neurons.

NEURON ANATOMY

The delicate **neurons** are the structural units of nervous tissue. They are highly specialized to transmit messages (nerve impulses) from one part of the body to another. Although neurons differ structurally, they have many identifiable features in common (Figure 15.2). All have a **cell body** from which slender processes or fibers extend. Although neuron cell bodies are typically found in the CNS (central nervous system: brain and spinal cord) in clusters called **nuclei,** occasionally they reside in **ganglia** (collections of neuron cell bodies outside the CNS). They make up the gray matter of the nervous system. Neuron processes running through the CNS form **tracts** of white matter; outside the CNS they form the peripheral **nerves.**

The neuron cell body contains a large round nucleus surrounded by cytoplasm (*neuroplasm*). The cytoplasm is riddled with neurofibrils and with darkly staining structures called Nissl bodies. **Neurofibrils,** the cytoskeletal elements of the neuron, have a support and intracellular transport function. **Nissl bodies,** an elaborate type of rough endoplasmic reticulum, are involved in the metabolic activities of the cell.

152

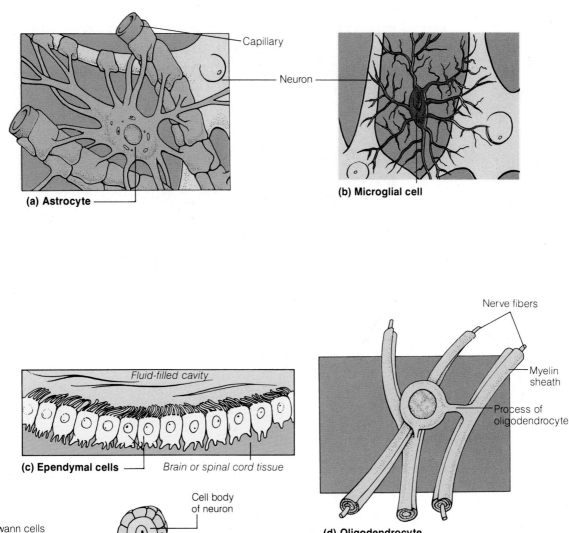

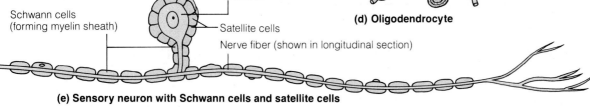

F15.1

Supporting cells of nervous tissue. (a) Astrocyte. **(b)** Microglial cell. **(c)** Ependymal cells. **(d)** Oligodendrocyte. **(e)** Neuron with Schwann cells and satellite cells.

According to the older, traditional scheme, neuron processes that conduct electrical currents *toward* the cell body are called **dendrites;** and those that carry impulses *away from* the nerve cell body are called **axons.** When it was discovered that this functional scheme had pitfalls (some axons carry impulses *both* toward and away from the cell body), a newer functional definition of neuron processes was adopted. According to this scheme, dendrites are *receptive regions* (they bear receptors for neurotransmitters released by other neu-

rons), whereas axons are *nerve impulse generators* and *transmitters.* Neurons have only one axon (which may branch into **collaterals**) but may have many dendrites, depending on the neuron type. Notice that the term *nerve fiber* is a synonym for axon and is, thus, quite specific.

In general, a neuron is excited by other neurons when their axons release neurotransmitters close to its dendrites or cell body. The electrical current generated travels across the cell body and down the axon. As Fig-

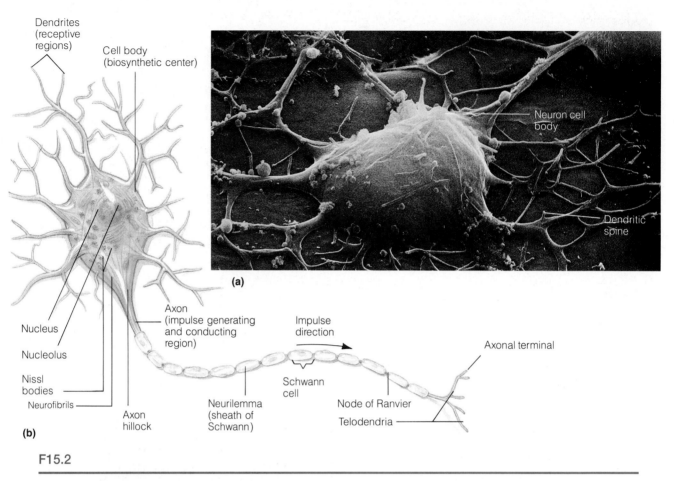

Dendrites
(receptive
regions)

Cell body
(biosynthetic center)

Neuron cell
body

Dendritic
spine

(a)

Nucleus

Nucleolus

Nissl
bodies

Neurofibrils

Axon
hillock

Axon
(impulse generating
and conducting
region)

Neurilemma
(sheath of
Schwann)

Schwann
cell

Impulse
direction

Node of Ranvier

Telodendria

Axonal terminal

(b)

F15.2

Structure of a typical motor neuron. (a) Photomicrograph showing the neuron cell body and dendrites with obvious dendritic spines, which are synapse sites (5000×). **(b)** Diagrammatic view.

ure 15.2 shows, the axon (in motor neurons) begins at a slightly enlarged cell body structure called the **axon hillock** and ends in many small structures called **axonal terminals,** or synaptic knobs. These terminals store the neurotransmitter chemical in tiny vesicles. Each axonal terminal is separated from the cell body or dendrites of the next (postsynaptic) neuron by a tiny gap called the **synaptic cleft.** Thus, although they are close, there is no actual physical contact between neurons. When an impulse reaches the axonal terminals, some of the synaptic vesicles rupture and release neurotransmitter into the synaptic cleft. The neurotransmitter then diffuses across the synaptic cleft to bind to membrane receptors on the next neuron, initiating the action potential.*

Many drugs can influence the transmission of impulses at synapses. Some, like caffeine, are stimulants which decrease the receptor neuron's threshold and make it more irritable. Others block transmission by

binding competitively with the receptor sites or by interfering with the release of neurotransmitter by the axonal terminals. As might be anticipated, some of these drugs are used as painkillers or tranquilizers. ■

Most long nerve fibers are covered with a fatty material called *myelin,* and such fibers are referred to as **myelinated fibers.** Axons in the peripheral nervous system are typically heavily myelinated by special cells called **Schwann cells,** which wrap themselves tightly around the axon jelly-roll fashion (Figure 15.3). During the wrapping process, the cytoplasm is squeezed from between adjacent layers of the Schwann cell membranes, so that when the process is completed a tight core of plasma membrane material (protein-lipoid material) encompasses the axon. This wrapping is the **myelin sheath.** The Schwann cell nucleus and the bulk of its cytoplasm end up just beneath the outermost portion of its plasma membrane. This peripheral part of the Schwann cell and its plasma membrane is referred to as the **neurilemma.** Since the myelin sheath is formed by many individual Schwann cells, it is a discontinuous sheath; the gaps or indentations in the sheath are called **nodes of Ranvier** (see Figure 15.2).

* Specialized synapses in skeletal muscle are called neuromuscular junctions. They are discussed in Exercise 13.

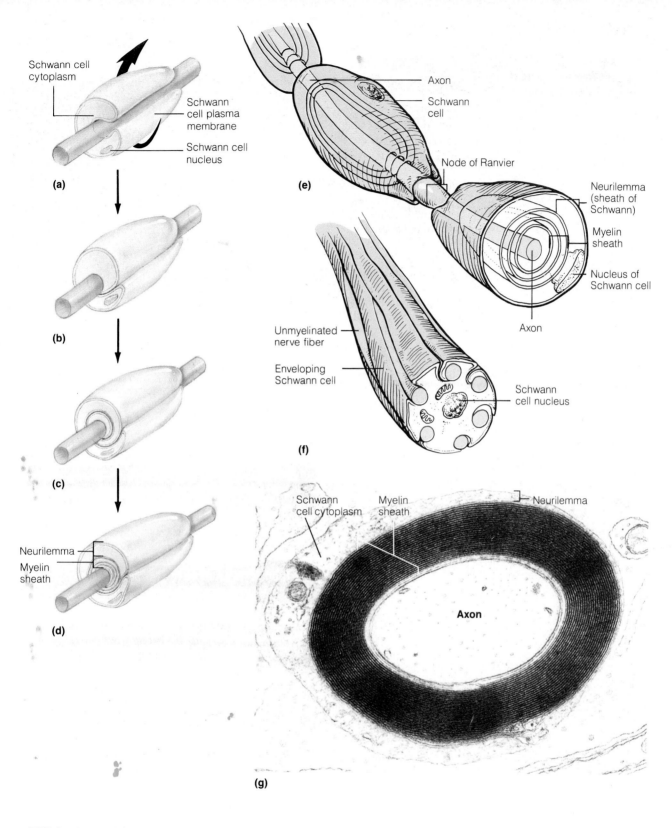

F15.3

Myelination of neuron processes by individual Schwann cells. (a–d) A Schwann cell becomes apposed to an axon and envelops it in a trough. It then begins to rotate around the axon, wrapping it loosely in successive layers of its plasma membrane. Eventually, the Schwann cell cytoplasm is forced from between the membranes and comes to lie peripherally just beneath the exposed portion of the Schwann cell membrane. The tight membrane wrappings surrounding the axon form the myelin sheath. The area of Schwann cell cytoplasm and its exposed membrane are referred to as the neurilemma or sheath of Schwann.
(e) Longitudinal view of myelinated axon showing portions of adjacent Schwann cells and the node of Ranvier between them.
(f) Unmyelinated fibers. Schwann cells may associate loosely with several axons, which they partially invest. In such cases, Schwann cell coiling around the axons does not occur. **(g)** Photomicrograph of a myelinated axon, cross-sectional view (20,000✕).

Within the CNS, myelination is accomplished by glial cells called **oligodendrocytes** (see Figure 15.1d). These CNS sheaths do not exhibit the neurilemma seen in fibers myelinated by Schwann cells. Because of its chemical composition, myelin insulates the fibers and greatly increases the speed of neurotransmission by neuron fibers.

 1. Study the typical motor neuron shown in Figure 15.2, noting the structural details described above, and then identify these structures on a neuron model.

2. Obtain a prepared slide of the ox spinal cord smear, which has large, easily identifiable neurons. Study one representative neuron under oil immersion and identify the cell body; the nucleus; the large, prominent "owl's eye" nucleolus; and the granular Nissl bodies. If possible, distinguish the axon from the many dendrites. Sketch the cell in the space provided here, and label the important anatomical details you have observed. Compare your sketch to Plate 4 of the Histology Atlas. Also examine Plate 5, which differentiates the neuronal processes more clearly.

3. Obtain a prepared slide of teased myelinated nerve fibers. Using Plate 6 of the Histology Atlas as a guide, identify the following: nodes of Ranvier, neurilemma, axis cylinder (the axon itself), Schwann cell nuclei, and myelin sheath.

Do the nodes seem to occur at consistent intervals, or are they irregularly distributed?

_____

Explain the significance of this finding: _____

_____

_____

_____

Sketch a portion of a myelinated nerve fiber in the space provided here, illustrating two or three nodes of Ranvier. Label the axon, myelin sheath, nodes, and neurilemma.

NEURON CLASSIFICATION

Neurons may be classified on the basis of structure or of function. Figure 15.4 depicts both schemes.

Basis of Structure

Structurally, neurons may be differentiated according to the number of processes attached to the cell body. In **unipolar neurons,** one very short process, which divides into *peripheral* and *central processes,* extends from the cell body. Functionally, only the most distal portions of the peripheral process act as dendrites; the rest acts as an axon along with the central process. Nearly all neurons that conduct impulses toward the CNS are unipolar.

    Bipolar neurons have two processes—one axon and one dendrite—attached to the cell body. This neuron type is quite rare, typically found only as part of the receptor apparatus of the eye, ear, and olfactory mucosa.

    Many processes issue from the cell body of **multipolar** neurons, all classified as dendrites except for a single axon. Most neurons in the brain and spinal cord (CNS neurons) and those whose axons carry impulses away from the CNS fall into this last category.

 Obtain prepared slides of Purkinje cells of the cerebellar cortex, pyramidal cells of the cerebral cortex, and a dorsal root ganglion. As you observe them under the microscope, try to pick out the anatomical details depicted in Figure 15.5. Notice that the neurons of the cerebral and cerebellar tissues (both brain tissues) are extensively branched; in contrast, the neurons of the dorsal root ganglion are more rounded. You may also be able to identify astrocytes (a type of neuroglia) in the brain tissue slides if you examine them closely.

Which of these neuron types would be classified as multipolar neurons?

_____

Which as unipolar? _____

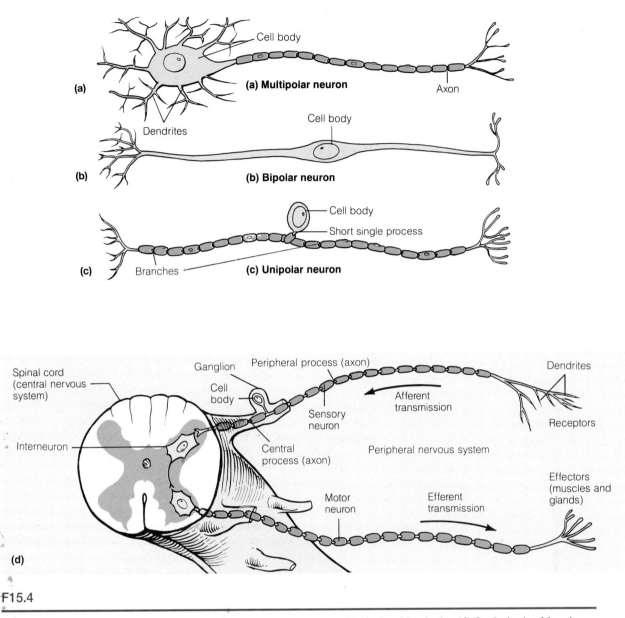

F15.4

Classification of neurons. (a–c) On the basis of structure: **(a)** multipolar; **(b)** bipolar; **(c)** unipolar. **(d)** On the basis of function, there are sensory, motor, and association neurons. Sensory (afferent) neurons conduct impulses from the body's sensory receptors to the central nervous system; most are unipolar neurons with their nerve cell bodies in ganglia in the peripheral nervous system (PNS). Motor (efferent) neurons transmit impulses from the CNS to effectors such as muscles and glands. Association neurons (interneurons) complete the communication line between sensory and motor neurons. They are typically multipolar and their cell bodies reside in the CNS.

Basis of Function

In general, neurons carrying impulses from the sensory receptors in the internal organs (viscera) or in the skin are termed **sensory,** or **afferent, neurons** (see Figure 15.4d). The dendritic endings of sensory neurons are often equipped with specialized receptors that are stimulated by specific changes in their immediate environment. The cell bodies of sensory neurons are always found in a ganglion outside the CNS, and these neurons are typically unipolar.

Neurons carrying activating impulses from the CNS to the viscera and/or body muscles and glands are termed **motor,** or **efferent, neurons.** Motor neurons are most often multipolar and their cell bodies are almost always located in the CNS.

The third functional category of neurons is the **association neurons,** or **interneurons,** which are situated between and contribute to pathways that connect sensory and motor neurons. Their cell bodies are always located within the CNS and they are multipolar neurons structurally.

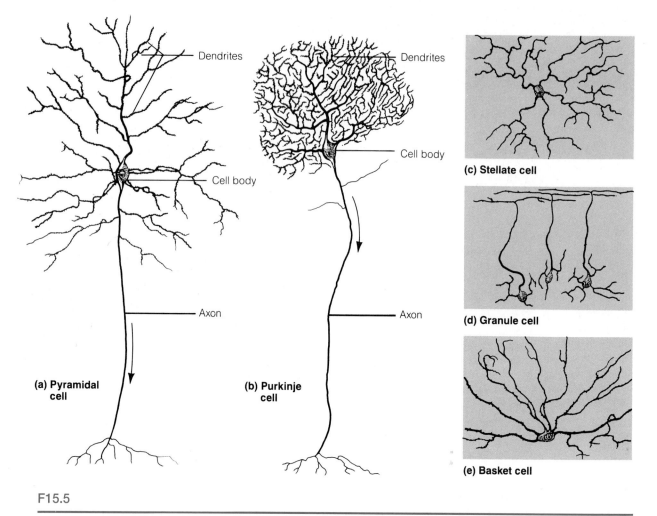

F15.5

Structure of selected neurons. (**a**) Pyramidal cell of the cerebral cortex and (**b**) Purkinje cells of the cerebellum have extremely long axons, but they are easily distinguished by their dendrite-branching patterns, whereas cerebellar neurons such as (**c**) stellate cells, (**d**) granule cells, and (**e**) basket cells have short (or even absent) axons and profuse dendrites.

STRUCTURE OF A NERVE

A nerve is a bundle of neuron fibers or processes wrapped in connective tissue coverings that extends to and/or from the CNS and visceral organs or structures of the body periphery (such as skeletal muscles, glands, and skin).

Within a nerve, each fiber is surrounded by a delicate connective tissue sheath called an **endoneurium,** which insulates it from the other neuron processes adjacent to it. (The endoneurium is often mistaken for the myelin sheath; it is instead an additional sheath that surrounds the myelin sheath.) Groups of fibers are bound by a coarser connective tissue, called the **perineurium,** to form bundles of fibers called **fascicles.** Finally, all the fascicles are bound together by a tough, white, fibrous connective tissue sheath called the **epineurium,** forming the cordlike nerve (Figure 15.6). In addition to the connective tissue wrappings, blood vessels and lym-

phatic vessels serving the fibers also travel within a nerve.

Like neurons, nerves are classified according to the direction in which they transmit impulses. Nerves carrying both sensory (afferent) and motor (efferent) fibers are called **mixed nerves;** all spinal nerves are mixed nerves. Nerves that carry only sensory processes and conduct impulses only toward the CNS are referred to as **sensory,** or **afferent, nerves.** A few of the cranial nerves are pure sensory nerves, but the majority are mixed nerves. The ventral roots of the spinal cord, which carry only motor fibers, can be considered **motor,** or **efferent, nerves.**

Examine under the compound microscope a prepared cross section of a peripheral nerve. Identify nerve fibers, myelin sheaths, fascicles, and endoneurium, perineurium, and epineurium sheaths. Compare your observations to Figure 15.6b.

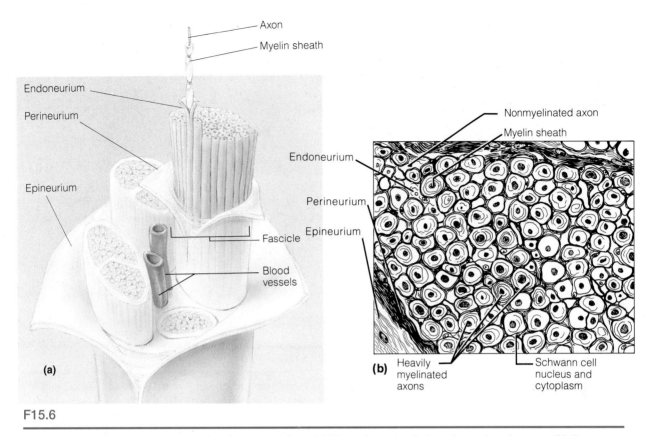

Axon
Myelin sheath

Endoneurium
Perineurium

Epineurium

Fascicle

Blood vessels

(a)

Nonmyelinated axon
Myelin sheath

Endoneurium

Perineurium

Epineurium

(b)

Heavily myelinated axons

Schwann cell nucleus and cytoplasm

F15.6

Structure of a nerve showing connective tissue wrappings. (a) Three-dimensional view of a portion of a nerve. **(b)** Cross-sectional view corresponding to Plate 8 of the Histology Atlas.

STRUCTURE OF GENERAL SENSORY RECEPTORS

You cannot become aware of changes in the environment unless your sensory neurons and their receptors are operating properly. **Sensory receptors** are modified dendritic endings (or specialized cells associated with the dendrites) that are sensitive to certain environmental stimuli. They react to such stimuli by initiating a nerve impulse.

The receptors of the *special sense organs* are complex and deserve considerable study. Thus the special senses (vision, hearing, equilibrium, taste, and smell) are covered separately in Exercises 18 through 20. Only the anatomically simpler **general sensory receptors**—cutaneous receptors and proprioceptors—will be studied in this section. Cutaneous receptors reside in the skin. Proprioceptors are located in muscles, tendons, and joint capsules where they report on the degree of stretch those structures are subjected to. Although many references link each type of receptor to specific stimuli, there is still considerable controversy about the precise qualitative function of each receptor. It may be that the responses of all these receptors overlap considerably. Certainly, intense stimulation of any of them is always interpreted as pain.

The least specialized of the cutaneous receptors are the **free dendritic endings** of sensory neurons (Figure 15.7 and Plate 10 of the Histology Atlas), which re-

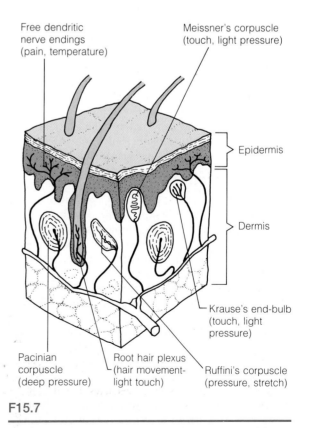

Free dendritic nerve endings (pain, temperature)

Meissner's corpuscle (touch, light pressure)

Epidermis

Dermis

Pacinian corpuscle (deep pressure)

Root hair plexus (hair movement-light touch)

Krause's end-bulb (touch, light pressure)

Ruffini's corpuscle (pressure, stretch)

F15.7

Cutaneous receptors. See also Plates 9, 10, and 11 of the Histology Atlas.

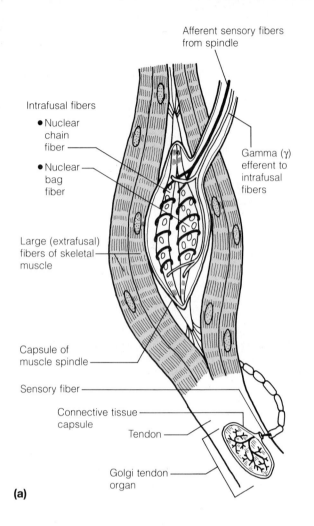

Afferent sensory fibers from spindle

Intrafusal fibers
- Nuclear chain fiber
- Nuclear bag fiber

Gamma (γ) efferent to intrafusal fibers

Large (extrafusal) fibers of skeletal muscle

Capsule of muscle spindle

Sensory fiber

Connective tissue capsule

Tendon

Golgi tendon organ

(a)

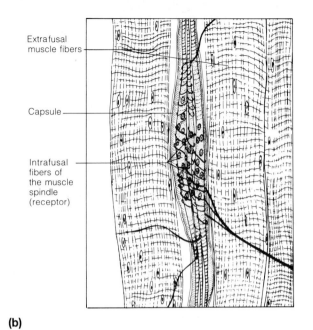

Extrafusal muscle fibers

Capsule

Intrafusal fibers of the muscle spindle (receptor)

(b)

F15.8

Proprioceptors. (a) Diagrammatic view of a muscle spindle and Golgi tendon organ. **(b)** Drawing of a photomicrograph of a muscle spindle. See corresponding Plate 12 in the Histology Atlas.

spond chiefly to pain and temperature. (The pain receptors are widespread in the skin and make up a sizable portion of the visceral interoceptors.) Certain free dendritic endings associate with specific epidermal cells to form **Merkel discs,** or entwine in hair follicles to form **root hair plexuses.** Both Merkel discs and root hair plexuses function as light touch receptors.

The other cutaneous receptors are a bit more complex, with the dendritic endings *encapsulated* by connective tissue cells. **Meissner's corpuscles,** commonly referred to as *tactile receptors* because they respond to light touch, are located in the dermal papillae of the skin. Although **Krause's end bulbs** and **Ruffini's corpuscles** were presumed to be thermoreceptors, now most authorities classify Krause's end bulbs with Meissner's corpuscles as a special kind of touch, or light pressure, receptor. Ruffini's corpuscles appear to respond to pressure and stretch stimuli. As you inspect Figure 15.7, note that all of the encapsulated receptors are quite similar, with the possible exception of the **Pacinian corpuscles** (deep pressure receptors), which are anatomically more distinctive and lie deepest in the dermis.

1. Obtain histologic slides of Pacinian and Meissner's corpuscles. Locate, under low power, a Meissner's corpuscle in the dermal layer of the skin. As mentioned above, these are usually found in the dermal papillae. Then, switch to the oil im-

mersion lens for a detailed study. Notice that the naked dendritic fibers within the capsule are aligned parallel to the skin surface. Compare your observations to Figure 15.7 and Plate 9 of the Histology Atlas.

2. Next observe a Pacinian corpuscle located much deeper in the dermis. Try to identify the slender naked dendrite ending in the center of the receptor and the heavy capsule of connective tissue surrounding it (which looks rather like an onion cut lengthwise). Also, note how much larger the Pacinian corpuscles are than the Meissner's corpuscles. Compare your observations to the views shown in Figure 15.7 and Plate 11 of the Histology Atlas.

3. Obtain slides of muscle spindles and Golgi tendon organs, the two major types of proprioceptors (see Figure 15.8). In the slide of **muscle spindles,** note that minute extensions of the dendrites of the sensory neurons coil around specialized slender skeletal muscle cells called **intrafusal cells,** or **fibers.** The **Golgi tendon organs** are composed of dendrites that ramify through the tendon tissue close to the muscle tendon attachment. Stretching of the muscle or tendon excites both types of receptors, which then transmit impulses that ultimately reach the cerebellum for interpretation. Compare your observations to Figure 15.8.

Gross Anatomy of the Brain and Cranial Nerves

OBJECTIVES

1. To identify the following brain structures on a dissected specimen, human brain model (or slices), or appropriate diagram, and to state their functions:

   - *Cerebral hemisphere structures:* lobes, important fissures, lateral ventricles, basal nuclei, corpus callosum, fornix, septum pellucidum
   - *Diencephalon structures:* thalamus, intermediate mass, hypothalamus, optic chiasma, pituitary gland, mammillary bodies, pineal body, choroid plexus of the third ventricle, interventricular foramen
   - *Brain stem structures:* corpora quadrigemina, cerebral aqueduct, cerebral peduncles of the midbrain, pons, medulla, fourth ventricle
   - *Cerebellum structures:* cerebellar hemispheres, vermis, arbor vitae

2. To describe the composition of gray and white matter.

3. To locate the well-recognized functional areas of the human cerebral hemispheres.

4. To define *gyri* and *fissures* (*sulci*).

5. To identify the three meningeal layers and state their function, and to locate the falx cerebri, falx cerebelli, and tentorium cerebelli.

6. To state the function of the arachnoid villi and dural sinuses.

7. To discuss the formation, circulation, and drainage of cerebrospinal fluid.

8. To identify at least four pertinent anatomical differences between the human brain and that of the sheep (or other mammal).

9. To identify the cranial nerves by number and name on an appropriate model or diagram, stating the origin and function of each.

MATERIALS

Human brain model (dissectible)
3-D model of ventricles
Preserved human brain (if available)
Coronally sectioned human brain slice (if available)
Preserved sheep brain (meninges and cranial nerves intact)
Dissecting tray and instruments
Protective skin cream or disposable gloves
Materials as needed for cranial nerve testing
The Human Nervous System videotape*

 See Appendix B, Exercise 16 for links to A.D.A.M. Standard.

 See Appendix C, Exercise 16 for links to *Anatomy and PhysioShow: The Videodisc.*

*Available to qualified adopters from Benjamin/Cummings.

When viewed alongside all nature's animals, humans are indeed unique, and the key to their uniqueness is found in the brain. Only in humans has the brain region called the cerebrum become so elaborated and grown out of proportion that it overshadows other brain areas. Other animals are primarily concerned with informational input and response for the sake of survival and preservation of the species, but human beings devote considerable time to nonsurvival ends. They are the only animals who manipulate abstract ideas and search for knowledge for its own sake, who are capable of emotional response and artistic creativity, or who can anticipate the future and guide their lives according to ethical and moral values. For all this, humans can thank their overgrown cerebrum (cerebral hemispheres).

We can be considered composite reflections of our brain's experience. If all past sensory input could mysteriously and suddenly be "erased," we would be unable to walk, talk, or communicate in any manner. Spontaneous movement would occur, as in a fetus, but no voluntary integrated function of any type would be possible. Clearly we would cease to be the same individuals.

Because of the complexity of the nervous system, its anatomical structures are usually considered in terms of two principal divisions: the **central nervous system (CNS)** and the **peripheral nervous system (PNS).** The central nervous system consists of the brain and spinal cord, which primarily interpret incoming sensory information and issue instructions based on past experience. The peripheral nervous system consists of the cranial and spinal nerves, ganglia, and sensory receptors. These structures serve as communication lines as they carry impulses—from the sensory receptors to the CNS and from the CNS to the appropriate glands or muscles.

In this exercise both CNS (brain) and PNS (cranial nerves) structures will be studied because of their close anatomical relationship.

F16.1

Embryonic development of the human brain. (a) The neural tube becomes subdivided into (b) the primary brain vesicles, which subsequently form (c) the secondary brain vesicles, which differentiate into (d) the adult brain structures. (e) The adult structures derived from the neural canal.

THE HUMAN BRAIN

During embryonic development of all vertebrates, the CNS first makes its appearance as a simple tubelike structure, the **neural tube,** that extends down the dorsal median plane. By the fourth week, the human brain begins to form as an expansion of the anterior or rostral end of the neural tube (the end toward the head). Shortly thereafter, constrictions appear, dividing the developing brain into three major regions—**forebrain, midbrain, and hindbrain** (Figure 16.1). The remainder of the neural tube becomes the spinal cord.

During fetal development, two anterior outpocketings extend from the forebrain and grow rapidly to form the cerebral hemispheres. Because of space restrictions imposed by the skull, the cerebral hemispheres are forced to grow posteriorly and inferiorly, and finally end up enveloping and obscuring the rest of the forebrain and most midbrain structures. Somewhat later in development, the dorsal portion of the hindbrain also enlarges to produce the cerebellum. The central canal of the neural tube, which remains continuous throughout the brain and cord, becomes enlarged in four regions of the brain, forming chambers called **ventricles** (see Figure 16.8a, p. 170).

External Anatomy

Generally, the brain is considered in terms of four major regions: the cerebral hemispheres, diencephalon, brain stem, and cerebellum. The relationship between these four anatomical regions and the structures of the forebrain, midbrain, and hindbrain is also outlined in Figure 16.1.

CEREBRAL HEMISPHERES   The **cerebral hemispheres** are the most superior portion of the brain (Figure 16.2). Their entire surface is thrown into elevated ridges of tissue called **gyri** that are separated by depressed areas called **fissures** or **sulci.** Of the two types of depressions, the fissures are deeper. Many of the fissures and gyri are important anatomical landmarks.

The cerebral hemispheres are divided by a single deep fissure, the **longitudinal fissure.** The **central sulcus** divides the **frontal lobe** from the **parietal lobe,** and the **lateral sulcus** separates the **temporal lobe** from the parietal lobe. The **parieto-occipital sulcus,** which divides the **occipital lobe** from the parietal lobe, is not visible externally. Notice that the cerebral hemisphere lobes are named for the cranial bones that lie over them.

Some important functional areas of the cerebral hemispheres have also been located (Figure 16.2d). The **primary somatosensory area** is located in the **postcentral gyrus** of the parietal lobe. Impulses traveling from the body's sensory receptors (such as those for pressure, pain, and temperature) are localized in this area of the brain. ("This information is from my big toe.") Immediately posterior to the primary somatosensory area is the **somatosensory association area,** in which the meaning of incoming stimuli is analyzed. ("Ouch! I have a *pain* there.") Thus, the somatosensory association area allows you to become aware of pain, coldness, a light touch, and the like.

Impulses from the special sense organs are interpreted in other specific areas also noted in Figure 16.2d. For example, the visual areas are in the posterior portion of the occipital lobe and the auditory area is located in the temporal lobe in the gyrus bordering the lateral sulcus. The olfactory area is deep within the temporal lobe

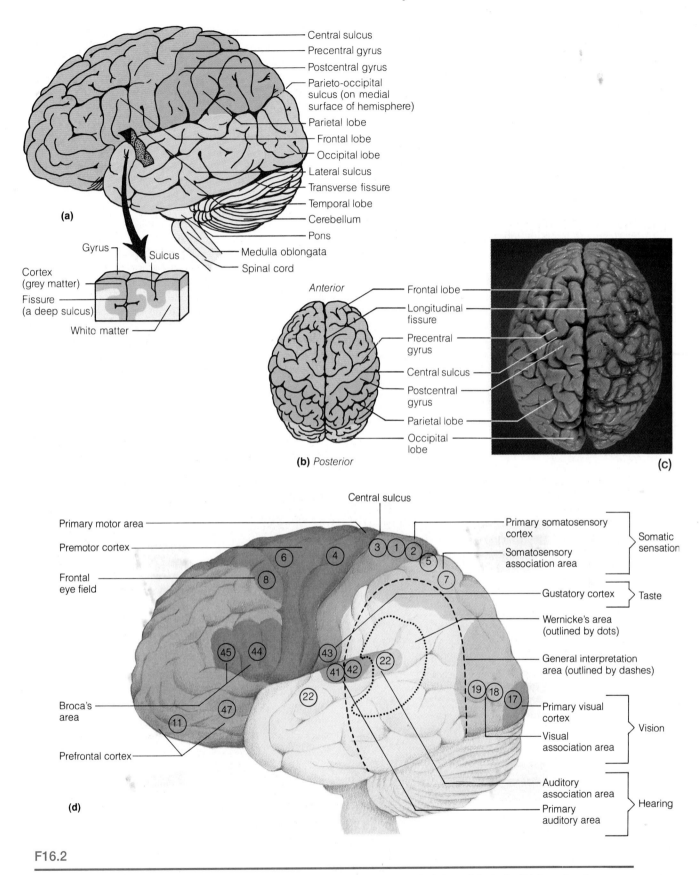

F16.2

External structure (lobes and fissures) of the cerebral hemispheres. (a) Left lateral view of the brain. **(b)** Superior view. **(c)** Photograph of the superior aspect of the human brain. **(d)** Functional areas of the left cerebral cortex. The olfactory area, which is deep to the temporal lobe on the medial hemispheric surface, is not identified. Numbers indicate brain regions plotted by the Brodman system.

along its medial surface, in a region called the **uncus** (see Figure 16.4a).

The **primary motor area,** which is responsible for conscious or voluntary movement of the skeletal muscles, is located in the **precentral gyrus** of the frontal lobe. A specialized motor speech area called **Broca's area** is found at the base of the precentral gyrus just above the lateral sulcus. Damage to this area (which is located only in one cerebral hemisphere, usually the left) reduces or eliminates the ability to articulate words. Areas involved in intellect, complex reasoning, and personality lie in the anterior portions of the frontal lobes, in a region called the **prefrontal cortex.**

A rather poorly defined region at the junction of the parietal and temporal lobes is Wernicke's area, an area in which unfamiliar words are sounded out. Like Broca's area, Wernicke's area is located in one cerebral hemisphere only, typically the left.

Although there are many similar functional areas in both cerebral hemispheres, such as motor and sensory areas, each hemisphere is also a "specialist" in certain ways. For example, the left hemisphere is the "language brain" in most of us, because it houses centers associated with language skills and speech. The right hemisphere is more specifically concerned with abstract, conceptual, or spatial processes—skills associated with artistic or creative pursuits.

The cell bodies of cerebral neurons involved in these functions are found only in the outermost gray matter of the cerebrum, the area called the **cerebral cortex.** Most of the balance of cerebral tissue—the deeper **cerebral white matter**—is composed of fiber tracts carrying impulses to or from the cortex.

 Using a model of the human brain (and a preserved human brain, if available), identify the areas and structures of the cerebral hemispheres described above.

DIENCEPHALON   The **diencephalon,** sometimes considered the most superior portion of the brain stem, is embryologically part of the forebrain, along with the cerebral hemispheres.

Turn the brain model so the ventral surface of the brain can be viewed. Using Figure 16.3 as a guide, start superiorly and identify the externally visible structures that mark the position of the floor of the diencephalon. These are the **olfactory bulbs** and **tracts, optic nerves, optic chiasma** (where the fibers of the optic nerves partially cross over), **optic tracts, pituitary gland,** and **mammillary bodies.**

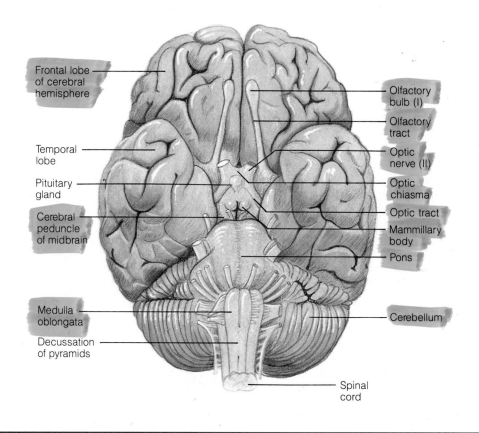

F16.3

Ventral aspect of the human brain, showing the three regions of the brain stem. Only a small portion of the midbrain can be seen; the rest is surrounded by other brain regions.

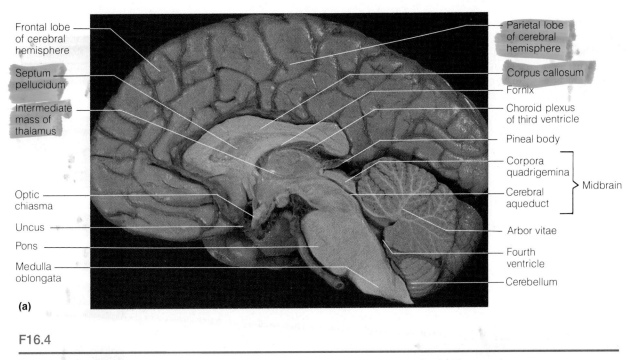

Frontal lobe of cerebral hemisphere

Septum pellucidum

Intermediate mass of thalamus

Optic chiasma

Uncus

Pons

Medulla oblongata

(a)

Parietal lobe of cerebral hemisphere

Corpus callosum

Fornix

Choroid plexus of third ventricle

Pineal body

Corpora quadrigemina

Cerebral aqueduct

} Midbrain

Arbor vitae

Fourth ventricle

Cerebellum

F16.4

Diencephalon and brain stem structures as seen in a midsagittal section of the brain. (a) Photograph.

BRAIN STEM   Continue inferiorly to identify the **brain stem** structures—the **cerebral peduncles** (fiber tracts in the **midbrain** connecting the pons below with cerebrum above), the pons, and the medulla oblongata. *Pons* means "bridge," and the **pons** consists primarily of motor and sensory fiber tracts connecting the brain with lower CNS centers. The lowest brain stem region, the **medulla,** is also composed primarily of fiber tracts. You can see the **decussation of pyramids,** a crossover point for the major motor tract (pyramidal tract) descending from the motor areas of the cerebrum to the cord, on the medulla's anterior surface. The medulla also houses many vital autonomic centers involved in the control of heart rate, respiratory rhythm, and blood pressure as well as involuntary centers involved in the initiation of vomiting, swallowing, and so on.

CEREBELLUM

1.   Turn the brain model so you can see the dorsal aspect. Identify the large cauliflowerlike **cerebellum,** which projects dorsally from under the occipital lobe of the cerebrum. Note that, like the cerebrum, the cerebellum has two major hemispheres and a convoluted surface (see Figure 16.6). It also has an outer cortex made up of gray matter with an inner region of white matter.

2.   Remove the cerebellum to view the **corpora quadrigemina,** located on the posterior aspect of the midbrain, a brain stem structure. The two superior prominences are the **superior colliculi** (visual reflex centers); the two smaller inferior prominences are the **inferior colliculi** (auditory reflex centers).

Internal Anatomy

The deeper structures of the brain have also been well mapped. Like the external structures, these can be studied in terms of the four major regions.

CEREBRAL HEMISPHERES

1.   Take the brain model apart so you can see a median sagittal view of the internal brain structures (Figure 16.4). Observe the model closely to see the extent of the outer cortex (gray matter), which contains the cell bodies of cerebral neurons. (The pyramidal cells of the cerebral motor cortex [studied in Exercise 15, pp. 156 and 158] are representative of the neurons seen in the precentral gyrus.)

2.   Observe the deeper area of white matter, which is composed of fiber tracts (Figure 16.5). The fiber tracts found in the cerebral hemisphere white matter are called *association tracts* if they connect two portions of the same hemisphere, *projection tracts* if they run between the cerebral cortex and the lower brain or spinal cord, and *commissures* if they run from one hemisphere to another. Observe the large **corpus callosum,** the major commissure connecting the cerebral hemispheres. The corpus callosum arches above the structures of the diencephalon and roofs over the lateral ventricles. Note also the **fornix,** a bandlike fiber tract concerned with olfaction and limbic system functions, and the membranous **septum pellucidum,** which separates the lateral ventricles of the cerebral hemispheres (see Figure 16.4b).

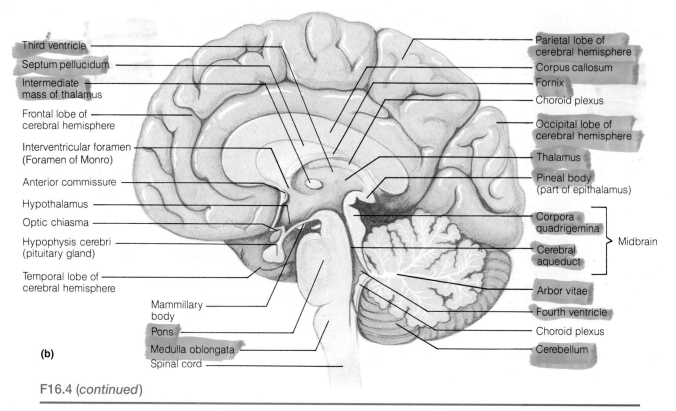

Third ventricle
Septum pellucidum
Intermediate mass of thalamus
Frontal lobe of cerebral hemisphere
Interventricular foramen (Foramen of Monro)
Anterior commissure
Hypothalamus
Optic chiasma
Hypophysis cerebri (pituitary gland)
Temporal lobe of cerebral hemisphere
Mammillary body
Pons
Medulla oblongata
Spinal cord

Parietal lobe of cerebral hemisphere
Corpus callosum
Fornix
Choroid plexus
Occipital lobe of cerebral hemisphere
Thalamus
Pineal body (part of epithalamus)
Corpora quadrigemina
Cerebral aqueduct
} Midbrain
Arbor vitae
Fourth ventricle
Choroid plexus
Cerebellum

(b)

F16.4 *(continued)*

Diencephalon and brain stem structures in a midsagittal section of the brain. (b) Diagrammatic view.

3. In addition to the gray matter of the cerebral cortex, there are several "islands" of gray matter (clusters of neuron cell bodies) called **nuclei** buried deep within the white matter of the cerebral hemispheres. One important group of cerebral nuclei, called the **basal nuclei,** flank the lateral and third ventricles. You can see the basal nuclei if you have an appropriate dissectible model or a coronally or cross-sectioned human brain slice. Otherwise, Figure 16.5 will suffice.

The basal nuclei, which are important subcortical motor nuclei (and part of the so-called *extrapyramidal system*), are involved in regulating voluntary motor activities. The most important of them are the arching, comma-shaped **caudate nucleus,** the **claustrum,** the **amygdaloid nucleus** (located at the tip of the caudate nucleus), and the **lentiform nucleus,** which is composed of the **putamen** and **globus pallidus nuclei.** The **corona radiata,** a spray of projection fibers coursing down from the precentral (motor) gyrus, combines with sensory fibers traveling to the sensory cortex to form a broad band of fibrous material called the **internal capsule.** The internal capsule passes between the diencephalon and the basal nuclei, and gives these basal nuclei a striped appearance. This is why the caudate nucleus and the lentiform nucleus are sometimes referred to collectively as the **corpus striatum,** or "striped body" (see Figure 16.5c).

4. Examine the relationship of the lateral ventricles and corpus callosum to the diencephalon structures; that is, hypothalamus, thalamus, and third ventricle—from the frontal-section viewpoint (Figure 16.5).

DIENCEPHALON

1. The major internal structures of the diencephalon are the thalamus, hypothalamus, and epithalamus (Figures 16.4 and 16.5). The **thalamus** consists of two large lobes of gray matter that laterally enclose the shallow third ventricle of the brain. A slender stalk of thalamic tissue, the **intermediate mass,** or **massa intermedia,** connects the two thalamic lobes and bridges the ventricle. The thalamus is a major integrating and relay station for sensory impulses passing upward to the cortical sensory areas for localization and interpretation. Locate also the **interventricular foramen** *(foramen of Monro),* a tiny orifice connecting the third ventricle with the lateral ventricle on the same side.

2. The **hypothalamus** makes up the floor and the inferolateral walls of the third ventricle. It is an important autonomic center involved in regulation of body temperature, water balance, and fat and carbohydrate metabolism as well as in many other activities and drives (sex, hunger, thirst). Locate again the pituitary gland, or **hypophysis,** which hangs from the anterior floor of the hypothalamus by a slender stalk, the **infundibulum.** (The pituitary gland is usually not present in preserved brain specimens.) In life, the pituitary rests in the hypophyseal fossa of the sella turcica of the sphenoid bone. Its function is discussed in Exercise 21.

Anterior to the pituitary, identify the optic chiasma portion of the optic pathway to the brain. The **mammillary bodies,** relay stations for olfaction, bulge exteriorly from the floor of the hypothalamus just posterior to the pituitary gland.

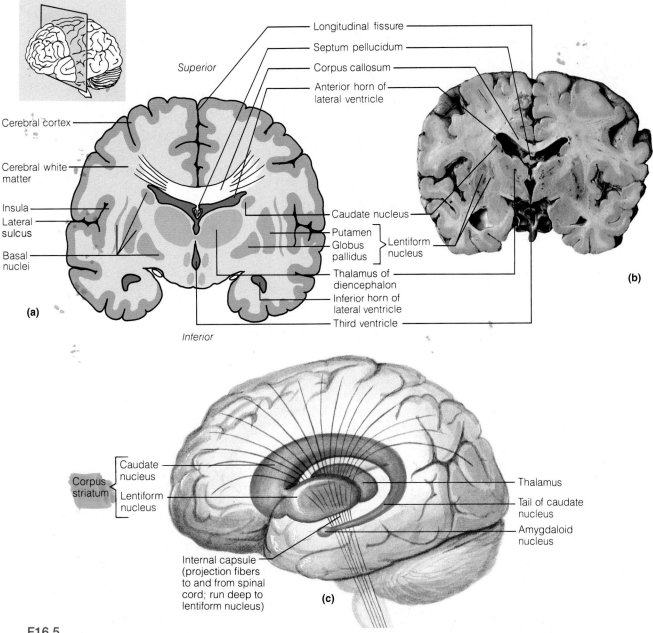

F16.5

Internal structure of the forebrain. (a and b) The cerebrum is sectioned frontally to show the positions of the outer cerebral cortex, the intermediate white matter, and the deep basal nuclei. Since the cerebral hemispheres enclose the diencephalon, that region is also evident. Compare the diagram in **(a)** to the photo in **(b)**. **(c)** Three-dimensional view of the basal (cerebral) nuclei showing their position within the cerebrum.

3.   The **epithalamus** forms the roof of the third ventricle and is the most dorsal portion of the diencephalon. Important structures in the epithalamus are the **pineal body,** or **gland** (a neuroendocrine structure), and the **choroid plexus** of the third ventricle. The choroid plexuses, knotlike collections of capillaries within each ventricle, form the cerebrospinal fluid.

BRAIN STEM

1.   Now trace the short midbrain from the mammillary bodies to the rounded pons below. Continue to refer to Figure 16.4. The **cerebral aqueduct** is a slender canal traveling through the midbrain; it connects the third ventricle to the fourth ventricle in the hindbrain below. The cerebral peduncles and the rounded corpora quadrigemina make up the midbrain tissue anterior and posterior (respectively) to the cerebral aqueduct.

2.   Locate the hindbrain structures. Trace the rounded pons to the medulla oblongata below, and identify the fourth ventricle posterior to these structures. Attempt to identify the single *median aperture* and the two *lateral apertures,* three orifices found in the walls of the fourth ventricle. These apertures serve as conduits for cerebrospinal fluid to circulate into the subarachnoid space from the fourth ventricle.

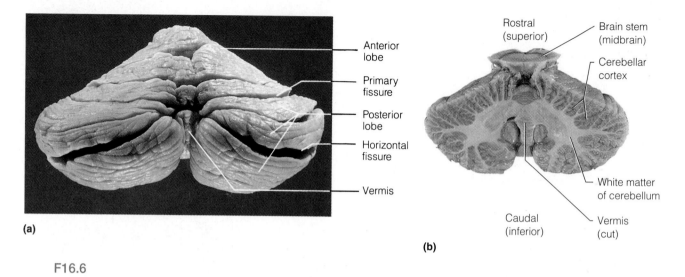

(a)

(b)

F16.6

Cerebellum. (a) Posterior (dorsal) view. **(b)** The cerebellum, sectioned to reveal its cortex and medullary regions. (Note that the cerebellum is sectioned frontally and the brain stem is sectioned horizontally in this posterior view.)

CEREBELLUM   Examine the cerebellum. Notice that it is composed of two lateral hemispheres connected by a midline lobe called the **vermis** (Figure 16.6). Each hemisphere has three lobes—*anterior, posterior,* and a deep *flocculonodular.* As in the cerebral hemispheres, the cerebellum has an outer cortical area of gray matter and an inner area of white matter. The distinctive tree-like branching of the cerebellar white matter is referred to as the **arbor vitae,** or tree of life. The cerebellum is concerned with unconscious coordination of skeletal muscle activity and control of balance and equilibrium. Fibers converge on the cerebellum from the equilibrium apparatus of the inner ear, visual pathways, propriocep-tors of the tendons and skeletal muscles, and from many other areas. Thus the cerebellum remains constantly aware of the position and state of tension of the various body parts.

Meninges of the Brain

The brain (and spinal cord) are covered and protected by three connective tissue membranes called **meninges** (Figure 16.7). The outermost meninx is the leathery **dura mater,** a double-layered membrane. One of its lay-ers (the *periosteal layer*) is attached to the inner surface of the skull, forming the periosteum. The other (the *meningeal layer*) forms the outermost brain covering and is continuous with the dura mater of the spinal cord.

The dural layers are fused together except in three places where the inner membrane extends inward to form a septum that secures the brain to structures inside the cranial cavity. One such extension, the **falx cerebri,** dips into the longitudinal fissure between the cerebral hemispheres to attach to the crista galli of the ethmoid bone of the skull. The cavity created at this point is the large **superior sagittal sinus,** which collects blood draining from the brain tissue. The **falx cerebelli,** sepa-rating the two cerebellar hemispheres, and the **tento-rium cerebelli,** separating the cerebrum from the cere-bellum below, are two other important inward folds of the inner dural membrane.

The middle meninx, the weblike **arachnoid mater,** or simply the **arachnoid,** underlies the dura mater and is partially separated from it by the **subdural space.** Threadlike projections bridge the **subarachnoid space** to attach the arachnoid to the innermost meninx, the **pia mater.** The delicate pia mater is highly vascular and clings tenaciously to the surface of the brain, following its convolutions.

In life, the subarachnoid space is filled with cere-brospinal fluid. Specialized projections of the arachnoid tissue called **arachnoid villi** protrude through the dura mater to allow the cerebrospinal fluid to drain back into the venous circulation via the superior sagittal sinus and other dural sinuses.

Meningitis, inflammation of the meninges, is a se-rious threat to the brain because of the intimate association between the brain and meninges. Should in-fection spread to the neural tissue of the brain itself, life-threatening **encephalitis** may occur. Meningitis is often diagnosed by taking a sample of cerebrospinal fluid from the subarachnoid space. ■

(Text continues on p. 171)

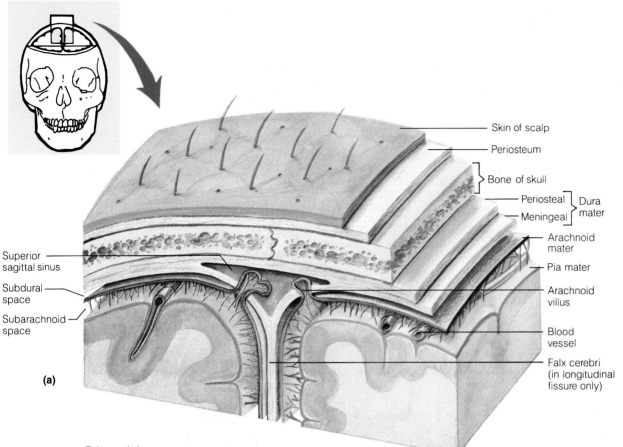

Skin of scalp

Periosteum

Bone of skull

Periosteal ⎱ Dura
Meningeal ⎰ mater

Arachnoid mater

Pia mater

Arachnoid villus

Blood vessel

Falx cerebri (in longitudinal fissure only)

Superior sagittal sinus

Subdural space

Subarachnoid space

(a)

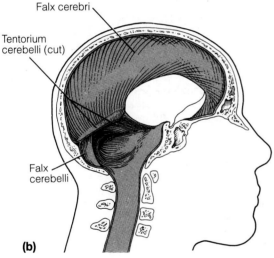

Falx cerebri

Tentorium cerebelli (cut)

Falx cerebelli

(b)

F16.7

Meninges of the brain. (a) Three-dimensional frontal section showing the relationship of the dura mater, arachnoid mater, and pia mater. The meningeal dura forms the falx cerebri fold, which extends into the longitudinal fissure and attaches the brain to the ethmoid bone of the skull. A dural sinus, the superior sagittal sinus, is enclosed by the dural membranes superiorly. Arachnoid villi, which return cerebrospinal fluid to the dural sinus, are also shown. (b) Position of the dural folds, the falx cerebri, tentorium cerebelli, and falx cerebelli. (c) Posterior view of the brain in place, surrounded by the dura mater.

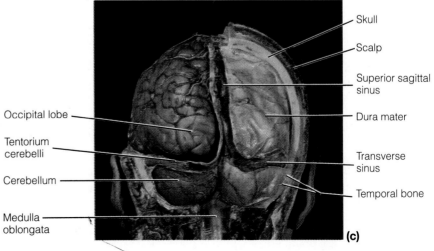

Occipital lobe

Tentorium cerebelli

Cerebellum

Medulla oblongata

Skull

Scalp

Superior sagittal sinus

Dura mater

Transverse sinus

Temporal bone

(c)

169

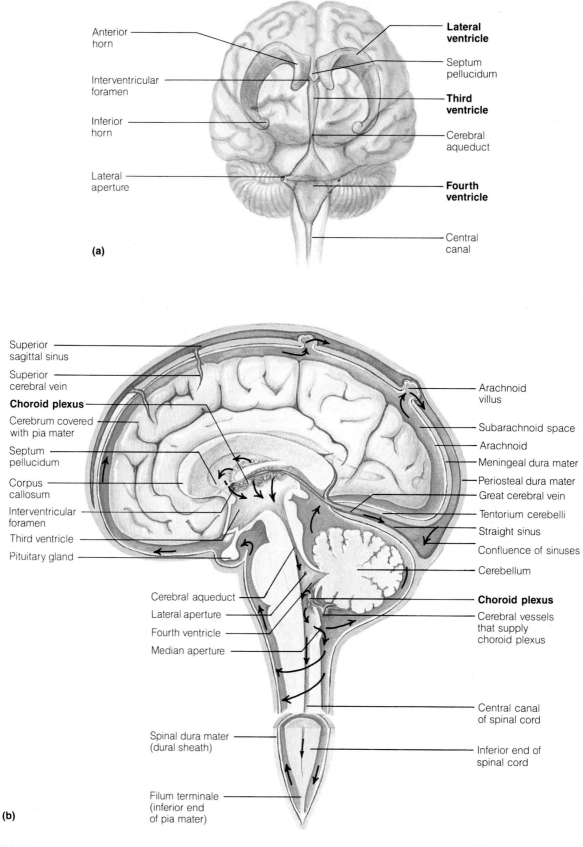

(a)

(b)

F16.8

Location and circulatory pattern of cerebrospinal fluid. (a) Anterior view; note that different regions of the large lateral ventricles are indicated by the terms *anterior horn, posterior horn,* and *inferior horn.* **(b)** The cerebrospinal fluid flows from the lateral ventricles, through the interventricular foramina, into the third ventricle, and then into the fourth ventricle via the cerebral aqueduct. (The relative position of the right lateral ventricle is indicated by the pale blue area deep to the corpus callosum and septum pellucidum.)

170

Cerebrospinal Fluid

The cerebrospinal fluid, much like plasma in composition, is continually formed by the **choroid plexuses,** small capillary knots hanging from the roof of the ventricles of the brain. The cerebrospinal fluid in and around the brain forms a watery cushion that protects the delicate brain tissue against blows to the head.

Within the brain, the cerebrospinal fluid circulates from the two lateral ventricles (in the cerebral hemispheres) into the third ventricle via the **interventricular foramina,** and then through the cerebral aqueduct of the midbrain into the fourth ventricle in the hindbrain (Figure 16.8). Some of the fluid reaching the fourth ventricle continues down the central canal of the spinal cord, but the bulk of it circulates into the subarachnoid space, exiting through the three foramina in the walls of the fourth ventricle (the two lateral and the single median apertures). The fluid returns to the blood in the dural sinuses via the arachnoid villi.

▲ Ordinarily, cerebrospinal fluid forms and drains at a constant rate. However, under certain conditions—for example, obstructed drainage or circulation resulting from tumors or anatomical deviations—the cerebrospinal fluid accumulates and exerts increasing pressure on the brain which, uncorrected, causes neurological damage in adults. In infants, **hydrocephalus** (literally, water on the brain) is indicated by a gradually enlarging head. Since the infant's skull is still flexible and contains fontanels, it can expand to accommodate the increasing size of the brain. ■

CRANIAL NERVES

The **cranial nerves** are part of the peripheral nervous system and not part of the brain proper, but they are most appropriately identified in conjunction with the study of brain anatomy. The 12 pairs of cranial nerves primarily serve the head and neck. Only one pair, the vagus nerves, extends into the thoracic and abdominal cavities. All but the first two pairs (olfactory and optic nerves) arise from the brain stem and pass through foramina in the base of the skull to reach their destination.

The cranial nerves are numbered consecutively, and in most cases their names reflect the major structures they control. The cranial nerves are described by name, number (Roman numeral), origin, course, and function in Table 16.1. This information should be committed to memory. A mnemonic device that might be helpful for remembering the cranial nerves in order is "*O*n occasion, *o*ur *t*rusty *t*ruck *a*cts *f*unny—*v*ery *g*ood *v*ehicle *a*ny*h*ow." The first letter of each word and the "a" and "h" of the final word "anyhow" will remind you of the first letter of the cranial nerve name.

Most cranial nerves are mixed nerves (containing both motor and sensory fibers). However, close scrutiny of Table 16.1 will reveal that three pairs of cranial nerves (optic, olfactory, and vestibulocochlear) are purely sensory in function.

You may recall that the cell bodies of neurons are always located within the central nervous system (cortex or nuclei) or in specialized collections of cell bodies (ganglia) outside the CNS. Neuron cell bodies of the sensory cranial nerves are located in ganglia; those of the mixed cranial nerves are found both within the brain and in peripheral ganglia.

(Text continues on p. 174)

TABLE 16.1   The Cranial Nerves (see Figure 16.9)

Number and name	Origin and course	Function	Testing
I. Olfactory	Fibers arise from olfactory mucosa and run through cribriform plate of ethmoid bone to synapse with olfactory bulbs.	Purely sensory—carries impulses associated with sense of smell.	Person is asked to sniff aromatic substances, such as oil of cloves and vanilla, and to identify each.
II. Optic	Fibers arise from retina of eye and pass through optic foramen in sphenoid bone. Fibers of the two optic nerves then take part in forming optic chiasma (with partial crossover of fibers) after which they continue on to thalamus as the optic tracts. Final fibers of this pathway travel from the thalamus to the optic cortex as the optic radiation.	Purely sensory—carries impulses associated with vision.	Vision and visual field are determined with eye chart and by testing the point at which the person first sees an object (finger) moving into the visual field. Fundus of eye viewed with ophthalmoscope to detect papilledema (swelling of optic disc, or point at which optic nerve leaves the eye) and to observe blood vessels.
III. Oculomotor	Fibers emerge from midbrain and exit from skull via superior orbital fissure to run to eye.	Mixed—somatic motor fibers to inferior oblique and superior, inferior, and medial rectus muscles, which direct eyeball, and to levator palpebrae muscles of eyelid; parasympathetic fibers to iris and smooth muscle controlling lens shape (reflex responses to varying light intensity and focusing of eye for near vision); contains proprioceptive sensory fibers carrying impulses from extrinsic eye muscles.	Pupils are examined for size, shape, and equality. Pupillary reflex is tested with penlight (pupils should constrict when illuminated). Convergence for near vision is tested, as is subject's ability to follow objects up, down, side to side, and diagonally.
IV. Trochlear	Fibers emerge from midbrain and exit from skull via superior orbital fissure to run to eye.	Mixed—provides somatic motor fibers to superior oblique muscle (an extrinsic eye muscle); conveys proprioceptive impulses from same muscle to brain.	Tested in common with cranial nerve III.
V. Trigeminal	Fibers emerge from pons and form three divisions, which exit separately from skull: mandibular division through foramen ovale in sphenoid bone, maxillary division via foramen rotundum in sphenoid bone, and ophthalmic division through superior orbital fissure of eye socket.	Mixed—major sensory nerve of face; conducts sensory impulses from skin of face and anterior scalp, from mucosae of mouth and nose, and from surface of eyes; mandibular division also contains motor fibers that innervate muscles of mastication and muscles of floor of mouth.	Sensations of pain, touch, and temperature are tested with safety pin and hot and cold objects. Corneal reflex tested with wisp of cotton. Motor branch assessed by asking person to clench his teeth, open mouth against resistance, and move jaw side to side.
VI. Abducens	Fibers leave inferior region of pons and exit from skull via superior orbital fissure to run to eye.	Carries motor fibers to lateral rectus muscle of eye and proprioceptive fibers from same muscle to brain.	Tested in common with cranial nerve III.

(continued)

TABLE 16.1   (*Continued*)

Number and name	Origin and course	Function	Testing
VII. Facial	Fibers leave pons and travel through temporal bone via internal acoustic meatus, exiting via stylomastoid foramen to reach the face.	Mixed—supplies somatic motor fibers to muscles of facial expression and parasympathetic motor fibers to lacrimal and salivary glands; carries sensory fibers from taste receptors of anterior portion of tongue.	Anterior two-thirds of tongue is tested for ability to taste sweet (sugar), salty, sour (vinegar), and bitter (quinine) substances. Symmetry of face is checked. Subject is asked to close eyes, smile, whistle, and so on. Tearing is assessed with ammonia fumes.
VIII. Vestibulocochlear	Fibers run from inner-ear equilibrium and hearing apparatus, housed in temporal bone, through internal acoustic meatus to enter pons.	Purely sensory—vestibular branch transmits impulses associated with sense of equilibrium from vestibular apparatus and semicircular canals; cochlear branch transmits impulses associated with hearing from cochlea.	Hearing is checked by air and bone conduction using tuning fork.
IX. Glossopharyngeal	Fibers emerge from medulla and leave skull via jugular foramen to run to throat.	Mixed—somatic motor fibers serve pharyngeal muscles, and parasympathetic motor fibers serve salivary glands; sensory fibers carry impulses from pharynx, tonsils, posterior tongue (taste buds), and pressure receptors of carotid artery.	Position of the uvula is checked. Gag and swallowing reflexes are checked. Subject is asked to speak and cough. Posterior third of tongue may be tested for taste.
X. Vagus	Fibers emerge from medulla and pass through jugular foramen and descend through neck region into thorax and abdomen.	Mixed—fibers carry somatic motor impulses to pharynx and larynx and sensory fibers from same structures; very large portion is composed of parasympathetic motor fibers, which supply heart and smooth muscles of abdominal visceral organs; transmits sensory impulses from viscera.	As for cranial nerve IX (IX and X are tested in common, since they both innervate muscles of throat and mouth).
XI. Accessory	Fibers arise from medulla and superior aspect of spinal cord and travel through jugular foramen to reach muscles of neck and back.	Mixed—provides somatic motor fibers to sternocleidomastoid and trapezius muscles and to muscles of soft palate, pharynx, and larynx (spinal and medullary fibers respectively); proprioceptive impulses are conducted from these muscles to brain.	Sternocleidomastoid and trapezius muscles are checked for strength by asking person to rotate head and shoulders against resistance.
XII. Hypoglossal	Fibers arise from medulla and exit from skull via hypoglossal canal to travel to tongue.	Mixed—carries somatic motor fibers to muscles of tongue and proprioceptive impulses from tongue to brain.	Person is asked to protrude and retract tongue. Any deviations in position are noted.

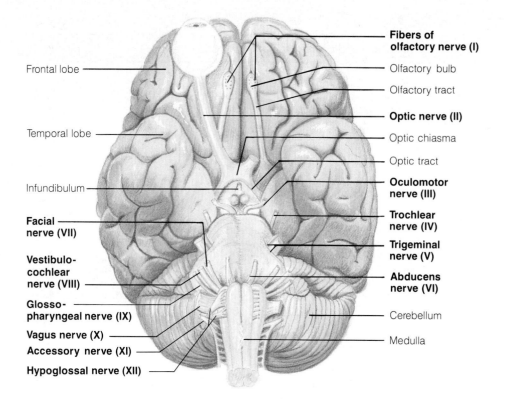

F16.9

Ventral aspect of the human brain, showing the cranial nerves.

1. Observe the anterior surface of the brain model to identify the cranial nerves. Figure 16.9 may also aid you in this study. Notice that the first (olfactory) cranial nerves are not visible on the model because they consist only of those short axons that run from the nasal mucosa through the cribriform plate of the ethmoid bone. (However, the synapse points of the first cranial nerves, the *olfactory bulbs,* are visible on the model.)

2. The last column of Table 16.1 describes techniques for testing cranial nerves, which is an important part of any neurologic examination. This information may help you understand cranial nerve function, especially as it pertains to some aspects of brain function. If materials are provided for cranial nerve testing, conduct tests of cranial nerve function following directions given in the "testing" column of the table.

3. Several cranial nerve ganglia are named here. *Using your textbook or an appropriate reference,* name the cranial nerve the ganglion is associated with and state its location.

Cranial nerve ganglion	Cranial nerve	Site of ganglion
trigeminal		
geniculate		
inferior		
superior		
spiral		
vestibular		

DISSECTION OF THE SHEEP BRAIN

The brain of any mammal is enough like the human brain to warrant comparison. Obtain a sheep brain, protective skin cream or disposable gloves, dissecting pan, and instruments, and bring them to your laboratory bench.

 1. Place the intact sheep brain ventral surface down on the dissecting pan and observe the dura mater. Feel its consistency and note its toughness. Cut through the dura mater along the line of the longitudinal fissure (which separates the cerebral hemispheres) to enter the superior sagittal sinus. Gently force the cerebral hemispheres apart laterally to expose the corpus callosum deep to the longitudinal fissure.

2. Carefully remove the dura mater and examine the superior surface of the brain. Notice that, like the human brain, its surface is thrown into convolutions (fissures and gyri). Locate the arachnoid mater, which appears on the brain surface as a delicate "cottony" material spanning the fissures. In contrast, the innermost meninx, the pia mater, closely follows the cerebral contours.

Dorsal Structures

1. Refer to Figures 16.10a and c as a guide in identifying the following structures. The cerebral hemispheres should be easy to locate. How do the size of the sheep's cerebral hemispheres and the depth of the fissures compare to those in the human brain?

_____

_____

_____

2. Carefully examine the cerebellum. Notice that, in contrast to the human cerebellum, it is not divided longitudinally, and that its fissures are oriented differently. What dural falx is missing that is present in humans?

_____

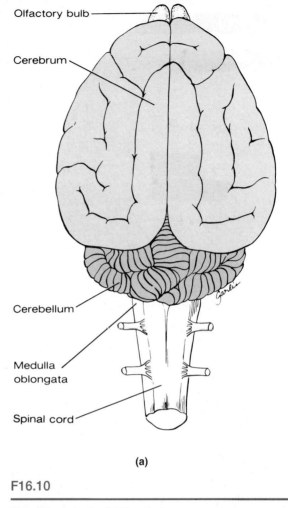

Olfactory bulb

Cerebrum

Cerebellum

Medulla oblongata

Spinal cord

(a)

F16.10

Intact sheep brain. (a) Dorsal view.

3. Locate the three pairs of cerebellar peduncles, fiber tracts that connect the cerebellum to other brain structures, by lifting the cerebellum dorsally away from the brain stem. The most posterior pair, the inferior cerebellar peduncles, connect the cerebellum to the medulla. The middle cerebellar peduncles attach the cerebellum to the pons, and the superior cerebellar peduncles run from the cerebellum to the midbrain.

(_Text continues on p. 177_)

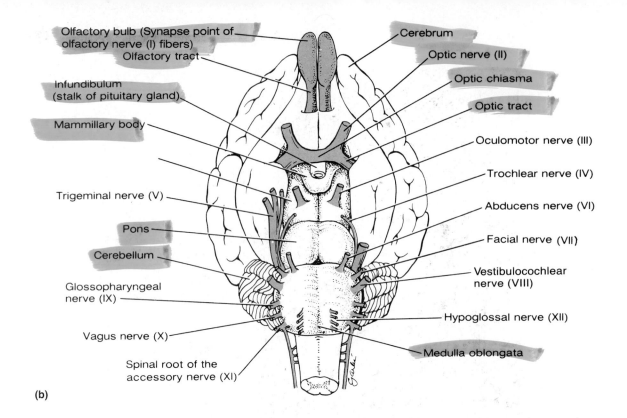

Olfactory bulb (Synapse point of olfactory nerve (I) fibers)

Olfactory tract

Infundibulum (stalk of pituitary gland)

Mammillary body

Trigeminal nerve (V)

Pons

Cerebellum

Glossopharyngeal nerve (IX)

Vagus nerve (X)

Spinal root of the accessory nerve (XI)

Cerebrum

Optic nerve (II)

Optic chiasma

Optic tract

Oculomotor nerve (III)

Trochlear nerve (IV)

Abducens nerve (VI)

Facial nerve (VII)

Vestibulocochlear nerve (VIII)

Hypoglossal nerve (XII)

Medulla oblongata

(b)

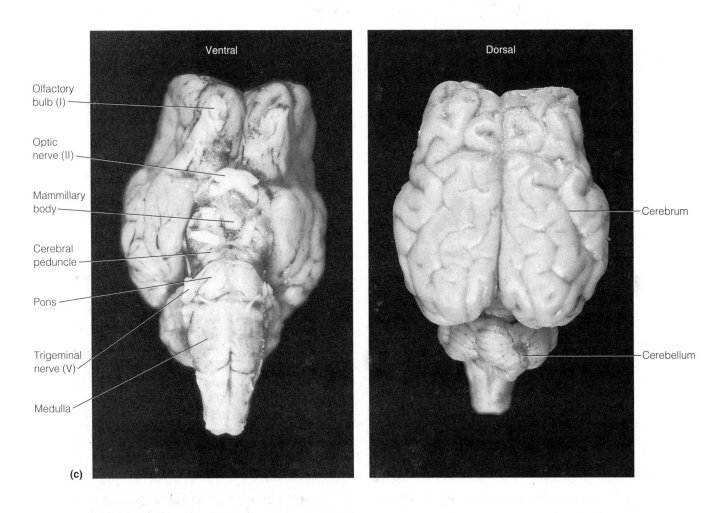

Ventral

Dorsal

Olfactory bulb (I)

Optic nerve (II)

Mammillary body

Cerebral peduncle

Pons

Trigeminal nerve (V)

Medulla

Cerebrum

Cerebellum

(c)

F16.10 (*continued*)

Intact sheep brain. (b) Ventral view. **(c)** Photographs showing ventral and dorsal views.

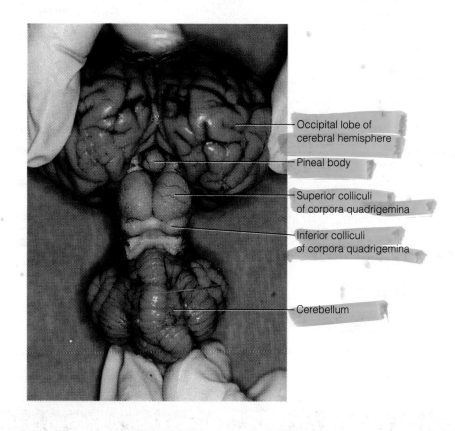

Occipital lobe of cerebral hemisphere

Pineal body

Superior colliculi of corpora quadrigemina

Inferior colliculi of corpora quadrigemina

Cerebellum

F16.11

Means of exposing the dorsal midbrain structures of the sheep brain.

4.   To expose the dorsal surface of the midbrain, gently spread the cerebrum and cerebellum apart, as shown in Figure 16.11. Identify the corpora quadrigemina, which appear as four rounded prominences on the dorsal midbrain surface. What is the function of the corpora quadrigemina?

Also locate the pineal body, which appears as a small oval protrusion in the midline just anterior to the corpora quadrigemina.

Ventral Structures

Figure 16.10b and c shows the important features of the ventral surface of the brain.

1.   Look for the clublike olfactory bulbs anteriorly, on the inferior surface of the frontal lobes of the cerebral hemispheres. Axons of olfactory neurons run from the nasal mucosa through the perforated cribriform plate of the ethmoid bone to synapse with the olfactory bulbs.

How does the size of these olfactory bulbs compare with those of humans?

Is the sense of smell more important as a protective and a food-getting sense in sheep or in humans?

2.   The optic nerve (II) carries sensory impulses from the retina of the eye. Thus this cranial nerve is involved in the sense of vision. Identify the optic nerves, optic chiasma, and optic tracts.

3.   Posterior to the optic chiasma, two structures protrude from the ventral aspect of the hypothalamus—the infundibulum (stalk of the pituitary gland) immediately posterior to the optic chiasma and the mammillary body. Notice that the sheep's mammillary body is a single rounded eminence; in humans it is a double structure.

4.   Identify the cerebral peduncles on the ventral aspect of the midbrain, just posterior to the mammillary body of the hypothalamus. The cerebral peduncles are fiber tracts connecting the cerebrum and medulla. Identify the large oculomotor nerves (III), which arise from the ventral midbrain surface, and the tiny trochlear nerves (IV), which can be seen at the junction of the midbrain and pons. Both of these cranial nerves provide motor fibers to extrinsic muscles of the eyeball.

5.   Move posteriorly from the midbrain to identify first the pons and then the medulla oblongata, both hindbrain structures composed primarily of ascending and descending fiber tracts.

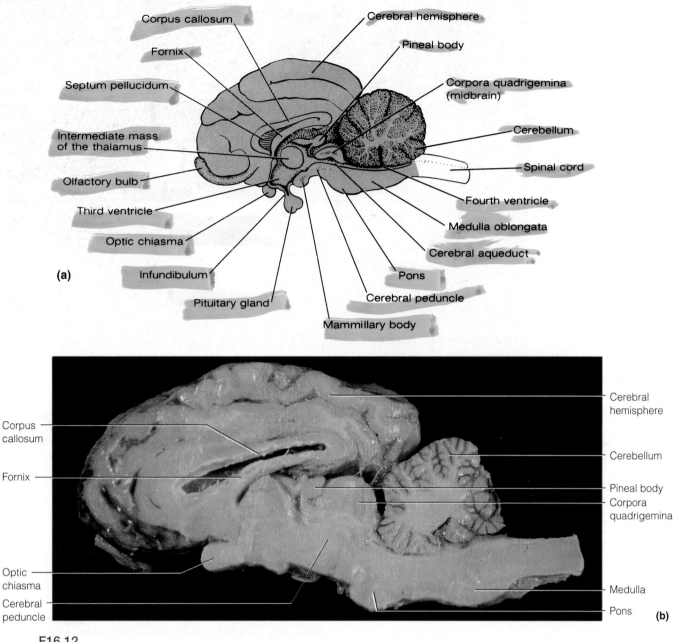

F16.12

Sagittal section of the sheep brain showing internal structures. (a) Diagrammatic view. **(b)** Photograph.

6. Return to the junction of the pons and midbrain and proceed posteriorly to identify the following cranial nerves, all arising from the pons:

- Trigeminal nerves (V), which are involved in chewing and sensations of the head and face
- Abducens nerves (VI), which abduct the eye (and thus work in conjunction with cranial nerves III and IV)
- Facial nerves (VII), large nerves involved in taste sensation, gland function (salivary and lacrimal glands), and facial expression

7. Continue posteriorly to identify the following:

- Vestibulocochlear nerves (VIII), purely sensory nerves which are involved with hearing and equilibrium
- Glossopharyngeal nerves (IX), which contain motor fibers innervating throat structures and sensory fibers transmitting taste stimuli (in conjunction with cranial nerve VII)
- Vagus nerves (X), often called "wanderers," which serve many organs of the head, thorax, and abdominal cavity
- Accessory nerves (XI), which serve muscles of the neck, larynx, and shoulder; notice that the accessory nerves arise from both the medulla and the spinal cord
- Hypoglossal nerves (XII), which stimulate tongue and neck muscles

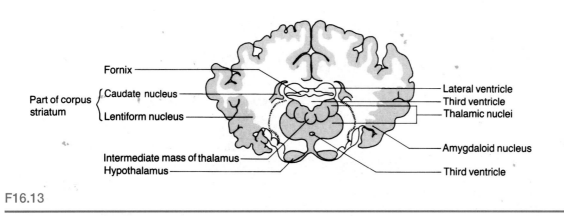

Fornix

Part of corpus striatum { Caudate nucleus

Lentiform nucleus }

Intermediate mass of thalamus
Hypothalamus

Lateral ventricle
Third ventricle
Thalamic nuclei

Amygdaloid nucleus

Third ventricle

F16.13

Frontal section of a sheep brain. Major structures revealed are the location of major basal nuclei deep in the interior, the thalamus, hypothalamus, and lateral and third ventricles.

Internal Structures

1.   The internal structure of the brain can only be examined after further dissection. Place the brain ventral side down on the dissecting pan and make a cut completely through it in a superior to inferior direction. Cut through the longitudinal fissure, corpus callosum, and midline of the cerebellum. Refer to Figure 16.12 as you work.

2.   The thin nervous tissue membrane immediately ventral to the corpus callosum that separates the lateral ventricles is the septum pellucidum. Pierce this membrane and probe the lateral ventricle cavity. The fiber tract ventral to the septum pellucidum and anterior to the third ventricle is the fornix.

How does the size of the fornix in this brain compare with the human fornix?

_____

_____

Why do you suppose this is so? (Hint: What is the function of this band of fibers?)

_____

_____

_____

3.   Identify the thalamus, which forms the walls of the third ventricle and is located posterior and ventral to the fornix. The intermediate mass spanning the ventricular cavity appears as an oval protrusion of the thalamic wall. Anterior to the intermediate mass, locate the interventricular foramen, a canal connecting the lateral ventricle on the same side with the third ventricle.

4.   The hypothalamus forms the floor of the third ventricle. Identify the optic chiasma, infundibulum, and mammillary body on its exterior surface. You can see the pineal body at the superoposterior end of the third ventricle, just beneath the junction of the corpus callosum and fornix.

5.   Locate the midbrain by identifying the corpora quadrigemina that form its dorsal roof. Follow the cerebral aqueduct (the narrow canal connecting the third and fourth ventricles) through the midbrain tissue to the fourth ventricle. Identify the cerebral peduncles, which form its anterior walls.

6.   Identify the pons and medulla, which lie anterior to the fourth ventricle. The medulla continues into the spinal cord without any obvious anatomical change, but the point at which the fourth ventricle narrows to a small canal is generally accepted as the beginning of the spinal cord.

7.   Identify the cerebellum posterior to the fourth ventricle. Note its internal treelike arrangement of white matter, the arbor vitae.

8.   If time allows, obtain another sheep brain and section it along the frontal plane so that the cut passes through the infundibulum. Compare your specimen to the diagrammatic view in Figure 16.13, and attempt to identify all the structures shown in the figure.

9.   Check with your instructor to determine if cow spinal cord sections (preserved) are available for the spinal cord studies in Exercise 17. If not, save the small portion of the spinal cord from your brain specimen. Otherwise, dispose of all the organic debris in the appropriate laboratory containers and clean the dissecting instruments and tray before leaving the laboratory.

Spinal Cord, Spinal Nerves, and the Autonomic Nervous System

OBJECTIVES

1. To identify important anatomical areas on a spinal cord model or appropriate diagram of the spinal cord, and to cite the neuron type found in these areas (where applicable).

2. To indicate two major areas where the spinal cord is enlarged, and to explain the reasons for this anatomical characteristic.

3. To define *conus medullaris, cauda equina,* and *filum terminale.*

4. To locate on a diagram the fiber tracts in the spinal cord and to state their functional importance.

5. To list two major functions of the spinal cord.

6. To name the meningeal coverings of the spinal cord and state their function.

7. To describe the origin, fiber composition, and distribution of the spinal nerves, differentiating between roots, the spinal nerve proper, and rami, and to discuss the result of transecting these structures.

8. To discuss the distribution of the dorsal rami and ventral rami of the spinal nerves.

9. To identify the four major nerve plexuses, the major nerves of each, and their distribution.

10. To identify on a dissected animal the musculocutaneous, radial, median, and ulnar nerves of the upper limb and the femoral, saphenous, sciatic, common peroneal, and tibial nerves of the lower limb.

11. To identify the site of origin and the function of the sympathetic and parasympathetic divisions of the autonomic nervous system, and to state how the autonomic nervous system differs from the somatic nervous system.

MATERIALS

Spinal cord model (cross section)
Laboratory charts of the spinal cord and spinal nerves and sympathetic chain
Red and blue pencils
Preserved cow spinal cord sections with meninges and nerve roots intact (or spinal cord segment saved from the brain dissection in Exercise 16)
Dissecting tray and instruments
Dissecting microscope
Histologic slide of spinal cord (cross section)
Protective skin cream or disposable gloves
Compound microscope
Animal specimen from previous dissections
The Human Nervous System videotape*

 See Appendix B, Exercise 17 for links to A.D.A.M. Standard.

 See Appendix C, Exercise 17 for links to *Anatomy and PhysioShow: The Videodisc.*

*Available to qualified adopters from Benjamin/Cummings

ANATOMY OF THE SPINAL CORD

The cylindrical **spinal cord,** a continuation of the brain stem, is an association and communication center. It plays a major role in spinal reflex activity and provides neural pathways to and from higher nervous centers. Enclosed within the vertebral canal of the spinal column, the spinal cord extends from the foramen magnum of the skull to the first or second lumbar vertebra, where it terminates in the cone-shaped **conus medullaris** (Figure 17.1). Like the brain, it is cushioned and protected by meninges. The dura mater and arachnoid meningeal coverings extend beyond the conus medullaris, approximately to the level of S_2, and a fibrous extension of the pia mater extends even farther (into the coccygeal canal) as the **filum terminale.**

The fact that the meninges, filled with cerebrospinal fluid, extend well beyond the end of the spinal cord provides an excellent site for removing cerebrospinal fluid for analysis (as when bacterial or viral infections of the spinal cord or its meningeal coverings are suspected) without endangering the delicate spinal cord. This procedure, called a *lumbar tap,* is usually performed below L_3. Additionally, "saddle block" or caudal anesthesia for childbirth is normally administered (injected) between L_3 and L_5.

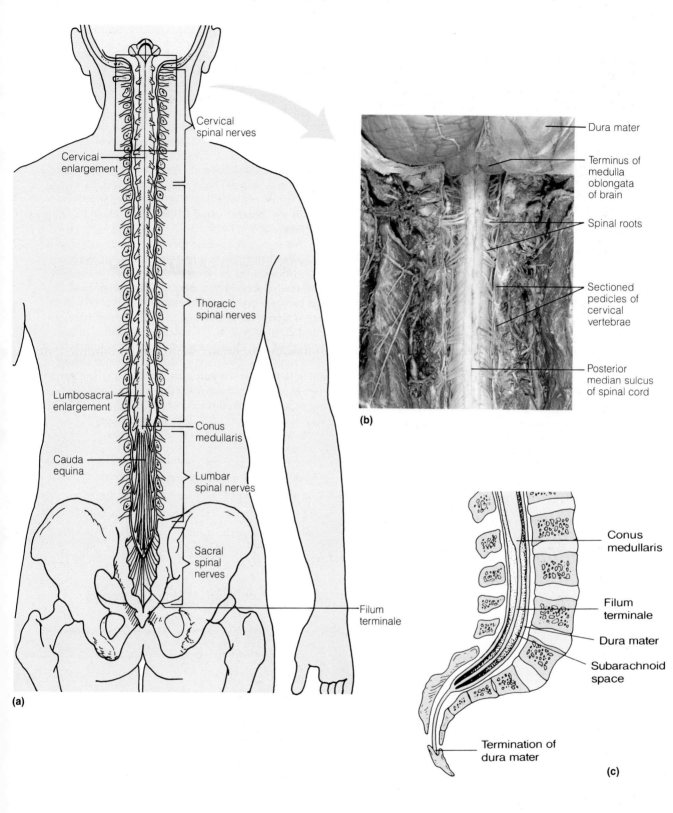

(a)

Cervical spinal nerves

Cervical enlargement

Thoracic spinal nerves

Lumbosacral enlargement

Conus medullaris

Cauda equina

Lumbar spinal nerves

Sacral spinal nerves

Filum terminale

(b)

Dura mater

Terminus of medulla oblongata of brain

Spinal roots

Sectioned pedicles of cervical vertebrae

Posterior median sulcus of spinal cord

(c)

Conus medullaris

Filum terminale

Dura mater

Subarachnoid space

Termination of dura mater

F17.1

Structure of the spinal cord. (a) The vertebral arches have been removed to show the dorsal aspect of the spinal cord (and its nerve roots). The dura mater is cut and reflected laterally. The various regions of the spinal cord are indicated in relation to the vertebral column as cervical, thoracic, lumbar, and sacral. **(b)** Photograph of the cervical region of the spinal cord; the meningeal coverings have been removed to show the spinal roots that give rise to the spinal nerves. **(c)** Lateral view, showing the extent of the filum terminale.

In humans, 31 pairs of spinal nerves arise from the spinal cord and pass through intervertebral foramina to serve the body area at their approximate level of emergence. The cord is about the size of a thumb in circumference for most of its length, but there are obvious enlargements in the cervical and lumbar areas where the nerves serving the upper and lower limbs issue from the cord.

Because the spinal cord does not extend to the end of the vertebral column, the spinal nerves emerging from the inferior end of the cord must travel through the vertebral canal for some distance before exiting at the appropriate intervertebral foramina. This collection of spinal nerves traversing the inferior end of the vertebral canal is called the **cauda equina** because of its similarity to a horse's tail (the literal translation of *cauda equina*).

Obtain a model of a cross section of a spinal cord and identify its structures as they are described next.

Gray Matter

In cross section, the **gray matter** of the spinal cord looks like a butterfly or the letter H (Figure 17.2). The two posterior projections are called the **posterior,** or **dorsal, horns;** the two anterior projections are the **anterior,** or **ventral, horns.** The tips of the anterior horns are broader and less tapered than those of the posterior horns. In the thoracic and lumbar regions of the cord, there is also a lateral outpocketing of gray matter on each side referred to as the **lateral horn.** The central area of gray matter connecting the two vertical regions is the **gray commissure.** The gray commissure surrounds the **central canal** of the cord, which contains cerebrospinal fluid.

Neurons with specific functions can be localized in the gray matter. The posterior horns, for instance, contain association neurons and sensory fibers that enter the cord from the body periphery via the **dorsal root.** The cell bodies of these sensory neurons are found in an enlarged area of the dorsal root called the **dorsal root ganglion.** The anterior horns contain cell bodies of motor neurons of the somatic nervous system (voluntary system), which send their axons out via the **ventral root** of the cord to enter the adjacent spinal nerve. The **spinal nerves** are formed from the fusion of the dorsal and ventral roots. The lateral horns, where present, contain cell

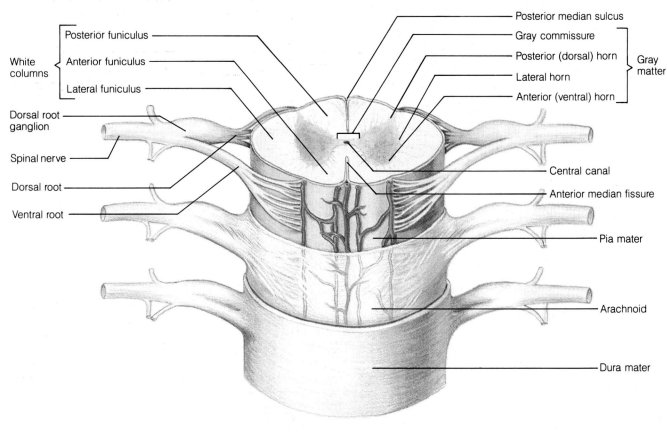

F17.2

Anatomy of the human spinal cord (three-dimensional view).

bodies of motor neurons of the autonomic nervous system (sympathetic division). Their axons also leave the cord via the ventral roots, along with those of the motor neurons of the anterior horns.

White Matter

The **white matter** of the spinal cord is nearly bisected by fissures (see Figure 17.2). The more open anterior fissure is the **anterior median fissure,** and the posterior one is the **posterior median sulcus.** The white matter is composed of myelinated fibers—some running to higher centers, some traveling from the brain to the cord, and some conducting impulses from one side of the cord to the other.

Because of the irregular shape of the gray matter, the white matter on each side of the cord can be divided into three primary regions or *white columns:* the **posterior, lateral,** and **anterior funiculi.** Each funiculus contains a number of fiber **tracts** composed of axons with the same origin, terminus, and function. Tracts conducting sensory impulses to the brain are called *ascending,* or *sensory, tracts;* those carrying impulses from the brain to the skeletal muscles are *descending,* or *motor, tracts.*

Because it serves as the transmission pathway between the brain and the body periphery, the spinal cord is an extremely important functional area. Even though it is protected by meninges and cerebrospinal fluid in the vertebral canal, it is highly vulnerable to traumatic injuries, such as might occur in an automobile accident.

When the cord is transected (or severely traumatized), both motor and sensory functions are lost in body areas normally served by that (and lower) regions of the spinal cord. Injury to certain spinal cord areas may even result in a permanent flaccid paralysis of both legs (paraplegia) or of all four limbs (quadriplegia). ■

 With the help of your textbook or a laboratory chart showing the tracts of the spinal cord, label Figure 17.3 with the tract names that follow. Since each tract is represented on both sides of the cord, for clarity you can label the motor tracts on the right side of the diagram and the sensory tracts on the left side of the diagram. *Color ascending tracts red and descending tracts blue.* Then fill in the functional importance of each tract beside its name below. As you work, try to be aware of how the naming of the tracts is related to their anatomical distribution.

Fasciculus gracilis _____

Fasciculus cuneatus _____

Dorsal spinocerebellar _____

Ventral spinocerebellar _____

Lateral spinothalamic _____

Ventral spinothalamic _____

Lateral corticospinal _____

Ventral corticospinal _____

Rubrospinal _____

Tectospinal _____

Vestibulospinal _____

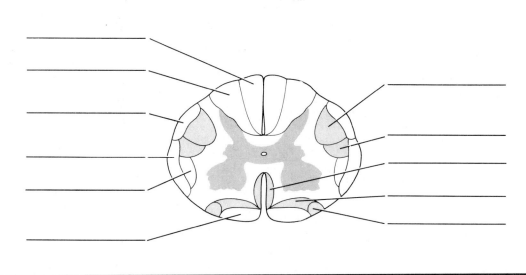

F17.3

Cross section of the spinal cord showing the relative positioning of its major tracts.

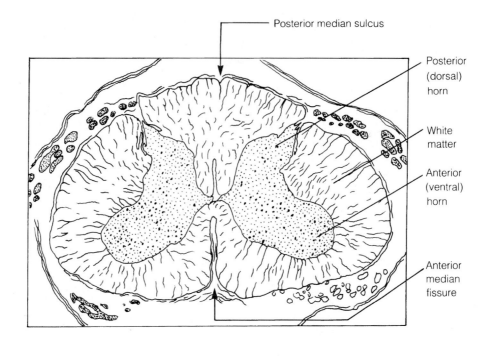

Posterior median sulcus

Posterior (dorsal) horn

White matter

Anterior (ventral) horn

Anterior median fissure

F17.4

Cross section of the spinal cord. See corresponding Plate 7 in the Histology Atlas.

Spinal Cord Dissection

1. Obtain a dissecting tray and instruments and a segment of preserved spinal cord (from a cow or saved from the brain specimen used in Exercise 16). Identify the tough outer meninx (dura mater) and the weblike arachnoid mater.

What name is given to the third meninx, and where is it found?

_____

_____

Peel back the dura mater and observe the fibers making up the dorsal and ventral roots. If possible, identify a dorsal root ganglion.

2. Cut a thin cross section of the cord and identify the anterior and posterior horns of the gray matter with the naked eye or with the aid of a dissecting microscope.

How can you be certain that you are correctly identifying the anterior and posterior horns?

_____

_____

Also identify the central canal, white matter, anterior median fissure, posterior median sulcus, and posterior, anterior, and lateral funiculi.

3. Obtain a prepared slide of the spinal cord (cross section) and a compound microscope. Refer to Figure 17.4 as you examine the slide carefully under low power. Observe the shape of the central canal.

Is it basically circular or oval? _____

Name the glial cell type that lines this canal. _____

What would you expect to find in this canal in the living animal?

_____

Can any neuron cell bodies be seen? _____

Where? _____

_____

What type of neurons would these most likely be—motor, association, or sensory?

_____

_____

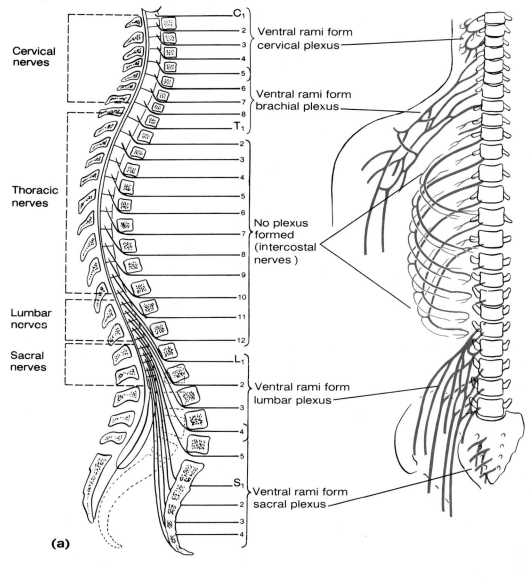

F17.5

Human spinal nerves. (a) Relationship of spinal nerves to vertebrae (areas of plexuses formed by the ventral rami are indicated).

SPINAL NERVES AND NERVE PLEXUSES

The 31 pairs of human spinal nerves arise from the fusions of the ventral and dorsal roots of the spinal cord. Figure 17.5a shows how the nerves are named according to their point of issue. Because the ventral roots contain myelinated axons of motor neurons located in the cord and the dorsal roots carry sensory fibers entering the cord, all spinal nerves are **mixed nerves.** The first pair of spinal nerves leaves the vertebral canal between the base of the occiput and the atlas, but all the rest exit via the intervertebral foramina. The first through seventh pairs of cervical nerves emerge *above* the vertebra for which they are named. C_8 emerges between C_7 and T_1. (Notice that there are 7 cervical vertebrae, but 8 pairs of cervical nerves.) The remaining spinal nerve pairs emerge from the spinal cord *below* the same-numbered vertebra.

Almost immediately after emerging, each nerve divides into **dorsal** and **ventral rami.** (Thus each spinal nerve is only about 1 or 2 cm long.) The rami, like the spinal nerves, contain both motor and sensory fibers. The smaller dorsal rami serve the skin and musculature

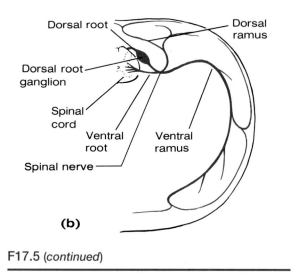

(b)

F17.5 (*continued*)

Human spinal nerves. (b) Relative distribution of the ventral and dorsal rami of a spinal nerve (cross section of left trunk).

of the posterior body trunk at their approximate level of emergence (Figure 17.5b). The ventral rami of spinal nerves T_2–T_{12} pass anteriorly as the **intercostal nerves** to supply the muscles of intercostal spaces, and the skin and muscles of the anterior and lateral trunk. The ventral rami of all other spinal nerves form complex networks of nerves called **plexuses.** These plexuses serve the motor and sensory needs of the muscles and skin of the limbs. The fibers of the ventral rami unite in the plexuses (with a few rami supplying fibers to more than one plexus). From the plexuses the fibers diverge again to form peripheral nerves, each of which contains fibers from more than one spinal nerve. The four major nerve plexuses and their chief peripheral nerves are illustrated in Figures 17.5 and 17.6 and are described below. Their names and site of origin should be committed to memory.

Cervical Plexus and the Neck

The **cervical plexus** arises from the ventral rami of C_1 through C_5 to supply muscles of the shoulder and neck. The major motor branch of this plexus is the **phrenic nerve,** which arises from C_3–C_4 (plus some fibers from C_5) and passes into the thoracic cavity in front of the first rib to innervate the diaphragm. The primary danger of a broken neck is that the phrenic nerve may be severed, leading to paralysis of the diaphragm and cessation of breathing. A jingle to help you remember the rami (roots) forming the phrenic nerves is "C_3, C_4, C_5 keep the diaphragm alive."

Brachial Plexus and the Upper Limb

The **brachial plexus** is large and complex, arising from the ventral rami of C_5 through C_8 and T_1. The plexus, after being rearranged consecutively into *trunks, divi-*

sions, and *cords,* finally becomes subdivided into five major *peripheral nerves.*

The **axillary nerve,** which serves the muscles and skin of the shoulder, has the most limited distribution. The large **radial nerve** passes down the posterolateral surface of the arm and forearm, supplying all the extensor muscles of the arm, forearm, and hand and the skin along its course. The radial nerve is often injured in the axillary region by the pressure of a crutch or by hanging one's arm over the back of a chair. The **median nerve** passes down the anteromedial surface of the arm to supply most of the flexor muscles in the forearm and several muscles in the hand (plus the skin of the lateral surface of the palm of the hand).

- Hyperextend your wrist to identify the long, obvious tendon of your palmaris longus muscle, which crosses the exact midline of the anterior wrist. Your median nerve lies immediately deep to that tendon, and the radial nerve lies just *lateral* to it.

The **musculocutaneous nerve** supplies the arm muscles that flex the forearm and the skin of the lateral surface of the forearm. The **ulnar nerve** travels down the posteromedial surface of the arm. It courses around the medial epicondyle of the humerus to supply the flexor carpi ulnaris, the ulnar head of the flexor digitorum profundus of the forearm, and all intrinsic muscles of the hand not served by the median nerve. It supplies the skin of the medial third of the hand, both the anterior and posterior surfaces. Trauma to the ulnar nerve, which often occurs when the elbow is hit, produces a smarting sensation commonly referred to as "hitting the funny bone."

Lumbosacral Plexus and the Lower Limb

The **lumbosacral plexus,** which serves the pelvic region of the trunk and the lower limbs, is actually a complex of two plexuses, the lumbar plexus and the sacral plexus (see Figure 17.6). The **lumbar plexus** arises from ventral rami of L_1 through L_4 (and sometimes T_{12}). Its nerves serve the lower abdominopelvic region and the anterior thigh. The largest nerve of this plexus is the **femoral nerve,** which passes beneath the inguinal ligament to innervate the anterior thigh muscles. The cutaneous branches of the femoral nerve (median and anterior femoral cutaneous and the saphenous nerves) supply the skin of the anteromedial surface of the entire lower limb.

Arising from L_4 through S_4, the nerves of the **sacral plexus** supply the buttock, the posterior surface of the thigh, and virtually all sensory and motor fibers of the leg and foot. The major peripheral nerve of this plexus is the **sciatic nerve,** the largest nerve in the body. The sciatic nerve leaves the pelvis through the greater sciatic notch and travels down the posterior thigh, serving its flexor muscles and skin. In the popliteal region, the sciatic nerve divides into the **common peroneal nerve** and the **tibial nerve,** which together supply the balance of the leg muscles and skin, both directly and via several branches.

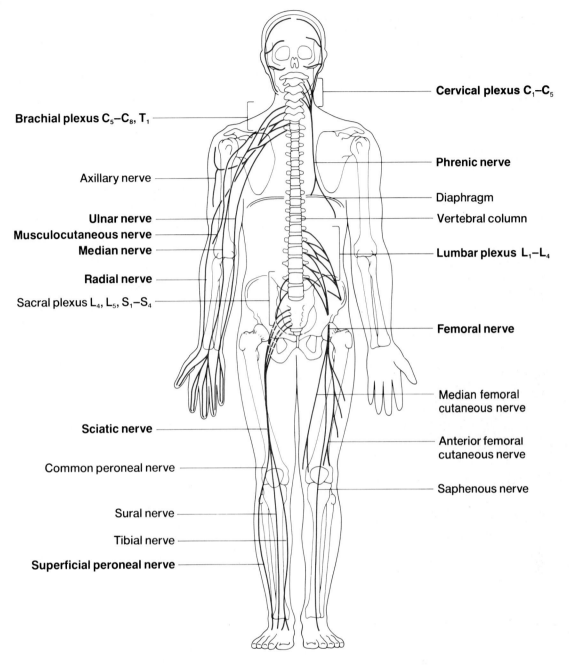

Cervical plexus C₁–C₅

Brachial plexus C₅–C₈, T₁

Axillary nerve

Ulnar nerve
Musculocutaneous nerve
Median nerve

Radial nerve

Sacral plexus L₄, L₅, S₁–S₄

Phrenic nerve

Diaphragm
Vertebral column

Lumbar plexus L₁–L₄

Femoral nerve

Median femoral
cutaneous nerve

Anterior femoral
cutaneous nerve

Saphenous nerve

Sciatic nerve

Common peroneal nerve

Sural nerve

Tibial nerve

Superficial peroneal nerve

F17.6

Nerve plexuses and major nerves arising from each. For clarity, each plexus is illustrated only on one side of the body.

Identify each of the four major nerve plexuses (and its major nerves) shown in Figure 17.6 on a large laboratory chart. Trace the course of the nerves.

THE AUTONOMIC NERVOUS SYSTEM

The **autonomic nervous system** is the subdivision of the PNS that regulates body activities that are generally not under conscious control. It is composed of a special group of motor neurons serving cardiac muscle (the heart), smooth muscle (found in the walls of the visceral organs and blood vessels), and internal glands. Because these structures typically function without conscious

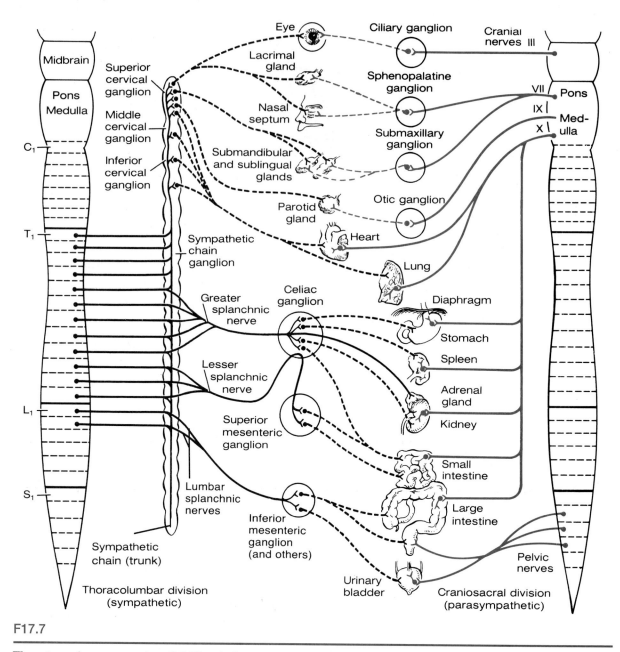

F17.7

The autonomic nervous system. Solid lines indicate preganglionic nerve fibers; dashed lines indicate postganglionic nerve fibers.

control, this system is often referred to as the *involuntary nervous system.*

There is a basic anatomical difference between the motor pathways of the **somatic** (voluntary) **nervous system,** which innervates the skeletal muscles, and those of the autonomic nervous system. In the somatic division, the cell bodies of the motor neurons reside in the CNS (spinal cord or brain), and their axons, sheathed in spinal nerves, extend all the way to the skeletal muscles they serve. However, the autonomic nervous system consists of chains of two motor neurons. The first motor neuron of each pair, called the *preganglionic neuron,* resides in the brain or cord. Its axon leaves the CNS to synapse with the second motor neuron (*postganglionic neuron*), whose cell body is located in a ganglion outside the CNS. The axon of the postganglionic neuron then extends to the organ it serves.

The autonomic nervous system has two major functional subdivisions (Figure 17.7). These, the sympathetic and parasympathetic divisions, serve most of the same organs, but generally cause opposing or antagonistic effects.

Parasympathetic Division

The preganglionic neurons of the **parasympathetic,** or **craniosacral,** division are located in brain nuclei of cranial nerves III, VII, IX, X and in the S_2 through S_4 level of the spinal cord. The axons of the preganglionic neurons of the cranial region travel in their respective cranial nerves to the *immediate area* of the head and neck organs to be stimulated. There they synapse with the postganglionic neuron in a **terminal,** or **intramural** (literally, "within the walls"), **ganglion.** The postgangli-

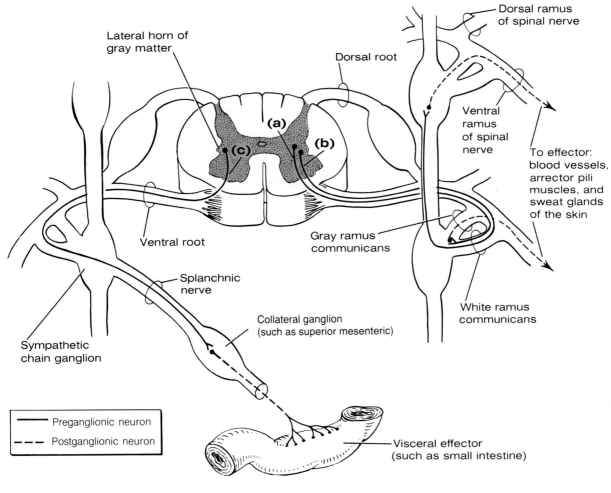

Lateral horn of gray matter

Dorsal ramus of spinal nerve

Dorsal root

(a)

(b)

(c)

Ventral ramus of spinal nerve

To effector: blood vessels, arrector pili muscles, and sweat glands of the skin

Ventral root

Gray ramus communicans

Splanchnic nerve

Collateral ganglion (such as superior mesenteric)

White ramus communicans

Sympathetic chain ganglion

—— Preganglionic neuron
--- Postganglionic neuron

Visceral effector (such as small intestine)

F17.8

Sympathetic pathways. (a) Synapse in a sympathetic chain (paravertebral) ganglion at the same level. **(b)** Synapse in a sympathetic chain ganglion at a different level. **(c)** Synapse in a collateral (prevertebral) ganglion.

onic neuron then sends out a very short axon to the organ it serves. In the sacral region, the preganglionic axons leave the ventral roots of the spinal cord and collectively form the **pelvic splanchnic nerves,** which travel to the pelvic cavity. In the pelvic cavity, the preganglionic axons synapse with the postganglionic neurons in ganglia located on or close to the organs served.

Sympathetic Division

The preganglionic neurons of the **sympathetic,** or **thoracolumbar,** division are in the lateral horns of the gray matter of the spinal cord from T_1 through L_2. The preganglionic axons leave the cord via the ventral root (in conjunction with the axons of the somatic motor neurons), enter the spinal nerve, and then travel briefly in the ventral ramus (Figure 17.8). From the ventral ramus, they pass through a small branch called the **white ramus communicans** to enter a **paravertebral ganglion** in the **sympathetic chain,** or **trunk,** which lies alongside the vertebral column (the literal meaning of *paravertebral*).

Having reached the ganglion, a preganglionic axon may take one of three main courses (see Figure 17.8). First, it may synapse with a postganglionic neuron in the sympathetic chain at that level. Second, the axon

may travel upward or downward through the sympathetic chain to synapse with a postganglionic neuron in a paravertebral ganglion at another level. In either of these two instances, the postganglionic axons then reenter the ventral or dorsal ramus of a spinal nerve via a **gray ramus communicans** and travel in the ramus to innervate skin structures (sweat glands, arrector pili muscles attached to hair follicles, and the smooth muscles of blood vessel walls). Third, the axon may pass through the ganglion without synapsing and form part of the **thoracic, lumbar,** and **sacral splanchnic nerves,** which travel to the viscera to synapse with a postganglionic neuron in a **prevertebral** or **collateral ganglion.** The major prevertebral ganglia—the *celiac, superior mesenteric, inferior mesenteric,* and *hypogastric ganglia*—supply the abdominal and pelvic visceral organs. The postganglionic axon then leaves the ganglion and travels to a nearby visceral organ which it innervates.

 Locate the sympathetic chain on the spinal nerve chart.

Autonomic Functioning

As noted earlier, most body organs served by the autonomic nervous system receive fibers from both the sympathetic and parasympathetic divisions. The only exceptions are the structures of the skin (sweat glands and arrector pili muscles attached to the hair follicles), the pancreas and liver, the adrenal medulla, and essentially all blood vessels except those of the external genitalia, all of which receive sympathetic innervation only. When both divisions serve an organ, they have antagonistic effects. This is because their postganglionic axons release different neurotransmitters. The parasympathetic fibers, called **cholinergic fibers,** release acetylcholine; the sympathetic postganglionic fibers, called **adrenergic fibers,** release norepinephrine. (However, there are isolated examples of postganglionic sympathetic fibers, such as those serving blood vessels in the skeletal muscles, that release acetylcholine.) The preganglionic fibers of both divisions release acetylcholine.

The parasympathetic division is often referred to as the housekeeping, or "resting and digesting," system because it maintains the visceral organs in a state most suitable for normal functions and internal homeostasis; that is, it promotes normal digestion and elimination. In contrast, activation of the sympathetic division is referred to as the "fight or flight" response because it readies the body to cope with situations that threaten homeostasis. Under such emergency conditions, the sympathetic nervous system induces an increase in heart rate and blood pressure, dilates the bronchioles of the lungs, increases blood sugar levels, and promotes many other effects that help the individual cope with a stressor.

As we grow older, our sympathetic nervous system gradually becomes less and less efficient, particularly in causing vasoconstriction of blood vessels. When elderly people stand up quickly after sitting or lying down, they often become light-headed or faint. This is because the sympathetic nervous system is not able to react quickly enough to counteract the pull of gravity by activating the vasoconstrictor fibers. So, blood pools in the feet. This condition, **orthostatic hypotension,** is a type of low blood pressure resulting from changes in body position as described. Orthostatic hypotension can be prevented to some degree if *slow* changes in position are made. This gives the sympathetic nervous system a little more time to react and adjust. ■

 Several body organs are listed in the chart below. *Using your textbook as a reference,* list the effect of the sympathetic and parasympathetic divisions on each.

Organ or function	Parasympathetic effect	Sympathetic effect
Heart		
Bronchioles of lungs		
Digestive tract activity		
Urinary bladder		
Iris of the eye		
Blood vessels (most)		
Penis/clitoris		
Sweat glands		
Adrenal medulla		
Pancreas		

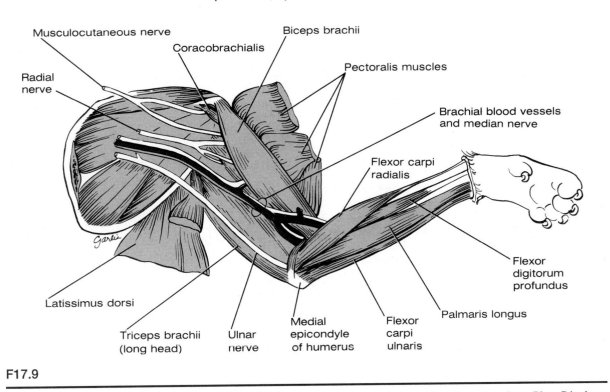

F17.9

Brachial plexus and major blood vessels of the left forelimb of the cat (ventral aspect). See dissection photo, Plate C in the Cat Anatomy Atlas.

DISSECTION OF CAT SPINAL NERVES

The cat has 38 or 39 pairs of spinal nerves (as compared to 31 in humans). Of these, 8 are cervical, 13 thoracic, 7 lumbar, 3 sacral, and 7 or 8 caudal.

A complete dissection of the cat's spinal nerves would be extraordinarily time-consuming and exacting, and is not warranted in a basic anatomy course. However, it is desirable for you to have some dissection work to complement your study of the anatomical charts. Thus at this point you will carry out a partial dissection of the brachial plexus and lumbosacral plexus and identify some of the major nerves.

Nerves of the Brachial Plexus

1. Place your cat specimen on the dissecting tray, dorsal side down. Reflect the cut ends of the left pectoralis muscles to expose the large brachial plexus in the axillary region (Figure 17.9 and Plate C in the Cat Anatomy Atlas). Carefully clean the exposed nerves as far back toward their points of origin as possible.

2. The *musculocutaneous nerve* is the most superior nerve of this group. It splits into two subdivisions that run under the margins of the coracobrachialis and biceps brachii muscles. Trace its fibers into the ventral muscles of the forelimb it serves.

3. Locate the large *radial nerve* inferior to the musculocutaneous nerve. The radial nerve serves the dorsal muscles of the arm and forearm. Follow it into the three heads of the triceps brachii muscle.

4. In the cat, the *median nerve* is closely associated with the brachial artery and vein. It courses through the arm to supply the ventral muscles of the forearm (with the exception of the flexor carpi ulnaris and the ulnar head of the flexor digitorum profundus). It also innervates some of the intrinsic hand muscles, as in humans.

5. The *ulnar nerve* is the most posterior large brachial plexus nerve. Follow it as it travels down the forelimb, passing over the medial epicondyle of the humerus, to supply the flexor carpi ulnaris and the ulnar head of the flexor digitorum profundus (and the hand muscles).

Nerves of the Lumbosacral Plexus

1. To locate the *femoral nerve* arising from the lumbar plexus, first identify the right femoral triangle, which is bordered by the sartorius and adductor muscles of the anterior thigh (Figure 17.10). The large femoral nerve travels through this region after emerging from the psoas major muscle in close association with the femoral artery and vein. Follow the nerve into the muscles and skin of the anterior thigh, which it supplies. Note also its cutaneous branch in the cat, the saphenous nerve, which continues down the anterior medial surface of the thigh (with the greater saphenous artery and vein) to supply the skin of the anterior shank and foot.

2. Turn the cat ventral side down so you can view the posterior aspect of the lower limb (Figure 17.11, p. 194). Reflect the ends of the transected biceps femoris muscle to view the large cordlike sciatic nerve. The *sciatic nerve* arises from the sacral plexus and serves the dorsal thigh muscles and all the muscles of the hindleg and foot. Follow the nerve as it travels down the posterior thigh lateral to the semimembranosus muscle. Note that just superior to the gastrocnemius muscle of the calf, it divides into its two major branches, which serve the leg.

3. Identify the *tibial nerve* medially and the *common peroneal nerve,* which curves over the lateral surface of the gastrocnemius.

4. When you have finished making your observations, wrap the cat for storage and clean all dissecting tools and equipment before leaving the laboratory.

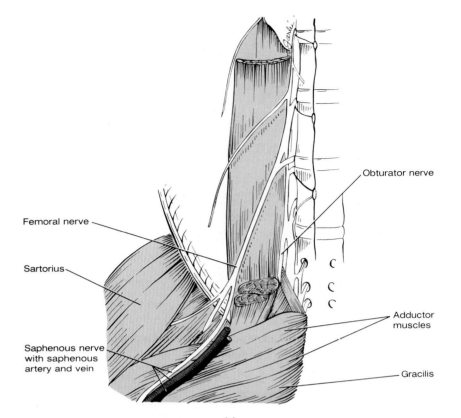

Femoral nerve

Sartorius

Saphenous nerve
with saphenous
artery and vein

Obturator nerve

Adductor
muscles

Gracilis

(a)

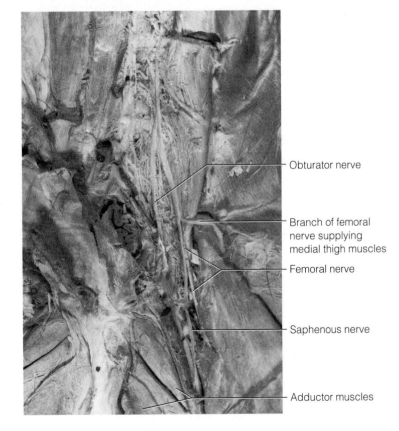

Obturator nerve

Branch of femoral
nerve supplying
medial thigh muscles

Femoral nerve

Saphenous nerve

Adductor muscles

(b)

F17.10

Lumbar plexus of the cat. Ventral views: **(a)** Diagrammatic view. **(b)** Photograph.

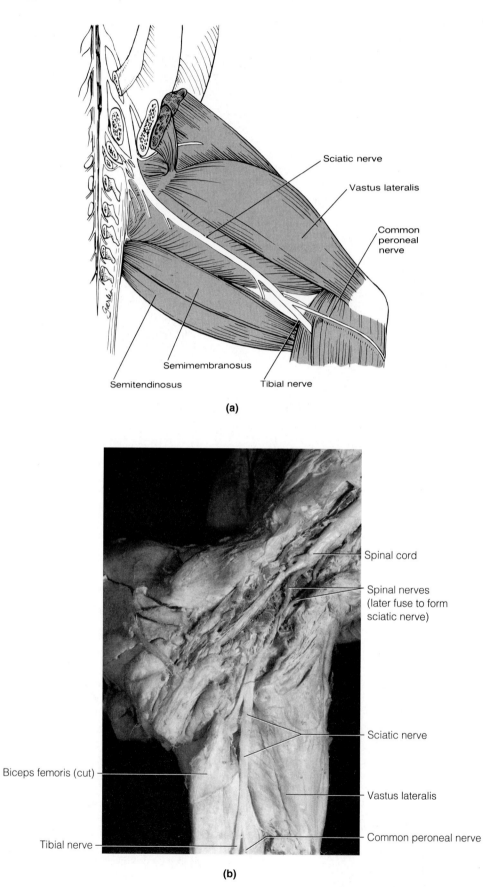

(a)

(b)

F17.11

Sacral plexus of the cat. Dorsal views: **(a)** Diagrammatic view. **(b)** Photograph.

Special Senses: Vision

OBJECTIVES

1. To describe the structure and function of the accessory visual structures.

2. To identify the structural components of the eye when provided with a model, an appropriate diagram, or a preserved sheep or cow eye, and list the function(s) of each.

3. To describe the cellular makeup of the retina.

4. To discuss the mechanism of image formation on the retina.

5. To trace the visual pathway to the optic cortex and note the effects of damage to various parts of this pathway.

6. To define the following terms:

   refraction   myopia
   accommodation   hyperopia
   convergence   cataract
   astigmatism   glaucoma
   emmetropia   conjunctivitis

7. To discuss the importance of the pupillary and convergence reflexes.

8. To explain the difference between rods and cones with respect to visual perception.

9. To state the importance of an ophthalmoscopic examination.

MATERIALS

Dissectible eye model
Chart of eye anatomy
Preserved cow or sheep eye
Dissecting pan and instruments
Protective skin cream or disposable gloves

Histologic section of an eye showing retinal layers
Compound microscope

Snellen eye chart (floor marked with chalk to indicate
   20-ft distance from posted Snellen chart)
Ishihara's color-blindness plates
Ophthalmoscope
Metric ruler; meter stick
Common straight pins

See Appendix B, Exercise 18 for links to
A.D.A.M. Standard.

See Appendix C, Exercise 18 for links to
Anatomy and PhysioShow: The Videodisc.

ANATOMY OF THE EYE

External Anatomy and Accessory Structures

The adult human eye is a sphere measuring about 2.5 cm (1 inch) in diameter. Only about one-sixth of the eye's anterior surface is observable; the remainder is enclosed and protected by a cushion of fat and the walls of the bony orbit.

Six **extrinsic eye muscles** attached to the exterior surface of each eyeball control eye movement and make it possible for the eye to follow a moving object. The names and positioning of these extrinsic muscles are noted in Figure 18.1. Their actions are given in the chart accompanying that figure.

The anterior surface of each eye is protected by the **eyelids,** or **palpebrae.** (See Figure 18.2.) The medial and lateral junctions of the upper and lower eyelids are referred to as the **medial** and **lateral canthus** (respectively). The **caruncle,** a fleshy elevation at the medial canthus, produces a whitish oily secretion. A mucous membrane, the **conjunctiva,** lines the internal surface of the eyelids (as the *palpebral conjunctiva*) and continues over the anterior surface of the eyeball to its junction with the corneal epithelium (as the *ocular,* or *bulbar, conjunctiva*). The conjunctiva secretes mucus, which aids in lubricating the eyeball. Inflammation of the conjunctiva, often accompanied by redness of the eye, is called **conjunctivitis.**

Name	Innervation (cranial nerve)	Action
Lateral rectus	VI	Moves eye horizontally (laterally)
Medial rectus	III	Moves eye horizontally (medially)
Superior rectus	III	Elevates eye
Inferior oblique	III	Elevates eye and turns it laterally
Inferior rectus	III	Depresses eye
Superior oblique	IV	Depresses eye and turns it laterally

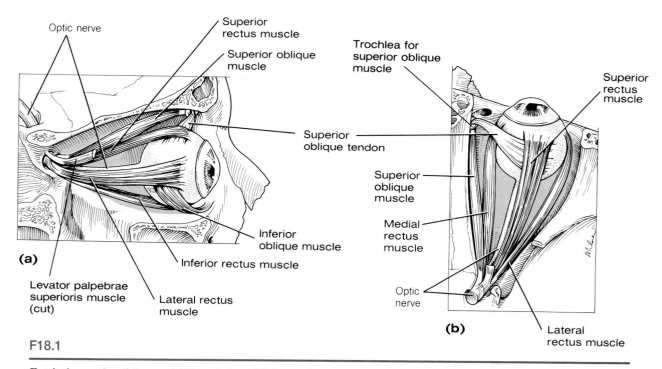

F18.1

Extrinsic muscles of the eye. (**a**) Lateral view of right eye. (**b**) Superior view of right eye.

Projecting from the border of each eyelid is a row of short hairs, the **eyelashes.** The **ciliary glands,** a type of sweat gland, lie between the eyelash hair follicles and help lubricate the eyeball. An inflammation of one of these glands is called a **sty.** Small sebaceous glands associated with the hair follicles and the larger **meibomian glands,** located posterior to the eyelashes, secrete an oily substance.

The **lacrimal apparatus** consists of the **lacrimal gland, lacrimal canals, lacrimal sac,** and the **nasolacrimal duct.** The lacrimal glands are situated superior to the lateral aspect of each eye. They continually liberate a dilute salt solution (tears) that flows onto the anterior surface of the eyeball through several small ducts. The tears flush across the eyeball into the lacrimal canals medially, then into the lacrimal sac, and finally into the nasolacrimal duct, which empties into the nasal cav-

ity. The lacrimal secretion also contains **lysozyme,** an antibacterial enzyme. Because it constantly flushes the eyeball, the lacrimal fluid cleanses and protects the eye surface as it moistens and lubricates it. As we age, our eyes tend to become dry due to decreased lacrimation, and thus are more vulnerable to bacterial invasion and irritation.

Observe the eyes of another student and identify as many of the accessory structures as possible. Ask the student to look to the left. What extrinsic eye muscles are responsible for this action?

Right eye _____

Left eye _____

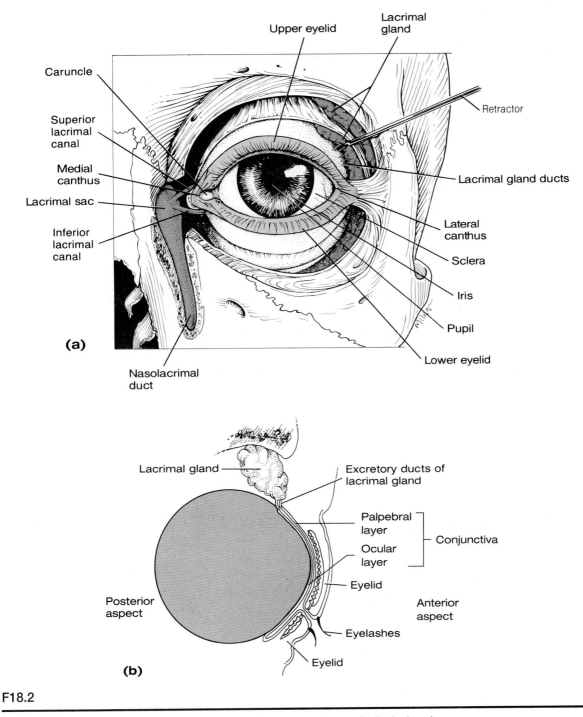

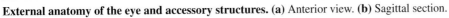

External anatomy of the eye and accessory structures. (**a**) Anterior view. (**b**) Sagittal section.

Internal Anatomy of the Eye

Obtain a dissectible eye model and identify its internal structures as they are described below. As you work, also refer to Figure 18.3.

Anatomically, the wall of the eye is constructed of three tunics, or coats. The outermost **fibrous tunic** is a protective layer composed of dense avascular connective tissue. It has two obviously different regions: The opaque white **sclera** forms the bulk of the fibrous tunic and is observable anteriorly as the "white of the eye." Its anteriormost portion is modified structurally to form the transparent **cornea,** through which light enters the eye.

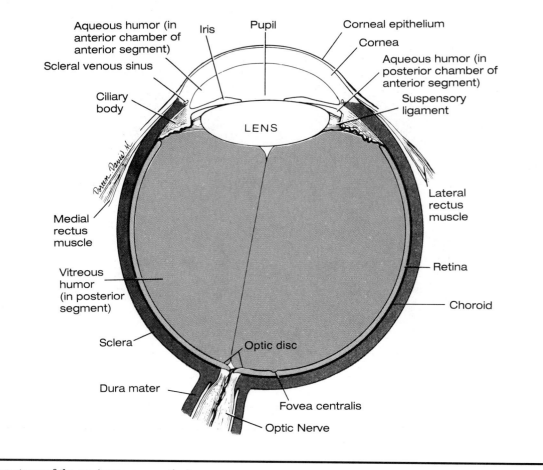

Aqueous humor (in anterior chamber of anterior segment)

Iris

Pupil

Corneal epithelium

Cornea

Aqueous humor (in posterior chamber of anterior segment)

Scleral venous sinus

Ciliary body

Suspensory ligament

LENS

Medial rectus muscle

Lateral rectus muscle

Vitreous humor (in posterior segment)

Retina

Choroid

Sclera

Optic disc

Dura mater

Fovea centralis

Optic Nerve

F18.3

Internal anatomy of the eye (transverse section).

The middle tunic, called the **uvea,** is the **vascular tunic.** Its posteriormost part, the **choroid,** is a richly vascular nutritive layer that contains a dark pigment that prevents light scattering within the eye. Anteriorly, the choroid is modified to form the **ciliary body,** to which the lens is attached, and then the pigmented **iris.** The iris is incomplete, resulting in a rounded opening, the **pupil,** through which light passes.

The iris is composed of circularly and radially arranged smooth muscle fibers and acts as a reflexively activated diaphragm to regulate the amount of light entering the eye. In close vision and bright light, the circular muscles of the iris contract, and the pupil constricts. In distant vision and in dim light, the radial fibers contract, enlarging (dilating) the pupil and allowing more light to enter the eye.

The innermost **sensory tunic** of the eye is the delicate, two-layered **retina.** The outer **pigmented epithelial layer** abuts and lines the entire uvea. The transparent inner **neural (nervous) layer** extends anteriorly only to the ciliary body. It contains the photoreceptors, **rods** and **cones,** which begin the chain of electrical events that ultimately result in the transduction of light energy into nerve impulses that are transmitted to the optic cortex of the brain. Vision is the result. The photoreceptor cells are distributed over the entire neural retina, except where the optic nerve leaves the eyeball. This site is called the **optic disc,** or blind spot. Lateral to

each blind spot, and directly posterior to the lens, is an area called the **macula lutea** (yellow spot), an area of high cone density. In its center is the **fovea centralis,** a minute pit about ½ mm in diameter, which contains only cones and is the area of greatest visual acuity. Focusing for discriminative vision occurs in the fovea centralis.

Light entering the eye is focused on the retina by the **lens,** a flexible crystalline structure held vertically in the eye's interior by the **suspensory ligament** attached to the ciliary body. Activity of the ciliary muscle, which accounts for the bulk of ciliary body tissue, changes lens thickness to allow light to be properly focused on the retina.

In the elderly the lens becomes increasingly hard and opaque. **Cataracts,** which often result from this process, cause vision to become hazy or entirely obstructed. ■

The lens divides the eye into two segments: the **anterior segment** anterior to the lens, which contains a clear watery fluid called the **aqueous humor,** and the **posterior segment** behind the lens, filled with a gel-like substance, the **vitreous humor,** or **vitreous body.** The anterior segment is further divided into **anterior** and **posterior chambers,** located before and after the iris, respectively. The aqueous humor is continually formed by the capillaries of the **ciliary processes** of the ciliary

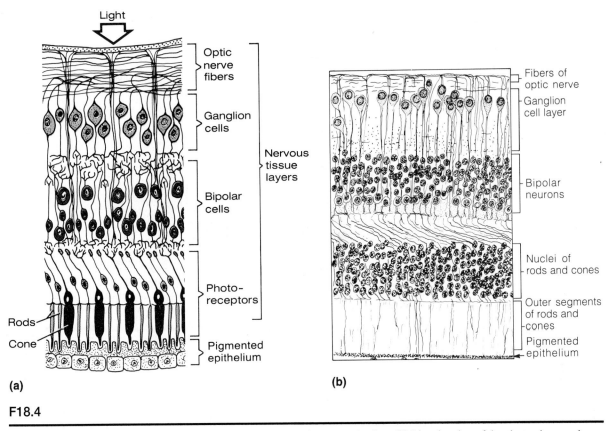

Light

Optic nerve fibers

Ganglion cells

Nervous tissue layers

Bipolar cells

Photo-receptors

Rods

Cone

Pigmented epithelium

(a)

Fibers of optic nerve

Ganglion cell layer

Bipolar neurons

Nuclei of rods and cones

Outer segments of rods and cones

Pigmented epithelium

(b)

F18.4

Microscopic anatomy of the cellular layers of the retina. (a) Diagrammatic view. (b) Line drawing of the photomicrograph provided in Plate 13 of the Histology Atlas.

body. It helps to maintain the intraocular pressure of the eye and provides nutrients for the avascular lens and cornea. The aqueous humor is reabsorbed into the **scleral venous sinus (canal of Schlemm).** The vitreous humor provides the major internal reinforcement of the posterior part of the eyeball, and helps to keep the neural layer of the retina pressed firmly against the wall of the eyeball. It is formed *only* before birth.

Anything that interferes with drainage of the aqueous fluid increases intraocular pressure. When intraocular pressure reaches dangerously high levels, the retina and optic nerve are compressed, resulting in pain and possible blindness, a condition called **glaucoma.** ■

MICROSCOPIC ANATOMY OF THE RETINA

As described above, the retina consists of two main types of cells: a pigmented *epithelial* layer, which abuts the choroid, and an inner cell layer composed of *neurons,* which is in contact with the vitreous humor (Figure 18.4). The inner nervous layer is composed of three major neuronal populations. These are, from outer to inner aspect, the **photoreceptors** (rods and cones), the **bipolar cells,** and the **ganglion cells.**

The **rods** are the specialized receptors for dim light. Visual interpretation of their activity is in gray tones. The **cones** are color receptors that permit high levels of visual acuity, but they function only under conditions of high light intensity; thus, for example, no color vision is possible in moonlight. Only cones are found in the fovea centralis, and their number decreases as the retinal periphery is approached. By contrast, rods are most numerous in the periphery, and their density decreases as the macula is approached.

Light must pass through the ganglion cell layer and the bipolar neuron layer to reach and excite the rods and cones. As a result of a light stimulus, the photoreceptors undergo changes in their membrane potential that ultimately influence the bipolar neurons. These in turn stimulate the ganglion cells, whose axons leave the retina in the tight bundle of fibers known as the optic nerve. The retinal layer is thickest where the optic nerve attaches to the eyeball because an increasing number of ganglion cell axons converge at this point. It thins as it approaches the ciliary body.

Obtain a histologic slide of a longitudinal section of the eye. Identify the retinal layers by comparing it to Figure 18.4.

VISUAL PATHWAYS TO THE BRAIN

The axons of the ganglion cells of the retina converge at the posterior aspect of the eyeball and exit from the eye as the optic nerve. At the **optic chiasma,** the fibers from the medial side of each eye cross over to the opposite side. The fiber tracts thus formed are called the **optic tracts.** Each optic tract contains fibers from the lateral side of the eye on the same side and from the medial side of the opposite eye.

The optic tract fibers synapse with neurons in the **lateral geniculate nucleus** of the thalamus, whose axons form the **optic radiation,** terminating in the **optic,** or **visual, cortex** in the occipital lobe of the brain. Here they synapse with the cortical cells, and visual interpretation occurs.

After examining Figure 18.5, determine what effects lesions in the following areas would have on vision:

In the right optic nerve _____

_____

Through the optic chiasma _____

_____

In the left optic tract _____

_____

In the right cerebral cortex (visual area) _____

_____

DISSECTION OF THE COW (SHEEP) EYE

1.   Obtain a preserved cow or sheep eye, dissecting instruments, and a dissecting pan. Apply protective skin cream or don disposable gloves if desired.

2.   Examine the external surface of the eye, noting the thick cushion of adipose tissue. Identify the optic nerve (cranial nerve II) as it leaves the eyeball, the remnants of the extrinsic eye muscles, the conjunctiva, the sclera, and the cornea. The normally transparent cornea is opalescent or opaque if the eye has been preserved. Refer to Figure 18.6 as you work.

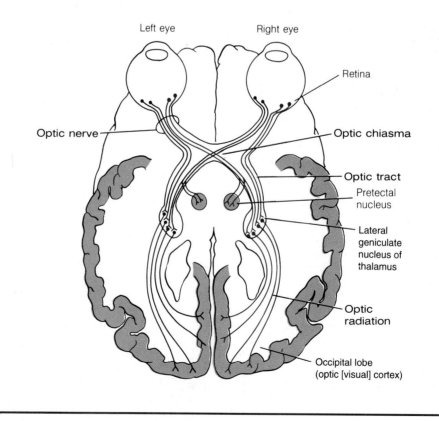

F18.5

Visual pathway to the brain. (Note that fibers from the lateral portion of each retinal field do not cross at the optic chiasma.)

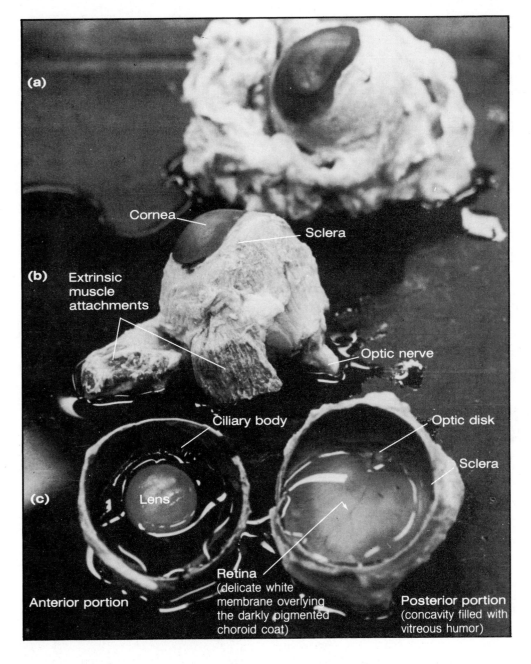

(a)

Cornea — Sclera

(b) Extrinsic muscle attachments

Optic nerve

Ciliary body — Optic disk

Sclera

(c) Lens

Retina (delicate white membrane overlying the darkly pigmented choroid coat)

Anterior portion

Posterior portion (concavity filled with vitreous humor)

F18.6

Anatomy of the cow eye. (a) Cow eye (entire) removed from orbit (notice the large amont of fat cushioning the eyeball). **(b)** Cow eye (entire) with fat removed to show the extrinsic muscle attachments and optic nerve. **(c)** Cow eye cut along the coronal plane to reveal internal structures.

3.   Trim away most of the fat and connective tissue, but leave the optic nerve intact. Holding the eye with the cornea facing downward, carefully make an incision with a sharp scalpel into the sclera about ¼ inch above the cornea. (The sclera of the preserved eyeball is *very* tough so you will have to apply substantial pressure to penetrate it.) Using scissors, complete the incision around the circumference of the eyeball paralleling the corneal edge.

4.   Carefully lift the anterior part of the eyeball away from the posterior portion. Conditions being proper, the vitreous body should remain with the posterior part of the eyeball.

5.   Examine the anterior part of the eye and identify the following structures:

Ciliary body: black pigmented body that appears to be a halo encircling the lens.

Lens: biconvex structure that is opaque in preserved specimens.

Suspensory ligament: a halo of delicate fibers attaching the lens to the ciliary body.

Carefully remove the lens and identify the adjacent structures:

Iris: anterior continuation of the ciliary body penetrated by the pupil.

Cornea: more convex anteriormost portion of the sclera; normally transparent but cloudy in preserved specimens.

6. Examine the posterior portion of the eyeball. Remove the vitreous humor, and identify the following structures:

Retina: the neural layer of the retina appears as a delicate white, probably crumpled membrane that separates easily from the pigmented choroid.

Note its point of attachment. What is this point called?

_____

Pigmented choroid coat: appears iridescent in the cow or sheep eye owing to a special reflecting surface called the **tapetum lucidum.** This specialized surface reflects the light within the eye and is found in the eyes of animals that live under conditions of low-intensity light. It is not found in humans.

OPHTHALMOSCOPIC EXAMINATION OF THE EYE (OPTIONAL)

The ophthalmoscope is an instrument used to examine the *fundus,* or eyeball interior, to determine visually the condition of the retina, optic disc, and internal blood vessels. Certain pathologic conditions such as diabetes mellitus, arteriosclerosis, and degenerative changes of the optic nerve and retina can be detected by such an examination. The ophthalmoscope consists of a set of lenses mounted on a rotating disc (the **lens selection disc**), a light source regulated by a **rheostat control,** and a mirror that reflects the light so that the eye interior can be illuminated (Figure 18.7a).

The lens selection disc is positioned in a small slit in the mirror, and the examiner views the eye interior through this slit, appropriately called the **viewing window.** The focal length of each lens is indicated in diopters preceded by a + sign if the lens is convex and by a − sign if the lens is concave. When the zero (0) is seen in the **diopter window,** there is no lens positioned in the slit. The depth of focus for viewing the eye interior is changed by changing the lens.

The light is turned on by depressing the red **rheostat lock button** and then rotating the rheostat control in the clockwise direction. The aperture selection disc on the front of the instrument allows the nature of the light beam to be altered. Generally, green light allows for clearest viewing of the blood vessels in the eye interior and is most comfortable for the subject.

Once you have examined the ophthalmoscope and have become familiar with it, you are ready to conduct an eye examination.

1. Conduct the examination in a dimly lit or darkened room with the subject comfortably seated and gazing straight ahead. To examine the right eye, sit face-to-face with the subject, and steady yourself by resting your left hand on the subject's head as shown in Figure 18.7b. Hold the instrument in your right hand, and use your right eye to view the eye interior. To view the left eye, use your left eye, hold the instrument in your left hand, and steady yourself with your right hand.

2. Begin the examination with the 0 (no lens) in position. Grasp the instrument so that the lens disc may be rotated with the index finger. Holding the ophthalmoscope about 6 inches from the subject's eye, direct the light into the pupil at a slight angle—through the pupil edge rather than directly through its center. You will see a red circular area that is the illuminated eye interior.

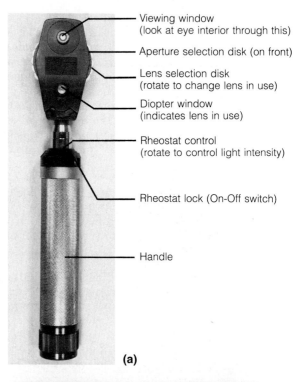

Viewing window
(look at eye interior through this)

Aperture selection disk (on front)

Lens selection disk
(rotate to change lens in use)

Diopter window
(indicates lens in use)

Rheostat control
(rotate to control light intensity)

Rheostat lock (On-Off switch)

Handle

(a)

(b)

F18.7

Structure and use of an ophthalmoscope. (a) Structure of an ophthalmoscope. **(b)** Proper position for examining the right eye with an ophthalmoscope.

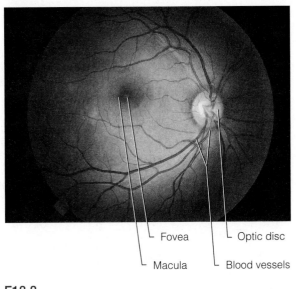

Fovea — Optic disc

Macula — Blood vessels

F18.8

Posterior portion of right retina. Photograph taken with slit-lamp camera.

3. Move in as close as possible to the subject's cornea (to within 2 in.) as you continue to observe the area. Steady your instrument-holding hand on the subject's cheek if necessary (see Figure 18.7b). If both your eye and that of the subject are normal, the fundus can be viewed clearly without further adjustment of the ophthalmoscope. If the fundus cannot be focused, slowly rotate the lens disc counterclockwise until the fundus can be clearly seen. When the ophthalmoscope is correctly set, the fundus of the right eye should appear as shown in Figure 18.8. (Note: If a positive [convex] lens is required and your eyes are normal, the subject has hyperopia. If a negative [concave] lens is necessary to view the fundus and your eyes are normal, the subject is myopic.)

When the examination is proceeding correctly, the subject can often see images of retinal vessels in his own eye that appear rather like cracked glass. If you are unable to achieve a sharp focus or to see the optic disc, move medially or laterally and begin again.

4. Examine the optic disc for color, elevation, and sharpness of outline, and observe the blood vessels radiating from near its center. Locate the macula, lateral to the optic disc. It is a darker area in which blood vessels are absent, and the fovea appears to be a slightly lighter area in its center. The macula is most easily seen when the subject looks directly into the light of the ophthalmoscope.

⚠️ Do not examine the macula for longer than 1 second at a time.

5. When you have finished examining your partner's retina, shut off the ophthalmoscope. Change places with your partner (become the subject) and repeat steps 1–4.

X ●

F18.9

Blind spot test figure.

☯ SELECTED VISUAL TESTS

Demonstration of the Blind Spot

1. Hold Figure 18.9 about 18 inches from your eyes. Close your left eye, and focus your right eye on the X, which should be positioned so that it is directly in line with your right eye. Move the figure slowly toward your face, keeping your right eye focused on the X. When the dot focuses on the blind spot, which lacks photoreceptors, it will disappear.

2. Have your laboratory partner record in metric units the distance at which this occurs. The dot will reappear as the figure is moved closer. Distance at which the dot disappears:

Right eye _____

Repeat the test for the left eye, this time closing the right eye and focusing the left eye on the dot. Record the distance at which the X disappears:

Left eye _____

Refraction, Tests for Visual Acuity, and Astigmatism

When light rays pass from one medium to another, their velocity changes, and the rays are bent or *refracted*. Thus the light rays in the visual field are refracted as they encounter the cornea, lens, and vitreous humor of the eye.

The refractive index (bending power) of the cornea and vitreous humor are constant. But the lens's refractive index, or strength, can be varied by changing the lens's shape—that is, by making it more or less convex so that the light is properly converged and focused on the retina. The greater the lens convexity, or bulge, the more the light will be bent and the stronger the lens. Conversely, the less the lens convexity (the flatter it is), the less it bends the light.

In general, light from a distant source (over 20 feet) approaches the eye as parallel rays, and no change in lens convexity is necessary for it to focus properly on the retina. However, light from a close source tends to diverge, and the convexity of the lens must increase to make close vision possible. To achieve this, the ciliary

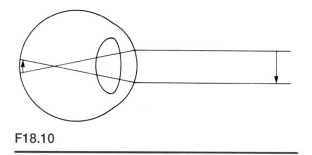

F18.10

Refraction of light in the eye, resulting in the production of a real image on the retina.

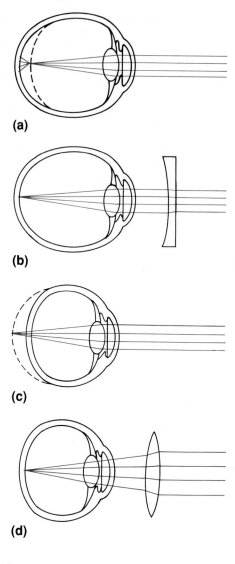

(a)

(b)

(c)

(d)

F18.11

Common refraction problems and their correction. In myopia: **(a)** light from a distant object focuses in front of the retina, and **(b)** correction involves the use of a concave lens that diverges the light before it enters the eye. In hyperopia: **(c)** light from a distant object focuses behind the retina, and **(d)** correction requires a convex lens that converges the light rays before they enter the eye.

muscle contracts, decreasing the tension on the suspensory ligament attached to the lens and allowing the elastic lens to "round up." Thus, a lens capable of bringing a *close* object into sharp focus is stronger (more convex) than a lens focusing on a more distant object. The ability of the eye to focus differentially for objects of near vision (less than 20 feet) is called **accommodation.** It should be noted that the image formed on the retina as a result of the refractory activity of the lens (see Figure 18.10) is a **real image** (reversed from left to right, inverted, and smaller than the object).

The normal or **emmetropic eye** is able to accommodate properly. However, visual problems may result from (1) lenses that are too strong or too "lazy" (overconverging and underconverging, respectively), (2) from structural problems such as an eyeball that is too long or too short to provide for proper focusing by the lens, or (3) a cornea or lens with improper curvatures.

▲ Individuals in whom the image normally focuses in front of the retina are said to have **myopia,** or "nearsightedness" (Figure 18.11a); they can see close objects without difficulty, but distant objects are blurred or seen indistinctly. Correction requires a concave lens, which causes the light reaching the eye to diverge (Figure 18.11b).

If the image focuses behind the retina, the individual is said to have **hyperopia** or farsightedness. Such persons have no problems with distant vision but need glasses with convex lenses to augment the converging power of the lens for close vision (Figure 18.11c and d).

Irregularities in the curvatures of the lens and/or the cornea lead to a blurred vision problem called **astigmatism.** Cylindrically ground lenses, which compensate for inequalities in the curvatures of the refracting surfaces, are prescribed to correct the condition. ■

TEST FOR NEAR-POINT ACCOMMODATION

The elasticity of the lens decreases dramatically with age, resulting in difficulty in focusing for near or close vision. This condition is called **presbyopia**—literally, old vision. Lens elasticity can be tested by measuring the **near point of accommodation.** The near point of vision is about 10 cm from the eye in young adults. It is closer in children and farther in old age.

 To determine your near point of accommodation, hold a common straight pin at arm's length in front of one eye. Slowly move the pin toward that eye until the pin image becomes distorted. Have your lab partner measure the distance from your eye to the pin at this point, and record the distance below. Repeat the procedure for the other eye.

Near point for right eye _____

Near point for left eye _____

TEST FOR VISUAL ACUITY   **Visual acuity,** or sharpness of vision, is generally tested with a Snellen eye chart, which consists of letters of various sizes printed on a white card. This test is based on the fact that letters of a certain size can be seen clearly by eyes with normal vision at a specific distance. The distance at which the normal, or emmetropic, eye can read a line of letters is printed at the end of that line.

1.   Have your partner stand 20 feet from the posted Snellen eye chart and cover one eye with a card or hand. As your partner reads each consecutive line aloud, check for accuracy. If this individual wears glasses, give the test twice—first with glasses off and then with glasses on.

2.   Record the number of the line with the smallest-sized letters read. If it is 20/20, the person's vision for that eye is normal. If it is 20/40, or any ratio with a value less than one, he or she has less than the normal visual acuity. (Such an individual is myopic.) If the visual acuity is 20/15, vision is better than normal, because this person can stand at 20 feet from the chart and read letters that are only discernible by the normal eye at 15 feet. Give your partner the number of the line corresponding to the smallest letters read, to record in step 4.

3.   Repeat the process for the other eye.

4.   Have your partner test and record your visual acuity. If you wear glasses, the test results *without* glasses should be recorded first.

Visual acuity, right eye _____

Visual acuity, left eye _____

TEST FOR ASTIGMATISM   The astigmatism chart (Figure 18.12) is designed to test for defects in the refracting surface of the lens and/or cornea.

View the chart first with one eye and then with the other, focusing on the center of the chart. If all the radiating lines appear equally dark and distinct, there is no distortion of your refracting surfaces. If some of the lines are blurred or appear less dark than others, at least some degree of astigmatism is present.

Is astigmatism present in your left eye? _____

Right eye? _____

Test for Color Blindness

Ishihara's color plates are designed to test for deficiencies in the cones or color photoreceptor cells. There are three cone types, each containing a different light-absorbing pigment. One type primarily absorbs the red

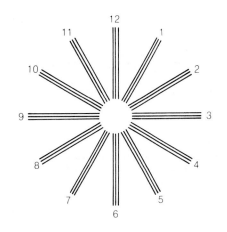

F18.12

Astigmatism testing chart.

wavelengths of the visible light spectrum, another the blue wavelengths, and a third the green wavelengths. Nerve impulses reaching the brain from these different photoreceptor types are then interpreted (seen) as red, blue, and green, respectively.

The interpretation of the intermediate colors of the visible light spectrum is a result of overlapping input from more than one cone type.

1.   View the various color plates in bright light or sunlight while holding them about 30 inches away and at right angles to your line of vision. Report to your laboratory partner what you see in each plate. (Take no more than 3 seconds for each decision.)

2.   Your partner is to write down your responses and then check their accuracy with the correct answers provided in the color plate book. Is there any indication that you have some degree of color blindness?

_____ If so, what type? _____

_____

Repeat the procedure to test your partner's color vision.

Special Senses: Hearing and Equilibrium

| OBJECTIVES |

1. To identify the anatomical structures of the outer, middle, and inner ear by appropriately labeling a diagram.

2. To describe the anatomy of the organ of hearing (organ of Corti in the cochlea) and explain its function in sound reception.

3. To describe the anatomy of the equilibrium organs of the inner ear (cristae ampullares and maculae), and to explain their relative function in maintaining equilibrium.

4. To define or explain *central deafness, conduction deafness,* and *nystagmus.*

5. To state the purpose of the Weber, Rinne, and Romberg tests.

6. To explain how one is able to localize the source of sounds.

7. To describe the effects of acceleration on the semicircular canals.

8. To explain the role of vision in maintaining equilibrium.

| MATERIALS |

Three-dimensional dissectible ear model and/or chart of ear anatomy
Histologic slides of the cochlea of the ear
Compound microscope

Demonstration: Microscope focused on a crista ampullaris receptor of a semicircular canal

Pocket watch or clock that ticks
Tuning forks (range of frequencies)
Rubber mallet
Absorbent cotton
Otoscope (if available)
Alcohol swabs
12-inch ruler

See Appendix C, Exercise 19 for links to *Anatomy and PhysioShow: The Videodisc.*

ANATOMY OF THE EAR

Gross Anatomy

The ear is a complex structure containing sensory receptors for hearing and equilibrium. The ear is divided into three major areas: the *outer ear,* the *middle ear,* and the *inner ear* (Figure 19.1). The outer and middle ear structures serve the needs of the sense of hearing *only,* while inner ear structures function both in equilibrium and hearing reception.

Obtain a dissectible ear model and identify the structures described below. Refer to Figure 19.1 as you work.

The **outer,** or **external, ear** is composed primarily of the **pinna,** or **auricle,** and the **external auditory canal.** The pinna is the skin-covered cartilaginous structure encircling the auditory canal opening. In many animals, it collects and directs sound waves into the auditory canal. In humans this function of the pinna is largely lost.

The external auditory canal is a short, narrow (about 1 inch long by ¼ inch wide) chamber carved into the temporal bone. In its skin-lined walls are wax-secreting glands called **ceruminous glands.** The sound waves that enter the external auditory canal eventually encounter the **tympanic membrane,** or **eardrum,** which vibrates at exactly the same frequency as the sound wave(s) hitting it. The membranous eardrum separates the outer from the middle ear.

The **middle ear** is essentially a small chamber—the **tympanic cavity**—found within the temporal bone. The cavity is spanned by three small bones, collectively called the **ossicles** (hammer, anvil, and stirrup),* which articulate to form a lever system that transmits the vibratory motion of the eardrum to the fluids of the inner ear via the **oval window.**

Connecting the middle ear chamber with the nasopharynx is the **pharyngotympanic,** or **auditory, tube.**

---

* The ossicles are often referred to by their Latin names, that is, **malleus, incus,** and **stapes,** respectively.

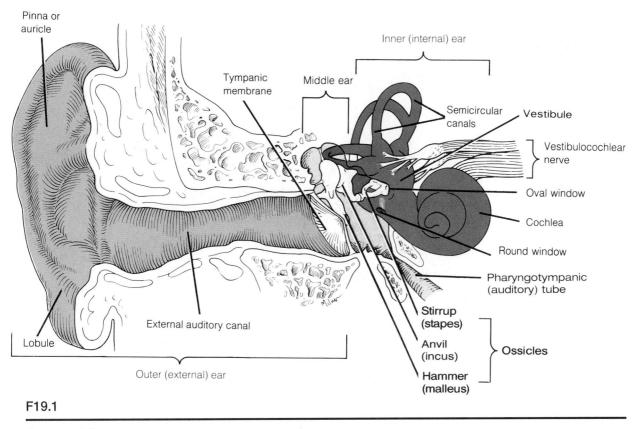

F19.1

Anatomy of the ear.

Normally this tube is flattened and closed, but swallowing or yawning can cause it to open temporarily to equalize the pressure of the middle ear cavity with external air pressure. This is an important function. The eardrum does not vibrate properly unless the pressure on both of its surfaces is the same.

Because the mucosal membranes of the middle ear cavity and nasopharynx are continuous through the pharyngotympanic tube, **otitis media,** or inflammation of the middle ear, is a fairly common condition, especially among youngsters prone to sore throats. In cases where large amounts of fluid or pus accumulate in the middle ear cavity, an emergency *myringotomy* (lancing of the eardrum) may be necessary to relieve the pressure. Frequently, tiny ventilating tubes are put in during the procedure. ∎

The **inner,** or **internal, ear** consists of a system of bony and rather tortuous chambers called the **osseous,** or **bony, labyrinth,** which is filled with an aqueous fluid called **perilymph** (Figure 19.2). Suspended in the perilymph is the **membranous labyrinth,** a system that mostly follows the contours of the osseous labyrinth. The membranous labyrinth is filled with a more viscous fluid called **endolymph.** The three subdivisions of the bony labyrinth are the **cochlea,** the **vestibule,** and the **semicircular canals,** with the vestibule situated between the cochlea and semicircular canals.

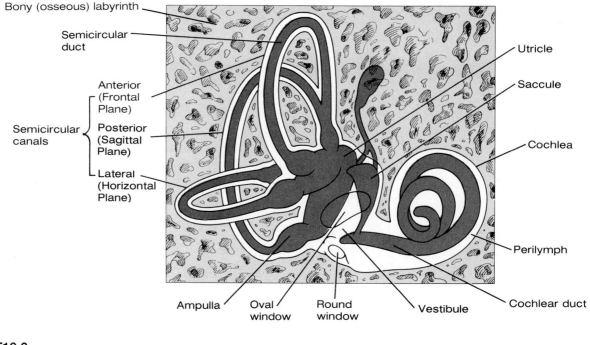

Bony (osseous) labyrinth

Semicircular duct

Anterior (Frontal Plane)

Posterior (Sagittal Plane)

Semicircular canals

Lateral (Horizontal Plane)

Utricle

Saccule

Cochlea

Perilymph

Cochlear duct

Ampulla   Oval window   Round window   Vestibule

F19.2

Inner ear. Right membranous labyrinth shown within the bony labyrinth.

The snail-like cochlea (see Figures 19.2 and 19.3) contains the sensory receptors for hearing. The cochlear membranous labyrinth, the **cochlear duct,** is a soft wormlike tube about 1½ inches long. It winds through the full two and three-quarter turns of the cochlea and separates the perilymph-containing cochlear cavity into upper and lower chambers, the **scala vestibuli** and **scala tympani,** respectively. The scala vestibuli terminates at the oval window, which "seats" the foot plate of the stirrup located laterally in the tympanic cavity. The scala tympani is bounded by a membranous area called the **round window.** The cochlear duct, itself filled with endolymph, supports the **organ of Corti,** which contains the receptors for hearing—the sensory hair cells and nerve endings of the cochlear division of the vestibulocochlear nerve (VIII).

Otoscopic Examination of the Ear (Optional)

1.   Obtain an otoscope and two alcohol swabs. Inspect your partner's ear canal and then select the largest-*diameter* (not length!) speculum that will fit comfortably into his or her ear to permit full visibility. Clean the speculum thoroughly with an alcohol swab, and then attach it to the battery-containing otoscope handle. Before beginning, check that the oto-scope light beam is strong. (If not, obtain another otoscope or new batteries.)

2.   Hold the lighted otoscope securely between your thumb and forefinger (like a pencil), and rest the little finger of the otoscope-holding hand against your partner's head when you are ready to begin the examination. This maneuver forms a brace that allows the speculum to move as your partner moves and prevents the speculum from penetrating too deeply into the ear canal during unexpected movements.

3.   Grasp the ear pinna firmly and pull it up, back, and slightly laterally. If your partner experiences pain or discomfort when the pinna is manipulated, an inflammation or infection of the external ear may be present. If this occurs, do not attempt to examine the ear canal.

4.   Carefully insert the speculum of the otoscope into the external auditory canal in a downward and forward direction only far enough to permit examination of the tympanic membrane or eardrum. Note its shape, color, and vascular network. The healthy tympanic membrane is pearly white. During the examination, notice if there is any discharge or redness in the canal and identify earwax.

5.   After the examination, thoroughly clean the speculum with the second alcohol swab before returning the otoscope to the supply area.

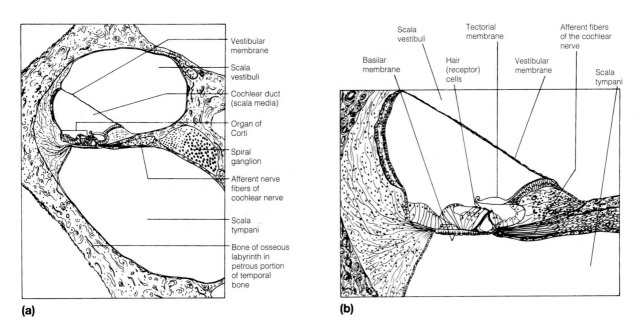

(a)

(b)

F19.3

Organ of Corti. (a) Cross section through one turn of the cochlea showing the position of the organ of Corti. **(b)** Enlarged view of the organ of Corti. Corresponding photomicrographs are Plates 14 and 15 respectively in the Histology Atlas.

Microscopic Anatomy of the Organ of Corti and the Mechanism of Hearing

The anatomical details of the organ of Corti are shown in Figure 19.3. The hair (auditory receptor) cells rest on the **basilar membrane,** which forms the floor of the cochlear duct, and their "hairs" (stereocilia) project into a gelatinous membrane, the **tectorial membrane,** that overlies them. The roof of the cochlear duct is called the **vestibular membrane.** The endolymph-filled chamber of the cochlear duct is the **scala media.**

Obtain a compound microscope and a prepared microscope slide of the cochlea and identify the areas shown in Figure 19.3a and b.

The mechanism of hearing begins as sound waves pass through the external auditory canal and through the middle ear into the inner ear, where the vibration eventually reaches the organ of Corti, which contains the receptors for hearing. Many theories have attempted to explain how the organ of Corti actually responds to sound.

The popular "traveling wave" hypothesis of Von Békésy suggests that vibration of the stirrup at the oval window initiates traveling waves that cause maximal displacements of the basilar membrane where they peak and stimulate the hair cells of the organ of Corti in that region. Since the area at which the traveling waves peak is a high-pressure area, the vestibular membrane is compressed at this point and, in turn, compresses the endolymph and the basilar membrane of the cochlear duct. The resulting pressure on the perilymph in the scala tympani causes the membrane of the round window to

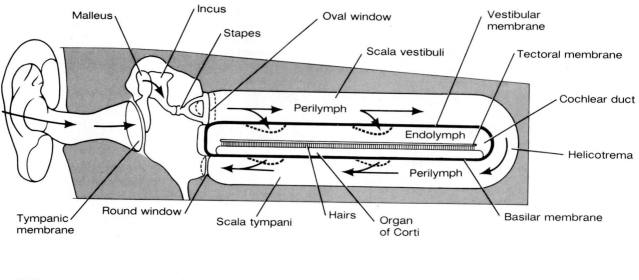

F19.4

Fluid movement in the cochlea following the stirrup thrust at the oval window. Compressional wave created causes the round window to bulge into the middle ear.

bulge outward into the middle ear chamber, thus acting as a relief valve for the compressional wave (Figure 19.4). Von Békésy found that high-frequency waves (high-pitched sounds) peaked close to the oval window and that low-frequency waves (low-pitched sounds) peaked farther up the basilar membrane near the apex of the cochlea. Although the mechanism of sound reception by the organ of Corti is not completely understood, we do know that hair cells on the basilar membrane are uniquely stimulated by sounds of various frequencies and amplitude and that once stimulated they depolarize and begin the chain of nervous impulses to the auditory centers of the temporal lobe cortex. This series of events results in the phenomenon we call hearing.

By the time most people are in their 60s, a gradual deterioration and atrophy of the organ of Corti begins, and leads to a loss in the ability to hear high tones and speech sounds. This condition, **presbycusis,** is a type of sensorineural deafness. Because many elderly people refuse to accept their hearing loss and resist using hearing aids, they begin to rely more and more on their vision for clues as to what is going on around them, and may be accused of ignoring people.

Although presbycusis is considered to be a disability of old age, it is becoming much more common in younger people as our world grows noisier. The damage (breakage of the "hairs" of the hair cells) caused by excessively loud sounds is progressive and cumulative. Each assault causes a bit more damage. Rock music played and listened to at deafening levels is definitely a contributing factor to the deterioration of hearing receptors. ∎

Laboratory Tests: Hearing

Perform the following hearing tests in a quiet area.

ACUITY TEST   Have your lab partner pack one ear with cotton and sit quietly with eyes closed. Obtain a ticking clock or pocket watch and hold it very close to his or her *unpacked* ear. Then slowly move it away from the ear until your partner signals that the ticking is no longer audible. Record the distance in inches at which ticking becomes inaudible.

Right ear _____ Left ear _____

Is the threshold of audibility sharp or indefinite?

_____

SOUND LOCALIZATION   Ask your partner to close both eyes. Hold the pocket watch at an audible distance (about 6 inches) from his or her ear, and move it to various locations (front, back, sides, and above his or her head). Have your partner locate the position by pointing in each instance. Can the sound be localized equally well at all positions?

_____ If not, at what position(s) was the sound less easily located?

_____

_____

The ability to localize the source of a sound depends on two factors—the difference in the loudness of the sound reaching each ear and the time of arrival of the sound at each ear. How does this information help to explain your findings?

_____

_____

FREQUENCY RANGE OF HEARING   Obtain three tuning forks: one with a low frequency (75 to 100 Hz [cps]), one with a frequency of approximately 1000 Hz, and one with a frequency of 4000 to 5000 Hz. Strike the lowest-frequency fork on the heel of your hand or with a rubber mallet, and hold it close to your partner's ear. Repeat with the other two forks.

Which fork was heard most clearly and comfortably?

_____ Hz

Which was heard least well? _____ Hz

WEBER TEST TO DETERMINE CONDUCTIVE AND SENSORINEURAL DEAFNESS   Strike a tuning fork and place the handle of the tuning fork medially on your forehead (see Figure 19.5a). Is the tone equally loud in both ears, or is it louder in one ear?

_____

If it is equally loud in both ears, you have equal hearing or equal loss of hearing in both ears. If sensorineural

deafness is present in one ear, the tone will be heard in the unaffected ear, but not in the ear with sensorineural deafness. If conduction deafness is present, the sound will be heard more strongly in the ear in which there is a hearing loss. Conduction deafness can be simulated by plugging one ear with cotton to interfere with the conduction of sound to the inner ear.

RINNE TEST FOR COMPARING BONE- AND AIR-CONDUCTION HEARING

1. Strike the tuning fork, and place its handle on your partner's mastoid process (Figure 19.5b).
2. When your partner indicates that the sound is no longer audible, hold the still-vibrating prongs close to his or her auditory canal (Figure 19.5c). If your partner hears the fork again (by air conduction) when it is moved to that position, hearing is not impaired and the test result is to be recorded as positive (+). (Record below.)
3. Repeat the test, but this time test air conduction hearing first.
4. After the tone is no longer heard by air conduction, hold the handle of the tuning fork on the bony mastoid process. If the subject hears the tone again by bone conduction after hearing by air conduction is lost, there is some conductive deafness and the result is recorded as negative (−).
5. Repeat the sequence for the opposite ear.

    Right ear _____ Left ear _____

    Does the subject hear better by bone or by air conduction?

    _____

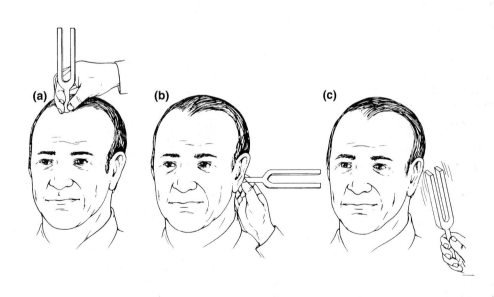

F19.5

The Weber and Rinne tuning fork tests. (a) The Weber test to evaluate whether the sound remains centralized (normal) or lateralizes to one side or the other (indicative of some degree of conductive or sensorineural deafness). **(b and c)** The Rinne test to compare bone conduction and air conduction.

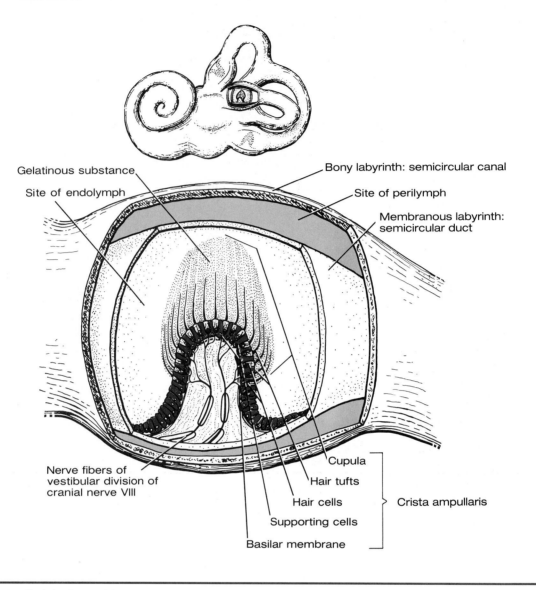

Gelatinous substance

Site of endolymph

Bony labyrinth: semicircular canal

Site of perilymph

Membranous labyrinth: semicircular duct

Nerve fibers of vestibular division of cranial nerve VIII

Cupula

Hair tufts

Hair cells

Supporting cells

Basilar membrane

Crista ampullaris

F19.6

Crista ampullaris in the semicircular canal.

Microscopic Anatomy of the Equilibrium Apparatus and Mechanisms of Equilibrium

The equilibrium apparatus of the inner ear is in the vestibular and semicircular canal portions of the bony labyrinth. Their chambers are filled with perilymph, in which membranous labyrinth structures are suspended. The vestibule contains the saclike **utricle** and **saccule,** and the semicircular chambers contain membranous **semicircular ducts.** Like the cochlear duct, these membranes are filled with endolymph and contain receptor cells that are activated by disturbance of their cilia.

The semicircular canals are centrally involved in the **mechanism of dynamic equilibrium.** They are about ½ inch in circumference and are oriented in three planes—horizontal, frontal, and sagittal. At the base of each semicircular duct is an enlarged region, the **ampulla,** which communicates with the utricle of the ves-

tibule. Within each ampulla is a receptor region called a **crista ampullaris,** which consists of a tuft of hair cells covered with a gelatinous cap, or **cupula** (Figure 19.6). When your head position changes in an angular direction, as when twirling on the dance floor or when taking a rough boat ride, the endolymph in the canal lags behind, pushing the cupula—like a swinging door—in a direction opposite to that of the angular motion. Depending on the ear, this movement depolarizes or hyperpolarizes the hair cells resulting in enhanced or reduced impulse transmission up the vestibular division of the eighth cranial nerve to the brain. Likewise, when the angular motion stops suddenly, the inertia of the endolymph causes it to continue to move, pushing the cupula in the same direction as the previous motion. This movement again initiates electrical changes in the hair cells. (This phenomenon accounts for the reversed motion sensation you feel when you stop suddenly after twirling.) If you begin to move at a constant rate of mo-

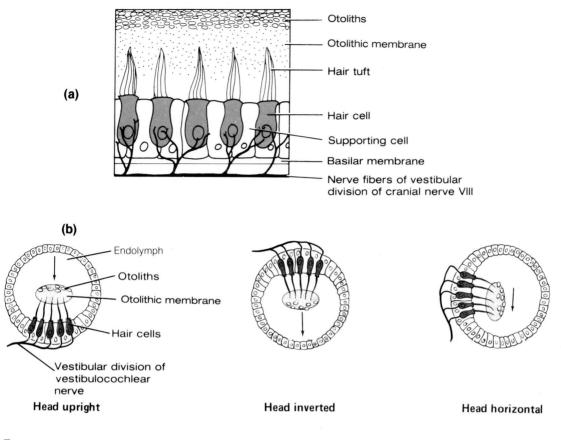

(a)

- Otoliths
- Otolithic membrane
- Hair tuft
- Hair cell
- Supporting cell
- Basilar membrane
- Nerve fibers of vestibular division of cranial nerve VIII

(b)

- Endolymph
- Otoliths
- Otolithic membrane
- Hair cells
- Vestibular division of vestibulocochlear nerve

Head upright   **Head inverted**   **Head horizontal**

F19.7

Structure and function of static equilibrium receptors (maculae). (a) Diagrammatic view of a portion of a macula. **(b)** Stimulation of the maculae by movement of otoliths in the gelatinous otolithic membrane creates a pull on the hair cells. Arrows indicate direction of gravitational pull.

tion, the cupula gradually returns to its original position. The hair cells, no longer bent, send no new signals, and you lose the sensation of spinning. Thus the response of these dynamic equilibrium receptors is a reaction to *changes* in angular motion rather than to motion itself.

 Go to the demonstration area and examine the slide of a crista ampullaris. Identify the areas depicted in Figure 19.6.

The vestibule contains **maculae,** receptors that are essential to the **mechanism of static equilibrium.** The maculae respond to gravitational pull, thus providing information on which way is up or down, and to linear or straightforward changes in speed. They are located on the walls of the saccule and utricle. The **otolithic membrane,** a gelatinous material containing small grains of calcium carbonate (**otoliths**), overrides the hair cells in each macula. As the head moves, the otoliths move in response to variations in gravitational pull (Figure 19.7). As they deflect different hair cells, they trigger electrical changes in the hair cells that modify their rate of impulse transmission along the vestibular nerve.

Although the receptors of the semicircular canals and the vestibule are responsible for dynamic and static equilibrium respectively, they rarely act independently. Complex interaction of many of the receptors is the rule. Also, the information these equilibrium, or balance, senses provide is significantly augmented by the proprioceptors and sight, as some of the following laboratory experiments demonstrate.

Laboratory Tests: Equilibrium

The function of the semicircular canals and vestibule are not routinely tested in the laboratory, but the following simple tests should serve to illustrate normal equilibrium apparatus functioning.

BALANCE TEST   Have your partner walk a straight line, placing one foot directly in front of the other.

Is he or she able to walk without undue wobbling from side to side?

Did he or she experience any dizziness? _____

The ability to walk with balance and without dizziness, unless subject to rotational forces, indicates normal function of the equilibrium apparatus.

Was nystagmus* present? _____

ROMBERG TEST   The Romberg test determines the integrity of the dorsal white column of the spinal cord, which transmits impulses to the brain from the proprioceptors involved with posture.

1.   Have your partner stand with his or her back to the blackboard.

2.   Draw one line parallel to each side of your partner's body. He or she should stand erect, with eyes open and staring straight ahead for 2 minutes while you observe any movements. Did you see any gross swaying movements?

_____

3.   Repeat the test. This time the subject's eyes should be closed. Note and record the degree of side-to-side movement.

_____

4.   Repeat the test with the subject's eyes first open and then closed. This time, however, the subject should be positioned with his or her left shoulder toward, but not touching, the board so that you may observe and record the degree of front-to-back swaying.

_____

Do you think the equilibrium apparatus of the inner ear was operating equally well in all these tests?

_____

The proprioceptors? _____

Why was the observed degree of swaying greater when the eyes were closed?

_____

_____

What conclusions can you draw regarding the factors necessary for maintaining body equilibrium and balance?

_____

_____

_____

_____

ROLE OF VISION IN MAINTAINING EQUILIB-RIUM   To further demonstrate the role of vision in maintaining equilibrium, perform the following experiment. (Ask your lab partner to record observations and act as a "spotter.") Stand erect, with your eyes open. Raise your left foot approximately 1 foot off the floor, and hold it there for 1 minute.

Record the observations: _____

_____

_____

Rest for 1 or 2 minutes; and then repeat the experiment with the same foot raised, but with your eyes closed. Record the observations:

_____

_____

_____

* **Nystagmus** is the involuntary rolling of the eyes in any direction or the trailing of the eyes slowly in one direction, followed by their rapid movement in the opposite direction. It is normal after rotation; abnormal otherwise. The direction of nystagmus is that of its quick phase on acceleration.

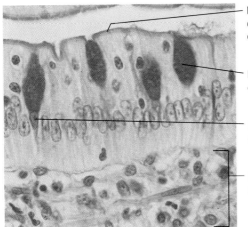

PLATE 1   Simple columnar epithelium. Mucus in goblet cells stains pink in this view (543X)

Microvilli of columnar epithelial cells

Mucus in goblet cell

Nucleus of goblet cell

Underlying connective tissue

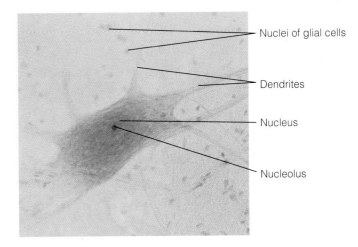

PLATE 4   Multipolar neuron in spinal cord smear (212X)

Nuclei of glial cells

Dendrites

Nucleus

Nucleolus

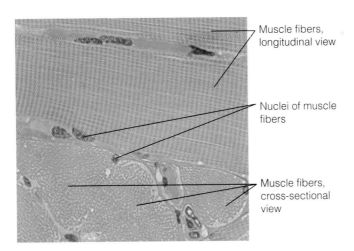

PLATE 2   Skeletal muscle, transverse and longitudinal views shown (543X)

Muscle fibers, longitudinal view

Nuclei of muscle fibers

Muscle fibers, cross-sectional view

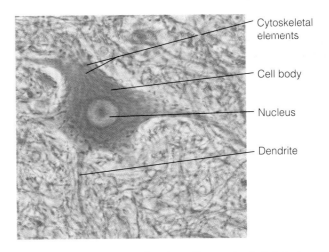

PLATE 5   Neuron stained to allow cytoskeletal elements in their processes to be seen (265X)

Cytoskeletal elements

Cell body

Nucleus

Dendrite

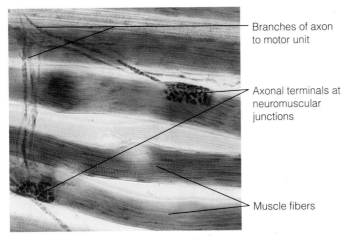

PLATE 3   Part of a motor unit (265X)

Branches of axon to motor unit

Axonal terminals at neuromuscular junctions

Muscle fibers

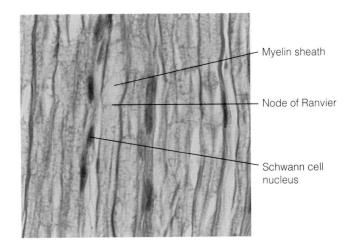

PLATE 6   Longitudinal view of myelinated axons (543X). Myelin sheaths appear "bubbly" because most of the fatty myelin is dissolved during slide preparation

Myelin sheath

Node of Ranvier

Schwann cell nucleus

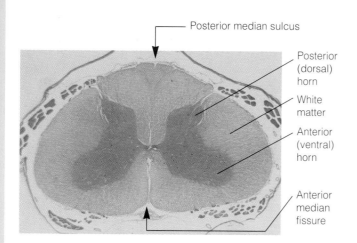

PLATE 7   Adult spinal cord, cross-sectional view (12X)

Labels: Posterior median sulcus; Posterior (dorsal) horn; White matter; Anterior (ventral) horn; Anterior median fissure

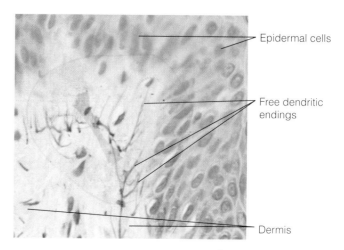

PLATE 10   Free dendritic endings at the dermal–epidermal junction (480X)

Labels: Epidermal cells; Free dendritic endings; Dermis

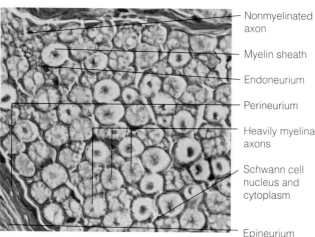

PLATE 8   Cross section of a portion of a peripheral nerve (543X). Heavily myelinated fibers are identified by a centrally-located axon, surrounded by an unstained ring of myelin, and then a peripheral rim of pink-staining Schwann cell

Labels: Nonmyelinated axon; Myelin sheath; Endoneurium; Perineurium; Heavily myelinated axons; Schwann cell nucleus and cytoplasm; Epineurium

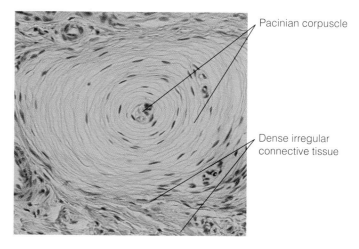

PLATE 11   Pacinian corpuscle in the hypodermis (212X)

Labels: Pacinian corpuscle; Dense irregular connective tissue

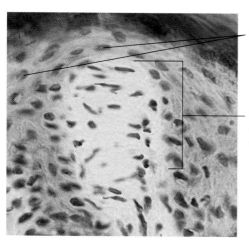

PLATE 9   Meissner's corpuscle in a dermal papilla (559X)

Labels: Epidermal cells; Meissner's corpuscle in dermal papilla

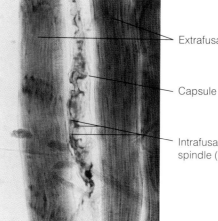

PLATE 12   Longitudinal section of a muscle spindle (305X)

Labels: Extrafusal muscle fibers; Capsule; Intrafusal fibers of the muscle spindle (receptor)

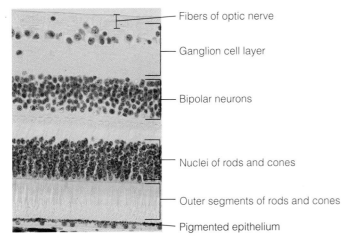

PLATE 13 Structure of the retina of the eye (353X)

- Fibers of optic nerve
- Ganglion cell layer
- Bipolar neurons
- Nuclei of rods and cones
- Outer segments of rods and cones
- Pigmented epithelium

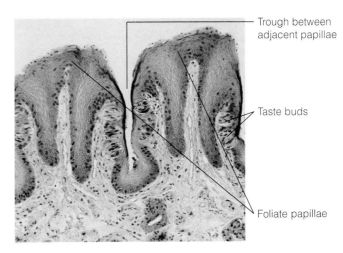

PLATE 16 Location of the taste buds on the lateral aspects of foliate papillae of the tongue (133X)

- Trough between adjacent papillae
- Taste buds
- Foliate papillae

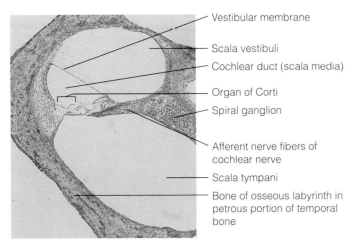

PLATE 14 Cross-sectional view of one turn of the cochlea showing the three scalas, and the location of the organ of Corti (36X)

- Vestibular membrane
- Scala vestibuli
- Cochlear duct (scala media)
- Organ of Corti
- Spiral ganglion
- Afferent nerve fibers of cochlear nerve
- Scala tympani
- Bone of osseous labyrinth in petrous portion of temporal bone

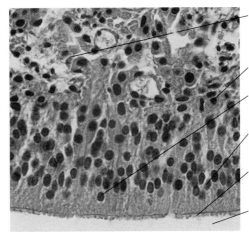

PLATE 17 Olfactory epithelium. From the lamina propria to the nasal cavity, the general arrangement of cells in this pseudostratified epithelium is basal cells, olfactory receptor cells, and supporting cells (543X)

- Lamina propria containing mucous secreting glands
- Basal cell nucleus
- Supporting cell nucleus
- Olfactory cell nucleus
- Cilia of olfactory receptor cells
- Lumen of nasal cavity

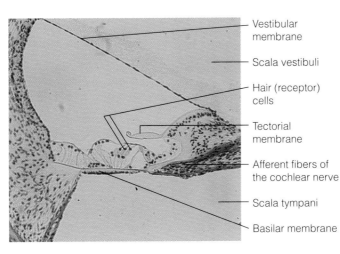

PLATE 15 The organ of Corti (106X)

- Vestibular membrane
- Scala vestibuli
- Hair (receptor) cells
- Tectorial membrane
- Afferent fibers of the cochlear nerve
- Scala tympani
- Basilar membrane

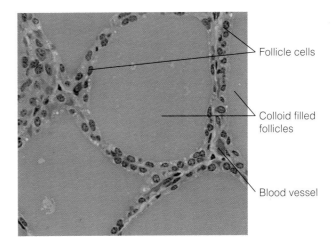

PLATE 18 The thyroid gland (345X)

- Follicle cells
- Colloid filled follicles
- Blood vessel

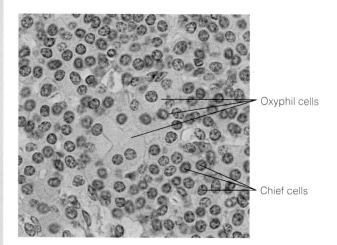

PLATE 19   Parathyroid gland tissue (543X)

Oxyphil cells

Chief cells

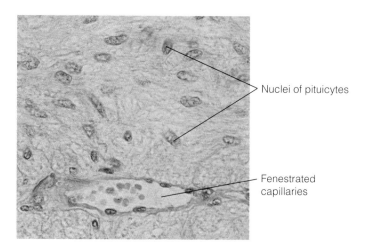

PLATE 22   Posterior pituitary (543X). The axons of the neurosecretory cells are indistinguishable from the cytoplasm of the pituicytes.

Nuclei of pituicytes

Fenestrated capillaries

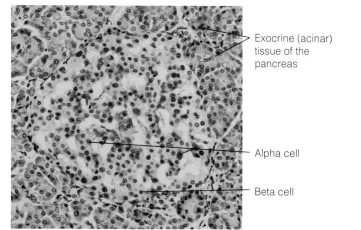

PLATE 20   Pancreatic islet stained differentially to allow identification of the glucagon-secreting alpha cells and the insulin-secreting beta cells (199X)

Exocrine (acinar) tissue of the pancreas

Alpha cell

Beta cell

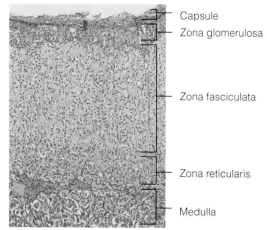

PLATE 23   Histologically distinct regions of the adrenal gland (56X)

Capsule
Zona glomerulosa

Zona fasciculata

Zona reticularis

Medulla

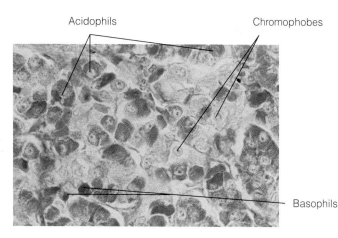

PLATE 21   Anterior pituitary gland. Differentially staining allows the acidophils, basophils, and chromophobes to be distinguished (434X)

Acidophils

Chromophobes

Basophils

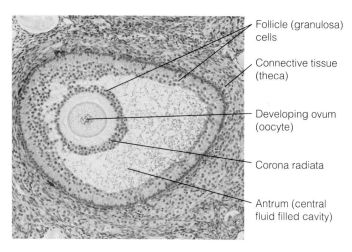

PLATE 24   A vesicular follicle of the ovary (106X)

Follicle (granulosa) cells

Connective tissue (theca)

Developing ovum (oocyte)

Corona radiata

Antrum (central fluid filled cavity)

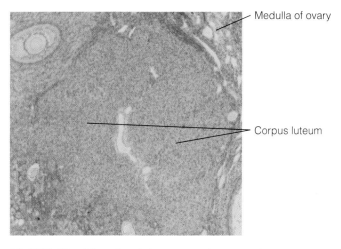

PLATE 25 The glandular corpus luteum of an ovary (45X)

Medulla of ovary

Corpus luteum

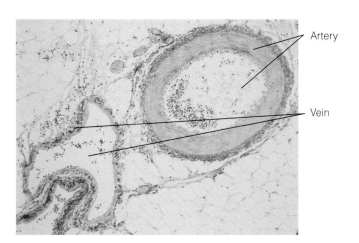

PLATE 26 Cross-sectional view of a small artery and vein (158X)

Artery

Vein

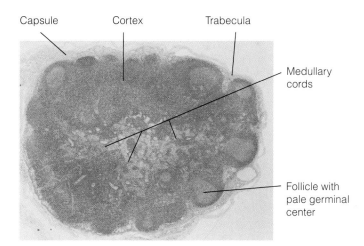

Capsule   Cortex   Trabecula

Medullary cords

Follicle with pale germinal center

PLATE 27 Main structural features of a lymph node (18X)

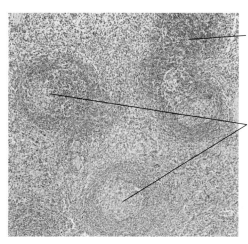

PLATE 28 Microscopic anatomy of a portion of the spleen showing the red and white pulp regions (56X)

Blood filled red pulp

White pulp containing lymphoid tissue

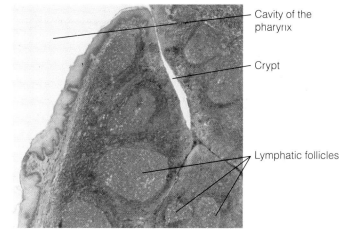

PLATE 29 Histology of a palatine tonsil. The luminal surface is covered with epithelium which invaginates deeply to form crypts (27X)

Cavity of the pharynx

Crypt

Lymphatic follicles

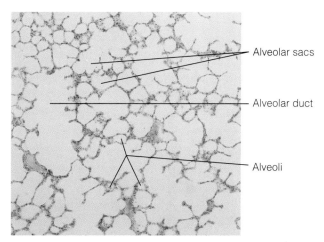

PLATE 30 Photomicrograph of part of the lung showing alveoli and alveolar ducts and sacs (34X)

Alveolar sacs

Alveolar duct

Alveoli

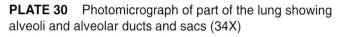

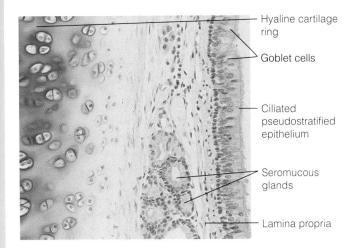

PLATE 31 Cross-section through the trachea showing the pseudostratified ciliated epithelium, glands, and part of the supporting ring of hyaline cartilage (159X)

Labels: Hyaline cartilage ring; Goblet cells; Ciliated pseudostratified epithelium; Seromucous glands; Lamina propria

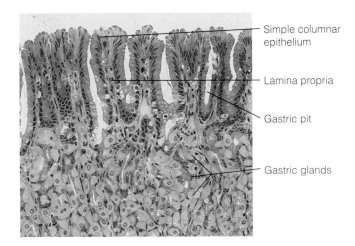

PLATE 34 Detailed structure of the gastric glands and pits (212X)

Labels: Simple columnar epithelium; Lamina propria; Gastric pit; Gastric glands

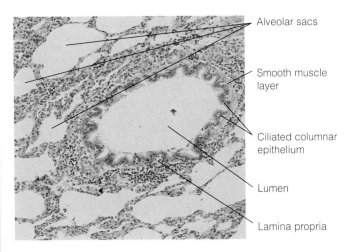

PLATE 32 Bronchiole, cross-sectional view (106X)

Labels: Alveolar sacs; Smooth muscle layer; Ciliated columnar epithelium; Lumen; Lamina propria

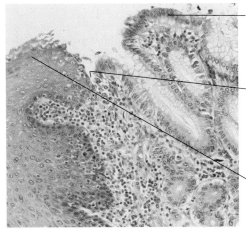

PLATE 35 Gastroesophageal junction showing the meeting of the simple columnar epithelium of the stomach and the stratified squamous epithelium of the esophagus (133X)

Labels: Simple columnar epithelium of the stomach; Cardio-esophageal junction; Stratified squamous epithelium of the esophagus

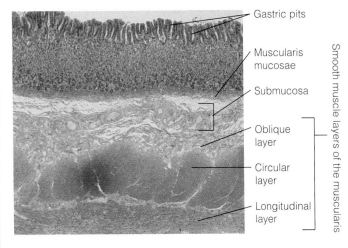

PLATE 33 Stomach. Low-power cross-sectional view through its wall showing three tunics (34X)

Labels: Gastric pits; Muscularis mucosae; Submucosa; Oblique layer; Circular layer; Longitudinal layer; Smooth muscle layers of the muscularis

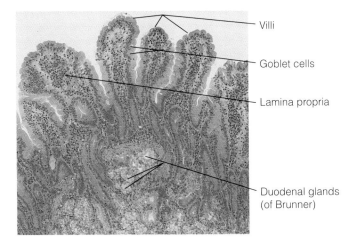

PLATE 36 Cross-sectional view of the duodenum showing villi and duodenal glands (66X)

Labels: Villi; Goblet cells; Lamina propria; Duodenal glands (of Brunner)

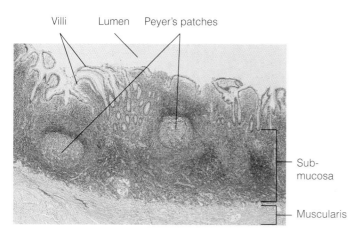

PLATE 37 Ileum, showing Peyer's patches (36X)

Villi — Lumen — Peyer's patches — Sub-mucosa — Muscularis

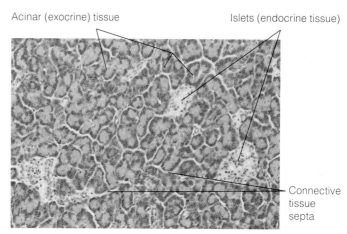

PLATE 40 Pancreas tissue. Exocrine and endocrine (islets) areas clearly visible (106X)

Acinar (exocrine) tissue — Islets (endocrine tissue) — Connective tissue septa

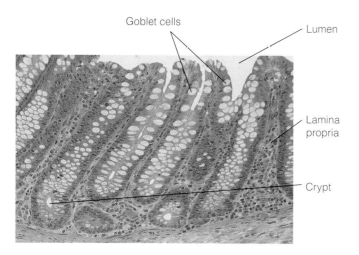

PLATE 38 Large intestine. Cross-sectional view showing the abundant goblet cells of the mucosa (127X)

Goblet cells — Lumen — Lamina propria — Crypt

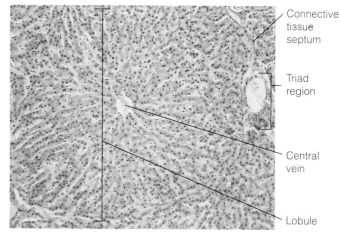

PLATE 41 Pig liver. Structure of the liver lobules (66X)

Connective tissue septum — Triad region — Central vein — Lobule

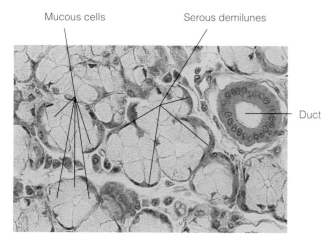

PLATE 39 Sublingual salivary glands (350X)

Mucous cells — Serous demilunes — Duct

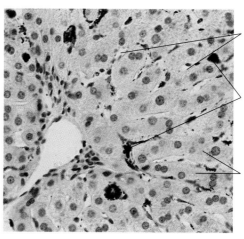

PLATE 42 Liver stained to show the location of the phagocytic cells (Kupffer cells) lining the sinusoids (265X)

Sinusoids — Kupffer cells containing dark deposits — Liver parenchyma (hepatocytes)

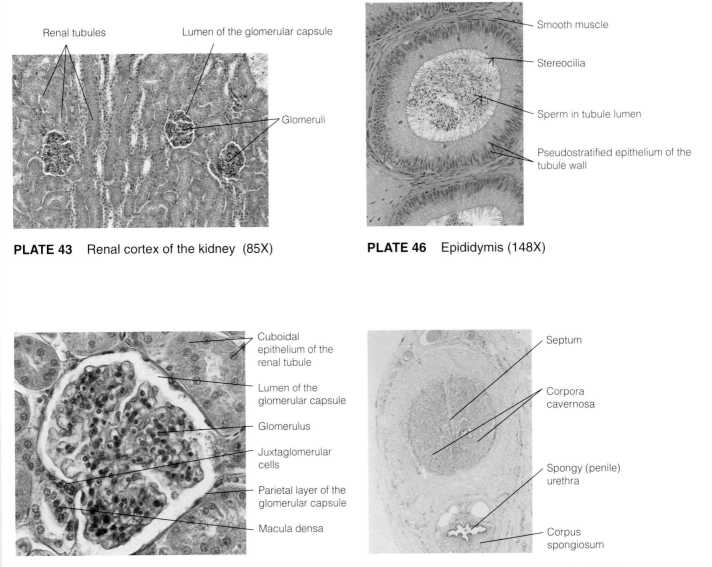

Renal tubules   Lumen of the glomerular capsule

Glomeruli

PLATE 43   Renal cortex of the kidney (85X)

Smooth muscle

Stereocilia

Sperm in tubule lumen

Pseudostratified epithelium of the tubule wall

PLATE 46   Epididymis (148X)

Cuboidal epithelium of the renal tubule

Lumen of the glomerular capsule

Glomerulus

Juxtaglomerular cells

Parietal layer of the glomerular capsule

Macula densa

PLATE 44   Detailed structure of a glomerulus (345X)

Septum

Corpora cavernosa

Spongy (penile) urethra

Corpus spongiosum

PLATE 47   Penis, transverse section (13X)

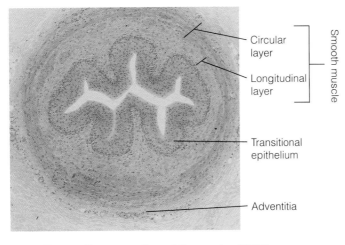

Circular layer

Longitudinal layer

Smooth muscle

Transitional epithelium

Adventitia

PLATE 45   Cross-section of the ureter (45X)

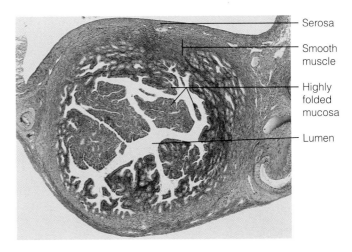

Serosa

Smooth muscle

Highly folded mucosa

Lumen

PLATE 48   Cross-sectional view of the uterine tube (34X)

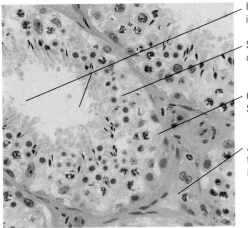

Immature sperm in lumen

Spermatogenic cells in tubule wall

Cytoplasm of Sertoli cell

Areolar connective tissue containing interstitial cells

PLATE 49 Parts of two seminiferous tubules and the intervening connective tissue (345X)

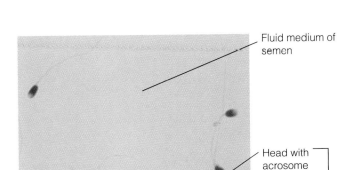

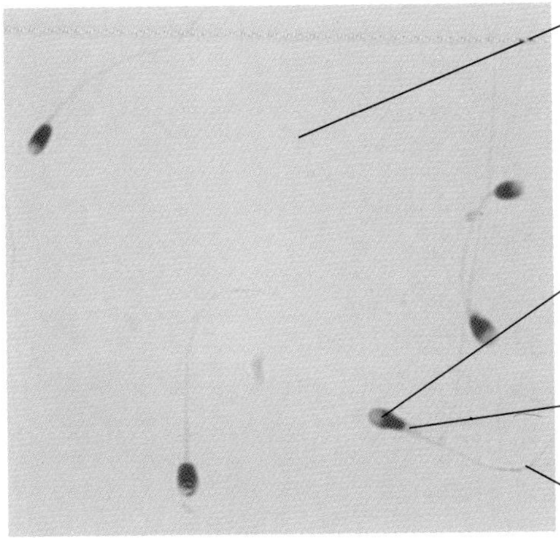

Fluid medium of semen

Head with acrosome

Midpiece

Tail

A sperm

PLATE 50 Semen, the product of ejaculation, consisting of sperm and fluids secreted by the accessory glands (particularly the prostate and seminal vesicles) (489X)

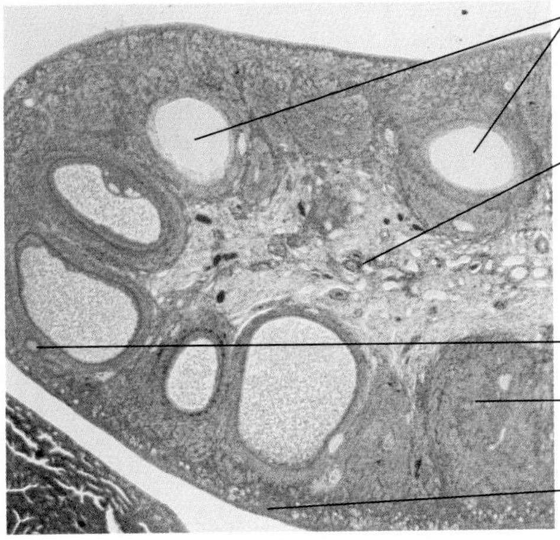

Vesicular follicles

Well vascularized medulla

Growing follicle

Corpus luteum

Tunica albuginea

PLATE 51 The ovary, showing its follicles in various stages of development (17X)

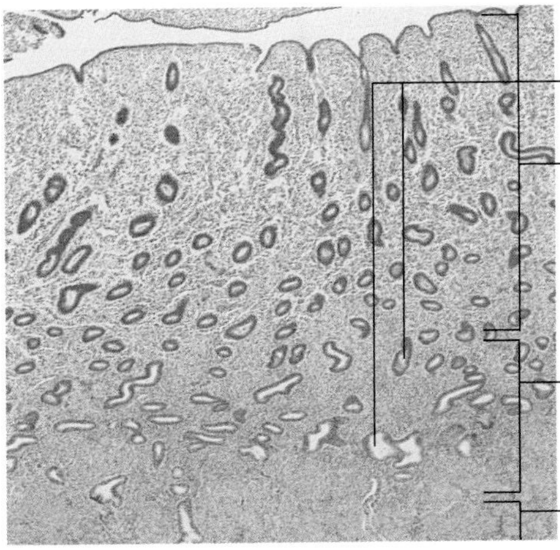

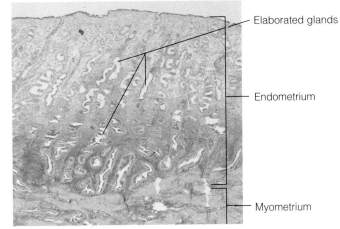

Glands

Functional layer

Endometrium

Basal layer

Myometrium

PLATE 52 Uterus: proliferative phase (17X)

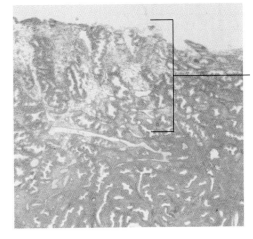

Elaborated glands

Endometrium

Myometrium

PLATE 53 Uterus: secretory phase (17X)

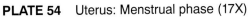

Necrotic (areas of dead and dying cells) fragments of functional layer of endometrium

PLATE 54 Uterus: Menstrual phase (17X)

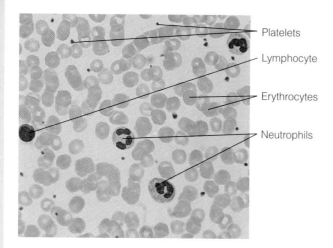

PLATE 55 Human blood smear (543X)

Platelets

Lymphocyte

Erythrocytes

Neutrophils

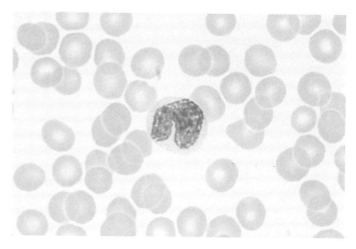

PLATE 58 A monocyte surrounded by erythrocytes (848X)

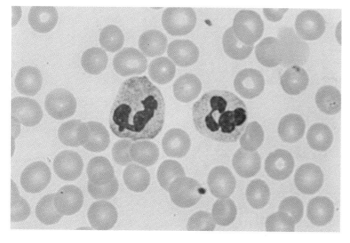

PLATE 56 Two neutrophils surrounded by erythrocytes (848X)

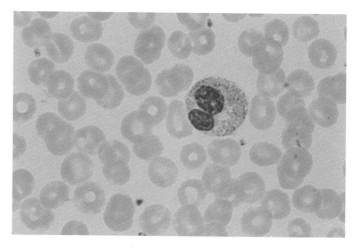

PLATE 59 An eosinophil surrounded by erythrocytes (848X)

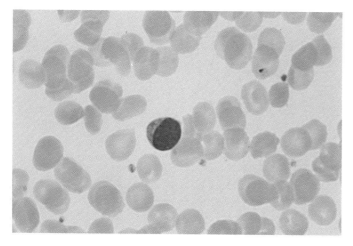

PLATE 57 A lymphocyte surrounded by erythrocytes (848X)

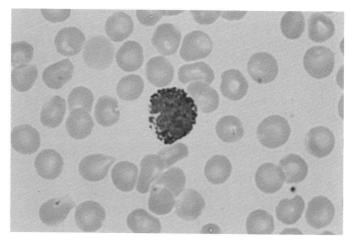

PLATE 60 A basophil surrounded by erythrocytes (848X)

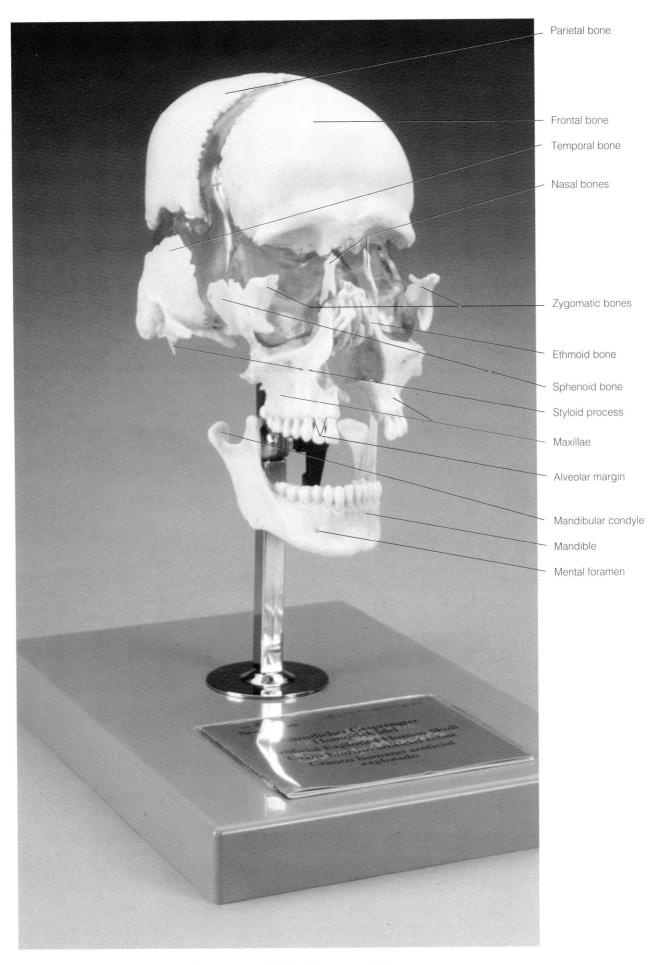

Parietal bone

Frontal bone

Temporal bone

Nasal bones

Zygomatic bones

Ethmoid bone

Sphenoid bone

Styloid process

Maxillae

Alveolar margin

Mandibular condyle

Mandible

Mental foramen

Beauchene skull, frontal view. See pages 65–70. (Somso model)

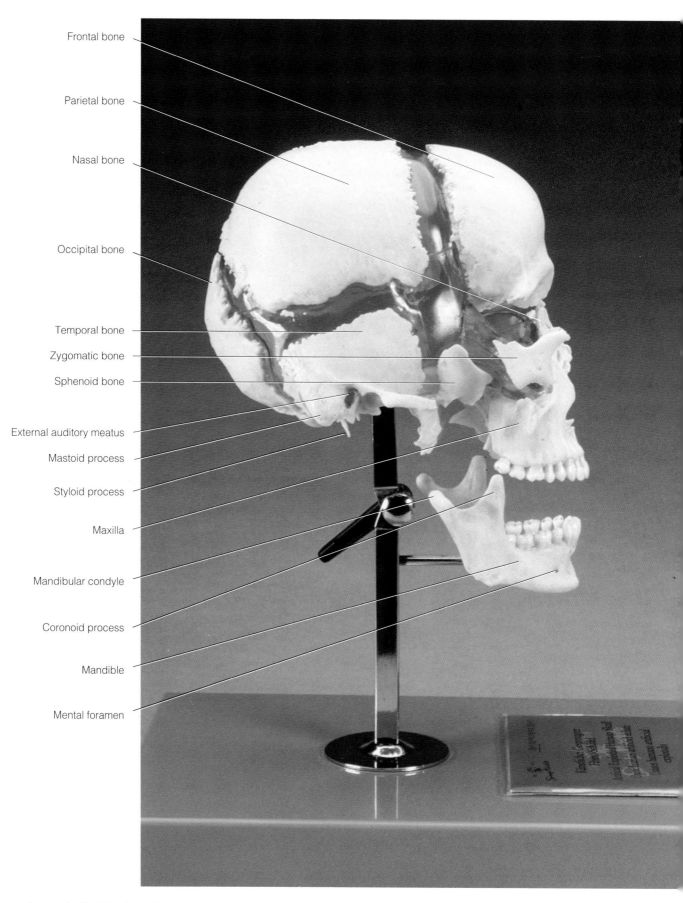

Frontal bone

Parietal bone

Nasal bone

Occipital bone

Temporal bone

Zygomatic bone

Sphenoid bone

External auditory meatus

Mastoid process

Styloid process

Maxilla

Mandibular condyle

Coronoid process

Mandible

Mental foramen

Beauchene skull, side view. See pages 65–70. (Somso model)

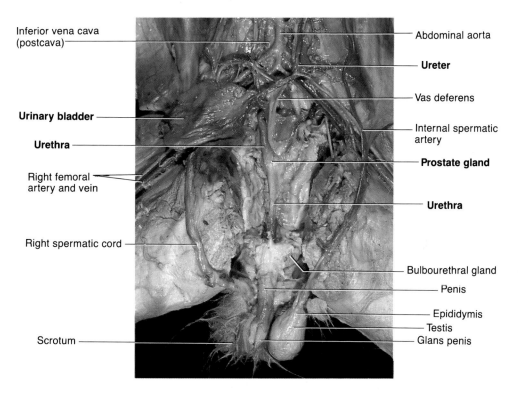

Inferior vena cava (postcava)

Abdominal aorta

Ureter

Vas deferens

Urinary bladder

Internal spermatic artery

Urethra

Prostate gland

Right femoral artery and vein

Urethra

Right spermatic cord

Bulbourethral gland

Penis

Epididymis

Testis

Scrotum

Glans penis

PLATE A   Reproductive organs of the male cat. See pages 334–335.

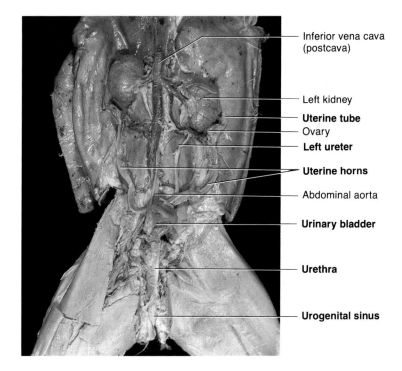

Inferior vena cava (postcava)

Left kidney

Uterine tube

Ovary

Left ureter

Uterine horns

Abdominal aorta

Urinary bladder

Urethra

Urogenital sinus

PLATE B   Urogenital system of the female cat, with emphasis on the urinary system. See pages 321–323.

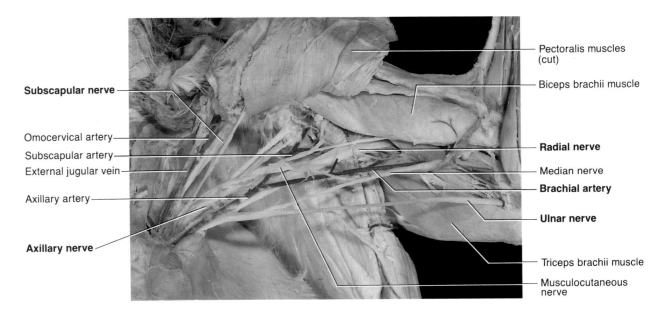

Subscapular nerve

Omocervical artery
Subscapular artery
External jugular vein

Axillary artery

Axillary nerve

Pectoralis muscles (cut)

Biceps brachii muscle

Radial nerve

Median nerve

Brachial artery

Ulnar nerve

Triceps brachii muscle

Musculocutaneous nerve

PLATE C   Brachial plexus and major blood vessels of the left forelimb of the cat. See pages 191 and 275–276.

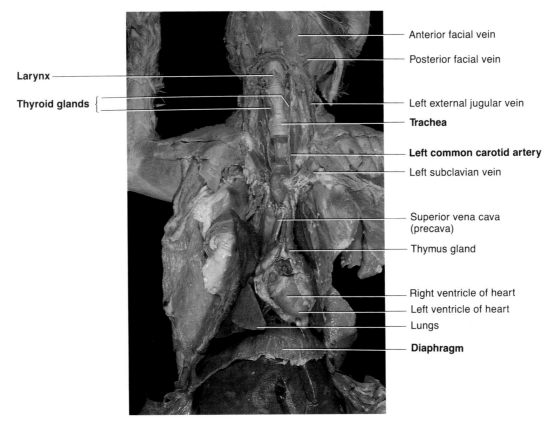

Larynx

Thyroid glands {

Anterior facial vein

Posterior facial vein

Left external jugular vein

Trachea

Left common carotid artery

Left subclavian vein

Superior vena cava (precava)

Thymus gland

Right ventricle of heart
Left ventricle of heart
Lungs

Diaphragm

PLATE D   Respiratory system of the cat. See pages 292–293.

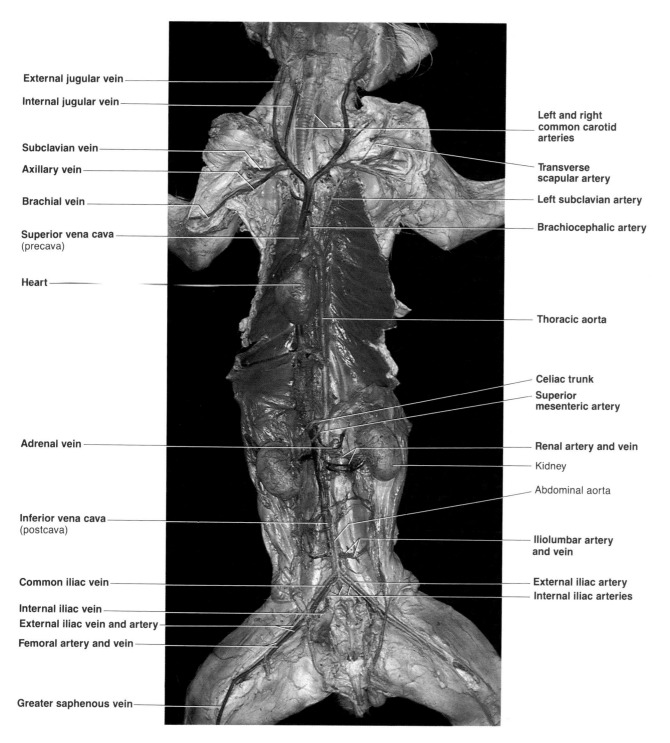

External jugular vein

Internal jugular vein

Subclavian vein

Axillary vein

Brachial vein

Superior vena cava
(precava)

Heart

Adrenal vein

Inferior vena cava
(postcava)

Common iliac vein

Internal iliac vein

External iliac vein and artery

Femoral artery and vein

Greater saphenous vein

Left and right
common carotid
arteries

Transverse
scapular artery

Left subclavian artery

Brachiocephalic artery

Thoracic aorta

Celiac trunk

Superior
mesenteric artery

Renal artery and vein

Kidney

Abdominal aorta

Iliolumbar artery
and vein

External iliac artery

Internal iliac arteries

PLATE E   Cat dissected to reveal major blood vessels. See pages 275–276.

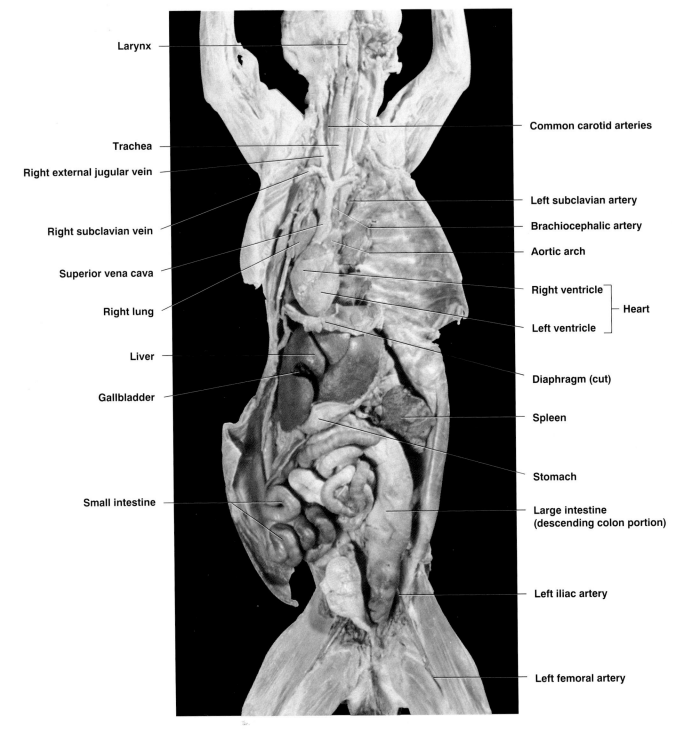

Larynx

Common carotid arteries

Trachea

Right external jugular vein

Left subclavian artery

Right subclavian vein

Brachiocephalic artery

Aortic arch

Superior vena cava

Right ventricle ⎤

Right lung

⎫ Heart

Left ventricle ⎦

Liver

Diaphragm (cut)

Gallbladder

Spleen

Stomach

Small intestine

Large intestine
(descending colon portion)

Left iliac artery

Left femoral artery

PLATE F   Ventral cavity of the cat. See pages 273–275.

Special Senses: Taste and Olfaction

OBJECTIVES

1. To describe the structure and function of the taste receptors.

2. To describe the location and cellular composition of the olfactory epithelium.

3. To name the four basic qualities of taste sensation and list the chemical substances that elicit these sensations.

4. To point out on a diagram of the tongue the predominant location of the basic types of taste receptors (salty, sweet, sour, bitter).

5. To explain the interdependence between the senses of smell and taste.

6. To name two factors other than olfaction that influence taste appreciation of foods.

7. To define *olfactory adaptation*.

MATERIALS

Prepared histologic slides: the tongue showing taste buds; nasal olfactory epithelium (longitudinal section)

Compound microscope

Paper towels

Small mirror

Granulated sugar

Cotton-tipped swabs

Disposable autoclave bag

Paper cups; paper plates

Prepared vials of 10% NaCl, 0.1% quinine or Epsom salt solution, 5% sucrose solution, and 1% acetic acid

Beaker containing 10% bleach solution

Prepared dropper bottles of oil of cloves, oil of peppermint, and oil of wintergreen or corresponding flavors found in the condiment section of a supermarket

See Appendix C, Exercise 20 for links to *Anatomy and PhysioShow: The Videodisc.*

The receptors for taste and olfaction are classified as **chemoreceptors** because they respond to chemicals or volatile substances in solution. Although four relatively specific types of taste receptors have been identified, the olfactory receptors are considered sensitive to a much wider range of chemical sensations. The sense of smell is the least understood of the special senses.

LOCALIZATION AND ANATOMY OF TASTE BUDS

The **taste buds,** specific receptors for the sense of taste, are widely but not uniformly distributed in the oral cavity. Most are located on the dorsal surface of the tongue (as described below). A few are found on the soft palate, epiglottis, and inner surface of the cheeks. The afferent fibers from the taste buds to the sensory cortex in the postcentral gyrus of the brain are carried in three cranial nerves: the *facial nerve* (*VII*) serves the anterior two-thirds of the tongue; the *glossopharyngeal nerve* (*IX*)

serves the posterior third of the tongue; and the *vagus nerve* (*X*) carries a few fibers from the pharyngeal region.

The dorsal tongue surface is covered with small projections, or **papillae,** of three major types: sharp *filiform papillae* and the rounded *fungiform* and *circumvallate papillae*. The taste buds are located primarily on the sides of the circumvallate papillae (arranged in a V-formation on the posterior surface of the tongue) and on the more numerous fungiform papillae. The latter look rather like minute mushrooms and are widely distributed on the tongue. (See Figure 20.1.)

● Use a mirror to examine your tongue. Can you pick out the various papillae types?

_____ If so, which? _____

_____

Each taste bud consists largely of a globular arrangement of two types of modified epithelial cells: the

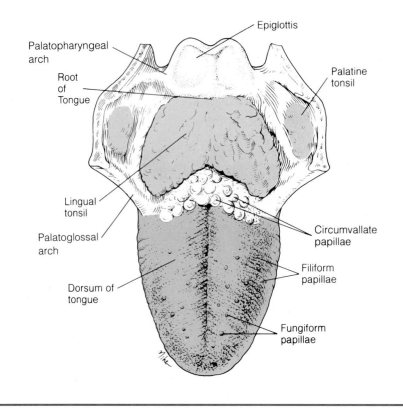

Dorsal surface of the tongue, showing major structures.

gustatory, or **taste, cells,** which are the actual receptor cells, and **supporting cells.** Several nerve fibers enter each taste bud and supply sensory nerve endings to each of the taste cells. The long microvilli of the receptor cells penetrate the epithelial surface through an opening called the **taste pore.** When these microvilli, called **gustatory hairs,** contact specific chemicals in the solution, the taste cells depolarize.

Obtain a microscope and a prepared slide of a tongue cross section. Use Figure 20.2 as a guide to aid you in locating the taste buds on the tongue papillae. Make a detailed study of one taste bud. Identify the taste pore and gustatory hairs if observed.

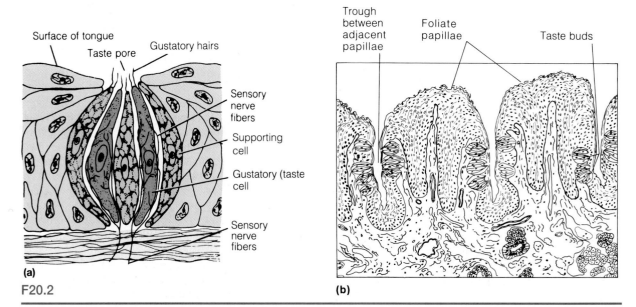

Taste bud anatomy and localization. (a) Diagrammatic view of taste bud. **(b)** View of tongue papillae, showing position of taste buds. See Plate 16 in the Histology Atlas for corresponding photomicrograph.

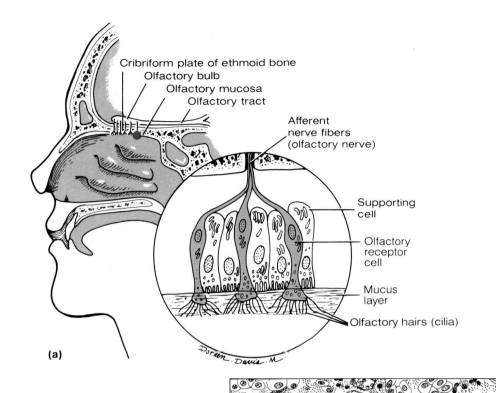

(a)

F20.3

Cellular composition of olfactory epithelium.
(**a**) Diagrammatic representation. (**b**) Line drawing
corresponding to the photomicrograph in Plate 17 in the
Histology Atlas.

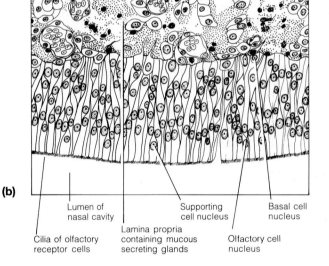

(b)

LOCALIZATION AND ANATOMY OF THE OLFACTORY RECEPTORS

The **olfactory epithelium** (organ of smell) occupies an area of about 2.5 cm in the roof of each nasal cavity. Since the air entering the human nasal cavity must make a hairpin turn to enter the respiratory passages below, the nasal epithelium is in a rather poor position for performing its function. This is why sniffing, which brings more air into contact with the receptors, intensifies the sense of smell.

The specialized receptor cells in the olfactory epithelium are surrounded by **supporting cells,** nonsensory epithelial cells. The **olfactory receptor cells** are bipolar neurons whose **olfactory hairs** (actually cilia) extend outward from the epithelium. Axonal nerve fibers emerging from their basal ends penetrate the cribriform plate of the ethmoid bone and proceed as the *olfactory nerves* to synapse in the olfactory bulbs lying on either side of the crista galli of the ethmoid bone. Impulses from neurons of the olfactory bulbs are then conveyed to the olfactory portion of the cortex (uncus).

Obtain a longitudinal section of olfactory epithelium. Examine it closely, comparing it to Figure 20.3.

 LABORATORY EXPERIMENTS

Plotting Taste Bud Distribution

When taste is tested with pure chemical compounds, taste sensations can all be grouped into one of four basic qualities—sweet, sour, bitter, or salty. Although all taste buds are believed to respond in some degree to all four classes of chemical stimuli, each type responds optimally to only one. This characteristic makes it possible to map the tongue to show the relative density of each type of taste bud.

The *sweet receptors* respond to a number of seemingly unrelated compounds such as sugars (fructose, sucrose, glucose), saccharine, and some amino acids. Some believe the common factor is the hydroxyl (OH^-) group. *Sour receptors* respond to hydrogen ions (H^+) or the acidity of the solution, *bitter receptors* to alkaloids, and *salty receptors* to metallic ions in solution.

 1.   Prepare to make a taste sensation map of your lab partner's tongue by obtaining the following: cotton-tipped swabs; one vial each of NaCl, quinine or Epsom salt solution, sucrose solution, and acetic acid; paper cups; a flask of distilled or tap water, and a disposable autoclave bag.

2.   Before each test, the subject should rinse his or her mouth thoroughly with water and lightly dry his or her tongue with a paper towel.

 Dispose of used paper towels in the autoclave bag.

3.   Generously moisten a swab with 5% sucrose solution and touch it to the center, back, tip, and sides of the dorsal surface of the subject's tongue.

4.   Map, with O's on the tongue outline below, the location of the sweet receptors.

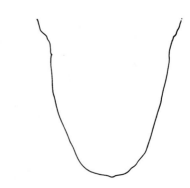

Put the used swab in the autoclave bag. Do not redip the swab into the sucrose solution.

5.   Repeat the procedure with quinine (or Epsom salt solution) to map the location of the bitter receptors (use the symbol B), with NaCl to map the salt receptors (symbol +), and with acetic acid to map the sour receptors (symbol −).

Use a fresh swab for each test, and properly dispose of the swabs immediately after use.

What area of the tongue dorsum seems to lack taste receptors?

_____

How closely does your localization of the different taste receptors coincide with the information in your textbook?

_____

_____

Olfactory Stimulation and Taste

What is commonly referred to as taste depends heavily on stimulation of the olfactory receptors, particularly in the case of strongly odoriferous substances. The following experiment should illustrate this.

1. Obtain vials of oil of wintergreen, peppermint, and cloves and some fresh cotton-tipped swabs. Ask the subject to sit so that he or she cannot see which vial is being used, and to dry the tongue and close the nostrils.

2. Apply a drop of one of the oils to the subject's tongue. Can he or she distinguish the flavor?

_____

3. Have the subject open the nostrils, and record the change in sensation he or she reports.

_____

4. Have the subject rinse the mouth well and dry the tongue.

5. Prepare two swabs, each with one of the two remaining oils.

6. Hold one swab under the subject's open nostrils, while touching the second swab to the tongue.

Record the reported sensations. _____

_____

_____

7. Appropriately dispose of the used swabs and paper towels in the autoclave bag.

Which sense, taste or smell, appears to be more important in the proper identification of a strongly flavored volatile substance?

_____

Functional Anatomy
of the Endocrine Glands

OBJECTIVES

1. To identify and name the major endocrine glands and tissues of the body when provided with an appropriate diagram.

2. To list the hormones produced by the endocrine glands and discuss the general function of each.

3. To indicate the means by which hormones contribute to body homeostasis by giving appropriate examples of hormonal actions.

4. To describe the structural and functional relationship between the hypothalamus and the pituitary.

5. To correctly identify the histologic structure of the anterior and posterior pituitary, thyroid, parathyroid, adrenal cortex and medulla, pancreas, testis, and ovary by microscopic inspection or when presented with an appropriate photomicrograph or diagram. (Optional)

6. To name and point out the specialized hormone-secreting cells in the above tissues as studied in the laboratory. (Optional)

MATERIALS

Human torso model
Anatomical chart of the human endocrine system
Compound microscopes
Colored pencils
Histologic slides of the anterior pituitary,* posterior pituitary, thyroid gland, parathyroid glands, adrenal gland, pancreas,* ovary, and testis tissue
Dissection animal, dissection tray, and instruments (if desired by the instructor)
Protective skin cream or disposable plastic gloves

See Appendix B, Exercise 21 for links to A.D.A.M. Standard.

 See Appendix C, Exercise 21 for links to *Anatomy and PhysioShow: The Videodisc.*

*Differential-staining if possible

The **endocrine system** is the second major controlling system of the body. Acting with the nervous system, it helps coordinate and integrate the activity of the body's cells. However, the nervous system employs electrochemical impulses to bring about rapid control, while the more slowly acting endocrine system employs chemical "messengers," or **hormones,** which are released into the blood to be transported throughout the body.

The term *hormone* comes from a Greek word meaning "to arouse." The body's hormones, which are steroids or amino acid–based molecules, arouse the body's tissues and cells by stimulating changes in their metabolic activity. These changes lead to growth and development and to the physiologic homeostasis of many body systems. Although all hormones are blood-borne, a given hormone affects only the biochemical activity of a specific organ or organs. Organs that respond to a particular hormone are referred to as the **target organs** of that hormone. The ability of the target tissue to respond seems to depend on the ability of the hormone to bind with specific receptors (proteins) occurring on the cells' plasma membrane or within the cells.

Although the function of some hormone-producing glands (the anterior pituitary, thyroid, adrenals, parathyroids) is purely endocrine, the function of others (the pancreas and gonads) is mixed—both endocrine and exocrine. Both types of glands are derived from epithelium, but the endocrine, or ductless, glands release their product (always hormonal) directly into the blood. The exocrine glands release their products at the body's surface or outside an epithelial membrane via ducts. In addition, there are varied numbers of hormone-producing cells within the intestine, stomach, kidney, and placenta, organs whose functions are primarily nonendocrine. Only the major endocrine organs are considered here.

You will use anatomical charts and models to localize the endocrine organs in this exercise and wait to identify the endocrine glands of the laboratory animals along with their other major organ systems in subsequent units. Partial dissection would cause the laboratory animals to become desiccated. However, if your instructor wishes you to perform an animal dissection to identify the endocrine organs, Figure 21.1 may be used as a guide for this purpose. Your instructor will provide methods of making incisions that minimize desiccation and damage to other organs.

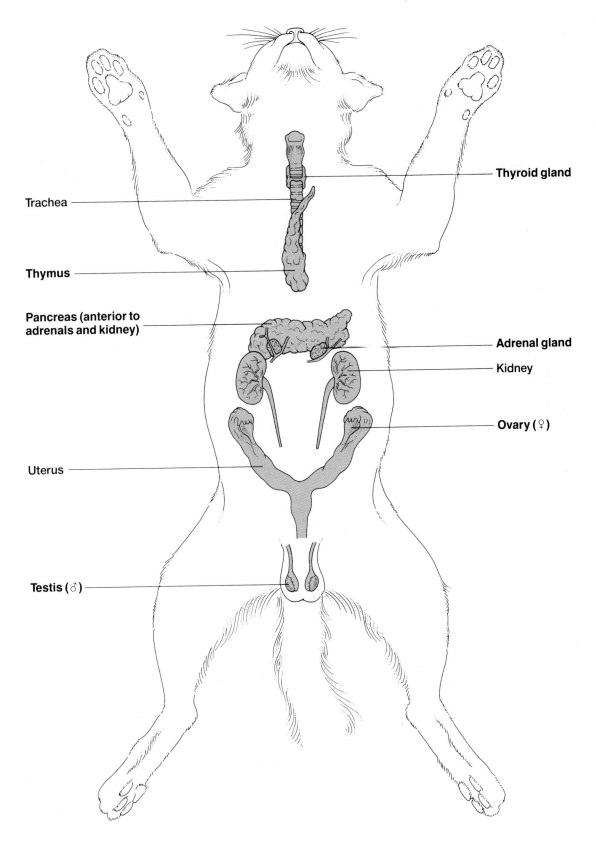

Thyroid gland

Trachea

Thymus

Pancreas (anterior to adrenals and kidney)

Adrenal gland

Kidney

Ovary (♀)

Uterus

Testis (♂)

F21.1

Endocrine organs in the cat.

GROSS ANATOMY AND BASIC FUNCTION OF THE ENDOCRINE GLANDS

As the endocrine organs are described, *locate and identify them by name* on Figure 21.2. When you have completed the descriptive material, also locate the organs on the anatomical charts or torso.

Pituitary Gland (Hypophysis)

The *pituitary gland,* or *hypophysis,* is located in the concavity of the sella turcica of the sphenoid bone. It consists largely of two functional areas, the **adenohypophysis,** or **anterior pituitary,** and the **neurohypophysis,** or **posterior pituitary,** and is attached to the hypothalamus by a stalk called the **infundibulum.**

ADENOHYPOPHYSEAL HORMONES   The adenohypophysis secretes a number of hormones. Four of these are **tropic hormones.** In each case, a tropic hor-

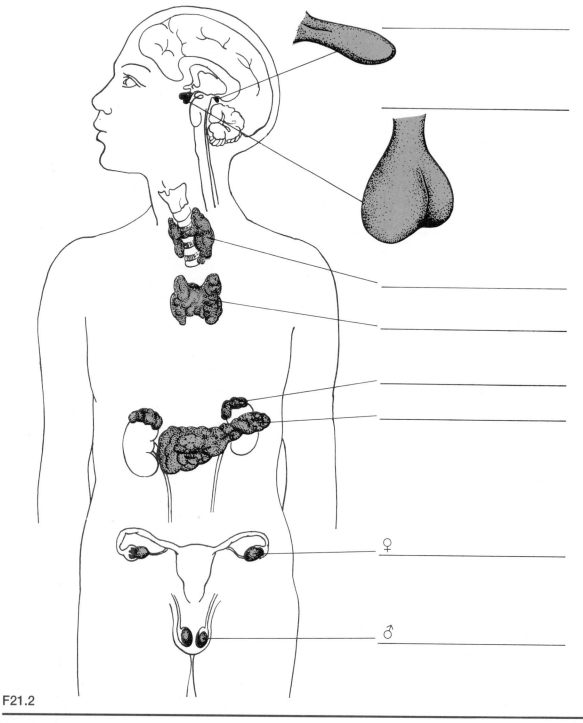

F21.2

Human endocrine organs.

mone released by the anterior pituitary stimulates its target organ, which is also an endocrine gland, to secrete its hormones. Target organ hormones then exert their effects on other body organs and tissues. The tropic hormones include:

- The **gonadotropins—follicle-stimulating hormone (FSH)** and **luteinizing hormone (LH)**—regulate the hormonal activity of the gonads (ovaries and testes).
- **Adrenocorticotropic hormone (ACTH)** regulates the endocrine activity of the cortex portion of the adrenal gland.
- **Thyrotropic hormone (TSH)** influences the growth and activity of the thyroid gland.

The three other hormones produced by the anterior pituitary are not directly involved in the regulation of other endocrine glands of the body.

- **Growth hormone (GH)** is a general metabolic hormone that plays an important role in determining body size. It affects many tissues of the body; however, its major effects are exerted on the growth of muscle and the long bones of the body.
- **Prolactin (PRL)** stimulates breast development and promotes and maintains lactation by the mammary glands after childbirth. Its function in males is unknown.
- **Melanocyte-stimulating hormone (MSH),** which stimulates melanocytes to increase their synthesis of melanin pigment, does not appear to be of major significance in humans.

The anterior pituitary controls the activity of so many other endocrine glands that it has often been called the *master endocrine gland.* However, the anterior pituitary is not autonomous in its control because release of the anterior pituitary hormones is controlled by neurosecretions, *releasing* or *inhibiting hormones,* produced by the hypothalamus. These hypothalamic hormones are liberated into the **hypophyseal portal system,** which serves the circulatory needs of the anterior pituitary (Figure 21.3).

NEUROHYPOPHYSEAL HORMONES The neurohypophysis, or posterior pituitary, is not an endocrine gland in a strict sense, because it does not synthesize the hormones it releases. (This relationship is also indicated in Figure 21.3.) Instead, it acts as a storage area for two hormones transported to it from the paraventricular and supraoptic nuclei of the hypothalamus. The first of these hormones is **oxytocin,** which stimulates powerful uterine contractions during birth and coitus and also causes milk ejection in the lactating mother. The second, **antidiuretic hormone (ADH),** causes the distal and collecting tubules of the kidneys to reabsorb more water from the urinary filtrate, thereby reducing urine output and conserving body water. It also plays a minor role in increasing blood pressure because of its vasoconstrictor effect on the arterioles.

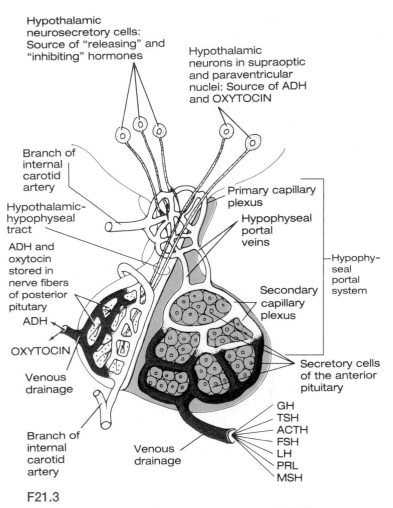

F21.3

Neural and vascular relationships between the hypothalamus and the anterior and posterior lobes of the pituitary.

Hyposecretion of ADH results in dehydration from excessive urine output, a condition called **diabetes insipidus.** Individuals with this condition experience an insatiable thirst. ■

Thyroid Gland

The *thyroid gland* is composed of two lobes joined by a central mass, or isthmus. It is located in the throat, just inferior to the larynx. It produces two major hormones, thyroid hormone and calcitonin.

Thyroid hormone (TH) is actually two physiologically active hormones known as T_4 (thyroxine) and T_3 (triiodothyronine). Because its primary function is to control the rate of body metabolism and cellular oxidation, TH affects virtually every cell in the body.

Hyposecretion of thyroxine leads to a condition of mental and physical sluggishness, which is called **myxedema** in the adult. ■

Calcitonin (also called **thyrocalcitonin**) decreases blood calcium levels by stimulating calcium deposit in the bones. It acts antagonistically to parathyroid hormone, the hormonal product of the parathyroid glands.

- Try to palpate your thyroid gland by placing your fingers against your windpipe. As you swallow, the thyroid gland will move up and down on the sides and front of the windpipe.

Parathyroid Glands

The *parathyroid glands* are found embedded in the posterior surface of the thyroid gland. Typically, there are two small oval glands on each lobe, but there may be more and some may be located in other regions of the neck. They secrete **parathyroid hormone (PTH),** the most important regulator of calcium-phosphate ion homeostasis of the blood. When blood calcium levels decrease below a certain critical level, the parathyroids release PTH, which causes release of calcium from bone matrix and prods the kidney to reabsorb more calcium and less phosphate from the filtrate. If blood calcium levels fall too low, **tetany** results and may be fatal.

Adrenal Glands

The two bean-shaped *adrenal,* or *suprarenal, glands* are located atop or close to the kidneys. Anatomically, the **adrenal medulla** develops from neural crest tissue and it is directly controlled by sympathetic nervous system neurons. The medullary cells respond to this stimulation by releasing **epinephrine** (80%) or **norepinephrine** (20%), which act in conjunction with the sympathetic nervous system to elicit the "flight or fight" response to stressors.

The **adrenal cortex** produces three major groups of steroid hormones, collectively called the **corticosteroids.** The **mineralocorticoids,** chiefly **aldosterone,** regulate water and electrolyte balance in the extracellular fluids, mainly by regulating sodium ion reabsorption by kidney tubules. The **glucocorticoids** (cortisone, hydrocortisone, and corticosterone) enable the body to resist long-term stressors, primarily by increasing blood glucose levels. The **gonadocorticoids,** or **sex hormones,** produced by the adrenal cortex are chiefly androgens (male sex hormones), but some estrogens (female sex hormones) are also formed. The gonadocorticoids are produced throughout life in relatively insignificant amounts; however, hypersecretion of these hormones produces abnormal hairiness (**hirsutism**), and masculinization occurs.

Pancreas

The *pancreas,* which functions as both an endocrine and exocrine gland, produces digestive enzymes as well as insulin and glucagon, important hormones concerned with the regulation of blood sugar levels. The pancreas is located partially behind the stomach in the abdomen.

Elevated blood glucose levels stimulate release of **insulin,** which decreases blood sugar levels, primarily by accelerating the transport of glucose into the body cells, where it is oxidized for energy or converted to glycogen or fat for storage.

Hyposecretion of insulin or some deficiency in the insulin receptors leads to **diabetes mellitus,** which is characterized by the inability of body cells to utilize glucose and the subsequent loss of glucose in the urine. Alterations of protein and fat metabolism also occur, but these are probably secondary to derangements in carbohydrate metabolism. ■

Glucagon acts antagonistically to insulin. Its release is stimulated by low blood glucose levels, and its action is basically hyperglycemic. It stimulates the liver, its primary target organ, to break down its glycogen stores to glucose and subsequently to release the glucose to the blood.

The Gonads

The female *gonads,* or *ovaries,* are paired, almond-sized organs located in the pelvic cavity. In addition to producing the female sex cells (ova), the ovaries produce two steroid hormone groups, the estrogens and progesterone. The endocrine and exocrine functions of the ovaries do not begin until the onset of puberty, when the anterior pituitary gonadotropic hormones prod the ovary into action that produces rhythmic ovarian cycles in which ova develop and hormonal levels rise and fall. The **estrogens** are responsible for the development of the secondary sex characteristics of the female at puberty (primarily maturation of the reproductive organs and development of the breasts) and act with progesterone to bring about cyclic changes of the uterine lining that occur during the menstrual cycle. The estrogens also help prepare the mammary glands for lactation.

Progesterone, as already noted, acts with estrogen to bring about the menstrual cycle. During pregnancy it maintains the uterine musculature in a quiescent state and helps to prepare the breast tissue for lactation.

The paired oval *testes* of the male are suspended in a pouchlike sac, the scrotum, outside the pelvic cavity. In addition to the male sex cells, sperm, the testes produce the male sex hormone, **testosterone.** Testosterone promotes the maturation of the reproductive system accessory structures, brings about the development of the secondary sex characteristics, and is responsible for the male sexual drive, or libido. Both the endocrine and exocrine functions of the testes begin at puberty under the influence of the anterior pituitary gonadotropins.

Two glands not mentioned earlier as major endocrine glands should also be briefly considered here, the thymus and the pineal gland.

Thymus

The *thymus* is a bilobed gland situated in the superior thorax, posterior to the sternum and anterior to the heart and lungs. Conspicuous in the infant, it begins to atrophy at puberty, and by old age it is relatively inconspicuous. The thymus produces a hormone called **thymosin** which helps direct the maturation and specialization of a unique population of white blood cells called T lymphocytes or T cells. T lymphocytes are responsible for the cellular immunity aspect of body defense; that is, rejection of foreign grafts, tumors, or virus-infected cells.

Pineal Body

The *pineal body,* or *epiphysis cerebri,* is a small cone-shaped gland located in the roof of the third ventricle of the brain. Its major endocrine product is **melatonin.**

The endocrine role of the pineal body in humans is still controversial, but it is known to play a role in the biological rhythms (particularly mating and migratory behavior) of other animals. In humans, melatonin appears to exert some inhibitory effect on the reproductive system that prevents precocious sexual maturation.

MICROSCOPIC ANATOMY OF SELECTED ENDOCRINE GLANDS (OPTIONAL)

 To prepare for the histologic study of the endocrine glands, obtain a microscope, one each of the slides listed in the list of materials, and colored pencils. We will study only organs in which it is possible to identify the endocrine-producing cells. Compare your observations with the line drawings in Figure 21.4a–f of the endocrine tissue photomicrographs.

Thyroid Gland

1.   Scan the thyroid under low power, noting the **follicles,** spherical sacs containing a pink-stained material (*colloid*). Stored T_3 and T_4 are attached to the protein colloidal material stored in the follicles as **thyroglobulin** and are released gradually to the blood. Compare the tissue viewed to Figure 21.4a and Plate 18 in the Histology Atlas.

2.   Observe the tissue under high power. Note that the walls of the follicles are formed by simple cuboidal or squamous epithelial cells that synthesize the follicular products. The **parafollicular,** or **C, cells** you see between the follicles are responsible for calcitonin production.

3.   Color appropriately two or three follicles in Figure 21.4a. Label the colloid and parafollicular cells.

When the thyroid gland is actively secreting, the follicles appear small, and the colloidal material has a ruffled border. When the thyroid is hypoactive or inactive, the follicles are large and plump and the follicular epithelium is squamouslike. What is the physiologic state of the tissue you have been viewing?

_____

_____

Parathyroid Glands

1.   Observe the parathyroid tissue under low power to view its two major cell types, the **chief cells** and the **oxyphil cells.** Compare your observations to the view in Plate 19 of the Histology Atlas. The chief cells, which synthesize parathyroid hormone (PTH), are small and abundant, and arranged in thick branching cords. The function of the scattered, much larger oxyphil cells is unknown.

2.   Color a small portion of the parathyroid tissue in Figure 21.4b. Label the chief cells, oxyphil cells, and the connective tissue matrix.

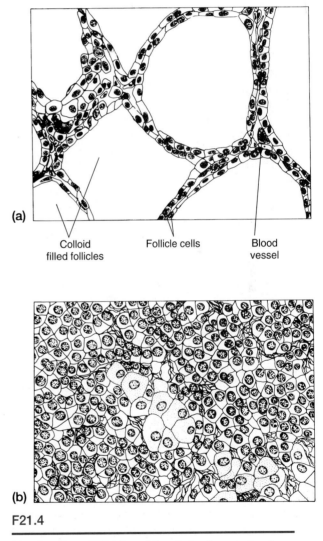

(a)

Colloid filled follicles   Follicle cells   Blood vessel

(b)

F21.4

Line drawings of Histology Atlas photomicrographs of selected endocrine organs. (a) Thyroid (see corresponding Plate 18 in the Histology Atlas); **(b)** parathyroid (see Plate 19).

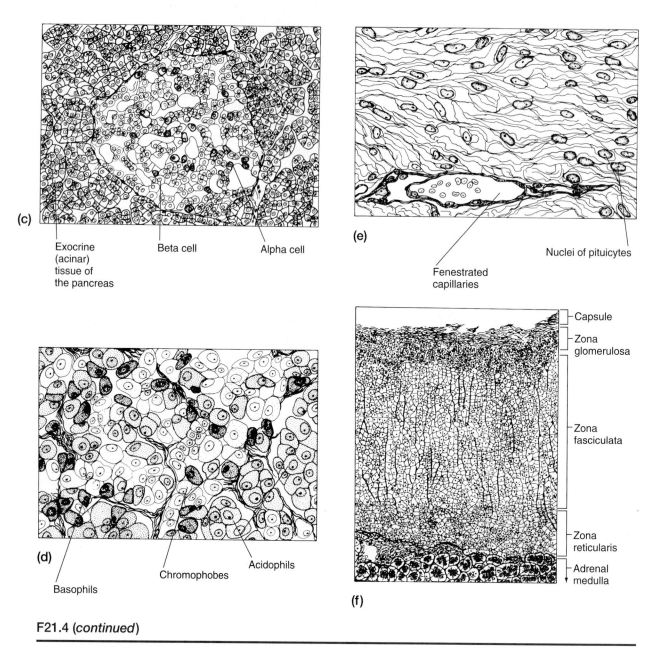

(c)

Exocrine (acinar) tissue of the pancreas   Beta cell   Alpha cell

(e)

Fenestrated capillaries   Nuclei of pituicytes

(d)

Basophils   Chromophobes   Acidophils

(f)

Capsule
Zona glomerulosa
Zona fasciculata
Zona reticularis
Adrenal medulla

F21.4 (*continued*)

Line drawings of Histology Atlas photomicrographs of selected endocrine organs. (**c**) Pancreas showing a pancreatic islet (see Plate 20); (**d**) anterior pituitary (see Plate 21); (**e**) posterior pituitary (see Plate 22); (**f**) adrenal gland (see Plate 23).

Pancreas

1. Observe pancreas tissue under low power to identify the roughly circular **pancreatic islets (islets of Langerhans),** the endocrine portions of the pancreas. The islets are scattered amid the more numerous acinar cells and stain differently (usually lighter), which makes their identification possible [Figure 27.15 (p. 310) and Plate 40 in the Histology Atlas].

2. Focus on an islet and examine its cells under high power. Notice that the islet cells are densely packed and have no definite arrangement. In contrast, the cuboidal acinar cells are arranged around secretory ducts. Unless special stains are used, it will not be possible to distinguish the **alpha cells,** which tend to cluster at the periphery of the islets and produce glucagon, from the **beta cells,** which synthesize insulin. With these specific stains, the beta cells are larger and stain gray-blue; and the alpha cells are smaller and appear bright pink, as hinted at in Figure 21.4c and shown in Plate 20 in the Histology Atlas. What is the product of the acinar cells?

3.   Draw a section of the pancreas in the space below. Label the islets and the acinar cells. If possible, differentiate the alpha and beta cells of the islets by color.

Pituitary Gland

1.   Observe the general structure of the pituitary gland under low power to differentiate between the glandular anterior pituitary and the neural posterior pituitary. Figure 21.4d and e should help you get started.

2.   Using the high-power lens, focus on the nests of cells of the anterior pituitary. It is possible to identify the specialized cell types that secrete the specific hormones when differential stains are used. Using Plate 21 in the Histology Atlas as a guide, locate the reddish-brown stained **acidophil cells,** which produce growth hormone and prolactin, and the **basophil cells,** whose deep-blue granules are responsible for the production of the tropic hormones (TSH, ACTH, FSH, and LH). **Chromophobes,** the third cellular population, do not take up the stain and appear rather dull and colorless. The role of the chromophobes is controversial, but they apparently are not directly involved in hormone production.

3.   Use appropriately colored pencils to identify acidophils, basophils, and chromophobes in Figure 21.4d.

4.   Switch your focus to the posterior pituitary. Observe the nerve fibers (axons of hypophyseal neurons) that compose most of this portion of the pituitary. Also note the **pituicytes,** glial cells which are randomly distributed among the nerve fibers. Refer to Plate 22 in the Histology Atlas as you scan the slide.

What two hormones are stored here?

_____

What is their source?

_____

Adrenal Gland

1.   Hold the slide of the adrenal gland up to the light to distinguish the outer cortex and inner medulla areas. Then scan the cortex under low power to distinguish the differences in cell appearance and arrangement in the three cortical areas. Refer to Figure 21.4f, Plate 23 in the Histology Atlas, and the descriptions below to identify the following cortical areas:

- Connective tissue capsule of the adrenal gland.
- The outermost **zona glomerulosa,** where most mineralocorticoid production occurs and where the tightly packed cells are arranged in spherical clusters.
- The deeper intermediate **zona fasciculata,** which produces glucocorticoids. This is the thickest part of the cortex. Its cells are arranged in parallel cords.
- The innermost cortical zone, the **zona reticularis** abutting the medulla, which produces sex hormones and some glucocorticoids. The cells here stain intensely and form a branching network.

2.   Switch focus to view the large, lightly stained cells of the adrenal medulla under high power. Note their clumped arrangement.

What hormones are produced by the medulla?

_____ and _____

3.   Draw a representative area of each of the adrenal regions, indicating in your sketch the differences in relative cell size and arrangement.

Zona glomerulosa   Zona fasciculata

Zona reticularis   Adrenal medulla

Ovary

Because you will consider the *ovary* in greater histologic detail when you study the reproductive system, the objective in this laboratory exercise is just to identify the endocrine-producing parts of the ovary.

1. Scan an ovary slide under low power, and look for a **vesicular (Graafian) follicle,** a circular arrangement of cells enclosing a central cavity. See Figure 29.8 (p. 332) and Plate 24 in the Histology Atlas. This structure synthesizes estrogens.
2. Examine the vesicular follicle under high power, identifying the follicular cells that produce estrogens, the antrum (fluid-filled cavity), and developing ovum (if present). The ovum will be the largest cell in the follicle.
3. Draw and color a vesicular follicle below, labeling the antrum, follicle cells, and developing ovum.

4. Switch to low power, and scan the slide to find a **corpus luteum,** a large amorphous-looking area that produces progesterone (and some estrogens). A corpus luteum is shown in Plate 25 in the Histology Atlas.

Testis

1. Examine a section of a *testis* under low power. Identify the seminiferous tubules, which produce sperm, and the **interstitial cells,** which produce testosterone. The interstitial cells are scattered between the seminiferous tubules in the connective tissue matrix. The photomicrograph of seminiferous tubules, Plate 49 in the Histology Atlas, will be helpful here.
2. Draw a representative area of the testis in the space provided. Label the seminiferous tubules and area of the interstitial cells.

Blood

OBJECTIVES

1. To name the two major components of blood and state their average percentages in whole blood.

2. To describe the composition and functional importance of plasma.

3. To define *formed elements* and list the cell types composing them, cite their relative percentages, and describe their major functions.

4. To identify red blood cells, basophils, eosinophils, monocytes, lymphocytes, and neutrophils when provided with a microscopic preparation or appropriate diagram.

5. To conduct the following blood test determinations in the laboratory, and to state their norms and the importance of each.

   hematocrit
   hemoglobin determination
   clotting time
   sedimentation rate
   differential white blood cell count
   ABO and Rh blood typing

6. To discuss the reason for transfusion reactions resulting from the administration of mismatched blood.

7. To define *anemia, polycythemia, leukopenia, leukocytosis,* and *leukemia* and to cite a possible reason for each condition.

MATERIALS

Compound microscope
Immersion oil
Models and charts of blood cells

Demonstration station:
Microscopes set up with prepared slides demonstrating the following blood (or bone marrow) conditions:
   Macrocytic, hypochromic anemia
   Microcytic, hypochromic anemia
   Sickle-cell anemia
   Lymphocytic leukemia (chronic)
   Eosinophilia

General supply area:
Plasma (obtained from an animal hospital or prepared by centrifuging animal (e.g., cattle or sheep) blood obtained from a biological supply house
Wide-range pH paper

Stained smears of human blood or, if desired by the instructor, heparinized blood obtained from an animal hospital (e.g., dog blood)
Clean microscope slides
Sterile lancets
Glass stirring rods
Alcohol swabs (wipes)
Absorbent cotton balls
Wright's stain in dropper bottle
Distilled water in dropper bottle
Test tubes
Test tube racks
Disposable gloves
Pipette cleaning solutions—(1) 10% household bleach solution, (2) distilled water, (3) 70% ethyl alcohol, (4) acetone
Bucket or large beaker containing 10% household bleach solution for slide and glassware disposal
Disposable autoclave bag
Spray bottles containing 10% bleach solution

Because many blood tests are to be conducted in this exercise, it seems advisable to set up a number of appropriately labeled supply areas for the various tests. These are designated below.

Differential count supply area:
Mechanical hand counters

Hematocrit supply area:
Heparinized capillary tubes
Microhematocrit centrifuge and reading gauge (if the reading gauge is not available, millimeter ruler may be used)
Seal-ease (Clay Adams Co.) or modeling clay

Hemoglobin determination supply area:
Tallquist hemoglobin scales and test paper or a hemoglobinometer, hemolysis applicator, and lens paper

(Materials list continues on next page)

Coagulation time supply area:
Capillary tubes (nonheparinized)
Fine triangular file

Blood typing supply area:
Blood typing sera (anti-A, anti-B, and anti-Rh [D])
Rh typing box
Wax marker

Toothpicks
Clean microscope slides

See Appendix C, Exercise 22 for links to
Anatomy and PhysioShow: The Videodisc.

Note to the Instructor: See directions for handling of soiled glass-ware and disposable items on p. 16.

In this exercise you will study plasma and formed elements of blood and conduct various hematological tests. These tests are extremely useful diagnostic tools for the physician because blood composition (number and types of blood cells, and chemical composition) reflects the status of many body functions and malfunctions.

ALERT: The decision to use animal blood for testing or to have students test their own blood will be made by the instructor in accordance with the educational purpose of the student group. For example, for students in the nursing or laboratory technician curricula, learning how to safely handle human blood or other human wastes is essential. If blood samples are provided and they are human blood samples, gloves should be worn while conducting the blood tests. If human blood is being tested, yours or that obtained from a clinical agency, precautions provided in the text for disposal of human waste **must be observed.** All soiled glassware is to be immersed in household bleach solution immediately after use, and disposable items (lancets, cotton balls, alcohol swabs, etc.) are to be placed in a disposable autoclave bag so that they can be sterilized before disposal. Safety glasses are to be worn throughout the laboratory session.

COMPOSITION OF BLOOD

The blood circulating to and from the body cells within the blood vessels is a rather viscous substance that varies from bright scarlet to a dull brick red, depending on the amount of oxygen it is carrying. The circulatory system of the average adult contains about 5.5 liters of blood.

Blood is classified as a type of connective tissue, because it consists of a nonliving fluid matrix (the **plasma**) in which living cells (**formed elements**) are suspended. The fibers typical of a connective tissue matrix become visible in blood only when clotting occurs. They then appear as fibrin threads, which form the structural basis for clot formation.

Over 100 different substances are dissolved or suspended in plasma (Figure 22.1), which is over 90% water. These include nutrients, gases, hormones, various wastes and metabolites, many types of proteins, and mineral salts. The composition of plasma varies continuously as cells remove or add substances to the blood.

Three types of formed elements are present in blood. The most numerous are the **erythrocytes,** or **red blood cells (RBCs),** which are literally sacs of hemoglobin molecules that transport the bulk of the oxygen carried in the blood (and a small percentage of the carbon dioxide). **Leukocytes,** or **white blood cells (WBCs),** are part of the body's nonspecific defenses and the immune system, and **platelets** function in hemostasis (blood clot formation). Formed elements normally constitute 45% of whole blood; plasma accounts for the remaining 55%.

Physical Characteristics of Plasma

Go to the general supply area and carefully pour a few milliliters of plasma into a test tube. Also obtain some wide-range pH paper and then return to your laboratory bench to make the following simple observations.

pH OF PLASMA   Test the pH of the plasma with wide-range pH paper. Record the pH observed.

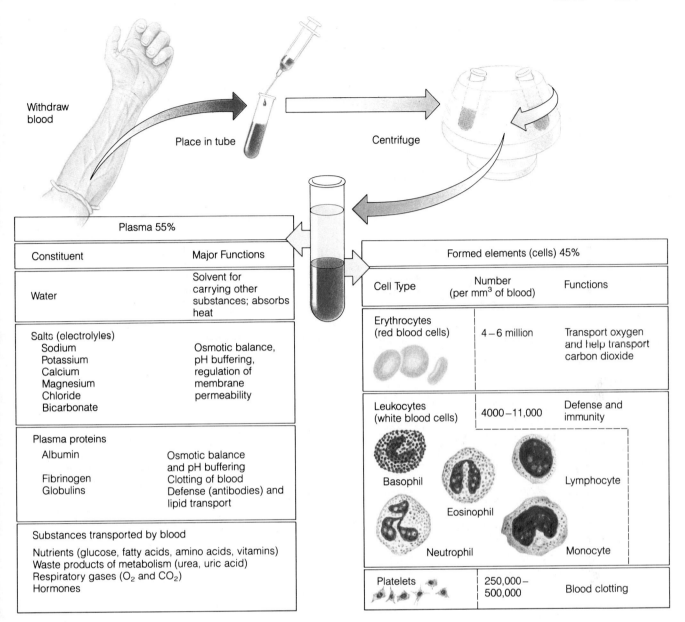

F22.1

The composition of blood.

COLOR AND CLARITY OF PLASMA   Hold the test tube up to a source of natural light. Note and record its color and degree of transparency. Is it clear, translucent, or opaque?

Color _____

Degree of transparency _____

CONSISTENCY   Dip your finger and thumb into the plasma and then press them firmly together for a few seconds. Gently pull them apart. How would you describe the consistency of plasma? Slippery, watery,

sticky, or granular? Record your observations.

Formed Elements of Blood

In this section, you will conduct your observations of blood cells on an already prepared (purchased) blood slide, or on a slide prepared from your own blood or blood provided by your instructor. Those using the purchased blood slide are to obtain a slide and begin their observations at step 6. Those testing blood provided by a biological supply source or an animal hospital are to obtain a tube of the supplied blood, disposable gloves, and the supplies listed in step 1, except for the lancets and alcohol swabs. After donning gloves, those students will jump to step 3b to begin their observations. If you are examining your own blood, you will perform all the steps described below *except* step 3b.

1.   Obtain two glass slides, a glass stirring rod, dropper bottles of Wright's stain and distilled water, two or three lancets, cotton balls, and alcohol swabs. Bring this equipment to the laboratory bench. Clean the slides thoroughly and dry them.

2.   Open the alcohol swab packet and scrub your third or fourth finger with the swab. (Because the pricked finger may be a little sore later, it is better to prepare a finger on the hand used less often.) Circumduct your hand (swing it in a cone-shaped path) for 10 to 15 seconds. This will dry the alcohol and cause your fingers to become engorged with blood. Then, open the lancet packet and grasp the lancet by its blunt end. Quickly jab the pointed end into the prepared finger to produce a free flow of blood. It is *not* a good idea to squeeze or "milk" the finger, as this forces out tissue fluid as well as blood. If the blood is not flowing freely, another puncture should be made.

⚠ *Under no circumstances is a lancet to be used for more than one puncture.* Dispose of the lancets in the disposable autoclave bag *immediately* after use.

3a.   With a cotton ball, wipe away the first drop of blood; then allow another large drop of blood to form. Touch the blood to one of the cleaned slides approximately ½ inch from the end. Then quickly (to prevent clotting) use the second slide to form a blood smear as shown in Figure 22.2. When properly prepared, the blood smear is uniformly thin. If the blood smear appears streaked, the blood probably began to clot or coagulate before the smear was made, and another slide should be prepared. Continue at step 4.

3b.   Dip a glass rod in the blood provided, and transfer a generous drop of blood to the end of a cleaned microscope slide. For the time being, lay the glass rod on a paper towel on the bench. Then, as described just above, use the second slide to make your blood smear.

4.   Dry the slide by waving it in the air. When it is completely dry, it will look dull. Place it on a paper towel, and flood it with Wright's stain. Count the number of drops of stain used. Allow the stain to remain on the slide for 3 to 4 minutes and then flood the slide with an equal number of drops of distilled water. Allow the water and Wright's stain mixture to remain on the slide for 4 or 5 minutes or until a metallic green film or scum is apparent on the fluid surface. Blow on the slide gently every minute or so to keep the water and stain mixed during this interval.

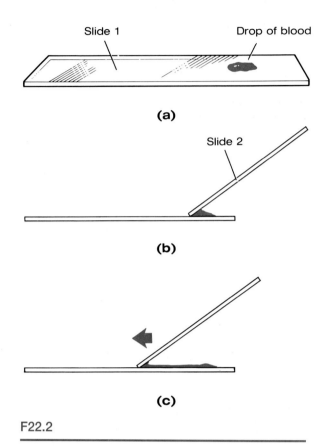

F22.2

Procedure for making a blood smear. (a) Place a drop of blood on slide 1 approximately ½ inch from one end. **(b)** Hold slide 2 at a 30° to 40° angle to slide 1 (it should touch the drop of blood) and allow blood to spread along entire bottom edge of angled slide. **(c)** Smoothly advance slide 2 to end of slide 1 (blood should run out before reaching the end of slide 1).

5.   Rinse the slide with a stream of distilled water. Then flood it with distilled water, and allow it to lie flat until the slide becomes translucent and takes on a pink cast. Then stand the slide on its long edge on the paper towel, and allow it to dry completely. Once the slide is dry, you can begin your observations.

6.   Obtain a microscope and scan the slide under low power to find the area where the blood smear is the thinnest. After scanning the slide in low power to find the areas with the largest numbers of nucleated WBCs, read the following descriptions of cell types, and find each one on Figure 22.1. (The formed elements are also shown in Plates 55 through 60 in the Histology Atlas.) Then, switch to the oil immersion lens and observe the slide carefully to identify each cell type.

ERYTHROCYTES   Erythrocytes, or red blood cells, which average 7.5 μm in diameter, vary in color from a salmon red color to pale pink, depending on the effectiveness of the stain. They have a distinctive biconcave disc shape and appear paler in the center than at the edge (see Plate 55 in the Histology Atlas).

As you observe the slide, notice that the red blood cells are by far the most numerous blood cells seen in the field. Their number averages 4.5 million to 5.0 million cells per cubic millimeter of blood (for women and men, respectively).

Red blood cells differ from the other blood cells because they are anucleate when mature and circulating in the blood. As a result, they are unable to reproduce and have a limited life span of 100 to 120 days, after which they begin to fragment and are destroyed in the spleen and other reticuloendothelial tissues of the body.

In various anemias, the red blood cells may appear pale (an indication of decreased hemoglobin content) or may be nucleated (an indication that the bone marrow is turning out cells prematurely). ■

LEUKOCYTES   Leukocytes, or white blood cells, are nucleated cells that are formed in the bone marrow from the same stem cell (*hemocytoblast*) as red blood cells. They are much less numerous than the red blood cells, averaging from 4000 to 11,000 cells per cubic millimeter. Basically, white blood cells are protective, pathogen-destroying cells that are transported to all parts of the body in the blood or lymph. Important to their protective function is their ability to move in and out of blood vessels, a process called **diapedesis,** and to wander through body tissues by **amoeboid motion** to reach sites of inflammation or tissue destruction. They are classified into two major groups, depending on whether or not they contain conspicuous granules in their cytoplasm.

Granulocytes comprise the first group. The granules in their cytoplasm stain differentially with Wright's stain, and they have peculiarly lobed nuclei, which often consist of expanded nuclear regions connected by thin strands of nucleoplasm. Additional information about the three types of granulocytes follows:

Neutrophil: The most abundant of the white blood cells (40% to 70% of the leukocyte population); nucleus consists of 3 to 7 lobes and the pale lilac cytoplasm contains fine cytoplasmic granules, which are generally indistinguishable and take up both the acidic (red) and basic (blue) dyes (*neutrophil* = neutral loving); functions as an active phagocyte. The number of neutrophils increases exponentially during acute infections. (See Plates 55 and 56.)

Eosinophil: Represents 1% to 4% of the leukocyte population; nucleus is generally figure 8 or bilobed in shape; contains large cytoplasmic granules (elaborate lysosomes) that stain red-orange with the acid dyes in Wright's stain (see Plate 59). Precise function is unknown, but they increase in number during allergies and parasite infections and may selectively phagocytize antigen-antibody complexes.

Basophil: Least abundant leukocyte type representing less than 1% of the population; large U- or S-shaped nucleus with two or more indentations. Cytoplasm contains coarse, sparse granules that are stained deep purple by the basic dyes in Wright's stain (see Plate 60). The granules contain several chemicals including histamine, a vasodilator which is discharged on exposure to antigens and helps mediate the inflammatory response.

The second group, **agranulocytes,** or **agranular leukocytes,** contains no observable cytoplasmic granules. Although found in the bloodstream, they are much more abundant in lymphoid tissues. Their nuclei tend to be closer to the norm, that is, spherical, oval, or kidney-shaped. Specific characteristics of the two types of agranulocytes are listed below.

Lymphocyte: The smallest of the leukocytes, approximately the size of a red blood cell (see Plates 55 and 57). The nucleus stains dark blue to purple, is generally spherical or slightly indented, and accounts for most of the cell mass. Sparse cytoplasm appears as a thin blue rim around the nucleus. Concerned with immunologic responses in body; one population, the B lymphocytes, oversees the production of antibodies that are released to blood. The second population, T lymphocytes, plays a regulatory role and destroys grafts, tumors, and virus-infected cells. Represents 20% to 45% of the WBC population.

Monocyte: The largest of the leukocytes; approximately twice the size of red blood cells (see Plate 58). Represents 4% to 8% of the leukocyte population. Dark blue nucleus is generally kidney-shaped; abundant cytoplasm stains gray-blue. Functions as an active phagocyte (the "long-term cleanup team"), increasing dramatically in number during chronic infections such as tuberculosis.

Students are often asked to list the leukocytes in order from the most abundant to the least abundant. The following silly phrase may help you with this task: **N**ever **l**et **m**onkeys **e**at **b**ananas (neutrophils, lymphocytes, monocytes, eosinophils, basophils).

PLATELETS   Platelets are cell fragments of large multinucleate cells (**megakaryocytes**) formed in the bone marrow. They appear as darkly staining, irregularly shaped bodies interspersed among the blood cells (see Plate 55). The normal platelet count in blood ranges from 250,000 to 500,000 per cubic millimeter. Platelets are instrumental in the clotting process that occurs in plasma when blood vessels are ruptured.

After you have identified these cell types on your slide, observe three-dimensional models of blood cells if these are available. *Do not dispose of your slide,* as it will be used later for the differential white blood cell count.

HEMATOLOGIC TESTS

When someone enters a hospital as a patient, several hematologic tests are routinely done to determine general level of health as well as the presence of pathologic conditions. You will be conducting a few of the most common of these tests in this exercise.

Materials such as cotton balls, lancets, and alcohol swabs are used in nearly all of the following diagnostic tests. These supplies are at the general supply area and should be properly disposed of (glassware to the "bleach bucket" and disposable items to the autoclave bag) immediately after use.

Other necessary supplies and equipment are at specific supply areas marked according to the test with which they are used. Since nearly all of the tests require a finger stab, if you will be using your own blood it might be wise to quickly read through the tests to determine in which instances more than one preparation can be done from the same finger stab. For example, the hematocrit capillary tubes and coagulation time samples might be prepared at the same time. A little preplanning will save you the discomfort of a multiply punctured finger.

An alternative to using blood obtained from the finger stab technique is using heparinized blood samples supplied by your instructor. The purpose of using heparinized tubes is to prevent the blood from clotting. Thus blood collected and stored in such tubes will be suitable for all tests except coagulation time testing.

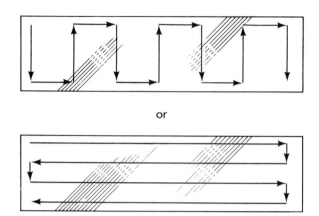

or

F22.3

Alternative methods of moving the slide for a differential WBC count.

Cell type	Number observed
Neutrophils	
Eosinophils	
Basophils	
Lymphocytes	
Monocytes	

Differential White Blood Cell Count

To make a **differential white blood cell count,** 100 WBCs are counted and classified according to type. Such a count is routine in a physical examination and in diagnosing illness, since any abnormality or significant elevation in percentages of WBC types may indicate a problem or the source of pathology. Use the slide prepared for the identification of the blood cells (pp. 231–232) for the count.

1. Begin at the edge of the smear and move the slide in a systematic manner on the microscope stage—either up and down or from side to side as indicated in Figure 22.3.

2. Record each type of white blood cell you observe by making a count on the chart below Figure 22.3 (for example, ||||| || = 7 cells) until you have observed and recorded a total of 100 WBCs. Using the equation below, compute the percentage of each WBC type counted, and record the percentages on the data sheet at the end of this exercise.

$$\text{Percent (\%)} = \frac{\text{\# observed}}{\text{Total \# counted (100)}} \times 100$$

How does your differential white blood cell count correlate with the percentages given for each type on page 233?

Total White and Red Blood Cell Counts

Total RBC and WBC counts are a routine part of any physical exam, and most clinical agencies use computers to conduct these counts. Since the hand counting technique is outdated, total RBC and WBC counts will not be done here, but the importance of such counts (both normal and abnormal values) is described briefly next.

TOTAL WHITE BLOOD CELL COUNT   Since white blood cells are an important part of the body's defense system, it is essential to note any abnormalities in them. **Leukocytosis,** an abnormally high WBC count, may indicate bacterial or viral infection, metabolic disease, hemorrhage, or poisoning by drugs or chemicals. A decrease in the white cell number below 4000/mm^3 (**leukopenia**) may indicate typhoid fever, measles, infectious hepatitis or cirrhosis, tuberculosis, or excessive antibiotic or X-ray therapy. A person with leukopenia lacks the usual protective mechanisms.

TOTAL RED BLOOD CELL COUNT   The red blood cell count, like the white blood cell count, determines the total number of this cell type per unit volume of blood. Since RBCs are absolutely necessary for oxygen transport, a doctor typically investigates any excessive change in their number immediately. An increase in the number of RBCs (**polycythemia**) may result from bone marrow cancer or from living at high altitudes where less oxygen is available. A decrease in the number of RBCs results in anemia. (The term **anemia** simply indicates a decreased oxygen-carrying capacity of blood that may result from a decrease in RBC number or size or a decreased hemoglobin content of the RBCs.) A decrease in RBCs may result suddenly from hemorrhage or more gradually from conditions that destroy RBCs or hinder RBC production.

Hematocrit

The **hematocrit,** or **packed cell volume (PCV),** is routinely determined when anemia is suspected. Centrifuging whole blood spins the formed elements to the bottom of the tube, with plasma forming the top layer (see Figure 22.1). Since the blood cell population is primarily RBCs, the PCV is generally considered equivalent to the RBC volume, and this is the only value reported. However, the relative percentage of WBCs can be differentiated, and both WBC and plasma volume will be reported here. Normal hematocrit values for the male and female, respectively, are 47.0 ± 7 and 42.0 ± 5.

The hematocrit is determined by the micromethod, so only a drop of blood is needed. If possible, all members of the class should prepare their capillary tubes at the same time so the centrifuge can be properly balanced and run only once.

1.   Obtain two heparinized capillary tubes, Seal-ease or modeling clay, a lancet, alcohol swabs, and some cotton balls.

2.   Cleanse the finger, and allow the blood to flow freely. Wipe away the first few drops and, holding the red-line-marked end of the capillary tube to the blood drop, allow the tube to fill at least three-fourths full by capillary action (Figure 22.4a). If the blood is not flowing freely, the end of the capillary tube will not be completely submerged in the blood during filling, air will enter, and you will have to prepare another sample.

3.   Plug the blood-containing end by pressing it into the Seal-ease or clay (Figure 22.4b). Prepare a second tube in the same manner.

4.   Place the prepared tubes opposite one another in the radial grooves of the microhematocrit centrifuge with the sealed ends abutting the rubber gasket at the centrifuge periphery (Figure 22.4c). This loading procedure balances the centrifuge and prevents blood from spraying everywhere by centrifugal force. *Make a note of the*

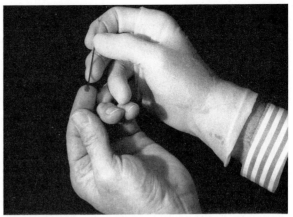
(a)

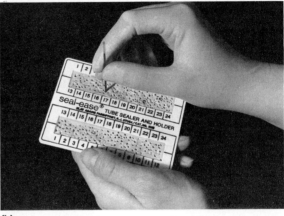
(b)

(c)

F22.4
___

Steps in a hematocrit determination. (a) Load a heparinized capillary tube with blood. **(b)** Plug the blood-containing end of the tube with clay. **(c)** Place the tube in a microhematocrit centrifuge. (Centrifuge must be balanced.)

numbers of the grooves your tubes are in. When all the tubes have been loaded, make sure the centrifuge is properly balanced, and secure the centrifuge cover. Turn the centrifuge on, and set the timer for 4 or 5 minutes.

5. Determine the percentage of RBCs, WBCs, and plasma by using the microhematocrit reader. The RBCs are the bottom layer, the plasma is the top layer, and the WBCs are the buff-colored layer between the two. If the reader is not available, use a millimeter ruler to measure the length of the filled capillary tube occupied by each element, and compute its percentage by using the following formula:

$$\frac{\text{Height of the column composed of the element (mm)}}{\text{Height of the original column of whole blood (mm)}} \times 100$$

Record your calculations below and on the data sheet.

% RBC _____ % WBC _____ % plasma _____

Usually WBCs constitute 1% of the total blood volume. How do your blood values compare to this figure and to the normal percentages for RBCs and plasma? (See p. 230.)

_____

As a rule, a hematocrit is considered a more accurate test for determining the RBC composition of the blood than the total RBC count. A hematocrit within the normal range generally indicates a normal RBC number, whereas an abnormally high or low hematocrit is cause for concern.

Hemoglobin Concentration Determination

As noted earlier, a person can be anemic even with a normal RBC count. Since hemoglobin is the RBC protein responsible for oxygen transport, perhaps the most accurate way of measuring the oxygen-carrying capacity of the blood is to determine its hemoglobin content. Oxygen, which combines reversibly with the heme (iron-containing portion) of the hemoglobin molecule, is picked up by the blood cells in the lungs and unloaded in the tissues. Thus, the more hemoglobin molecules the RBCs contain, the more oxygen they will be able to transport. Normal blood contains 12 to 16 g hemoglobin per 100 ml blood. Hemoglobin content in men is slightly higher (14 to 18 g) than in women (12 to 16 g).

Several techniques have been developed to estimate the hemoglobin content of blood, ranging from the old, rather inaccurate Tallquist method to expensive colorimeters, which are precisely calibrated and yield highly accurate results. Directions for both the Tallquist method and a hemoglobinometer are provided here.

TALLQUIST METHOD

1. Obtain a Tallquist hemoglobin scale, lancets, alcohol swabs, and cotton balls.

2. Use instructor-provided blood or prepare the finger as previously described. (For best results, make sure the alcohol evaporates before puncturing your finger.) Place one good-sized drop of blood on the special absorbent paper provided with the color chart. The blood stain should be larger than the holes on the color chart.

3. As soon as the blood has dried and loses its glossy appearance, match its color, under natural light, with the color standards by moving the specimen under the comparison chart so that the blood stain appears at all the various apertures. (The blood should not be allowed to dry to a brown color, as this will result in an inaccurate reading.) Because the colors on the chart represent 1% variations in hemoglobin content, it may be necessary to estimate the percentage if the color of your blood sample is intermediate between two color standards.

4. On the data sheet on page 240, record your results as the percentage of hemoglobin concentration and as grams per 100 ml of blood.

HEMOGLOBINOMETER DETERMINATION

1. Obtain a hemoglobinometer, hemolysis applicator stick, alcohol swab, and lens paper and bring them to your bench. Test the hemoglobinometer light source to make sure it is working; if not, request new batteries before proceeding and test it again.

2. Remove the blood chamber from the slot in the side of the hemoglobinometer and disassemble the blood chamber by separating the glass plates from the metal clip. Notice as you do this that the larger glass plate has an H-shaped depression cut into it that acts as a moat to hold the blood, whereas the smaller glass piece is flat and serves as a coverslip.

3. Clean the glass plates with an alcohol swab and then wipe dry with lens paper. Hold the plates by their sides to prevent smearing during the wiping process.

4. Reassemble the blood chamber (remember: larger glass piece on the bottom with the moat up), but leave the moat plate about halfway out to provide adequate exposed surface to charge it with blood.

5. Obtain a drop of blood (from the provided sample or from your fingertip as before), and place it on the depressed area of the moat plate that is closest to you (Figure 22.5a).

6. Using the wood hemolysis applicator, stir or agitate the blood to rupture (lyse) the RBCs (Figure 22.5b). This usually takes 35 to 45 seconds. Hemolysis is complete when the blood appears transparent rather than cloudy.

7. Push the blood-containing glass plate all the way into the metal clip and then firmly insert the charged blood chamber back into the slot on the side of the instrument (Figure 22.5c).

8. Hold the hemoglobinometer in your left hand with your left thumb resting on the light switch located on the underside of the instrument. Look into the eyepiece

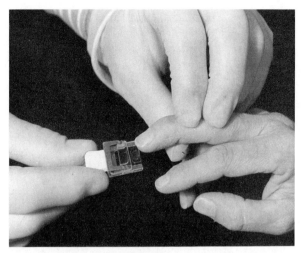

(a) A drop of blood is added to the moat plate of the blood chamber. The blood must flow freely.

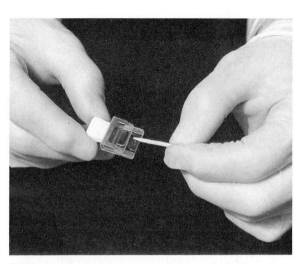

(b) The blood sample is hemolyzed with a wooden hemolysis applicator. Thirty-five to forty-five seconds are required for complete hemolysis.

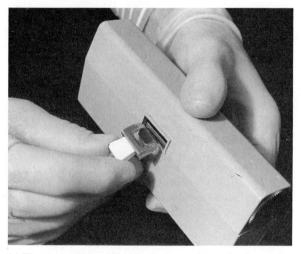

(c) The charged blood chamber is inserted into the slot on the side of the hemoglobinometer.

(d) The colors of the green split screen are found by moving the slide with right index finger. When the two colors match in density, the grams/100 ml and %Hb are read on the scale.

F22.5

Hemoglobin determination using a hemoglobinometer.

and notice that there is a green area divided into two halves (a split field).

9. With the index finger of your right hand, slowly move the slide on the right side of the hemoglobinometer back and forth until the two halves of the green field match (Figure 22.5d).

10. Note and record on the data sheet on page 240 the grams Hb (hemoglobin)/100 ml blood indicated on the uppermost scale by the index mark on the slide. Also record % Hb, indicated by one of the lower scales.

11. Disassemble the blood chamber once again, and carefully place its parts (glass plates and clip) into a bleach-containing beaker.

Generally speaking, the relationship between the PCV and grams of hemoglobin per 100 ml blood is 3:1. How do your values compare?

Record, on the data sheet, the value obtained from your data.

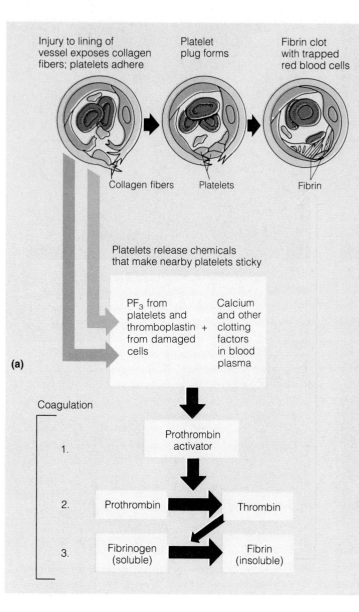

(a)

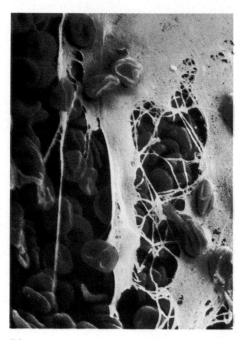

(b)

F22.6

Events of hemostasis and blood clotting. (a)
Simple schematic of events. Steps numbered 1–3
represent the major events of coagulation.
(b) Photomicrograph of RBCs trapped in a fibrin
mesh (1,600×).

Coagulation Time

Blood clotting, or **coagulation,** is a protective mechanism that minimizes blood loss when blood vessels are
ruptured. This process requires the interaction of many
substances normally present in the plasma (clotting factors, or procoagulants) as well as some released by
platelets and injured tissues. Basically hemostasis pro-
ceeds as follows (Figure 22.6a): The injured tissues and
platelets release **thromboplastin** and **PF$_3$** respectively,
which trigger the clotting mechanism, or cascade.
Thromboplastin and PF$_3$ interact with other blood protein clotting factors and calcium ions to form **pro-
thrombin activator,** which in turn converts **prothrom-
bin** (present in plasma) to **thrombin.** Thrombin then
acts enzymatically to polymerize the soluble **fibrinogen**
proteins (present in plasma) into insoluble **fibrin,** which
forms a meshwork of strands that traps the RBCs and
forms the basis of the clot (Figure 22.6b). Normally,
blood removed from the body clots within 2 to 6
minutes.

1.   Obtain a *nonheparinized* capillary tube, a lancet,
cotton balls, a triangular file, and alcohol swabs.

2.   Use instructor-supplied blood or clean and prick the
finger to produce a free flow of blood.

3.   Place one end of the capillary tube in the blood
drop, and hold the opposite end at a lower level to collect the sample.

4.   Lay the capillary tube on a paper towel.

Record the time. _____

5.   At 30-sec intervals, make a small nick on the tube
close to one end with the triangular file, and then carefully break the tube. Slowly separate the ends to see if a
gel-like thread of fibrin spans the gap. When this occurs,
record below and on the data sheet the time for coagulation to occur. Are your results within the normal time
range?

_____

6.   Dispose of the capillary tube and used supplies in
the disposable autoclave bag.

ABO blood type	Antigens present on RBC membranes	Antibodies present in plasma	% of U.S. Population		
			White	Black	Asian
A	A	Anti-B	40	27	28
B	B	Anti-A	11	20	27
AB	A and B	None	4	4	5
O	Neither	Anti-A and anti-B	45	49	40

Blood Typing

Blood typing is a system of blood classification based on the presence of specific glycoproteins on the outer surface of the RBC plasma membrane. Such proteins are called **antigens,** or **agglutinogens,** and are genetically determined. In many cases, these antigens are accompanied by plasma proteins, **antibodies** or **agglutinins,** that react with RBCs bearing different antigens, causing them to be clumped, agglutinated, and eventually hemolyzed. It is because of this phenomenon that a person's blood must be carefully typed before a whole blood or packed cell transfusion.

Several blood typing systems exist, based on the various possible antigens, but the factors routinely typed for are antigens of the ABO and Rh blood groups which are most commonly involved in transfusion reactions. Other blood factors, such as Kell, Lewis, M, and N, are not routinely typed for unless the individual will require multiple transfusions. The basis of the ABO typing is shown in the chart at the top of this page.

Individuals whose red blood cells carry the Rh antigen are Rh positive (approximately 85% of the U.S. population); those lacking the antigen are Rh negative. Unlike ABO blood groups neither the blood of the Rh-positive (Rh$^+$) nor Rh-negative (Rh$^-$) individuals carries preformed anti-Rh antibodies. This is understandable in the case of the Rh-positive individual. However, Rh-negative persons who receive transfusions of Rh-positive blood become sensitized by the Rh antigens of the donor RBCs and their systems begin to produce anti-Rh antibodies. On subsequent exposures to Rh-positive blood, typical transfusion reactions occur, resulting in the clumping and hemolysis of the donor blood cells.

ALERT: Although the blood of dogs and other mammals does react with some of the human agglutinins (present in the antisera), the reaction is not as pronounced and varies with the animal blood used. Hence, the most accurate and predictable blood typing results are obtained with human blood.

1.   Obtain two clean microscope slides, a wax marking pencil, anti-A, anti-B, and anti-Rh typing sera, toothpicks, lancets, alcohol swabs, and the Rh typing box.

2.   Divide slide 1 into two equal halves with the wax marking pencil. Label the lower left-hand corner "anti-A" and the lower right-hand corner "anti-B." Mark the bottom of slide 2 "anti-Rh."

3.   Place one drop of anti-A serum on the *left* side of slide 1. Place one drop of anti-B serum on the *right* side of slide 1. Place one drop of anti-Rh serum in the center of slide 2.

4.   Cleanse your finger with an alcohol swab, pierce the finger with a lancet, and wipe away the first drop of blood. Obtain 3 drops of freely flowing blood, placing one drop on each side of slide 1 and a drop on slide 2.

5.   Quickly mix each blood-antiserum sample with a *fresh* toothpick. Then dispose of the toothpicks, lancet, and used alcohol swab in the autoclave bag.

6.   Place slide 2 on the Rh typing box and rock gently back and forth. (A slightly higher temperature is required for precise Rh typing than for ABO typing.)

7.   After 2 minutes, observe all three blood samples for evidence of clumping. The agglutination that occurs in the positive test for the Rh factor is very fine and difficult to perceive; thus if there is any question, observe the slide under the microscope. Record your observations in the chart below.

	Observed (+)	Not observed (−)
Presence of clumping with anti-A		
Presence of clumping with anti-B		
Presence of clumping with anti-Rh		

F22.7

Blood typing of ABO blood types. When serum containing anti-A or anti-B agglutinins is added to a blood sample, agglutination will occur between the agglutinin and the corresponding agglutinogen (A or B). As illustrated, agglutination occurs with both sera in blood group AB, with anti-B serum in blood group B, with anti-A serum in blood group A, and with neither serum in blood group O.

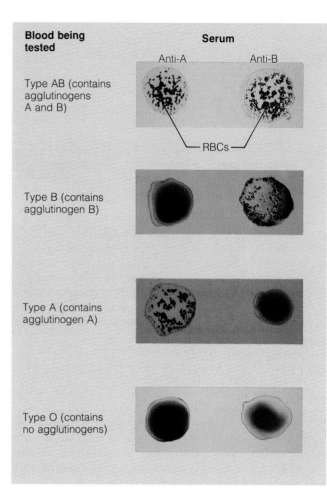

8. Interpret your ABO results in light of the information in Figure 22.7. If clumping was observed on slide 2, you are Rh positive. If not, you are Rh negative.

9. Record your blood type on the data sheet.

10. Put the used slides in the bleach-containing bucket at the general supply area; put disposable supplies in the autoclave bag.

Observation of Demonstration Slides

Before leaving the laboratory, take the time to look at the five slides that have been put on demonstration by your instructor. Record your observations in the appropriate section of the Review Sheets for Exercise 22. You can refer to your notes later to respond to questions about the blood pathologies represented on the slides.

Hematologic Test Data Sheet

Differential WBC count:

_____ % granulocytes   _____ % agranulocytes

_____ % neutrophils   _____ % lymphocytes

_____ % eosinophils   _____ % monocytes

_____ % basophils

Hematocrit (PCV):

RBC _____ % of blood volume

WBC _____ % of blood volume } not generally reported

Plasma _____ % of blood volume

Hemoglobin content:

Tallquist method:

_____ g/100 ml blood; _____ %

Hemoglobinometer (type: _____)

_____ g/100 ml blood; _____ %

Ratio (PCV/grams Hb in 100 ml blood): _____

Coagulation time _____

Blood typing:

ABO group _____ Rh factor _____

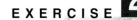

Anatomy of the Heart

OBJECTIVES

1. To describe the location of the heart.
2. To name and locate the major anatomical areas and structures of the heart when provided with an appropriate model, diagram, or dissected sheep heart, and to explain the function of each.
3. To trace the pathway of blood through the heart.
4. To explain why the heart is called a double pump, and to compare the pulmonary and systemic circuits.
5. To explain the operation of the atrioventricular and semilunar valves.
6. To name and follow the functional blood supply of the heart.
7. To describe the histologic structure of cardiac muscle, and to note the importance of its intercalated discs and the spiral arrangement of its fibers in the heart.

MATERIALS

Torso model or laboratory chart showing heart anatomy

Red and blue pencils

Three-dimensional models of cardiac and skeletal muscle

Heart model (three-dimensional)

X ray of the human thorax for observation of the position of the heart *in situ;* X ray viewing box

Preserved sheep heart, pericardial sacs intact (if possible)

Dissecting pan and instruments

Pointed glass rods for probes

Protective skin cream or disposable gloves

Compound microscope

Histologic slides of cardiac muscle (longitudinal section)

Human Cardiovascular System: The Heart videotape*

 See Appendix B, Exercise 23 for links to A.D.A.M. Standard.

See Appendix C, Exercise 23 for links to *Anatomy and PhysioShow: The Videodisc.*

* Available to qualified adopters from Benjamin/Cummings.

The major function of the **cardiovascular system** is transportation. Using blood as the transport vehicle, the system carries oxygen, digested foods, cell wastes, electrolytes, and many other substances vital to the body's homeostasis to and from the body cells. The system's propulsive force is the contracting heart, which can be compared to a muscular pump equipped with one-way valves. As the heart contracts, it forces blood into a closed system of large and small plumbing tubes (blood vessels) within which the blood is confined and circulated. This exercise deals with the structure of the heart or circulatory pump. The anatomy of the blood vessels is considered separately in Exercise 24.

GROSS ANATOMY OF THE HUMAN HEART

The **heart,** a cone-shaped organ approximately the size of a fist, is located within the mediastinum, or medial cavity, of the thorax. It is flanked laterally by the lungs, posteriorly by the vertebral column, and anteriorly by

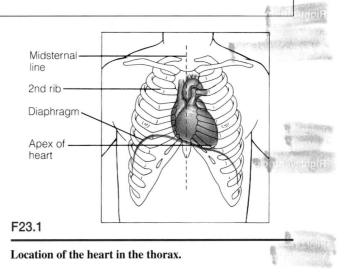

Midsternal line

2nd rib

Diaphragm

Apex of heart

F23.1

Location of the heart in the thorax.

the sternum (Figure 23.1). Its more pointed **apex** extends slightly to the left and rests on the diaphragm, approximately at the level of the fifth intercostal space. Its broader **base,** from which the great vessels emerge, lies

Brachiocephalic artery

Superior vena cava

Right pulmonary artery

Ascending aorta

Pulmonary trunk

Right pulmonary veins

Right atrium

Right coronary artery (in right atrioventricular groove)

Anterior cardiac vein

Right ventricle

Marginal artery

Small cardiac vein

Inferior vena cava

Left common carotid artery

Left subclavian artery

Aortic arch

Ligamentum arteriosum

Left pulmonary artery

Left pulmonary veins

Left atrium

Auricle

Circumflex artery

Left coronary artery (in left atrioventricular groove)

Left ventricle

Great cardiac vein

Anterior interventricular artery (in anterior interventricular sulcus)

Apex

(a)

Superior vena cava

Right pulmonary artery

Pulmonary trunk

Right atrium

Right pulmonary veins

Fossa ovalis

Pectinate muscles

Tricuspid valve

Right ventricle

Chordae tendineae

Trabeculae carneae

Inferior vena cava

Aorta

Left pulmonary artery

Left atrium

Left pulmonary veins

Pulmonary semilunar valve

Bicuspid (mitral) valve

Aortic semilunar valve

Left ventricle

Papillary muscle

Interventricular septum

Myocardium

Visceral pericardium

(b)

F23.2

Anatomy of the human heart. (a) External anterior view. **(b)** Frontal section.

beneath the second rib and points toward the right shoulder. *In situ,* the right ventricle of the heart forms most of its anterior surface.

If an X ray of a human thorax is available, verify the relationships described above; otherwise Figure 23.1 will suffice.

The heart is enclosed within a double-walled fibroserous sac called the pericardium. The thin **visceral pericardium,** or **epicardium,** is closely applied to the heart muscle. It reflects downward at the base of the heart to form its companion serous membrane, the outer, loosely applied **parietal pericardium,** which is attached at the heart apex to the diaphragm. Serous fluid produced by these membranes allows the heart to beat in a relatively frictionless environment. The serous parietal pericardium, in turn, lines the loosely fitting superficial **fibrous pericardium** composed of dense connective tissue.

Inflammation of the pericardium, **pericarditis,** causes painful adhesions between the serous pericardial layers. These adhesions interfere with heart movements. ■

The walls of the heart are composed primarily of cardiac muscle—the **myocardium**—which is reinforced internally by a dense fibrous connective tissue network. This network—the *fibrous skeleton of the heart*—is more elaborate and thicker in certain areas, for example, around the valves and at the base of the great vessels leaving the heart.

Figure 23.2 shows two views of the heart—an external anterior view and a frontal section. As its anatomical areas are described in the text, consult the figure. When you have pinpointed all the structures, observe the human heart model, and reidentify the same structures without reference to the figure.

Heart Chambers

The heart is divided into four chambers: two superior **atria** and two inferior **ventricles,** each lined by a thin serous membrane called the **endocardium.** The septum that divides the heart longitudinally is referred to as the **interatrial** or **interventricular septum,** depending on which chambers it partitions. Functionally, the atria are receiving chambers and are relatively ineffective as pumps. Blood flows into the atria under low pressure from the veins of the body. The right atrium receives relatively oxygen-poor blood from the body via the **superior** and **inferior venae cavae.** Four **pulmonary veins** deliver oxygen-rich blood from the lungs to the left atrium.

The inferior thick-walled ventricles, which form the bulk of the heart, are the discharging chambers. They force blood out of the heart into the large arteries that emerge from its base. The right ventricle pumps blood into the **pulmonary trunk,** which routes blood to the lungs to be oxygenated. The left ventricle discharges blood into the **aorta,** from which all systemic arteries of the body diverge to supply the body tissues. Discussions

of the heart's pumping action usually refer to ventricular activity.

Heart Valves

Four valves enforce a one-way blood flow through the heart chambers. The **atrioventricular (AV) valves,** located between the atrial and ventricular chambers on each side, prevent backflow into the atria when the ventricles are contracting. The left atrioventricular valve, also called the **mitral** or **bicuspid valve,** consists of two cusps, or flaps, of endocardium. The right atrioventricular valve, the **tricuspid valve,** has three cusps (Figure 23.3). Tiny white collagenic cords called the **chordae tendineae** (literally, "heart strings") anchor the cusps to the ventricular walls. The chordae tendineae originate from small bundles of cardiac muscle, called **papillary muscles,** that project from the myocardial wall (see Figure 23.2).

When blood is flowing passively into the atria and then into the ventricles during **diastole** (the period of ventricular relaxation), the AV valve flaps hang limply into the ventricular chambers and then are carried passively toward the atria by the accumulating blood. When the ventricles contract (**systole**) and compress the blood in their chambers, the intraventricular blood pressure rises, causing the valve flaps to be reflected superiorly, which closes the AV valves. The chordae tendineae, pulled taut by the contracting papillary muscles, anchor the flaps in a closed position that prevents backflow into the atria during ventricular contraction. If unanchored, the flaps would blow upward into the atria rather like an umbrella being turned inside out by a strong wind.

The second set of valves, the **pulmonary** and **aortic semilunar valves,** each composed of three pocketlike cusps, guards the bases of the two large arteries leaving the ventricular chambers. The valve cusps are forced open and flatten against the walls of the artery as the ventricles discharge their blood into the large arteries during systole. However, when the ventricles relax, blood flows backward toward the heart and the cusps fill with blood, closing the semilunar valves and preventing arterial blood from reentering the heart.

PULMONARY, SYSTEMIC, AND CARDIAC CIRCULATIONS

Pulmonary and Systemic Circulations

The heart functions as a double pump. The right side serves as the **pulmonary circulation** pump, shunting the carbon dioxide–rich blood entering its chambers to the lungs to unload carbon dioxide and pick up oxygen, and then back to the left side of the heart. The function of this circuit is strictly to provide for gas exchange. The second circuit, which carries oxygen-rich blood from the left heart through the body tissues and back to the right heart, is called the **systemic circulation.** It supplies the functional blood supply to all body tissues.

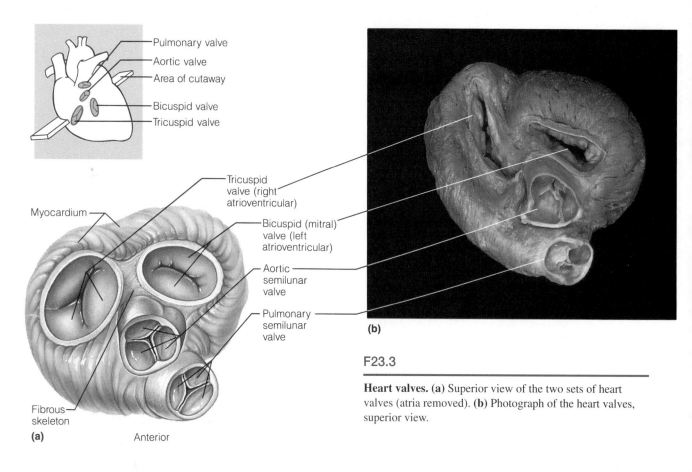

Pulmonary valve
Aortic valve
Area of cutaway
Bicuspid valve
Tricuspid valve

Tricuspid valve (right atrioventricular)

Myocardium

Bicuspid (mitral) valve (left atrioventricular)

Aortic semilunar valve

Pulmonary semilunar valve

Fibrous skeleton

(a)   Anterior

(b)

F23.3

Heart valves. (a) Superior view of the two sets of heart valves (atria removed). **(b)** Photograph of the heart valves, superior view.

Trace the pathway of blood through the heart by adding arrows to the frontal section diagram (see Figure 23.2b). Use red arrows for the oxygen-rich blood and blue arrows for the less oxygen-rich blood.

Cardiac Circulation

Even though the heart chambers are almost continually bathed with blood, this contained blood does not nourish the myocardium. The functional blood supply of the heart is provided by the right and left coronary arteries (see Figures 23.2 and 23.4). The **coronary arteries** issue from the base of the aorta just above the aortic semilunar valve and encircle the heart in the **atrioventricular groove** at the junction of the atria and ventricles. They then ramify over the heart's surface, the right coronary artery supplying the posterior surface of the ventricles and the lateral aspect of the right side of the heart, largely through its **posterior interventricular** and **marginal artery** branches. The left coronary artery supplies the anterior ventricular walls and the laterodorsal part of the left side of the heart via its two major branches, the **anterior interventricular artery** and the **circumflex artery.** The coronary arteries and their branches are compressed during systole and fill when the heart is relaxed. The myocardium is largely drained

by the **great, middle,** and **small cardiac veins,** which empty into the **coronary sinus.** The coronary sinus, in turn, empties into the right atrium. In addition, several **anterior cardiac veins** empty directly into the right atrium (Figure 23.4).

MICROSCOPIC ANATOMY OF CARDIAC MUSCLE

Cardiac muscle is found in only one place—the heart. The heart acts as a vascular pump, propelling blood to all tissues of the body; cardiac muscle is thus very important to life. Cardiac muscle is involuntary, ensuring a constant blood supply.

The cardiac cells, only sparingly invested in connective tissue, are arranged in spiral or figure-8-shaped bundles (Figure 23.5). When the heart contracts, its internal chambers become smaller (or are temporarily obliterated), forcing the blood into the large arteries leaving the heart.

1. Observe the three-dimensional model of cardiac muscle, examining its branching cells and the areas where the cells interdigitate, the **intercalated discs.** These two structural features provide a continuity to cardiac muscle not seen in other

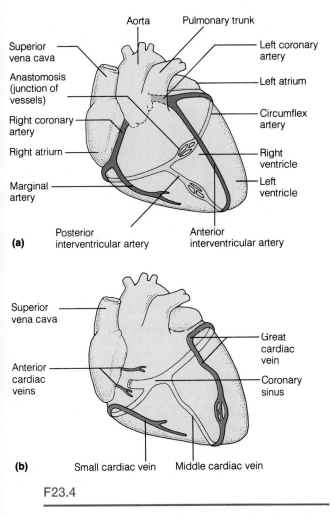

(a)

Aorta
Pulmonary trunk
Superior vena cava
Left coronary artery
Anastomosis (junction of vessels)
Left atrium
Right coronary artery
Circumflex artery
Right atrium
Right ventricle
Marginal artery
Left ventricle
Posterior interventricular artery
Anterior interventricular artery

(b)

Superior vena cava
Great cardiac vein
Anterior cardiac veins
Coronary sinus
Small cardiac vein
Middle cardiac vein

F23.4

Cardiac circulation. (a) Main coronary arteries. **(b)** Major cardiac veins.

muscle tissues and allow close coordination of heart activity.

2.   Compare the model of cardiac muscle to the model of skeletal muscle. Note the similarities and differences between the two kinds of muscle tissue.

3.   Obtain and observe a longitudinal section of cardiac muscle under high power. Identify the nucleus, striations, intercalated discs, and sarcolemma of the individual cells and then compare your observations to the view seen in Figure 23.6.

DISSECTION OF THE SHEEP HEART

Dissection of a sheep heart is valuable because it is similar in size and structure to the human heart. Also, a dissection experience allows you to view structures in a way not possible with models and diagrams. Refer to Figure 23.7 as you proceed with the dissection.

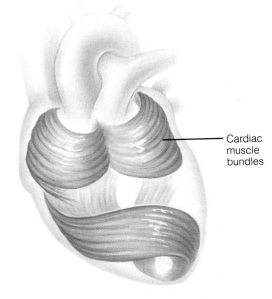

Cardiac muscle bundles

F23.5

Longitudinal view of the heart chambers showing the spiral arrangement of the cardiac muscle fibers.

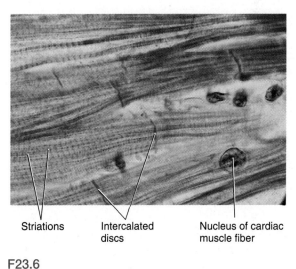

Striations
Intercalated discs
Nucleus of cardiac muscle fiber

F23.6

Photomicrograph of cardiac muscle (688×).

1.   Obtain a preserved sheep heart, a dissection tray, dissecting instruments and, if desired, protective skin cream or gloves. Rinse the sheep heart in cold water to remove excessive preservatives and to flush out any trapped blood clots. Now you are ready to make your observations.

2.   Observe the texture of the pericardium. Also, note its point of attachment to the heart. Where is it attached?

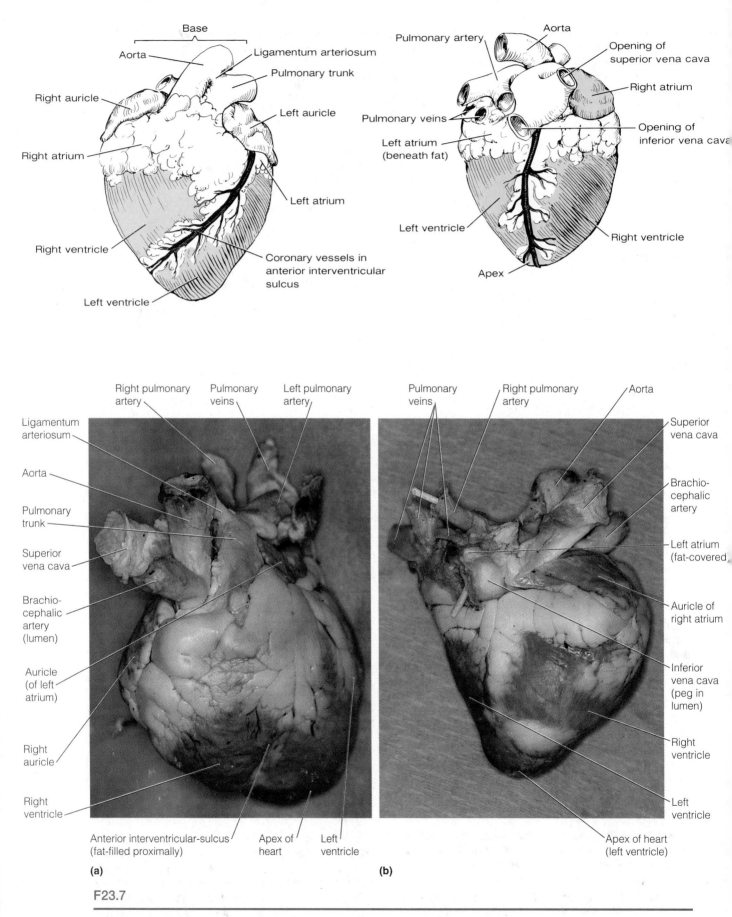

Anatomy of the sheep heart. (a) Anterior view. **(b)** Posterior view. Diagrammatic views at top; photographs at bottom.

F23.7

3. If the serous pericardial sac is still intact, slit open the parietal pericardium and cut it from its attachments. Observe the visceral pericardium (epicardium). Using a sharp scalpel, carefully pull a little of this serous membrane away from the myocardium. How does its position, thickness, and apposition to the heart differ from those of the parietal pericardium?

_____

_____

_____

4. Examine the external surface of the heart. Notice the accumulation of adipose tissue, which in many cases marks the separation of the chambers and the location of the coronary arteries that nourish the myocardium. Carefully scrape away some of the fat with a scalpel to expose the coronary blood vessels.

5. Identify the base and apex of the heart, and then identify the two wrinkled **auricles,** earlike flaps of tissue projecting from the atrial chambers. The balance of the heart muscle is ventricular tissue. To identify the left ventricle, compress the ventricular chambers on each side of the longitudinal fissures carrying the coronary blood vessels. The side that feels thicker and more solid is the left ventricle. The right ventricle feels much thinner and somewhat flabby when compressed. This difference reflects the greater demand placed on the left ventricle, which must pump blood through the much longer systemic circulation, a pathway with much higher resistance than the pulmonary circulation served by the right ventricle. Hold the heart in its anatomical position (Figure 23.7a), with the anterior surface uppermost. In this position the left ventricle composes the entire apex and the left side of the heart.

6. Identify the pulmonary trunk and the aorta extending from the superior aspect of the heart. The pulmonary trunk is the more anterior, and you may see its division into the right and left pulmonary arteries if it has not been cut too closely to the heart. The thicker-walled aorta, which branches almost immediately, is located just beneath the pulmonary trunk. The first branch of the sheep aorta, the **brachiocephalic artery,** is identifiable unless the aorta has been cut immediately as it leaves the heart. The brachiocephalic artery splits to form the right carotid and subclavian arteries, which supply the right side of the head and right forelimb, respectively.

Carefully clear away some of the fat between the pulmonary trunk and the aorta to expose the **ligamentum arteriosum,** a cordlike remnant of the **ductus arteriosus.** (In the fetus, the ductus arteriosus allows blood to pass directly from the pulmonary trunk to the aorta, thus bypassing the nonfunctional fetal lungs.)

7. Cut through the wall of the aorta until you see the aortic semilunar valve. Identify the two openings to the coronary arteries just above the valve. Insert a probe into one of these holes to see if you can follow the course of a coronary artery across the heart.

8. Turn the heart to view its posterior surface. The heart will appear as shown in Figure 23.7b. Notice that the right and left ventricles appear equal-sized in this view. Identify the four thin-walled pulmonary veins entering the left atrium. (It may or may not be possible to locate the pulmonary veins from this vantage point, depending on how they were cut as the heart was removed.) Identify the superior and inferior venae cavae entering the right atrium. Compare the approximate diameter of the superior vena cava with the diameter of the aorta.

Which is larger? _____

Which has thicker walls? _____

Why do you suppose these differences exist?

_____

_____

_____

9. Insert a probe into the superior vena cava and use scissors to cut through its wall so that you can view the interior of the right atrium. Do not extend your cut entirely through the right atrium or into the ventricle. Observe the right atrioventricular valve.

How many flaps does it have? _____

Pour some water into the right atrium and allow it to flow into the ventricle. Slowly and gently squeeze the right ventricle to watch the closing action of this valve. (If you squeeze too vigorously, you'll get a face full of water!) Drain the water from the heart before continuing.

10. Return to the pulmonary trunk and cut through its anterior wall until you can see the pulmonary semilunar valve. Pour some water into the base of the pulmonary trunk to observe the closing action of this valve. How does its action differ from that of the atrioventricular valve?

_____

_____

_____

After observing semilunar valve action, drain the heart once again. Return to the superior vena cava, and continue the cut made in its wall through the right atrium and right atrioventricular valve into the right ventricle. Parallel the anterior border of the interventricular septum until you "round the corner" to the dorsal aspect of the heart (Figure 23.8).

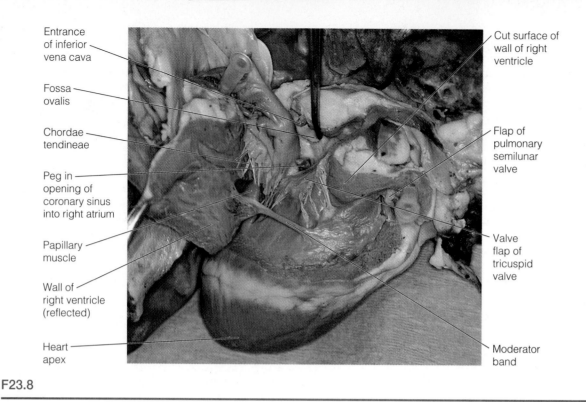

Entrance of inferior vena cava

Fossa ovalis

Chordae tendineae

Peg in opening of coronary sinus into right atrium

Papillary muscle

Wall of right ventricle (reflected)

Heart apex

Cut surface of wall of right ventricle

Flap of pulmonary semilunar valve

Valve flap of tricuspid valve

Moderator band

F23.8

Right side of the heart opened and reflected to reveal internal structures.

11. Reflect the cut edges of the superior vena cava, right atrium, and right ventricle to obtain the view seen in Figure 23.8. Observe the comblike ridges of muscle throughout most of the right atrium. This is called **pectinate muscle** (*pectin* means "comb"). Identify, on the ventral atrial wall, the large opening of the inferior vena cava and follow it to its external opening with a probe. Notice that the atrial walls in the vicinity of the venae cavae are smooth and lack the roughened appearance (pectinate musculature) of the other regions of the atrial walls. Just below the inferior vena caval opening, identify the opening of the **coronary sinus,** which returns venous blood of the coronary circulation to the right atrium. Nearby, locate an oval depression, the **fossa ovalis,** in the interatrial septum. This depression marks the site of an opening in the fetal heart, the **foramen ovale,** which allows blood to pass from the right to the left atrium, thus bypassing the fetal lungs.

12. Identify the papillary muscles in the right ventricle, and follow their attached chordae tendineae to the flaps of the tricuspid valve. Notice the pitted and ridged appearance (**trabeculae carneae**) of the inner ventricular muscle.

13. Make a longitudinal incision through the aorta and continue it into the left ventricle. Notice how much thicker the myocardium of the left ventricle is than that of the right ventricle. Compare the *shape* of the left ventricular cavity to the shape of the right ventricular cavity.

Are the papillary muscles and chordae tendineae observed in the right ventricle also present in the left ventricle?

Count the number of cusps in the left atrioventricular valve. How does this compare with the number seen in the right atrioventricular valve?

How do the sheep valves compare with their human counterparts?

14. Continue your incision from the left ventricle superiorly into the left atrium. Reflect the cut edges of the atrial wall, and attempt to locate the entry points of the pulmonary veins into the left atrium. Follow the pulmonary veins to the heart exterior with a probe. Notice how thin-walled these vessels are.

15. Properly dispose of the organic debris, and clean the dissecting tray and instruments before leaving the laboratory.

Anatomy of Blood Vessels

OBJECTIVES

1. To describe the tunics of blood vessel walls and state the function of each layer.

2. To correlate differences in artery, vein, and capillary structure with the functions these vessels perform.

3. To recognize a cross-sectional view of an artery and vein when provided with a microscopic view or appropriate diagram.

4. To list and/or identify the major arteries arising from the aorta, and to indicate the body region supplied by each.

5. To list and/or identify the major veins draining into the superior and inferior venae cavae, and to indicate the body regions drained.

6. To point out and/or discuss the unique features of special circulations (hepatic portal system, circle of Willis, pulmonary circulation, fetal circulation) in the body.

7. To point out anatomical differences between the vascular system of the human and the laboratory dissection specimen.

MATERIALS

Anatomical charts of human arteries and veins (or a three-dimensional model of the human circulatory system)

Anatomical charts of the following specialized circulations: pulmonary circulation, hepatic portal circulation, arterial supply and circle of Willis of the brain (or a brain model showing this circulation), fetal circulation

Compound microscope

Prepared microscope slides showing cross sections of an artery and vein

Dissection animals

Dissecting pans and instruments

Protective skin cream or disposable gloves

Human Cardiovascular System: The Blood Vessels videotape*

 See Appendix B, Exercise 24 for links to A.D.A.M. Standard.

See Appendix C, Exercise 24 for links to *Anatomy and PhysioShow: The Videodisc.*

*Available to qualified adopters from Benjamin/Cummings.

The blood vessels constitute a closed transport system. As the heart contracts, blood is propelled into the large arteries leaving the heart. It moves into successively smaller arteries and then to the arterioles, which feed the capillary beds in the tissues. Capillary beds are drained by the venules, which in turn empty into veins that ultimately converge on the great veins entering the heart. Thus arteries, carrying blood away from the heart, and veins, which drain the tissues and return blood to the heart, function simply as conducting vessels or conduits. Only the tiny capillaries that connect the arterioles and venules and ramify throughout the tissues directly serve the needs of the body's cells. It is through the capillary walls that exchanges are made between tissue cells and blood.

Respiratory gases, nutrients, and wastes move along diffusion gradients. Thus, oxygen and nutrients diffuse from the blood to the tissue cells, and carbon dioxide and metabolic wastes move from the cells to the blood.

In this exercise you will examine the microscopic structure of blood vessels and identify the major arteries and veins of the systemic and special circulations.

MICROSCOPIC STRUCTURE OF THE BLOOD VESSELS

Except for the microscopic capillaries, the walls of blood vessels are constructed of three coats, or *tunics* (Figure 24.1). The **tunica intima,** or **interna,** which lines the lumen of a vessel, is a single thin layer of *endothelium* (squamous cells underlain by a scant basal lamina) that is continuous with the endocardium of the heart. Its cells fit closely together, forming an extremely smooth blood vessel lining that helps decrease resistance to blood flow.

The **tunica media** is the more bulky middle coat and is composed primarily of smooth muscle and elastic tissue. The smooth muscle, under the control of the sympathetic nervous system, plays an active role in reducing or increasing the diameter of blood vessels, which in turn increases or decreases the peripheral resistance and blood pressure.

The **tunica externa,** or **adventitia,** the outermost tunic, is composed of areolar or fibrous connective tissue. Its function is basically supportive and protective.

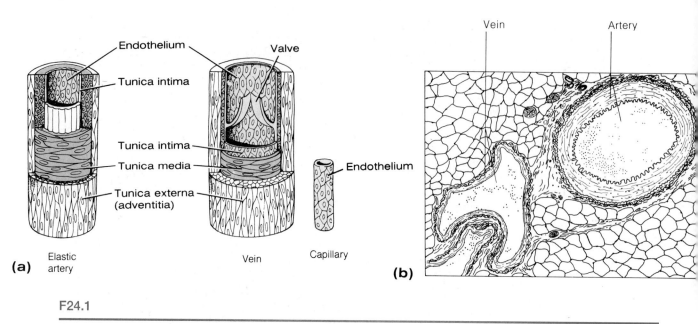

(a)

(b)

F24.1

Structure of arteries, veins, and capillaries. (a) Diagrammatic view. **(b)** Line drawing of a small artery (right) and vein (left), cross-sectional view, shown in Plate 26 in the Histology Atlas.

In general, the walls of arteries are thicker than those of veins. The tunica media in particular tends to be much heavier and contains substantially more smooth muscle and elastic tissue. This anatomical difference reflects a functional difference in the two types of vessels. Arteries, which are closer to the pumping action of the heart, must be able to expand as an increased volume of blood is propelled into them and then recoil passively as the blood flows off into the circulation during diastole. Their walls must be sufficiently strong and resilient to withstand such pressure fluctuations. Since these larger arteries have such large amounts of elastic tissue in their media, they are often referred to as *elastic arteries.* Smaller arteries, further along in the circulatory pathway, are exposed to less extreme pressure fluctuations. They have less elastic tissue but still have substantial amounts of smooth muscle in their media. For this reason, they are called *muscular arteries.*

By contrast, veins, which are far removed from the heart in the circulatory pathway, are not subjected to such pressure fluctuations and are essentially low-pressure vessels. Thus, veins may be thinner-walled without jeopardy. However, the low-pressure condition itself and the fact that blood returning to the heart often flows against gravity require structural modifications to ensure that venous return equals cardiac output. Thus, the lumens of veins tend to be substantially larger than those of corresponding arteries, and valves in larger veins function to prevent backflow of blood in much the same manner as the semilunar valves of the heart. The skeletal muscle "pump" also promotes venous return; as the skeletal muscles surrounding the veins contract and relax, the blood is milked through the veins toward the heart. (Anyone who has been standing relatively still for an extended time will be happy to show you their swollen ankles, caused by blood pooling in their feet during the period of muscle inactivity!) Pressure

changes that occur in the thorax during breathing also aid the return of blood to the heart.

The transparent walls of the tiny capillaries are only one cell layer thick, consisting of just the endothelium underlain by a basal lamina, that is, the tunica intima. Because of this exceptional thinness, exchanges are easily made between the blood and tissue cells.

1.   Obtain a slide showing a cross-sectional view of blood vessels and a microscope.

2.   Using Figure 24.1b and Plate 26 in the Histology Atlas as guides, scan the section to identify a thick-walled artery. Very often, but not always, its lumen will appear scalloped due to the constriction of its walls by the elastic tissue of the media.

3.   Identify a vein. Its lumen may appear elongated or irregularly shaped and collapsed, and its walls will be considerably thinner. Notice the difference in the relative amount of elastic fibers in the media of the two vessels. Also, note the thinness of the intima layer, which is composed of flat squamous-type cells.

MAJOR BLOOD VESSELS OF THE BODY

The next portion of this exercise provides tables (Tables 24.1–24.11) and figures (Figures 24.2–24.14) to help you locate and follow the major arteries and veins of the body on models or anatomical charts of the human vasculature.

(*Text continues on p. 271*)

TABLE 24.1   Pulmonary and Systemic Circulations

Pulmonary Circulation

The pulmonary circulation (Figure 24.2a) functions only to bring blood into close contact with the alveoli (air sacs) of the lungs so that gaseous exchanges can occur; it does not directly serve the metabolic needs of body tissues.

Oxygen-poor, dark red blood enters the pulmonary circulation as it is pumped from the right ventricle into the large **pulmonary trunk** (Figure 24.2b). The pulmonary trunk runs diagonally upward for about 8 cm and then divides abruptly to form the **right** and **left pulmonary arteries.** In the lungs, the pulmonary arteries subdivide into the **lobar arteries** (three in the right lung and two in the left lung), each of which serves one lung lobe. The lobar arteries accompany the main bronchi into the lungs and then branch profusely, forming arterioles and, finally, the dense networks of **pulmonary capillaries** that surround and cling to the delicate air sacs. It is here that oxygen loading and carbon dioxide unloading between the blood and alveolar air occur. As gas exchanges occur and oxygen content of the blood rises, the blood becomes bright red. The pulmonary capillary beds drain into venules, which in turn join to form the two **pulmonary veins** exiting from each lung. The four pulmonary veins complete the circuit by unloading their precious cargo into the left atrium of the heart. Note that any vessel with the term *pulmonary* or *lobar* in its name is part of the pulmonary circulation. All others are part of the systemic circulation.

Pulmonary arteries carry oxygen-poor, carbon dioxide–rich blood, and pulmonary veins carry oxygen-rich blood. This is exactly opposite the situation in the systemic circulation, where arteries carry oxygen-rich blood and veins carry carbon dioxide–rich, relatively oxygen-poor blood.

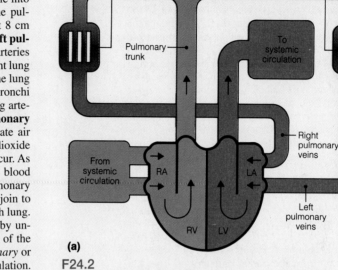

(a)

F24.2

Pulmonary circulation. (a) Schematic flow chart. **(b)** Illustration. The arterial system is shown in blue to indicate that the blood carried is oxygen-poor; the venous drainage is shown in red to indicate that the blood transported is oxygen-rich.

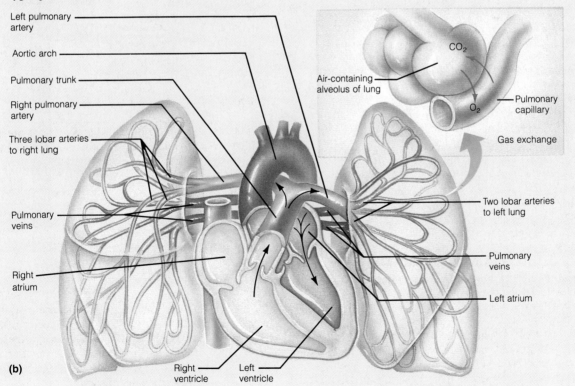

(b)

(Table continues on next page)

TABLE 24.1   (*continued*)

Systemic Circulation

The systemic circulation provides the *functional blood supply* to all body tissues; that is, it delivers oxygen, nutrients, and other needed substances while carrying away carbon dioxide and other metabolic wastes. Freshly oxygenated blood returning from the pulmonary circuit is pumped out of the left ventricle into the aorta (Figure 24.3). From the aorta, blood can take various routes, since essentially all systemic arteries branch from this single great vessel. The aorta arches upward from the heart and then curves and runs downward along the body midline to its terminus in the pelvis, where it splits to form the two large arteries serving the lower extremities. The various branches of the aorta continue to subdivide to produce the arterioles and, finally, the seemingly countless capillaries that ramify through the body organs. Venous blood draining from organs inferior to the diaphragm ultimately enters the inferior vena cava. Except for some thoracic venous drainage (which enters the azygos system of veins), body regions above the diaphragm are drained by the superior vena cava. The venae cavae empty the carbon dioxide–laden blood into the right atrium of the heart.

Two points concerning the two major circulations must be emphasized: (1) Blood passes from systemic veins to systemic arteries only after first moving through the pulmonary circuit (see Figure 24.2a), and (2) although the entire cardiac output of the right ventricle passes through the pulmonary circulation, only a small fraction of the output of the left ventricle flows through any single organ (Figure 24.3). The systemic circulation can be viewed as multiple circulatory channels functioning in parallel to distribute blood to all body organs.

As you examine tables that follow and locate the various systemic arteries and veins in the illustrations, be aware of cues that make your memorization task easier. In many cases, the name of a vessel reflects the body region traversed (axillary, brachial, femoral, etc.), the organ served (renal, hepatic, gonadal, etc.), or the bone followed (vertebral, radial, tibial, etc.). Also, notice that arteries and veins tend to run together side by side and, in many places, they also run with nerves. Finally, be alert to the fact that the systemic vessels do not always match on the right and left sides of the body. Some of the large, deep vessels of the trunk region are asymmetrical or unpaired (their initial symmetry is lost during embryonic development), while almost all vessels are bilaterally symmetrical in the head and limbs.

F24.3

Schematic flow chart showing an overview of the systemic circulation. The pulmonary circulation is shown in gray for comparison.

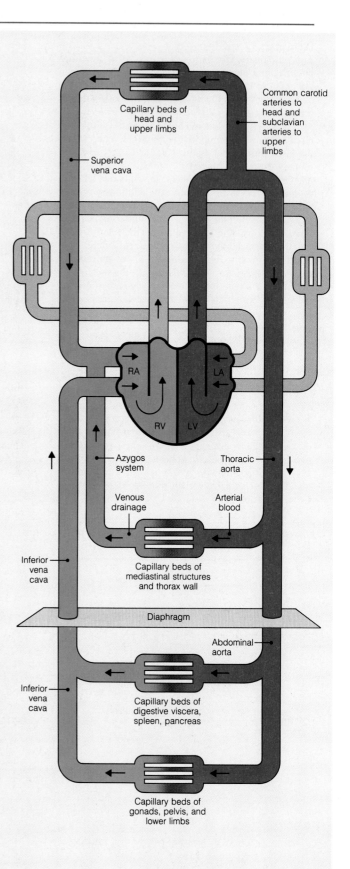

TABLE 24.2   The Aorta and Major Arteries of the Systemic Circulation

The major arteries of the systemic circulation are diagrammed in flow chart form in Figure 24.4. Notice that only the distribution of the aorta is described in detail in this figure. Fine points about the various vessels arising from the aorta are provided in Tables 24.3 through 24.6.

Description and Distribution

Aorta.   This vessel is the largest artery in the body. In adults, the aorta's internal diameter is 2.5 cm and its wall is about 2 mm thick where it issues from the left ventricle of the heart. It decreases in size only slightly as it runs to its terminus. The aortic semilunar valve guards the base of the aorta and prevents backflow of blood during diastole. Opposite each of the semilunar valve cusps is an *aortic sinus,* a slight enlargement containing baroreceptors important in reflex regulation of blood pressure.

Different portions of the aorta are named according to their relative shape or location. The first portion, the **ascending aorta,** runs posteriorly and to the right of the pulmonary trunk for only about 5 cm before curving to the left as the aortic arch. The only branches of the ascending aorta are the **right** and **left coronary arteries** (see p. 244). The **aortic arch,** deep to the sternum, begins and ends at the sternal angle (level of T_4). Its three major branches (R to L) are: (1) the **brachiocephalic** ("arm-head") **artery,** which passes under the right clavicle and branches into the **right common carotid artery** and the **right subclavian artery,** (2) the **left common carotid artery,** and (3) the **left subclavian artery.** These three vessels supply the head, neck, upper limbs, and part of the thorax wall. The **thoracic,** or **descending, aorta** runs along the anterior spine from T_5 to T_{12}, sending off small branches to the thorax wall and viscera before piercing the diaphragm to become the **abdominal aorta** in the abdominal cavity. The abdominal aorta supplies the abdominal walls and viscera and splits into the **right** and **left common iliac arteries** at the level of L_4.

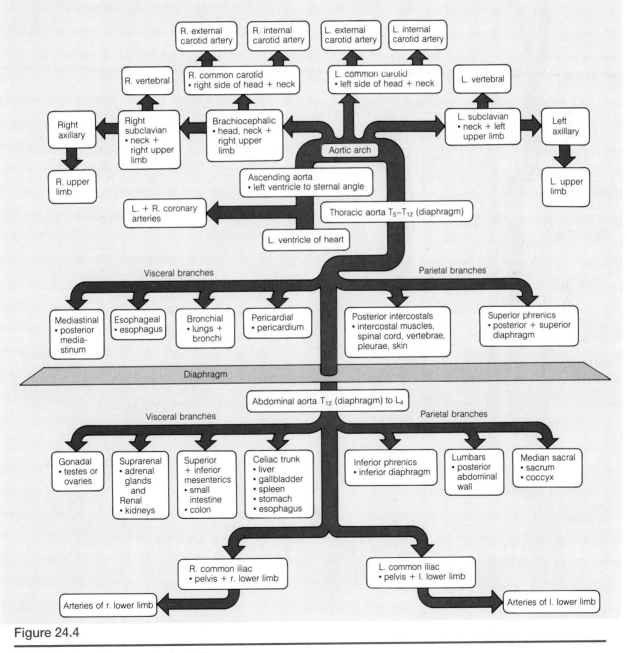

Figure 24.4

Major arteries of the systemic circulation.

TABLE 24.3   Arteries of the Head and Neck

Four paired arteries provide the functional blood supply of the head and neck. These are the common carotid arteries, the vertebral arteries, and the thyrocervical and costocervical trunks illustrated in Figure 24.5b. Of these, the common carotid arteries have the broadest distribution; these arteries and their branches are diagrammed in Figure 24.5a.

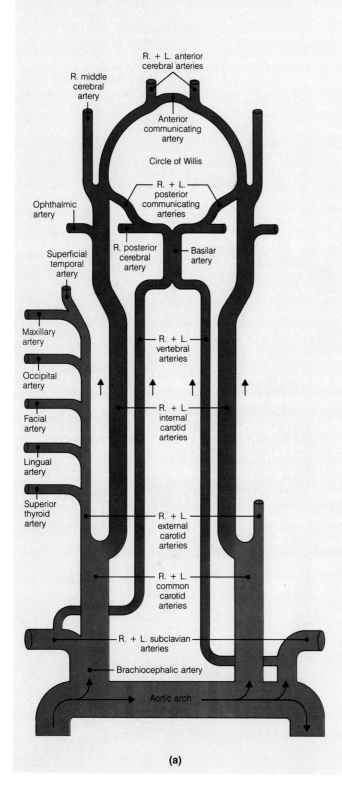

(a)

Each common carotid divides into 2 major branches (the internal and external carotid arteries) and, at the division point, each internal carotid artery has a slight dilation, the *carotid sinus,* that contains baroreceptors that assist in reflex blood pressure control. The *carotid bodies,* chemoreceptors involved in the control of respiratory rate, are located close by. Pressing on the neck in the area of the carotid sinuses can cause unconsciousness (*carot* = stupor) because it interferes with blood delivery to the brain.

Description and Distribution

Common carotid arteries.   The origins of these two arteries differ: The right common carotid artery arises from the brachiocephalic artery; the left is the second branch of the aortic arch. The common carotid arteries ascend through the lateral neck, and at the superior border of the larynx (the level of the "Adam's apple"), each divides into its two major branches, the *external* and *internal carotid arteries.*

The **external carotid arteries** supply most tissues of the head except for the brain and orbit. As each artery runs superiorly, it sends branches to the thyroid gland and larynx (**superior thyroid artery),** the tongue (**lingual artery),** the skin and muscles of the anterior face (**facial artery),** and the posterior scalp (**occipital artery).** Each external carotid artery terminates by splitting into **maxillary** and **superficial temporal arteries,** which together supply the upper and lower jaws and chewing muscles, the teeth, the nasal cavity, most of the scalp, lateral aspects of the face, and the dura mater.

The larger **internal carotid arteries** supply the orbits and most (over 80%) of the cerebrum. They assume a deep course and enter the skull through the carotid canals of the temporal bones. Inside the cranium, each gives off one main branch, the ophthalmic artery, and then divides into the anterior and middle cerebral arteries. The **ophthalmic arteries** supply the eyes, orbits, forehead, and nose. Each **anterior cerebral artery** supplies the medial surface of the cerebral hemisphere on its side and also anastomoses with its partner on the opposite side via a short arterial shunt called the **anterior communicating artery** (Figure 24.5d). The **middle cerebral arteries** run in the lateral fissures of their respective cerebral hemispheres and supply the lateral parts of the temporal and parietal lobes.

Vertebral arteries.   These vessels spring from the subclavian arteries at the root of the neck and ascend through foramina in the transverse processes of the cervical vertebrae to enter the skull through the foramen magnum. En route, they send branches to the cervical spinal cord and some deep structures of the neck. Within the cranium, the right and left vertebral arteries join to form the **basilar artery,** which ascends along the anterior aspect of the brain stem, giving off branches to the cerebellum, pons, and inner ear (see Figure 24.5b and d). At the pons-midbrain border, the basilar artery divides into a pair of **posterior cerebral arteries,** which supply the occipital lobes and the inferior parts of the temporal lobes of the cerebral hemispheres.

(Table continues on next page)

TABLE 24.3   *(continued)*

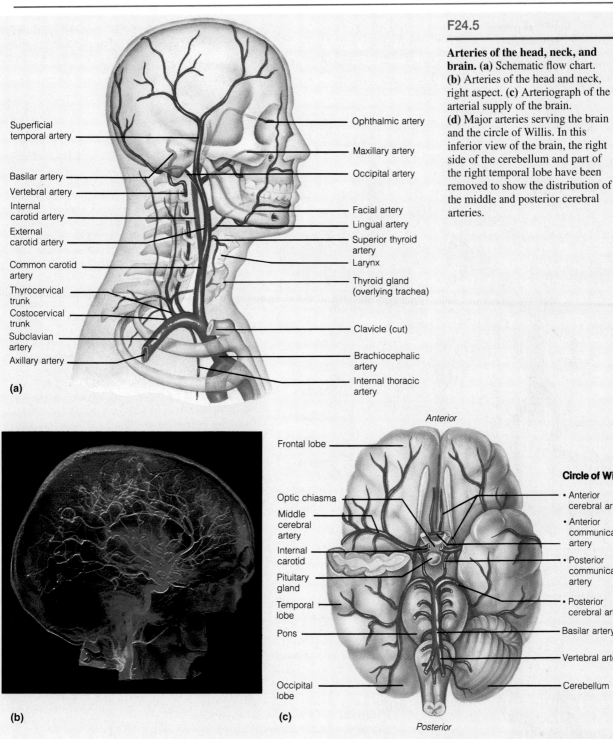

(a)

- Superficial temporal artery
- Basilar artery
- Vertebral artery
- Internal carotid artery
- External carotid artery
- Common carotid artery
- Thyrocervical trunk
- Costocervical trunk
- Subclavian artery
- Axillary artery
- Ophthalmic artery
- Maxillary artery
- Occipital artery
- Facial artery
- Lingual artery
- Superior thyroid artery
- Larynx
- Thyroid gland (overlying trachea)
- Clavicle (cut)
- Brachiocephalic artery
- Internal thoracic artery

F24.5

Arteries of the head, neck, and brain. (a) Schematic flow chart. **(b)** Arteries of the head and neck, right aspect. **(c)** Arteriograph of the arterial supply of the brain. **(d)** Major arteries serving the brain and the circle of Willis. In this inferior view of the brain, the right side of the cerebellum and part of the right temporal lobe have been removed to show the distribution of the middle and posterior cerebral arteries.

(b)

(c)

Anterior

- Frontal lobe
- Optic chiasma
- Middle cerebral artery
- Internal carotid
- Pituitary gland
- Temporal lobe
- Pons
- Occipital lobe

Circle of Willis:
- Anterior cerebral artery
- Anterior communicating artery
- Posterior communicating artery
- Posterior cerebral artery
- Basilar artery
- Vertebral artery
- Cerebellum

Posterior

Arterial shunts called **posterior communicating arteries** connect the posterior cerebral arteries to the middle cerebral arteries anteriorly. The two posterior and single anterior communicating arteries complete the formation of an arterial anastomosis called the **circle of Willis.** The circle of Willis encircles the pituitary gland and optic chiasma and unites the brain's anterior and posterior blood supplies.

Thyrocervical and costocervical trunks.   These short vessels arise from the subclavian artery just lateral to the vertebral arteries on each side (Figures 24.5b and 24.6). The thyrocervical trunk mainly supplies the thyroid gland and some scapular muscles. The costocervical trunk serves the deep neck and superior intercostal muscles.

TABLE 24.4   Arteries of the Upper Limbs and Thorax

The upper limbs are supplied entirely by arteries arising from the **subclavian arteries** (see Figure 24.6a). After giving off branches to the neck, each subclavian artery courses laterally between the clavicle and first rib to enter the axilla, where its name changes to axillary artery. The thorax wall is supplied by an array of vessels that arise either directly from the thoracic aorta or from branches of the subclavian arteries. Most visceral organs of the thorax receive their functional blood supply from small branches issuing from the thoracic aorta. Because these vessels are so small and tend to vary in number (except for the bronchial arteries), they are not illustrated in Figure 24.6a and b, but several of these are listed on p. 257.

Description and Distribution

Arteries of the Upper Limb

Axillary artery.   As it runs through the axilla accompanied by cords of the brachial plexus, each axillary artery gives off branches to the structures of the axilla, chest wall, and shoulder girdle. These include the **thoracoacromial trunk,** which supplies the superior shoulder (deltoid) and pectoral region; the **lateral thoracic artery,** which serves the lateral chest wall; the **subscapular artery** to the scapula, dorsal thorax wall, and latissimus dorsi muscle; and the **anterior** and **posterior circumflex arteries,** which wrap around the humeral neck and help supply the shoulder joint and the deltoid muscle. As the axillary artery emerges from the axilla, it becomes the brachial artery.

Brachial artery.   The brachial artery runs down the medial aspect of the humerus and supplies the anterior flexor muscles of the arm. One major branch, the **deep brachial artery,** serves the posterior triceps brachii muscle. Near the elbow, it gives off several small branches that contribute to an anastomosis serving the elbow joint and connecting it to the arteries of the forearm. As the brachial artery crosses the anterior midline aspect of the elbow, it provides an easily palpated pulse point (brachial pulse). Immediately beyond the elbow, the brachial artery splits to form the radial and ulnar arteries, which more or less follow the course of similarly named bones down the length of the anterior forearm.

Radial artery.   The radial artery, which runs from the median line of the cubital fossa to the styloid process of the radius, supplies the lateral muscles of the forearm, the wrist, and the thumb and index finger. At the root of the thumb, the radial artery provides a convenient site for taking the radial pulse.

Ulnar artery.   The ulnar artery supplies the medial aspect of the forearm, fingers 3–5, and the medial aspect of the index finger. Proximally, the ulnar artery gives off a short branch, the **common interosseous artery,** which runs between the radius and ulna to serve the deep flexors and extensors of the forearm. In the palm, branches of the radial and ulnar arteries anastomose to form two palmar arches, the **superficial** and **deep palmer arches.** The **metacarpal arteries** and the **digital arteries** that supply the fingers arise from these palmar arches.

Arteries of the Thorax Wall

Internal thoracic (mammary) arteries.   The internal thoracic (mammary) arteries, which arise from the subclavian arteries, supply blood to most of the anterior thorax wall. Each of these arteries descends lateral to the sternum and gives off **anterior intercostal arteries,** which supply the structures of the intercostal spaces anteriorly. The interior thoracic artery also supplies the mammary glands and terminates in twiglike branches to the anterior abdominal wall and diaphragm.

Posterior intercostal arteries.   Nine pairs of *posterior intercostal arteries* issue from the thoracic aorta and course around the rib cage to anastomose anteriorly with the *anterior intercostal arteries.* The superior two pairs are derived from the **costocervical trunk.** Inferior to the 12th rib, a pair of **subcostal arteries** emerges from the thoracic aorta (not illustrated). The posterior inter-

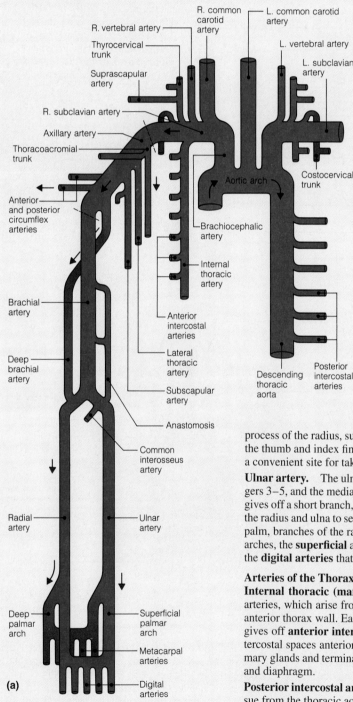

R. common carotid artery
L. common carotid artery
R. vertebral artery
Thyrocervical trunk
Suprascapular artery
L. vertebral artery
L. subclavian artery
R. subclavian artery
Axillary artery
Thoracoacromial trunk
Costocervical trunk
Anterior and posterior circumflex arteries
Aortic arch
Brachiocephalic artery
Brachial artery
Internal thoracic artery
Anterior intercostal arteries
Deep brachial artery
Lateral thoracic artery
Subscapular artery
Descending thoracic aorta
Posterior intercostal arteries
Anastomosis
Common interosseus artery
Radial artery
Ulnar artery
Deep palmar arch
Superficial palmar arch
Metacarpal arteries
(a)
Digital arteries

TABLE 24.4   (*continued*)

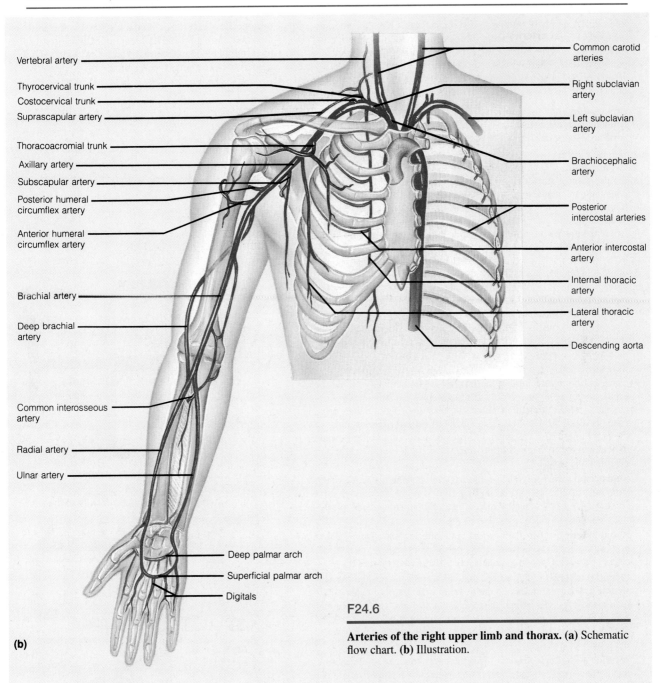

(b)

Vertebral artery

Thyrocervical trunk
Costocervical trunk
Suprascapular artery

Thoracoacromial trunk
Axillary artery
Subscapular artery
Posterior humeral circumflex artery

Anterior humeral circumflex artery

Brachial artery

Deep brachial artery

Common interosseous artery

Radial artery

Ulnar artery

Common carotid arteries
Right subclavian artery
Left subclavian artery
Brachiocephalic artery
Posterior intercostal arteries
Anterior intercostal artery
Internal thoracic artery
Lateral thoracic artery
Descending aorta

Deep palmar arch
Superficial palmar arch
Digitals

F24.6

Arteries of the right upper limb and thorax. (a) Schematic flow chart. **(b)** Illustration.

costal arteries supply the posterior intercostal spaces, the deep muscles of the back, the vertebrae, and the spinal cord. Together, the posterior and anterior intercostal arteries supply the intercostal muscles.

Superior phrenic arteries.   One or more paired superior phrenic arteries serve the posterior superior aspect of the diaphragm surface.

Arteries of the Thoracic Viscera
Pericardial arteries.   Several tiny branches supply the posterior pericardium.

Bronchial arteries.   Two left and one right bronchial arteries supply systemic (oxygen-rich) blood to the lungs, bronchi, and pleurae.

Esophageal arteries.   Four to five esophageal arteries supply the esophagus.

Mediastinal arteries.   Many small mediastinal arteries serve the contents of the posterior mediastinum.

TABLE 24.5   Arteries of the Abdomen

The arterial supply to the abdominal organs arises from the abdominal aorta (Figure 24.7a). Except for the celiac trunk, the superior and inferior mesenteric arteries, and the median sacral artery, all are paired vessels. These arteries supply the abdominal wall, the diaphragm, and visceral organs of the abdominopelvic cavity. The branches are given here in order of their issue.

Description and Distribution

Inferior phrenic arteries.   The inferior phrenics emerge from the aorta at T_{12}, just inferior to the diaphragm. They serve the inferior diaphragm surface.

Celiac trunk.   This large unpaired branch of the abdominal aorta divides almost immediately into three branches: the common hepatic, splenic, and left gastric arteries (Figure 24.7b). The **common hepatic artery** runs superiorly, giving off branches to the stomach, duodenum, and pancreas. It becomes the **hepatic artery** proper where the **gastroduodenal artery** branches off, then splits into right and left branches that serve the liver. The **splenic artery** passes deep to the stomach, sends branches to the pancreas and stomach, and terminates in branches to the spleen. The **left gastric** (*gaster* = stomach) **artery** supplies part of the stomach and the inferior esophagus. The **right** and **left gastroepiploic arteries,** branches of the gastroduodenal and splenic arteries, respectively, serve the left (greater) curvature of the stomach. A **right gastric artery,** which supplies the stomach's right (lesser) curvature, commonly arises from the common hepatic artery.

Superior mesenteric artery.   This large, unpaired artery arises from the abdominal aorta at the level of L_1 immediately below the celiac trunk (see Figure 24.7d). It runs deep to the pancreas and enters the mesentery, where its numerous anastomosing branches serve virtually all of the small intestine via the **intestinal arteries,** most of the large intestine—the appendix, cecum, ascending colon (via the **ileocolic artery**), and part of the transverse colon (via the **right** and **middle colic arteries**).

Suprarenal arteries.   The suprarenal arteries flank the origin of the superior mesenteric artery as they emerge from the abdominal aorta (Figure 24.7c). They supply the adrenal (suprarenal) glands overlying the kidneys.

Renal arteries.   The short, wide, renal arteries issue from the lateral surfaces of the aorta slightly below the superior mesenteric artery (between L_1 and L_2). Each serves the kidney on its side.

Gonadal arteries.   The paired gonadal arteries are called the **testicular arteries** in males and the **ovarian arteries** in females. In females, these vessels extend into the pelvis to serve the ovaries and part of the uterine tubes. In males the much longer testicular arteries descend through the pelvis to enter the scrotal sac, where they serve the testes.

Inferior mesenteric artery.   This final major branch of the abdominal aorta is unpaired and arises from the anterior aortic surface at the level of L_3. It serves the distal part of the large intestine via its **left colic, sigmoidal,** and **superior rectal branches.** Looping anastomoses between the superior and inferior mesenteric arteries help ensure blood delivery to the digestive viscera in cases of trauma to one of these abdominal arteries.

Lumbar arteries.   Four pairs of lumbar arteries arise from the posterolateral surface of the aorta in the lumbar region. These segmental arteries supply the posterior abdominal wall.

Median sacral artery.   The median sacral artery issues from the posterior surface of the abdominal aorta at its terminus. This tiny artery supplies the sacrum and coccyx.

Common iliac arteries.   At the level of L_4, the aorta splits into the right and left common iliac arteries, which supply blood to the lower abdominal wall, pelvic organs, and lower limbs (see Figure 24.8).

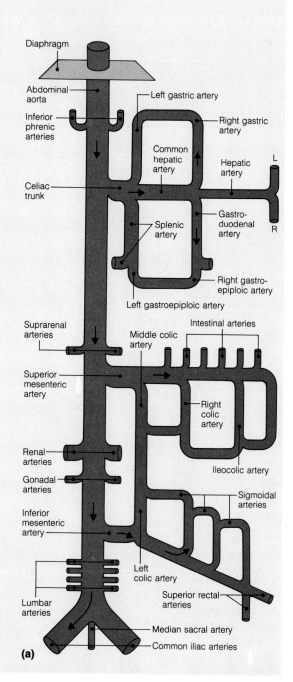

(a)

TABLE 24.5   (*continued*)

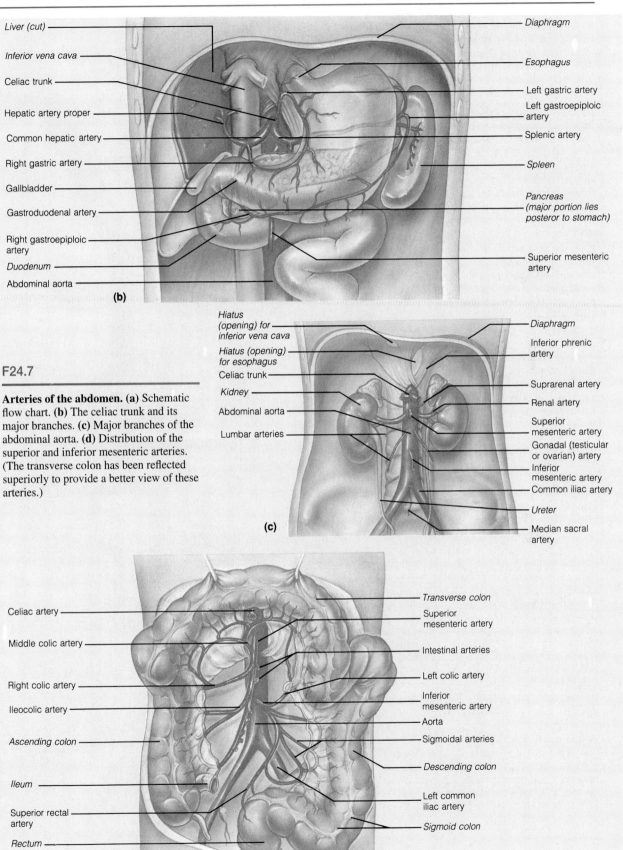

Liver (cut)
Inferior vena cava
Celiac trunk
Hepatic artery proper
Common hepatic artery
Right gastric artery
Gallbladder
Gastroduodenal artery
Right gastroepiploic artery
Duodenum
Abdominal aorta

Diaphragm
Esophagus
Left gastric artery
Left gastroepiploic artery
Splenic artery
Spleen
Pancreas (major portion lies posteror to stomach)
Superior mesenteric artery

(b)

F24.7

Arteries of the abdomen. (a) Schematic flow chart. **(b)** The celiac trunk and its major branches. **(c)** Major branches of the abdominal aorta. **(d)** Distribution of the superior and inferior mesenteric arteries. (The transverse colon has been reflected superiorly to provide a better view of these arteries.)

Hiatus (opening) for inferior vena cava
Hiatus (opening) for esophagus
Celiac trunk
Kidney
Abdominal aorta
Lumbar arteries

Diaphragm
Inferior phrenic artery
Suprarenal artery
Renal artery
Superior mesenteric artery
Gonadal (testicular or ovarian) artery
Inferior mesenteric artery
Common iliac artery
Ureter
Median sacral artery

(c)

Celiac artery
Middle colic artery
Right colic artery
Ileocolic artery
Ascending colon
Ileum
Superior rectal artery
Rectum

Transverse colon
Superior mesenteric artery
Intestinal arteries
Left colic artery
Inferior mesenteric artery
Aorta
Sigmoidal arteries
Descending colon
Left common iliac artery
Sigmoid colon

(d)

TABLE 24.6   Arteries of the Pelvis and Lower Limbs

At the level of the sacroiliac joints, the **common iliac arteries** divide into two major branches, the internal and external iliac arteries (Figure 24.8a). The internal iliacs distribute blood mainly to the pelvic region. The external iliacs serve the lower limbs and also send some branches to the abdominal wall.

Description and Distribution

Internal iliac arteries. These paired arteries run into the pelvis and distribute blood via several branches to the pelvic walls and viscera (bladder, rectum, uterus, and vagina in the female and prostate gland and ductus deferens in the male). They also serve the gluteal muscles via the **superior** and **inferior gluteal arteries,** adductor muscles of the medial thigh via the **obturator artery,** and external genitalia and perineum via the **internal pudendal artery** (Figure 24.8c).

External iliac arteries. These arteries supply the lower limbs (Figure 24.8b). As they course through the pelvis, they give off branches to the anterior abdominal wall. After passing under the inguinal ligaments to enter the thigh, they become the femoral arteries.

Femoral arteries. As each of these arteries passes down the anteromedial thigh, it gives off several branches to the muscles of the thigh. The largest of the deep branches is the **deep femoral artery** that serves the posterior (knee flexor) thigh muscles. Proximal branches of the deep femoral artery, the **lateral** and **medial femoral circumflex arteries,** encircle the neck of the femur. The medial femoral circumflex artery supplies most of the blood to the head and neck of the femur. Descending branches of the two circumflex arteries supply the hamstring muscles. As the femoral artery approaches the knee, it passes posteriorly and through a gap in the adductor magnus muscle, the *adductor hiatus,* to enter the popliteal fossa, where its name changes to popliteal artery.

Popliteal artery. This posterior vessel contributes to an arterial anastomosis that supplies the knee region and then splits into the anterior and posterior tibial arteries of the leg.

Anterior tibial artery. The anterior tibial artery runs through the anterior compartment of the leg, supplying the extensor muscles along the way. At the ankle, it becomes the **dorsalis pedis artery,** which supplies the ankle and dorsum of the foot, and continues as the **arcuate artery,** which gives off the metatarsal arteries to the metatarsus of the foot. The superficial dorsalis pedis artery provides a clinically important pulse point, the pedal pulse. (If the pedal pulse is easily felt, it is fairly certain that the blood supply to the leg is good.)

Posterior tibial artery. This large artery courses through the posteromedial part of the leg and supplies the flexor muscles. Proximally, it gives off a large branch, the **peroneal artery,** which supplies the lateral peroneal muscles of the leg. At the ankle, the posterior tibial artery divides into **lateral** and **medial plantar arteries** that serve the plantar surface of the foot. The **digital arteries** serving the toes arise from the plantar arch formed by the lateral plantar artery.

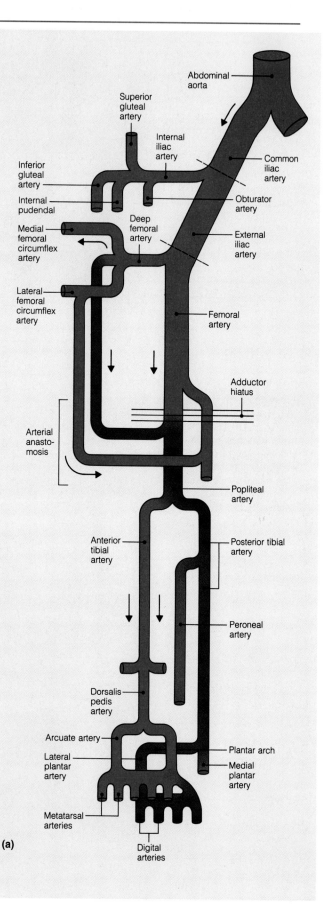

(a)

TABLE 24.6   (*continued*)

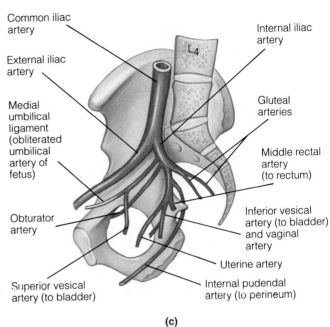

Common iliac artery

External iliac artery

Medial umbilical ligament (obliterated umbilical artery of fetus)

Obturator artery

Superior vesical artery (to bladder)

Internal iliac artery

Gluteal arteries

Middle rectal artery (to rectum)

Inferior vesical artery (to bladder) and vaginal artery

Uterine artery

Internal pudendal artery (to perineum)

L4

(c)

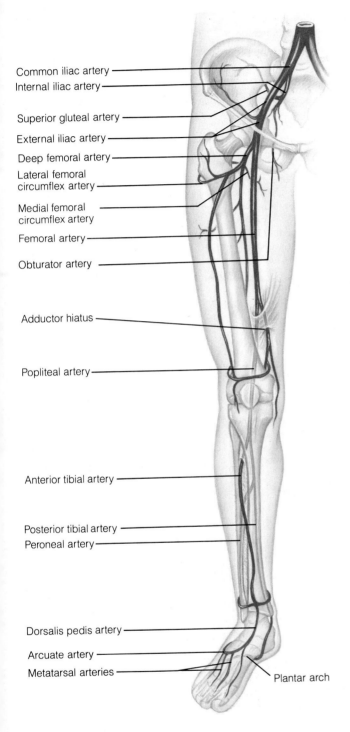

Common iliac artery

Internal iliac artery

Superior gluteal artery

External iliac artery

Deep femoral artery

Lateral femoral circumflex artery

Medial femoral circumflex artery

Femoral artery

Obturator artery

Adductor hiatus

Popliteal artery

Anterior tibial artery

Posterior tibial artery

Peroneal artery

Dorsalis pedis artery

Arcuate artery

Metatarsal arteries

Plantar arch

(b)

F24.8

Arteries of the right pelvis and lower limb. (a) Schematic flow chart. **(b)** Illustration. **(c)** Branches of the iliac artery in right half of the female pelvis (medial view). The branches of the internal iliac artery are variable, arising in somewhat different places in different individuals.

TABLE 24.7   The Venae Cavae and the Major Veins of the Systemic Circulation

Although there are many similarities between the systemic arteries and veins, there are also important differences:

1.   Whereas the heart pumps all of its blood into a single systemic artery—the aorta—blood returning to the heart is delivered largely by two terminal systemic veins, the superior and inferior venae cavae (see Figure 24.9). The single exception to this is the blood draining from the myocardium of the heart, which is collected by the cardiac veins and reenters the right atrium via the coronary sinus (see p. 244).

2.   All arteries run deep and are well protected by body tissues along most of their course, but both deep and superficial veins exist. Deep veins follow the course of the systemic arteries, and with a few exceptions, the naming of these veins is identical to that of their companion arteries. Superficial veins run just beneath the skin and are readily seen, especially in the limbs, face, and neck. Since there are no superficial arteries, the names of the superficial veins do not correspond to the names of any of the arteries.

3.   Unlike the fairly clear arterial pathways, venous pathways tend to have numerous interconnections, and many veins (particularly those in the limbs) are represented by not one but two similarly named vessels. As a result, venous pathways are more difficult to follow.

4.   In most body regions, there is a similar and predictable arterial supply and venous drainage. However, the pattern of venous drainage in at least two important body areas is unique. First, venous blood draining from the brain enters large *dural sinuses* rather than typical veins. Second, blood draining from the digestive organs enters a special subcirculation, called the *hepatic portal circulation,* and perfuses through the liver before it reenters the general systemic circulation.

In our survey of the systemic veins, the major tributaries (branches) of the venae cavae will be noted first, followed by a description in Tables 24.8 through 24.11 of the venous pattern of the various body regions. Since veins run toward the heart, the most distal veins are named first and those closest to the heart last. Deep veins are named, but since they generally drain the same areas served by their companion arteries, they are not described in detail. An overview of the systemic veins appears in Figure 24.10.

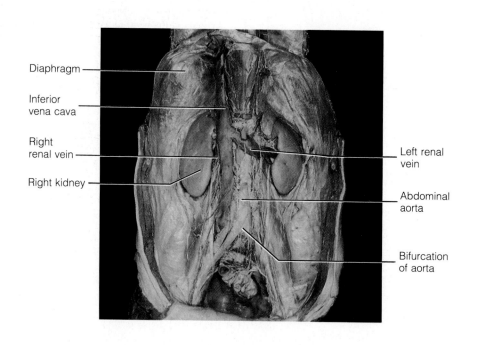

F24.9

Photograph of the posterior abdominal wall, revealing the position of the inferior vena cava. Note that this vessel lies just to the right of the abdominal aorta.

TABLE 24.7   (*continued*)

Description and Areas Drained

Superior vena cava.   This great vein receives blood draining from all areas superior to the diaphragm, except from the pulmonary circuit. The superior vena cava is formed by the union of the **right** and **left brachiocephalic veins** and empties into the superior aspect of the right atrium (Figure 24.10). Notice that there are two brachiocephalic veins, but only one brachiocephalic artery. Each brachiocephalic vein is formed in turn by the joining of the **internal jugular** and **subclavian veins** on its side. In most of the flow charts that follow, only the vessels draining blood from the right side of the body are followed (with the exception of vessels of the azygos circulation of the thorax).

Inferior vena cava.   This vein, the widest blood vessel in the body, is substantially longer than the superior vena cava. It returns blood to the heart from all body regions below the diaphragm and corresponds most closely to the abdominal aorta, which lies directly to its left (see Figure 24.9). The distal end of the inferior vena cava is formed by the junction of the paired **common iliac veins** at L_5. From this point, it courses superiorly along the anterior aspect of the spine, receiving venous blood draining from the abdominal walls, gonads, and kidneys. Just before it penetrates the diaphragm, it is joined by the hepatic veins, which transport blood from the liver. Immediately above the diaphragm, the inferior vena cava ends as it enters the inferior aspect of the right atrium of the heart.

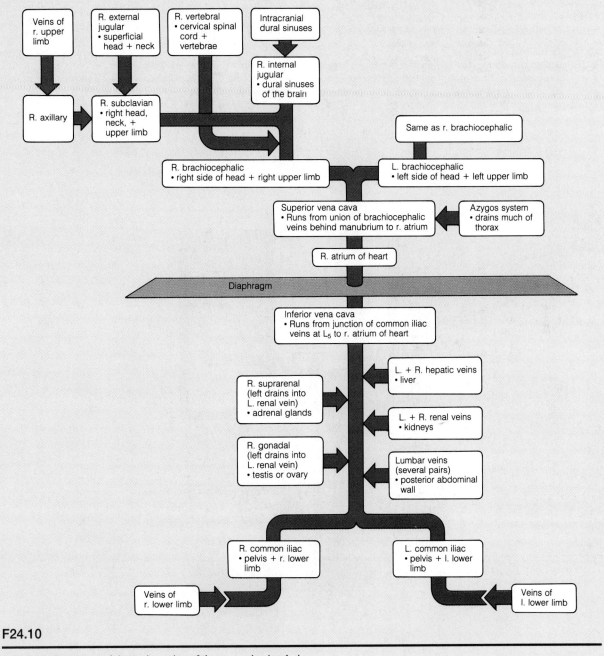

F24.10

Schematic flow chart of the major veins of the systemic circulation.

TABLE 24.8   Veins of the Head and Neck

Most blood draining from the head and neck is collected by three pairs of veins: the external jugular veins (which empty into the subclavians), the internal jugular veins and the vertebral veins, which all drain into the brachiocephalic vein (Figure 24.11a). Most of the extracranial veins have the same names as the extracranial arteries (for example, facial, ophthalmic, occipital, and superficial temporal), but their courses and interconnections differ substantially.

Most veins of the brain drain into the **dural sinuses,** an interconnected series of enlarged chambers located between the dura mater layers. The most important of these venous sinuses are the **superior** and **inferior sagittal sinuses** found within the falx cerebri, which dips down between the cerebral hemispheres, and the **cavernous sinuses,** which flank the sphenoid body (see Figure 24.11c). The cavernous sinus receives venous blood from the **ophthalmic vein** of the orbit and the facial vein, which drains the nose and upper lip area.

Description and Area Drained
External jugular veins.   The right and left external jugular veins drain superficial head (scalp and face) structures served by the external carotid arteries. However, their tributaries anastomose frequently, and some of the superficial drainage

from these regions enters the internal jugular veins as well. The external jugular veins descend through the lateral neck, passing obliquely over the sternocleidomastoid muscles, and then empty into the subclavian veins.

Vertebral veins.   Unlike the vertebral arteries, the vertebral veins do not serve much of the brain. Instead, they drain only the cervical vertebrae, the spinal cord, and some small neck muscles. The vertebral veins run inferiorly through the transverse foramina of the cervical vertebrae and join the brachiocephalic veins at the root of the neck.

Internal jugular veins.   The paired internal jugular veins, which receive the bulk of blood draining from the brain, are the largest of the paired veins draining the head and neck. They arise from the dural venous sinuses, exit from the skull via the *jugular foramina,* and then descend through the neck alongside the internal carotid arteries. As they move inferiorly, they receive blood from some of the deep veins of the face and neck—branches of the **facial** and **superficial temporal veins** (see Figure 24.11b). At the base of the neck, each internal jugular vein joins the subclavian vein on its own side to form a brachiocephalic vein. As already noted, the two brachiocephalic veins unite to form the superior vena cava.

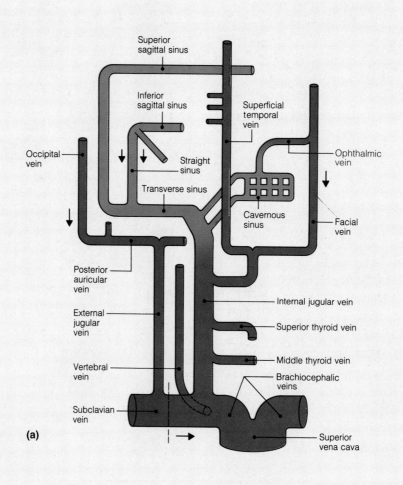

(a)

TABLE 24.8 (*continued*)

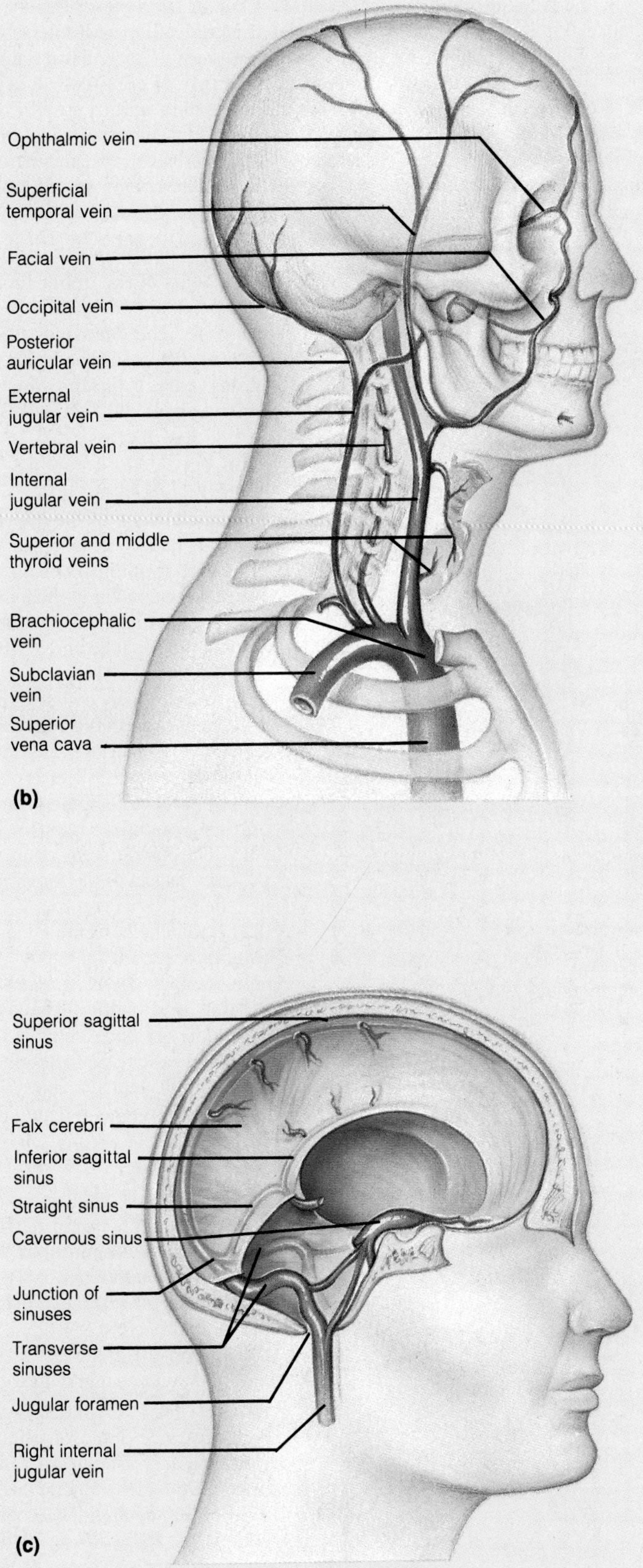

Ophthalmic vein

Superficial temporal vein

Facial vein

Occipital vein

Posterior auricular vein

External jugular vein

Vertebral vein

Internal jugular vein

Superior and middle thyroid veins

Brachiocephalic vein

Subclavian vein

Superior vena cava

(b)

Superior sagittal sinus

Falx cerebri

Inferior sagittal sinus

Straight sinus

Cavernous sinus

Junction of sinuses

Transverse sinuses

Jugular foramen

Right internal jugular vein

(c)

F24.11

Venous drainage of the head, neck, and brain. (a) Schematic flow chart. **(b)** Veins of the head and neck, right superficial aspect. **(c)** Dural sinuses of the brain, right aspect.

TABLE 24.9   Veins of the Upper Limbs and Thorax

The deep veins of the upper limbs follow the paths of their companion arteries and have the same names (Figure 24.12a). However, except for the largest, most are paired or double veins that flank their artery on both sides. The superficial veins of the upper limbs are larger than the deep veins and are easily seen just beneath the skin. The median cubital vein, crossing the anterior aspect of the elbow, is commonly used to obtain venous blood samples or administer intravenous medications or blood transfusions.

Blood draining from the mammary glands and the first two to three intercostal spaces enters the **brachiocephalic veins.** However, most thoracic tissues and the thorax wall are drained by a network of veins collectively called the **azygos system.** The branching azygos system provides a collateral circulation that drains the abdominal wall and other areas served by the inferior vena cava, and there are numerous anastomoses between the azygos system and the inferior vena cava.

Description and Areas Drained

Deep veins of the upper limb.   The most distal deep veins of the upper limb are the radial and ulnar veins. The **deep** and **superficial palmar venous arches** of the hand empty into the **radial** and **ulnar veins** of the forearm, which then unite to form the **brachial vein** of the arm. As the brachial vein enters the axilla, it becomes the **axillary vein,** which in turn becomes the **subclavian vein** at the level of the first rib.

Superficial veins of the upper limb.   The superficial venous system begins with the **dorsal venous arch** (not illustrated), a plexus of superficial veins in the dorsum of the hand. In the distal forearm, this plexus drains into three major superficial veins—the cephalic and basilic veins and the median vein of the forearm—that anastomose frequently as they course upward (see Figure 24.12b). The **cephalic vein** coils around the radius as it travels superiorly and then continues up the lateral superficial aspect of the arm to the shoulder, where it runs in the groove between the deltoid and pectoralis muscles to join the axillary vein. The **basilic vein** courses along the medial posterior aspect of the forearm, crosses the elbow, and then takes a deep course. In the axilla, it joins the brachial vein, forming the axillary vein. At the anterior aspect of the elbow, the **median cubital vein** connects the basilic and cephalic veins. The **median vein of the forearm** is located between the radial and ulnar veins in the forearm and terminates (variably) at the elbow by entering the basilic or the cephalic vein.

The azygos system.   The azygos system consists of the following vessels, which flank the vertebral column laterally:

Azygos vein.   Found against the right side of the vertebral column; originates in the abdomen from the **right ascending lumbar vein** that drains most of the right abdominal cavity wall and from the **right posterior intercostal veins** (except the first) that drain the chest muscles. At the level of T_4, it arches over the great vessels that run to the right lung and empties into the superior vena cava.

Hemiazygos ("half the azygos") **vein.**   Ascends on the left side of the vertebral column. Its origin, from the **left ascending lumbar vein** and the lower (9th–11th) **posterior intercostal veins,** mirrors that of the inferior portion of the azygos vein on the right. About midthorax, the hemiazygos vein passes in front of the vertebral column and empties into the azygos vein.

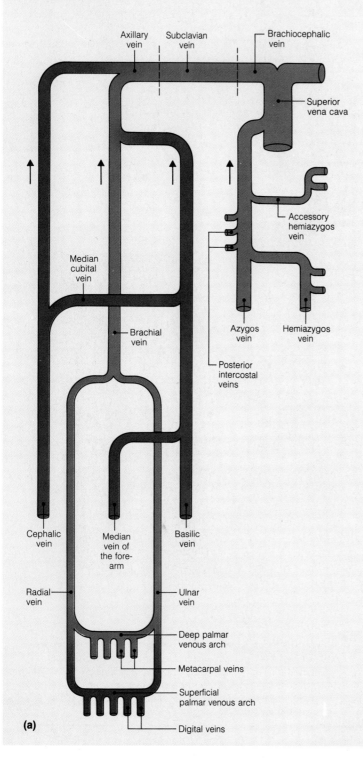

(a)

TABLE 24.9   (*continued*)

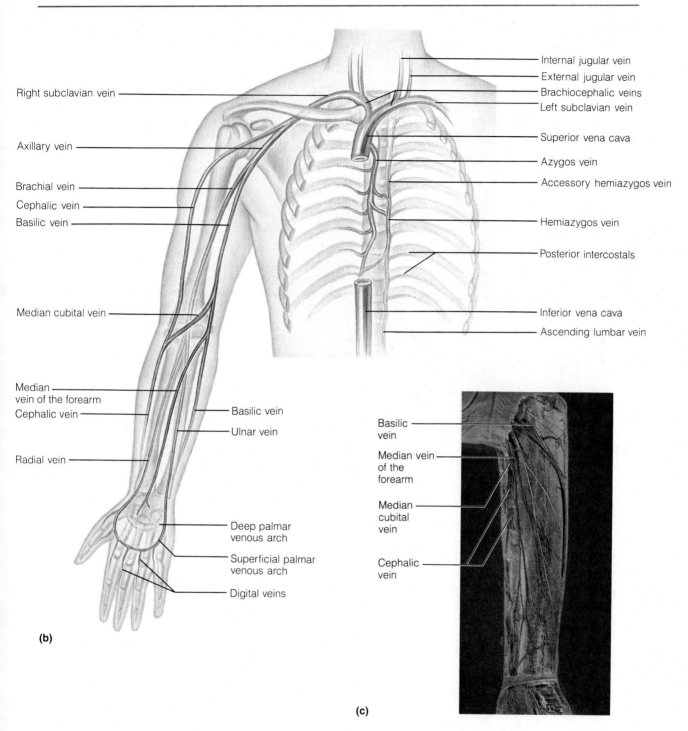

Right subclavian vein

Axillary vein

Brachial vein

Cephalic vein

Basilic vein

Median cubital vein

Median
vein of the forearm

Cephalic vein

Radial vein

Internal jugular vein
External jugular vein
Brachiocephalic veins
Left subclavian vein

Superior vena cava

Azygos vein

Accessory hemiazygos vein

Hemiazygos vein

Posterior intercostals

Inferior vena cava
Ascending lumbar vein

Basilic vein
Ulnar vein

Deep palmar
venous arch

Superficial palmar
venous arch

Digital veins

(b)

Basilic
vein

Median vein
of the
forearm

Median
cubital
vein

Cephalic
vein

(c)

Accessory hemiazygos vein.   The accessory hemiazygos completes the venous drainage of the left (middle) thorax and can be thought of as a superior continuation of the hemiazygos vein. It recieves blood from the 4th to 8th posterior intercostal veins and then crosses to the right to empty into the azygos vein. Like the azygos, it receives venous blood from the lungs (bronchial veins).

F24.12

Veins of the right upper limb and shoulder. (a) Schematic flow chart. **(b)** Illustration. For clarity, the abundant branching and anastomoses of these vessels are not shown. **(c)** Photograph of the superficial veins on the anteromedial aspect of the right forearm.

TABLE 24.10 Veins of the Abdomen

Blood draining from the abdominopelvic viscera and abdominal walls is returned to the heart by the **inferior vena cava** (Figures 24.13a and 24.9). Most of its venous tributaries have names that correspond to the arteries serving the abdominal organs.

Veins draining the digestive viscera empty their blood into a common vessel, the **hepatic portal vein,** which transports this venous blood into the liver before it is allowed to enter the major systemic circulation via the hepatic veins (Figure 24.13b). Such a venous system—veins to capillaries (or sinusoids) to veins—is called a **portal system.** Portal systems always serve very specific regional tissue needs. The **hepatic portal system** delivers blood laden with nutrients from the digestive organs to the liver. As the blood percolates slowly through the liver sinusoids, the hepatic parenchymal cells remove nutrients required for their various metabolic functions and phagocytic cells lining the sinusoids deftly rid the blood of bacteria and other foreign matter that has penetrated the digestive mucosa. The veins of the abdomen are listed below in inferior to superior order.

Description and Areas Drained

Lumbar veins. Several pairs of lumbar veins drain the posterior abdominal wall. They empty both directly into the inferior vena cava and into the ascending lumbar veins of the azygos system of the thorax.

Gonadal (testicular or ovarian) veins. The right gonadal vein drains the ovary or testis on the right side of the body and empties into the inferior vena cava. The left member drains into the left renal vein superiorly.

Renal veins. The right and left renal veins drain the kidneys.

Suprarenal veins. The right suprarenal vein drains the adrenal gland on the right and empties into the inferior vena cava. The left suprarenal vein drains into the left renal vein.

Hepatic portal system. The **hepatic portal vein** is a short vessel about 8 cm long; it begins at the level of L_2. Numerous tributaries from the stomach and pancreas contribute to the hepatic portal system (see Figure 24.13c), but the major vessels are as follows:

- **Superior mesenteric vein.** Drains the entire small intestine, part of the large intestine (ascending and transverse regions), and stomach.
- **Splenic vein.** Collects blood from the spleen, parts of the stomach, and pancreas and then joins the superior mesenteric vein to form the hepatic portal vein.
- **Inferior mesenteric vein.** Drains the distal portions of the large intestine and rectum and joins the splenic vein just before that vessel unites with the superior mesenteric vein.

Hepatic veins. The right and left hepatic veins carry venous blood from the liver to the inferior vena cava.

Cystic veins. The cystic veins drain the gallbladder and join the hepatic veins.

Inferior phrenic veins. The inferior phrenic veins drain the inferior surface of the diaphragm.

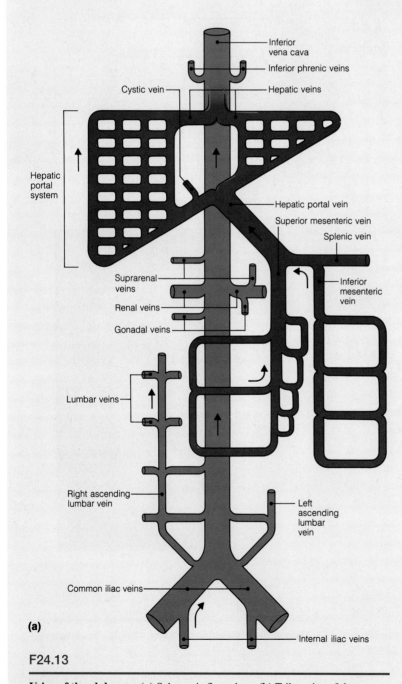

(a)

F24.13

Veins of the abdomen. (a) Schematic flow chart. **(b)** Tributaries of the inferior vena cava. Venous drainage of abdominal organs not drained by the hepatic portal vein. **(c)** The hepatic portal circulation.

TABLE 24.10 *(continued)*

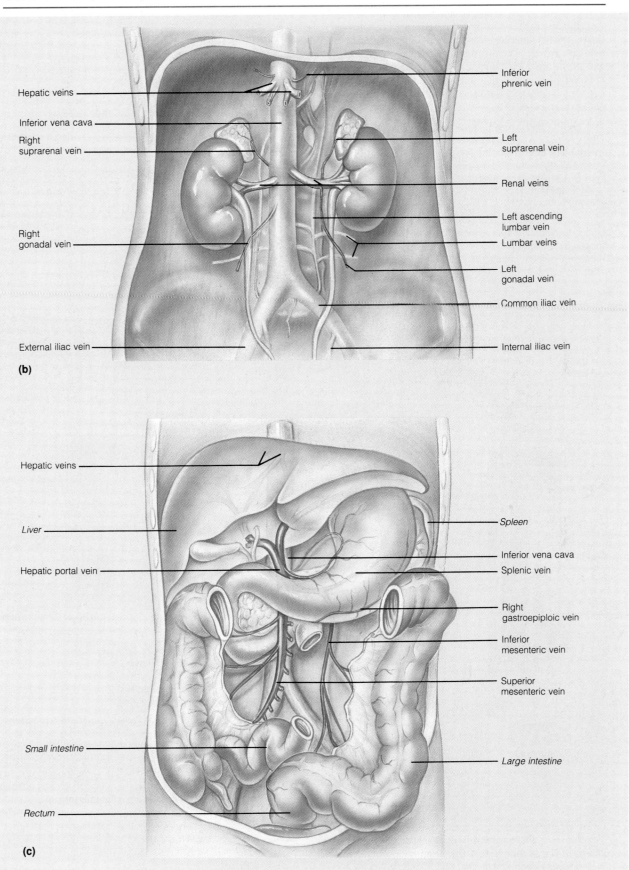

(b)

Hepatic veins

Inferior vena cava

Right suprarenal vein

Right gonadal vein

External iliac vein

Inferior phrenic vein

Left suprarenal vein

Renal veins

Left ascending lumbar vein

Lumbar veins

Left gonadal vein

Common iliac vein

Internal iliac vein

(c)

Hepatic veins

Liver

Hepatic portal vein

Small intestine

Rectum

Spleen

Inferior vena cava

Splenic vein

Right gastroepiploic vein

Inferior mesenteric vein

Superior mesenteric vein

Large intestine

TABLE 24.11   Veins of the Pelvis and Lower Limbs

As is the case in the upper limbs, most deep veins of the lower limbs have the same names as the arteries they accompany and many are double. The two superficial saphenous veins (great and small) are poorly supported by surrounding tissues, and are common sites of varicosities. The great saphenous (*saphenous* = obvious) vein is frequently excised and used as a coronary bypass vessel.

Description and Areas Drained
Deep veins.   After being formed by the union of the small **medial** and **lateral plantar veins**, the **posterior tibial vein** ascends deep within the calf muscle (Figure 24.14). The **anterior tibial vein** is the superior continuation of the **dorsalis pedis vein** of the foot. At the knee it unites with the posterior tibial vein to form the **popliteal vein,** which crosses the back of the knee. As the popliteal vein emerges from the knee, it becomes the **femoral vein,** which drains the deep structures of the thigh. The femoral vein becomes the **external iliac vein** as it enters the pelvis. In the pelvis, the external iliac vein unites with the **internal iliac vein** to form the **common iliac vein.**

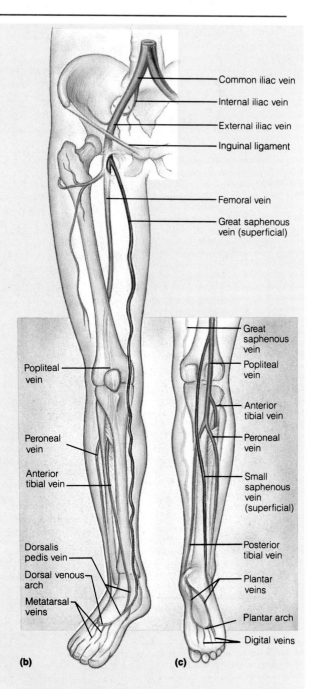

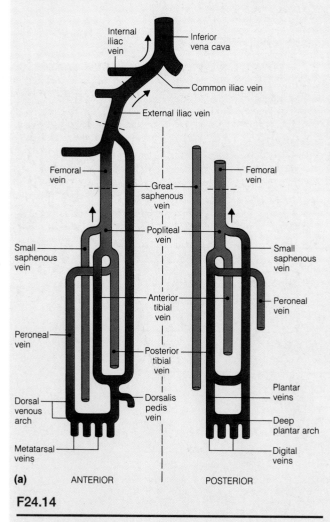

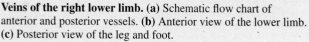

F24.14

Veins of the right lower limb. (a) Schematic flow chart of anterior and posterior vessels. **(b)** Anterior view of the lower limb. **(c)** Posterior view of the leg and foot.

The distribution of the internal iliac veins parallels that of the internal iliac arteries.

Superficial veins.   The **great** and **small saphenous veins** issue from the **dorsal venous arch** of the foot (Figure 24.14b). These veins anastomose frequently with each other and with the deep veins along their course. The great saphenous vein is the longest vein in the body. It travels superiorly along the medial aspect of the leg to the thigh, where it empties into the femoral vein just distal to the inguinal ligament. The small saphenous vein runs along the lateral aspect of the foot and then through the deep fascia of the calf muscles, which it drains. At the knee, it empties into the popliteal vein.

Re-identify the important arteries and veins on the large anatomical chart or model without referring to the figures.

SPECIAL CIRCULATIONS

There are several unique circulations in the body which are collectively referred to as *special circulations*. These include the pulmonary circulation, the arterial anastamosis called the circle of Willis, the hepatic portal circulation, and the fetal circulation. With the exception of the pulmonary circulation, all are part of the systemic circulation, and all except the fetal circulation have already been considered in the preceding tables. Thus our focus here will be simply to recap the important points of the special circulations already covered and to introduce the special blood vessels of the fetal circulation.

Pulmonary Circulation

The *pulmonary circulation* (described on p. 251) differs in many ways from the systemic circulation because it does not serve the metabolic needs of the body tissues with which it is associated (in this case, lung tissue). It functions instead to bring the blood into close contact with the alveoli of the lungs to permit gas exchanges that rid the blood of excess carbon dioxide and replenish its supply of vital oxygen.

The arteries of the pulmonary circulation are more like veins than arteries; and they create a low-pressure bed in the lungs. (If the arterial pressure in the systemic circulation is 120/80, the pressure in the pulmonary artery is likely to be approximately 25/10.)

Circle of Willis

A continuous blood supply to the brain is crucial, since oxygen deprivation for even a few minutes causes irreparable damage to the delicate brain tissue. The brain is supplied by two pairs of arteries arising from the region of the aortic arch—the *internal carotid arteries* and the *vertebral arteries*. Figure 24.5a and d (pp. 254–255) diagrams the brain's arterial supply. Branches

of the internal carotid arteries (the *anterior* and *middle cerebral arteries*), which supply the bulk of the cerebrum, also contribute to the formation of the *circle of Willis*, the arterial anastomosis at the base of the brain surrounding the pituitary gland and the optic chiasma, by forming a *posterior communicating artery* on each side. The circle is completed by the *anterior communicating artery*, a short shunt connecting the right and left anterior cerebral arteries.

The *basilar artery*, which results from the joining of the paired vertebral arteries, runs superiorly along the ventral aspect of the brain stem. At the base of the cerebrum, the basilar artery divides to form the *posterior cerebral arteries*, which contribute to the circle of Willis by joining with the posterior communicating arteries.

The uniting of the blood supply of the internal carotid arteries and the vertebral arteries via the circle of Willis is a protective device that theoretically provides an alternate set of pathways for blood to reach the brain tissue in the case of arterial occlusion or impaired blood flow anywhere in the system. In actuality, the communicating arteries are tiny, and in many cases the communicating system is defective.

Hepatic Portal Circulation

Blood vessels of the *hepatic portal circulation* drain the digestive viscera, spleen, and pancreas and deliver this blood to the liver for processing via the **hepatic portal vein.** If a meal has recently been eaten, the hepatic portal blood will be nutrient rich. The liver is the key body organ involved in maintaining proper sugar, fatty acid, and amino acid concentrations in the blood, and this system ensures that these substances pass through the liver before entering the systemic circulation. As blood percolates through the liver sinusoids, some of the nutrients are removed to be stored or processed in various ways for release to the general circulation. At the same time, the liver's hepatocytes are detoxifying alcohol and other possibly harmful chemicals present in the blood, and the liver's macrophages are removing bacteria and other debris from the passing blood. The liver in turn is drained by the hepatic veins that enter the inferior vena cava.

As indicated in Figure 24.13, the crucial veins draining the digestive viscera and contributing to the hepatic portal vein drainage are the splenic, *inferior mesenteric*, and *superior mesenteric veins*.

Fetal Circulation

In a developing fetus, the lungs and digestive system are not yet functional, and all nutrient, excretory, and gaseous exchanges occur through the placenta (see Figure 24.15a). Nutrients and oxygen move across placental barriers from the mother's blood into fetal blood, and carbon dioxide and other metabolic wastes move from the fetal blood supply to the mother's blood.

Fetal blood travels through the umbilical cord, which contains three blood vessels: two smaller umbilical arteries and one large umbilical vein. The **umbilical vein** carries blood rich in nutrients and oxygen to the fetus; the **umbilical arteries** carry carbon dioxide and waste-laden blood from the fetus to the placenta. The umbilical arteries, which transport blood away from the fetal heart, meet the umbilical vein at the *umbilicus* (navel, or belly button) and wrap around the vein within the cord en route to their placental attachments. Newly oxygenated blood flows in the umbilical vein superiorly toward the fetal heart. En route, some of this blood perfuses the liver, but the larger proportion is ducted through the relatively nonfunctional liver to the inferior vena cava via a vessel called the **ductus venosus,** which carries the blood to the right atrium of the heart.

Because fetal lungs are nonfunctional and collapsed, two shunting mechanisms ensure that blood almost entirely bypasses the lungs. Much of the blood entering the right atrium is shunted into the left atrium through the **foramen ovale,** a flaplike opening in the interatrial septum. The left ventricle then pumps the blood out the aorta to the systemic circulation. Blood that does enter the right ventricle and is pumped out of the pulmonary trunk encounters a second shunt, the **ductus arteriosus,** a short vessel connecting the pulmonary trunk and the aorta. Because the collapsed lungs present an extremely high-resistance pathway, blood more readily enters the systemic circulation through the ductus arteriosus.

The aorta carries blood to the tissues of the body; this blood ultimately finds its way back to the placenta via the umbilical arteries. The only fetal vessel that carries highly oxygenated blood is the umbilical vein. All other vessels contain varying degrees of oxygenated and deoxygenated blood.

At birth, or shortly after, several changes occur in the fetal cardiovascular system (Figure 24.15b). The foramen ovale closes and becomes the **fossa ovalis,** and the ductus arteriosus collapses and is converted to the fibrous **ligamentum arteriosum.** Lack of blood flow through the umbilical vessels leads to their eventual obliteration, and the circulatory pattern becomes that of the adult. Remnants of the umbilical arteries persist as the **medial umbilical ligaments** on the inner surface of the anterior abdominal wall, of the umbilical vein as the **ligamentum teres** or **round ligament** of the liver, and of the ductus venosus as a fibrous band called the **ligamentum venosum** on the inferior surface of the liver.

The pathway of fetal blood flow is indicated with arrows on Figure 24.15. Identify all specialized fetal circulatory structures and trace the blood flow to and from the placenta.

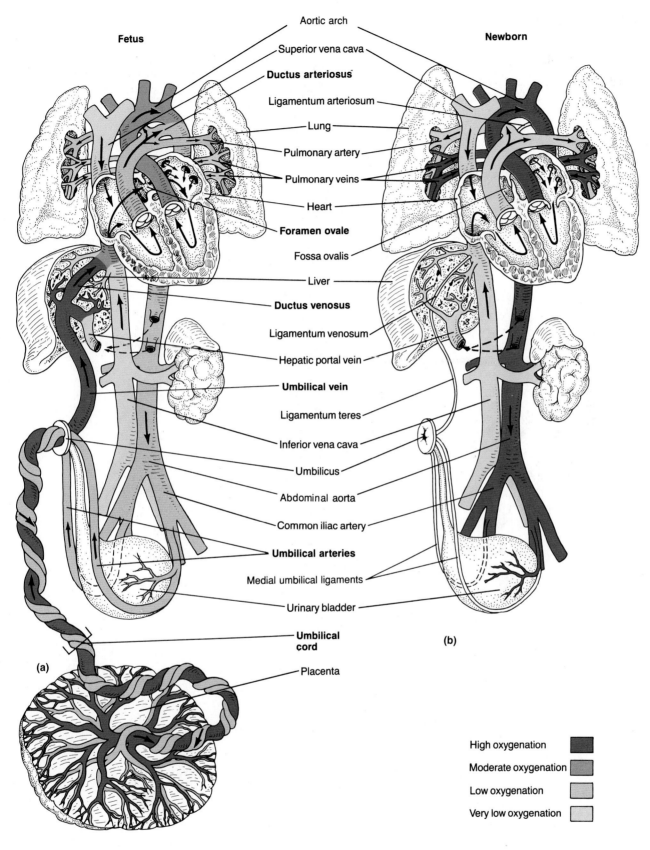

Fetus

Aortic arch

Superior vena cava

Ductus arteriosus

Ligamentum arteriosum

Lung

Pulmonary artery

Pulmonary veins

Heart

Foramen ovale

Fossa ovalis

Liver

Ductus venosus

Ligamentum venosum

Hepatic portal vein

Umbilical vein

Ligamentum teres

Inferior vena cava

Umbilicus

Abdominal aorta

Common iliac artery

Umbilical arteries

Medial umbilical ligaments

Urinary bladder

Umbilical cord

Placenta

Newborn

(b)

(a)

High oxygenation

Moderate oxygenation

Low oxygenation

Very low oxygenation

F24.15

Fetal circulation. Arrows indicate the direction of blood flow. **(a)** Special adaptations for fetal life. The umbilical vein carries oxygen- and nutrient-rich blood from the placenta to the fetus. The umbilical arteries carry waste-laden, low-oxygen blood from the fetus to the placenta. The ductus arteriosus and foramen ovale shunt blood away from the nonfunctional lungs, and the ductus venosus enables much of the blood to bypass the liver circulation. The ductus venosus lies in a deep cleft between the two major lobes of the liver and is therefore a shunt *around* the liver. **(b)** Changes in the cardiovascular system at birth. The umbilical vessels close, as do the ductus venosus, ductus arteriosus, and foramen ovale.

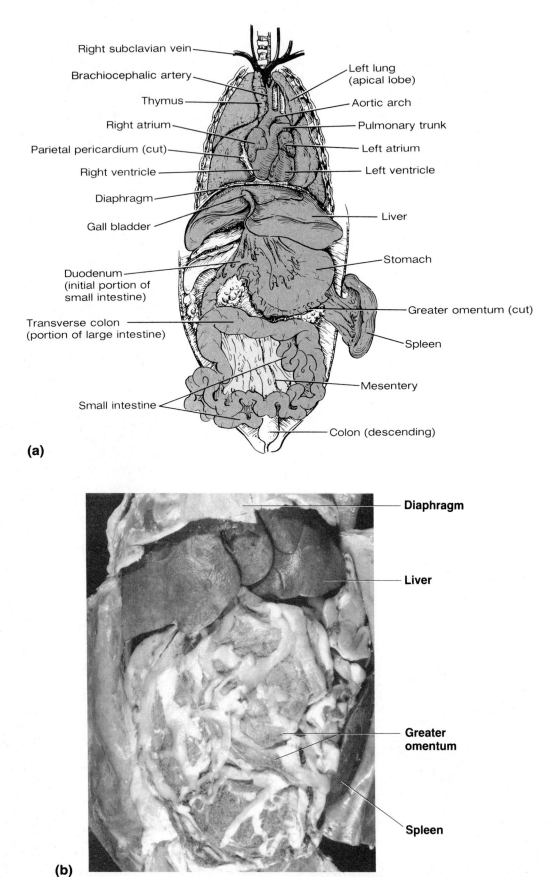

Right subclavian vein

Brachiocephalic artery

Thymus

Right atrium

Parietal pericardium (cut)

Right ventricle

Diaphragm

Gall bladder

Duodenum
(initial portion of
small intestine)

Transverse colon
(portion of large intestine)

Small intestine

Left lung
(apical lobe)

Aortic arch

Pulmonary trunk

Left atrium

Left ventricle

Liver

Stomach

Greater omentum (cut)

Spleen

Mesentery

Colon (descending)

(a)

Diaphragm

Liver

Greater
omentum

Spleen

(b)

F24.16

Ventral body cavity organs of the cat. (a) Diagrammatic view, greater omentum removed. **(b)** Photograph of abdominal cavity with greater omentum intact and in its normal position. (Also see Plate F in the Cat Anatomy Atlas.)

DISSECTION OF THE BLOOD VESSELS OF THE CAT

Opening the Ventral Body Cavity

1. Using scissors, make a longitudinal incision through the ventral body wall. Begin just superiorly to the midline of the pubic bone, and continue the cut anteriorly to the rib cage.

2. Angle the scissors slightly (½ inch) to the right or left of the sternum, and continue the cut through the rib cartilages just lateral to the body midline, to the base of the throat.

3. Make two lateral cuts on either side of the ventral body surface, anterior and posterior to the diaphragm. Leave the diaphragm intact. Spread the thoracic walls laterally to expose the thoracic organs.

Preliminary Organ Identification

A helpful prelude to identifying and tracing the blood supply of the various organs of the cat is a preliminary identification of ventral body cavity organs shown in Figure 24.16. Since you will study the organ systems contained in the ventral cavity in later units, the objective here is simply to identify the most important organs. Using Figure 24.16 as a guide, identify the following body cavity organs:

THORACIC CAVITY ORGANS

Heart: in the mediastinum enclosed by the pericardium
Lungs: flanking the heart
Thymus: superior to and partially covering the heart

ABDOMINAL CAVITY ORGANS

Liver: posterior to the diaphragm

Lift the large, drapelike, fat-infiltrated greater omentum covering the abdominal organs to expose the following:

Stomach: dorsally located and to the left side of the liver
Spleen: a flattened, brown organ curving around the lateral aspect of the stomach
Small intestine: continuing posteriorly from the stomach
Large intestine: taking a U-shaped course around the small intestine and terminating in the rectum

Preparing to Identify the Blood Vessels

1. Carefully clear away any thymus tissue or fat obscuring the heart and the large vessels associated with the heart. Before identifying the blood vessels, try to locate the *phrenic nerve* (from the cervical plexus), which innervates the diaphragm.

The phrenic nerves lie ventral to the root of the lung on each side, as they pass to the diaphragm. Also attempt to locate the *vagus nerve* (cranial nerve X) passing laterally along the trachea and dorsal to the root of the lung.

2. Slit the parietal pericardium and reflect it superiorly. Then, cut it away from its heart attachments. Review the structures of the heart. Notice its pointed inferior end (apex) and its broader superior portion. Identify the two atria, which appear darker than the inferior ventricles. Identify the coronary arteries in the sulcus on the ventral surface of the heart; these should be injected with red latex. (As an aid to blood vessel identification, the arteries of laboratory dissection specimens are injected with red latex; the veins are injected with blue latex. Exceptions to this will be noted as they are encountered.)

3. Identify the two large venae cavae—the superior and inferior venae cavae—entering the right atrium. The superior vena cava is the largest dark-colored vessel entering the base of the heart. These vessels are called the **precava** and **postcava,** respectively, in the cat. The caval veins drain the same relative body areas as in humans. Also identify the pulmonary trunk (usually injected with blue latex) extending anteriorly from the right ventricle and the right and left pulmonary arteries. Trace the pulmonary arteries until they enter the lungs. Locate the pulmonary veins entering the left atrium and the ascending aorta arising from the left ventricle and running dorsally to the precava and to the left of the body midline.

Arteries of the Cat

Refer to Figure 24.17 and Plate E in the Cat Anatomy Atlas as you study the arterial system of the cat.

1. Reidentify the aorta as it emerges from the left ventricle. As you noted in the dissection of the sheep heart, the first branches of the aorta are the **coronary arteries,** which supply the myocardium. The coronary arteries emerge from the base of the aorta and can be seen on the surface of the heart. Follow the aorta as it arches (aortic arch), and identify its major branches. In the cat, the aortic arch gives off two large vessels, the **right brachiocephalic artery** and the **left subclavian artery.** The right brachiocephalic artery has three major branches, the right subclavian artery and the right and left common carotid arteries. (Note that in humans, the left common carotid artery is a direct branch off the aortic arch.)

2. Follow the **right common carotid artery** along the right side of the trachea as it moves anteriorly, giving off branches to the neck muscles, thyroid gland, and trachea. At the level of the larynx, it branches to form **external** and **internal carotid arteries.** The size of the internal carotid is much reduced in the cat and may be difficult to locate. It may even be absent. The distribution of the carotid arteries parallels that in humans.

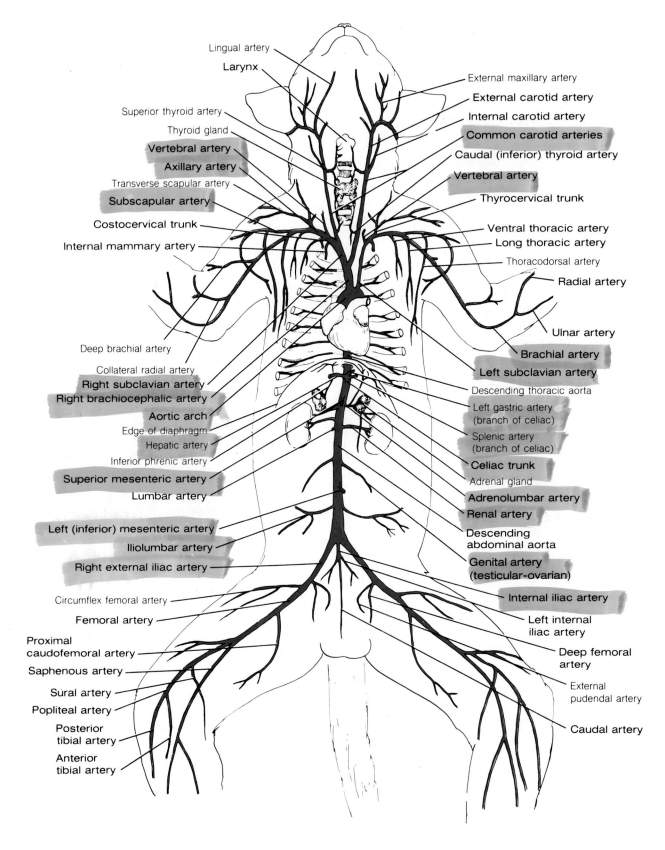

Lingual artery
Larynx
Superior thyroid artery
Thyroid gland
Vertebral artery
Axillary artery
Transverse scapular artery
Subscapular artery
Costocervical trunk
Internal mammary artery

External maxillary artery
External carotid artery
Internal carotid artery
Common carotid arteries
Caudal (inferior) thyroid artery
Vertebral artery
Thyrocervical trunk
Ventral thoracic artery
Long thoracic artery
Thoracodorsal artery
Radial artery

Deep brachial artery
Collateral radial artery
Right subclavian artery
Right brachiocephalic artery
Aortic arch
Edge of diaphragm
Hepatic artery
Inferior phrenic artery
Superior mesenteric artery
Lumbar artery
Left (inferior) mesenteric artery
Iliolumbar artery
Right external iliac artery

Ulnar artery
Brachial artery
Left subclavian artery
Descending thoracic aorta
Left gastric artery
(branch of celiac)
Splenic artery
(branch of celiac)
Celiac trunk
Adrenal gland
Adrenolumbar artery
Renal artery
Descending
abdominal aorta
Genital artery
(testicular-ovarian)
Internal iliac artery

Circumflex femoral artery
Femoral artery
Proximal
caudofemoral artery
Saphenous artery
Sural artery
Popliteal artery
Posterior
tibial artery
Anterior
tibial artery

Left internal
iliac artery
Deep femoral
artery
External
pudendal artery
Caudal artery

F24.17

Arterial system of the cat. See blood vessel dissection, Plate E in the Cat Anatomy Atlas.

3. Follow the right **subclavian artery** laterally. It gives off four branches, the first being the tiny vertebral artery, which along with the internal carotid artery provides the arterial circulation of the brain. Other branches of the subclavian artery include the **costocervical trunk** (to the costal and cervical regions), the **thyrocervical trunk** (to the shoulder), and the **internal mammary artery** (serving the ventral thoracic wall). As the subclavian passes in front of the first rib it becomes the **axillary artery.** Its branches, which supply the trunk and shoulder muscles, are the **ventral thoracic artery** (the pectoral muscles), the **long thoracic artery** (pectoral muscles and latissimus dorsi), and the **subscapular artery** (the trunk muscles). As the axillary artery enters the arm, it is called the **brachial artery,** and it travels with the median nerve down the length of the humerus. At the elbow, the brachial artery branches to produce the two major arteries serving the forearm and hand, the **radial** and **ulnar arteries.**

4. Return to the thorax, lift the left lung, and follow the course of the **descending aorta** through the thoracic cavity. The esophagus overlies it along its course. Notice the paired intercostal arteries that branch laterally from the aorta in the thoracic region.

5. Follow the aorta through the diaphragm into the abdominal cavity. Carefully pull the peritoneum away from its ventral surface and identify the following vessels:

Celiac trunk: the first branch diverging from the aorta immediately as it enters the abdominal cavity; supplies the stomach, liver, gallbladder, pancreas, and spleen. (Trace as many of its branches to these organs as possible.)

Superior mesenteric artery: immediately posterior to the celiac trunk; supplies the small intestine and most of the large intestine. (Spread the mesentery of the small intestine to observe the branches of this artery as they run to supply the small intestine.)

Adrenolumbar arteries: paired arteries diverging from the aorta slightly posterior to the superior mesenteric artery; supply the muscles of the body wall and adrenal glands

Renal arteries: paired arteries running to the kidneys, which they supply

Genital arteries (testicular or ovarian): paired arteries supplying the gonads

Inferior mesenteric artery: an unpaired thin vessel arising from the ventral surface of the aorta posterior to the genital arteries; supplies the second half of the large intestine

Iliolumbar arteries: paired, rather large arteries that supply the body musculature in the iliolumbar region

6. The descending aorta ends posteriorly by dividing into three arteries: the two lateral **external iliac arteries,** which continue through the body wall and pass under the inguinal ligament to the hindlimb, and the single median **internal iliac artery.** (There is no common iliac artery in the cat.) The internal iliac persists only briefly before dividing into the right and left internal iliac arteries and the caudal artery, which supply the pelvic viscera.

7. Trace the external iliac artery into the thigh, where it becomes the **femoral artery.** The femoral artery is most easily identified in the **femoral triangle** at the medial surface of the upper thigh. Follow the femoral artery as it courses through the thigh (along with the femoral vein and nerve) and gives off branches to the thigh muscles. (These various branches are indicated on Figure 24.17.) As you approach the knee, the **saphenous artery** branches off the femoral artery to supply the medial portion of the leg. The femoral artery then descends deep to the knee to become the **popliteal artery** in the popliteal region. The popliteal artery in turn gives off two main branches, the **sural artery** and the **posterior tibial artery,** and continues as the **anterior tibial artery.** These branches supply the leg and foot.

Veins of the Cat

Refer to Figure 24.18 and to Plate E in the Cat Anatomy Atlas as you study the venous system of the cat. Keep in mind that the vessels are named for the region drained, not for the point of union with other veins. (As you continue with the dissection, notice that not all vessels shown on Figure 24.18 are discussed.)

 1. Reidentify the precava as it enters the right atrium. Trace it anteriorly to identify its branches:

Azygos vein: passes directly into the dorsal surface of the precava just anterior to its point of entry into the heart; receives venous blood from the intercostal veins, which drain the intercostal muscles of the thorax

Internal thoracic (mammary) veins: drain the chest and abdominal walls

Right vertebral vein: drains the spinal cord and brain; usually enters right side of precava approximately at the level of the internal mammary veins but may enter the brachiocephalic vein in your specimen

Right and left brachiocephalic veins: form the precava by their union

2. Reflect the pectoral muscles, and trace the brachiocephalic vein laterally. Identify the two large veins that unite to form it—the external jugular vein and the subclavian vein.

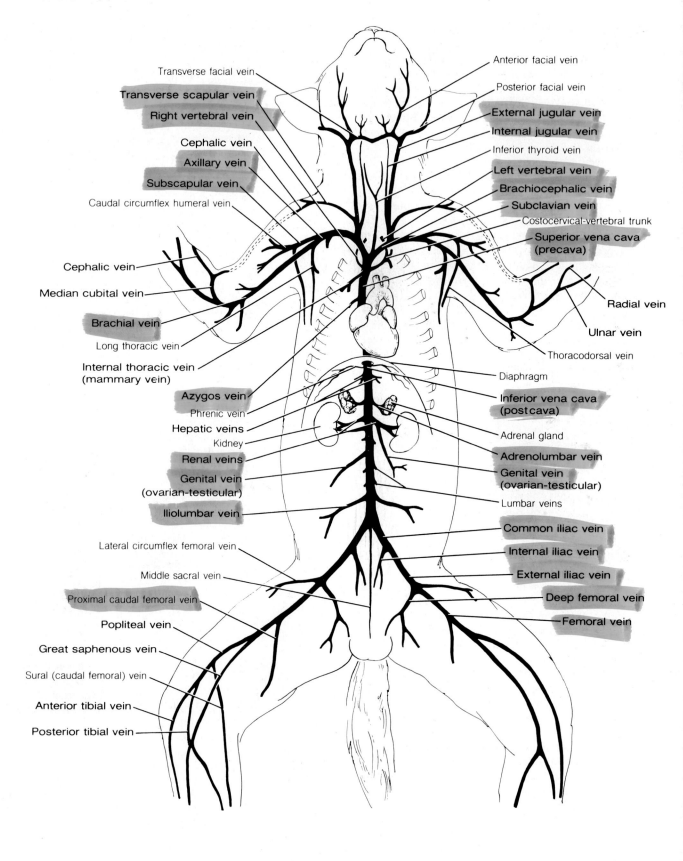

Transverse facial vein

Transverse scapular vein

Right vertebral vein

Cephalic vein

Axillary vein

Subscapular vein

Caudal circumflex humeral vein

Cephalic vein

Median cubital vein

Brachial vein

Long thoracic vein

Internal thoracic vein
(mammary vein)

Azygos vein

Phrenic vein

Hepatic veins

Kidney

Renal veins

Genital vein
(ovarian-testicular)

Iliolumbar vein

Lateral circumflex femoral vein

Middle sacral vein

Proximal caudal femoral vein

Popliteal vein

Great saphenous vein

Sural (caudal femoral) vein

Anterior tibial vein

Posterior tibial vein

Anterior facial vein

Posterior facial vein

External jugular vein

Internal jugular vein

Inferior thyroid vein

Left vertebral vein

Brachiocephalic vein

Subclavian vein

Costocervical-vertebral trunk

Superior vena cava
(precava)

Radial vein

Ulnar vein

Thoracodorsal vein

Diaphragm

Inferior vena cava
(post cava)

Adrenal gland

Adrenolumbar vein

Genital vein
(ovarian-testicular)

Lumbar veins

Common iliac vein

Internal iliac vein

External iliac vein

Deep femoral vein

Femoral vein

F24.18

Venous system of the cat. See blood vessel dissection, Plate E in the Cat Anatomy Atlas.

3. Follow the **external jugular vein** as it courses anteriorly along the side of the neck to the point where it is joined on its medial surface by the **internal jugular vein.** The internal jugular veins are small and may be difficult to identify in the cat. Note the difference in cat and human jugular veins. The internal jugular is considerably larger in humans and drains into the subclavian vein. In the cat, the external jugular is larger, and the interior jugular vein drains into it. Identify the common carotid artery, since it accompanies the internal jugular vein in this region. Also attempt to find the **sympathetic trunk,** which is located in the same area, running lateral to the trachea. Several other vessels drain into the external jugular vein (transverse scapular vein, facial veins, and others). These are not discussed here but are shown on the figure and may be traced if time allows.

4. Return to the shoulder region and follow the course of the **subclavian vein** as it moves laterally toward the arm. It becomes the **axillary vein** as it passes in front of the first rib and runs through the brachial plexus, giving off several branches, the first of which is the **subscapular vein.** The subscapular vein drains the proximal part of the arm and shoulder. The four other branches that receive drainage from the shoulder, pectoral, and latissimus dorsi muscles are shown in the figure but need not be identified in this dissection.

5. Follow the axillary vein into the arm, where it becomes the **brachial vein.** You can locate this vein on the medial side of the arm accompanying the brachial artery and nerve. Trace it to the point where it receives the **radial** and **ulnar veins** (which drain the forelimb) at the inner bend of the elbow. Also locate the superficial **cephalic vein** on the dorsal side of the arm. It communicates with the brachial vein via the median cubital vein in the elbow region and then enters the transverse scapular vein in the shoulder.

6. Reidentify the postcaval vein, and trace it to its passage through the diaphragm.

7. Attempt to identify the **hepatic veins** entering the postcava from the liver. These may be seen if some of the anterior liver tissue is scraped away where the postcava enters the liver.

8. Displace the intestines to the left side of the body cavity, and proceed posteriorly to identify the following veins in order. All of these veins empty into the postcava and drain the organs served by the same-named arteries. In the cat, variations in the connections of the veins to be located are common, and in some cases the postcaval vein may be double below the level of the renal veins. If you observe deviations, call them to the attention of your instructor.

Adrenolumbar veins: from the adrenal glands and body wall
Renal veins: from the kidneys (it is common to find two renal veins on the right side)

Genital veins (testicular or ovarian veins): the left vein of this venous pair enters the left renal vein anteriorly
Iliolumbar veins: drain muscles of the back
Common iliac veins: unite to form the postcava

The common iliac veins are formed in turn by the union of the **internal iliac** and **external iliac veins.** The more medial internal iliac veins receive branches from the pelvic organs and gluteal region whereas the external iliac vein receives venous drainage from the lower extremity. As the external iliac vein enters the thigh by running beneath the inguinal ligament, it receives the **deep femoral vein,** which drains the thigh and the external genital region. Just inferior to that point, the external iliac vein becomes the **femoral vein,** which receives blood from the thigh, leg, and foot. Follow the femoral vein down the thigh to identify the **great saphenous vein,** a superficial vein that courses up the inner aspect of the calf and across the inferior portion of the gracilis muscle (accompanied by the great saphenous artery and nerve) to enter the femoral vein. The femoral vein is formed by the union of this vein and the popliteal vein. The **popliteal vein** is located deep in the thigh beneath the semimembranosus and semitendinosus muscles in the popliteal space accompanying the popliteal artery. Trace the popliteal vein to its point of division into the **posterior** and **anterior tibial veins,** which drain the leg.

9. In your specimen, trace the hepatic portal drainage depicted in Figure 24.19. Locate the **hepatic portal vein** by removing the peritoneum between the first portion of the small intestine and the liver. It appears brown due to coagulated blood, and it is unlikely that it or any of the vessels of this circulation contain latex. In the cat, the hepatic portal vein is formed by the union of the **gastrosplenic** and **superior mesenteric veins.** (In the human, the hepatic portal vein is formed by the union of the splenic and superior mesenteric veins.) If possible, locate the following vessels, which empty into the hepatic portal vein:

Gastrosplenic vein: Carries blood from the spleen and stomach; located dorsal to the stomach.
Superior mesenteric vein: A large vein draining the small and large intestines and the pancreas.
Inferior mesenteric vein: Parallels the course of the inferior mesenteric artery.
Coronary vein: Drains the lesser curvature of the stomach.
Pancreaticoduodenal veins (anterior and posterior): The anterior branch empties into the hepatic portal vein; the posterior branch empties into the superior mesenteric vein. (In the human, both of these are branches of the superior mesenteric vein.)

(If the structures of the lymphatic system of the cat are to be studied during this laboratory session, turn to Exercise 25 for instructions to conduct the study. Otherwise, properly clean your dissecting instruments and dissecting pan, and wrap and tag your cat for storage.)

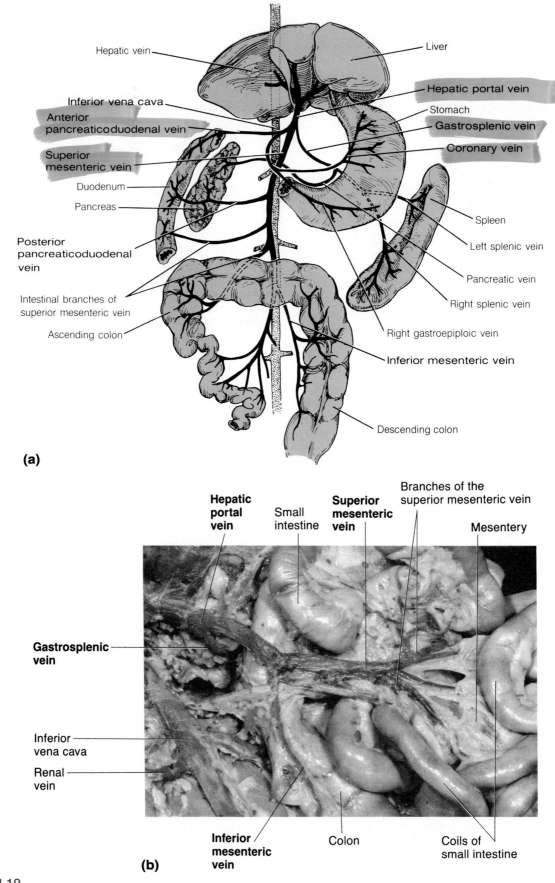

(a)

(b)

F24.19

Hepatic portal circulation of the cat. (a) Diagrammatic view. (b) Photograph of hepatic portal system of the cat, midline to left lateral view, just posterior to the liver and pancreas. Intestines have been pulled to the left side of the cat. The mesentery of the small intestine has been partially dissected to show the veins of the portal system.

The Lymphatic System

OBJECTIVES

1. To name the components of the lymphatic system.
2. To relate the function of the lymphatic system to that of the blood vascular system.
3. To describe the formation and composition of lymph, and to describe how it is transported through the lymphatic vessels.
4. To describe the structure and function of lymph nodes, and to indicate the localization of T cells, B cells, and macrophages in a typical lymph node.

MATERIALS

Large anatomical chart of the human lymphatic system
Dissection animal, tray, and dissection instruments
Prepared microscope slides of lymph nodes
Compound microscope

See Appendix B, Exercise 25 for links to A.D.A.M. Standard.

See Appendix C, Exercise 25 for links to *Anatomy and PhysioShow: The Videodisc.*

THE LYMPHATIC SYSTEM

General Description

The **lymphatic system** consists of a network of lymphatic vessels (lymphatics), lymph nodes, and a number of other lymphoid organs, such as the tonsils, thymus, and spleen (Figure 25.1). We will focus on the lymphatic vessels and lymph nodes in this section. The overall function of the lymphatic system is twofold. It transports tissue fluid (lymph) to the blood vessels. In addition, it protects the body by removing foreign material such as bacteria from the lymphatic stream and by serving as a site for lymphocyte "policing" of body fluids and lymphocyte multiplication. These white blood cells are the central actors in body immunity.

Distribution and Function of Lymphatic Vessels and Lymph Nodes

As blood circulates through the body, the hydrostatic and osmotic pressures operating at the capillary beds result in fluid outflow at the arterial end of the bed and in its return at the venous end. However, not all of the lost fluid is returned to the bloodstream by this mechanism, and the fluid that lags behind in the tissue spaces must eventually return to the blood if the vascular system is to operate properly. (If it does not, fluid accumulates in the tissues, producing a condition called edema.) It is the microscopic, blind-ended **lymphatic capillaries,** which ramify through nearly all the tissues of the body, that pick up this leaked fluid (primarily water and a small amount of dissolved proteins) and carry it through successively larger vessels—**lymphatic collecting vessels to lymphatic trunks**—until the lymph finally returns to the blood vascular system through one of the two large ducts in the thoracic region. The **right lymphatic duct** drains lymph from the right upper extremity, head, and thorax; the large **thoracic duct** receives lymph from the rest of the body (Figure 25.1b). In humans, both ducts empty the lymph into the venous circulation at the junction of the internal jugular vein and the subclavian vein, on their respective sides of the body. Notice that the lymphatic system, lacking both a contractile "heart" and arteries, is a one-way system; it carries lymph only toward the heart.

Like veins of the blood vascular system, the lymphatic collecting vessels have three tunics and are equipped with valves. However, the lymphatics tend to be thinner-walled, to have *more* valves, and to anastomose more than veins. Since the lymphatic system is a pumpless system, lymph transport depends largely on the milking action of the skeletal muscles and on pressure changes within the thorax during breathing.

As lymph is transported, it filters through bean-shaped **lymph nodes,** which cluster along the lymphatic vessels of the body. There are thousands of lymph nodes; but because they are usually embedded in connective tissue, they are not ordinarily seen. Within the lymph nodes are **macrophages,** phagocytes that destroy bacteria, cancer cells, and other foreign matter in the

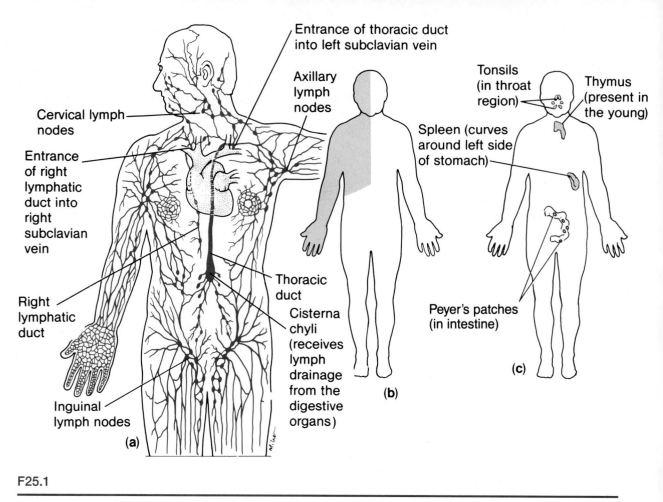

Entrance of thoracic duct into left subclavian vein

Axillary lymph nodes

Cervical lymph nodes

Entrance of right lymphatic duct into right subclavian vein

Right lymphatic duct

Inguinal lymph nodes

Thoracic duct

Cisterna chyli (receives lymph drainage from the digestive organs)

Tonsils (in throat region)

Thymus (present in the young)

Spleen (curves around left side of stomach)

Peyer's patches (in intestine)

(a)

(b)

(c)

F25.1

Lymphatic system. (a) Distribution of lymphatic vessels and lymph nodes. (b) Shaded area represents body area drained by the right lymphatic duct. (c) Body location of the tonsils, thymus, spleen, and Peyer's patches.

lymphatic stream, thus rendering many harmful substances or cells harmless before the lymph enters the bloodstream. Particularly large collections of lymph nodes are found in the inguinal, axillary, and cervical regions of the body. Although we are not usually aware of the filtering and protective nature of the lymph nodes, most of us have experienced "swollen glands" during an active infection. This swelling is a manifestation of the trapping function of the nodes.

The tonsils, thymus, and spleen are generally considered to be lymphoid organs because they resemble the lymph nodes histologically, and house similar cell populations (lymphocytes and macrophages).

 Study the large anatomical chart to observe the general plan of the lymphatic system. Notice the distribution of lymph nodes, various lymphatics, the lymphatic trunks, and the location of the right lymphatic duct and the thoracic duct. Also identify the **cisterna chyli,** the enlarged terminus of the thoracic duct that receives lymph from the digestive viscera.

MAIN LYMPHATIC DUCTS OF THE CAT

 1. Obtain your cat and a dissecting tray and instruments. Because lymphatic vessels are extremely thin-walled, it is difficult to locate them in a dissection. However, the large thoracic duct can be localized and identified.

2. Move the thoracic organs to the side to locate the **thoracic duct.** Typically it lies just to the left of the mid-dorsal line, abutting the dorsal aspect of the descending aorta. It is usually about the size of pencil lead and red-brown with a segmented or beaded appearance caused by the valves within it. Trace it anteriorly to the site where it passes behind the left brachiocephalic vein and then bends and enters the venous system at the junction of the left subclavian and external jugular veins. If the veins are well injected, some of the blue latex may have slipped past the valves and entered the first portion of the thoracic duct.

3. While in this region, also attempt to identify the short **right lymphatic duct** draining into the right subclavian vein, and notice the clustered lymph nodes in the axillary region.

282

4.   If the cat is triply injected and the lymphatic vessels have been injected with yellow or green latex, trace the thoracic duct posteriorly to identify the cisterna chyli, the saclike enlargement of its distal end. This structure, which receives fat-rich lymph from the intestine, begins at the level of the diaphragm and can be localized posterior to the left kidney.

5.   Clean the dissecting instruments and tray, and properly wrap and return the cat to storage before continuing with the laboratory exercise.

MICROSCOPIC ANATOMY OF A LYMPH NODE

Obtain a prepared slide of a lymph node and a compound microscope. As you examine the slide, notice the following anatomical features, depicted in Figure 25.2. The node is enclosed within a fibrous **capsule,** from which **trabeculae** (connective tissue septa) extend inward to divide the node into several compartments. Fine strands of reticular connective tissue issue from the trabeculae, forming the stroma ("mattress") of the gland within which cells are found.

In the outer region of the node, the **cortex,** some of the cells are arranged in globular masses, referred to as **germinal centers.** The germinal centers contain rapidly dividing B lymphocytes, commonly called *B cells.* The rest of the cortical cells are primarily T lymphocytes that circulate continuously, moving from the blood into the node and then exiting from the node in the lymphatic stream.

In the internal portion of the gland, the **medulla,** the cells are arranged in cordlike fashion. Most of the medullary cells are macrophages. Macrophages are important not only for their phagocytic function, but also because they play an essential role in "presenting" the antigens (foreign substances present in the body) to the T cells.

Lymph enters the node through a number of afferent vessels, circulates through sinuses within the node, and leaves the node through efferent vessels at the **hilus.** Since each node has fewer efferent than afferent vessels, the lymph flow stagnates somewhat within the node. This allows time for the generation of an immune response and for the macrophages to remove debris from the lymph before it reenters the blood vascular system.

In the space provided here, draw a pie-shaped section of a lymph node, showing the detail of cells in a germinal center, sinusoids, and afferent and efferent vessels. Label all elements.

Before leaving the topic of the lymphoid organs, compare and contrast the structure of the lymph node examined here with the microscopic anatomy of the spleen and tonsils illustrated in Plates 28 and 29 in the Histology Atlas.

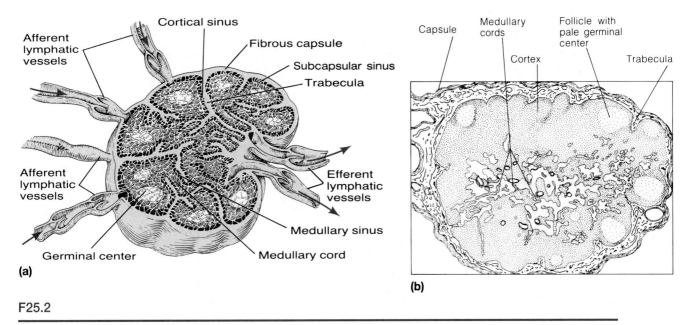

(a)

(b)

F25.2

Structure of lymph node. (a) Cross section of a lymph node, diagrammatic view. Notice that the afferent vessels outnumber the efferent vessels, which slows the rate of lymph flow. The arrows indicate the direction of the lymph flow. **(b)** Line drawing of a photomicrograph (Plate 27 in the Histology Atlas) showing a part of a lymph node.

Anatomy of the Respiratory System

OBJECTIVES

1. To define the following terms: *respiratory system, pulmonary ventilation, external respiration,* and *internal respiration.*

2. To label the major respiratory system structures on a diagram (or identify them on a model or dissection) and to describe the function of each.

3. To recognize the histologic structure of the trachea (cross section) and lung tissue on prepared slides, and describe the functions the observed structural modifications serve.

MATERIALS

Human torso model

Respiratory organ system model and/or chart of the respiratory system

Larynx model (if available)

Preserved inflatable lung preparation (obtained from a biological supply house) or sheep pluck fresh from the slaughterhouse

Source of compressed air*

2-foot length of laboratory rubber tubing

Dissection animals, trays, and instruments

Protective skin cream or disposable gloves

Histologic slides of the following (if available): trachea (cross section), lung tissue, both normal and pathologic specimens (e.g., sections taken from lung tissues exhibiting bronchitis, pneumonia, emphysema, or lung cancer)

Compound and dissecting microscopes

See Appendix B, Exercise 26 for links to A.D.A.M. Standard.

 See Appendix C, Exercise 26 for links to *Anatomy and PhysioShow: The Videodisc.*

* If a compressed air source is not available, cardboard mouthpieces that fit the cut end of the rubber tubing should be available for student use. Disposable autoclave bags should also be provided for discarding the mouthpiece.

Body cells require an abundant and continuous supply of oxygen. As the cells use oxygen, they release carbon dioxide, a waste product that the body must get rid of. These oxygen-using cellular processes, collectively referred to as *cellular respiration,* are more appropriately described in conjunction with the topic of cellular metabolism. The major role of the **respiratory system,** our focus in this exercise, is to supply the body with oxygen and dispose of carbon dioxide. For it to fulfill this role, at least four distinct processes, collectively referred to as **respiration,** must occur:

Pulmonary ventilation: the tidelike movement of air into and out of the lungs so that the gases in the alveoli are continuously changed and refreshed. Also more simply called *ventilation,* or *breathing.*

External respiration: the gas exchanges to and from the pulmonary circuit blood that occur in the lungs (oxygen loading/carbon dioxide unloading).

Transport of respiratory gases: the transport of respiratory gases between the lungs and tissue cells of the body accomplished by the cardiovascular system, using blood as the transport vehicle.

Internal respiration: exchange of gases to and from the blood capillaries of the systemic circulation (oxygen unloading and carbon dioxide loading).

Only the first two processes are the exclusive province of the respiratory system, but all four must occur for the respiratory system to "do its job." Hence, the respiratory and circulatory system are irreversibly linked. Should either system fail, the cells begin to die from oxygen starvation and the accumulation of carbon dioxide. If uncorrected, this situation soon causes death of the entire organism.

UPPER RESPIRATORY SYSTEM STRUCTURES

The upper respiratory system structures—the nose, pharynx, and larynx—are shown in Figure 26.1 and described below. As you read through the descriptions, identify each structure in the figure.

Air generally passes into the respiratory tract through the **external nares** (nostrils), and enters the **nasal cavity** (divided by the **nasal septum**). It then flows posteriorly over three pairs of lobelike structures, the **inferior, superior,** and **middle nasal conchae,** which increase the air turbulence. As the air passes through the nasal cavity, it is also warmed, moistened, and filtered by the nasal mucosa. The air that flows di-

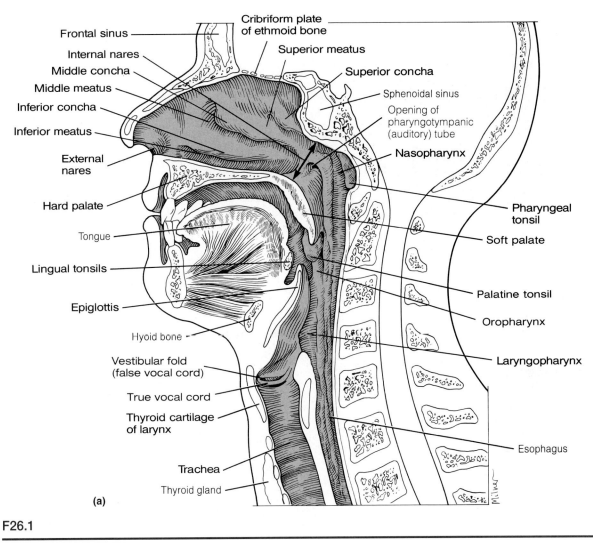

Frontal sinus

Internal nares

Middle concha

Middle meatus

Inferior concha

Inferior meatus

External nares

Hard palate

Tongue

Lingual tonsils

Epiglottis

Hyoid bone

Vestibular fold (false vocal cord)

True vocal cord

Thyroid cartilage of larynx

Trachea

Thyroid gland

Cribriform plate of ethmoid bone

Superior meatus

Superior concha

Sphenoidal sinus

Opening of pharyngotympanic (auditory) tube

Nasopharynx

Pharyngeal tonsil

Soft palate

Palatine tonsil

Oropharynx

Laryngopharynx

Esophagus

(a)

F26.1

Structures of the upper respiratory tract (sagittal section). (a) Diagrammatic view.

rectly beneath the upper part of the nasal cavity may chemically stimulate the olfactory receptors located in the mucosa of that region. The nasal cavity is surrounded by the **paranasal sinuses** in the frontal, sphenoid, ethmoid, and maxillary bones. These sinuses act as resonance chambers in speech, and their mucosae, like that of the nasal cavity, warm and moisten the incoming air.

The nasal passages are separated from the oral cavity below by a partition composed anteriorly of the **hard palate** and posteriorly by the **soft palate.**

The genetic defect called **cleft palate** (failure of the palatine bones and/or the palatine processes of the maxillary bones to fuse medially) causes difficulty in breathing and oral cavity functions such as sucking and, later, mastication and speech. ■

Needless to say, air may also enter the body via the mouth, and pass through the oral cavity to move into the pharynx posteriorly, where the oral and nasal cavities are joined temporarily.

Commonly called the *throat,* the funnel-shaped **pharynx** connects the nasal and oral cavities to the larynx and esophagus inferiorly. It has three named parts:

1.   The **nasopharynx** lies posterior to the nasal cavity and is continuous with it via the **internal nares.** It lies above the soft palate; hence, it serves only as an air passage. High on its posterior wall are the *pharyngeal tonsils,* paired masses of lymphoid tissue that help to protect the respiratory passages from invading pathogens. The *pharyngotympanic (auditory) tubes,* which allow middle ear pressure to become equalized to atmospheric pressure, drain into the lateral aspects of the nasopharynx.

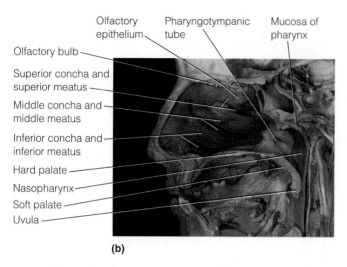

Olfactory epithelium

Pharyngotympanic tube

Mucosa of pharynx

Olfactory bulb

Superior concha and superior meatus

Middle concha and middle meatus

Inferior concha and inferior meatus

Hard palate

Nasopharynx

Soft palate

Uvula

(b)

F26.1 (*continued*)

Structures of the upper respiratory tract (sagittal section). (b) Photograph.

Because of the continuity of the middle ear and nasopharyngeal mucosae, nasal infections may invade the middle ear cavity causing *otitis media,* which is difficult to treat. ■

2.   The **oropharynx** is continuous posteriorly with the oral cavity. Since it extends from the soft palate to the epiglottis of the larynx inferiorly, it serves as a common conduit for food and air. In its lateral walls are the *palatine tonsils.* The *lingual tonsils* cover the base of the tongue.

3.   The **laryngopharynx,** like the oropharynx, accommodates both ingested food and air. It lies directly posterior to the upright epiglottis and extends to the larynx, where the common pathway divides into the respiratory and digestive channels. From the laryngopharynx, air enters the lower respiratory passageways by passing through the larynx (voice box) and into the trachea below.

The **larynx** (Figure 26.2) consists of nine cartilages. The two most prominent are the large shield-shaped **thyroid cartilage,** whose anterior medial prominence is commonly referred to as *Adam's apple,* and the inferiorly located, ring-shaped **cricoid cartilage,** whose widest dimension faces posteriorly. All the laryngeal cartilages are composed of hyaline cartilage except the flaplike **epiglottis,** a flexible elastic cartilage located superior to the opening of the larynx. The epiglottis, sometimes referred to as the "guardian of the airways," forms a lid over the larynx when we swallow. This closes off the respiratory passageways to incoming food or drink, which is routed into the posterior esophagus, or food chute.

●   Palpate your larynx by placing your hand on the anterior neck surface approximately halfway down its length. Swallow. Can you feel the cartilaginous larynx rising?

If anything other than air enters the larynx, a cough reflex attempts to expel the substance. Note that this reflex operates only when a person is conscious; thus you should never try to feed or pour liquids down the throat of an unconscious person.

The mucous membrane of the larynx is thrown into two pairs of folds—the upper **vestibular folds,** also

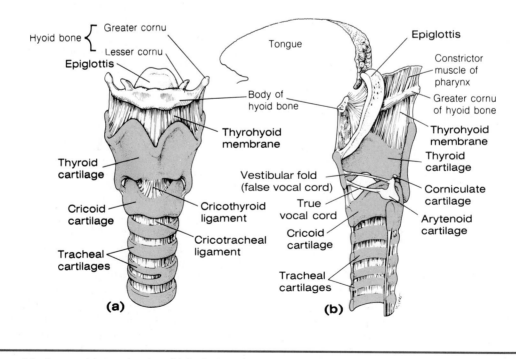

F26.2

Structure of the larynx. (a) Anterior view. **(b)** Sagittal section.

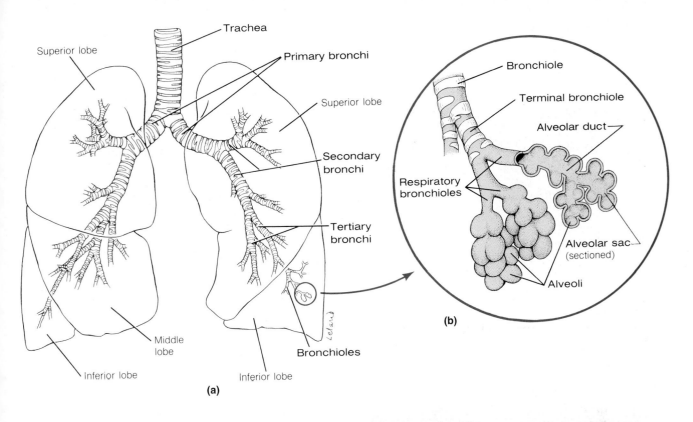

(a)

F26.3

Structures of the lower respiratory tract. (a) Diagrammatic view. **(b)** Inset shows enlarged view of alveoli. **(c)** Resin cast showing the extensive branching of the respiratory tree.

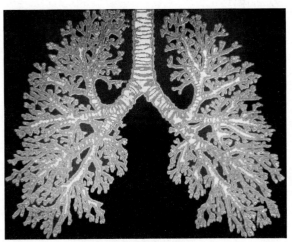

(c)

called the **false vocal cords,** and the lower **vocal folds,** or **true vocal cords,** which vibrate with expelled air for speech. The vocal cords are attached posterolaterally to the small triangular **arytenoid cartilages** by the *vocal ligaments.* The slitlike passageway between the folds is called the **glottis.**

LOWER RESPIRATORY SYSTEM STRUCTURES

Air entering the **trachea,** or windpipe, from the larynx travels down its length (about 11.0 cm) to the level of the *sternal angle* (or the disc between the fourth and fifth thoracic vertebrae). There the passageway divides into the right and left **primary bronchi** (Figure 26.3), which plunge into their respective lungs at an indented area called the **hilus** (see Figure 26.5b). The right primary bronchus is wider, shorter, and more vertical than the left. As a result, foreign objects that enter the respiratory passageways are more likely to become lodged in it.

The trachea is lined with a ciliated mucus-secreting, pseudostratified columnar epithelium, as are many of the other respiratory system passageways. The cilia propel mucus (produced by goblet cells) laden with dust particles, bacteria, and other debris away from the lungs and toward the throat, where it can be expectorated or swallowed. The walls of the trachea are reinforced with C-shaped cartilage rings, the incomplete portion located posteriorly. These C-shaped cartilages serve a double function: The incomplete parts allow the esophagus to expand anteriorly when a large food bolus is swallowed. The solid portions reinforce the trachea walls to maintain its open passageway regardless of the pressure changes that occur during breathing.

The primary bronchi further divide into smaller and smaller branches (the secondary, tertiary, on down), finally becoming the **bronchioles,** which have terminal branches called **respiratory bronchioles** (Figure 26.3b). All but the most minute branches have cartilaginous reinforcements in their walls, usually in the form

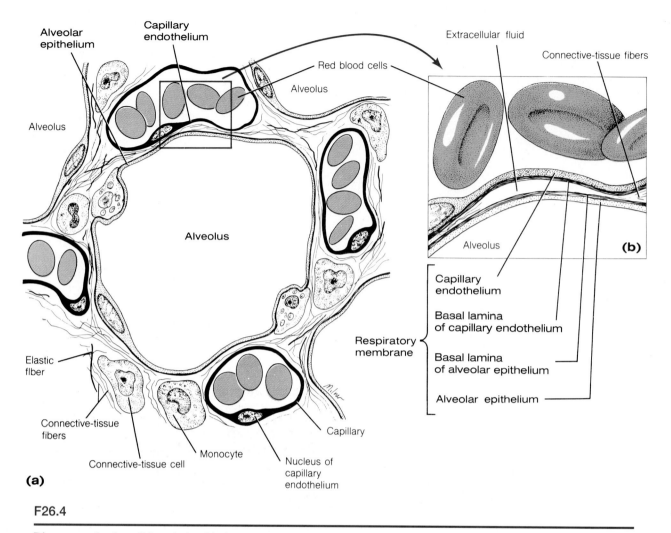

Alveolar epithelium

Capillary endothelium

Red blood cells

Alveolus

Alveolus

Alveolus

Extracellular fluid

Connective-tissue fibers

Alveolus

(b)

Capillary endothelium

Basal lamina of capillary endothelium

Basal lamina of alveolar epithelium

Alveolar epithelium

Respiratory membrane

Elastic fiber

Connective-tissue fibers

Connective-tissue cell

Monocyte

Capillary

Nucleus of capillary endothelium

(a)

F26.4

Diagrammatic view of the relationship between the alveoli and pulmonary capillaries involved in gas exchange. (a) One alveolus surrounded by capillaries. (b) Enlargement of the respiratory membrane.

of small plates of hyaline cartilage rather than cartilaginous rings. As the respiratory tubes get smaller and smaller, the relative amount of smooth muscle in their walls increases as the amount of cartilage declines and finally disappears. The complete layer of smooth muscle present in the bronchioles enables them to provide considerable resistance to air flow under certain conditions (asthma, hay fever, etc.). The continuous branching of the respiratory passageways in the lungs is often referred to as the **respiratory tree.** The comparison becomes much more meaningful if you observe a resin cast of the respiratory passages. (Do so, if one is available for observation in the laboratory; otherwise, refer to Figure 26.3c.)

The respiratory bronchioles in turn subdivide into several **alveolar ducts,** which terminate in alveolar sacs that rather resemble clusters of grapes. **Alveoli,** tiny balloonlike expansions along the alveolar sacs, and occasionally found protruding from the alveolar ducts and

respiratory bronchioles, are composed of a single thin layer of squamous epithelium overlying a wispy basal lamina. The external surfaces of the alveoli are densely spiderwebbed with a network of pulmonary capillaries (Figure 26.4). Together, the alveolar and capillary walls and their fused basal laminas form the **respiratory membrane.** Because gas exchanges occur by simple diffusion across the respiratory membrane—oxygen passing from the alveolar air to the capillary blood and carbon dioxide leaving the capillary blood to enter the alveolar air—the alveolar sacs, alveolar ducts, and respiratory bronchioles are referred to collectively as **respiratory zone structures.** All other respiratory passageways (from the nasal cavity to the bronchioles), which simply serve as access or exit routes to and from these gas exchange chambers, are called **conducting zone structures.** Because the conducting zone structures have no exchange function, they are also referred to as *anatomical dead space.*

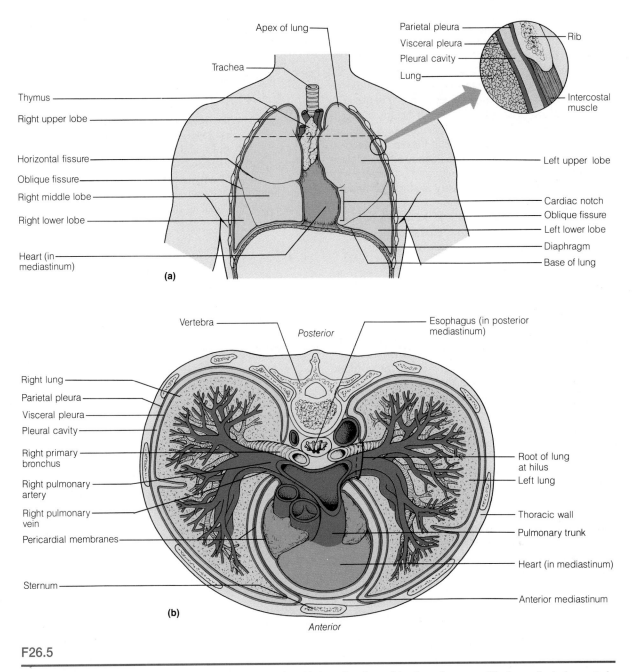

F26.5

Anatomical relationships of organs in the thoracic cavity. (a) Anterior view of the thoracic organs. The lungs flank the central mediastinum. The inset at upper right depicts the pleura and the pleural cavity. **(b)** Transverse section through the superior part of the thorax, showing the lungs and the main organs in the mediastinum. The plane of section is shown by the dotted line in part (a).

The Lungs and Their Pleural Coverings

The paired lungs are soft, spongy organs that occupy the entire thoracic cavity except for the *mediastinum,* which houses the heart, bronchi, esophagus, and other organs (Figure 26.5). Each lung is connected to the mediastinum by a *root* containing its vascular and bronchial attachments. The structures of the root enter (or leave) the lung via a medial indentation called the *hilus.* All structures distal to the primary bronchi are found within the lung substance. A lung's *apex,* the narrower superior aspect, lies just deep to the clavicle, and its *base,* the inferior concave surface, rests on the diaphragm. Anterior, lateral, and posterior lung surfaces are in close contact with the ribs. The medial surface of the left lung exhibits a concavity called the *cardiac notch,* which accommodates the heart where it extends left from the body midline. Fissures divide the lungs into a number of lobes—two in the left lung and three in the right. Other than the respiratory passageways and air spaces that make up the bulk of their volume, the lungs are mostly elastic connective tissue, which allows them to recoil passively during expiration.

Each lung is enclosed in a double-layered sac of serous membrane called the **pleura.** The outer layer, the **parietal pleura,** is attached to the thoracic walls and the **diaphragm.** The inner layer, covering the lung tissue, is the **visceral pleura.** The two pleural layers are separated by the *pleural space,* which is more of a potential space than an actual one. The pleural layers produce lubricating serous fluid that causes them to adhere closely to one another, holding the lungs to the thoracic wall and allowing them to move easily against one another during the movements of breathing.

Before proceeding, be sure to locate on the torso model, thoracic cavity structures model, or an anatomical chart all the respiratory structures described.

SHEEP PLUCK DEMONSTRATION

 A *sheep pluck* includes the larynx, trachea with attached lungs, the heart and pericardium, and portions of the major blood vessels found in the mediastinum (aorta, pulmonary artery and vein, venae cavae).

Don plastic gloves, obtain a fresh sheep pluck (or a preserved pluck of another animal), and identify the lower respiratory system organs. Once you have completed your observations, insert a hose from an air compressor (vacuum pump) into the trachea and alternately allow air to flow in and out of the lungs. Notice how the lungs inflate. This observation is educational in a preserved pluck but it is a spectacular sight in a fresh one. Another advantage of using a fresh pluck is that the lung pluck changes color (becomes redder) as hemoglobin in trapped RBCs becomes loaded with oxygen.

If air compressors are not available, the same effect may be obtained by using a length of laboratory rubber tubing to blow into the trachea. Obtain a cardboard mouthpiece and fit it into the cut end of the laboratory tubing before attempting to inflate the lungs.

 Dispose of the mouthpiece and gloves in the autoclave bag immediately after use.

EXAMINATION OF PREPARED SLIDES OF LUNG AND TRACHEA TISSUE

 1.   Obtain and examine a cross section of the trachea wall. Identify the smooth muscle layer, the hyaline cartilage supporting rings, and the pseudostratified ciliated epithelium. Using Figure 26.6a as a guide, also try to identify a few goblet cells in the epithelium. In the space below, draw a section of the trachea wall and label all tissue layers.

2.   Obtain a slide of lung tissue for examination. The alveolus is the main structural and functional unit of the lung and is the actual site of gas exchange. Identify a bronchiole (Figure 26.6b) and the thin squamous epithelium of the alveolar walls (Figure 26.6c). Draw your observations of a small section of the alveolar tissue in the space below. Label the alveoli.

3.   Examine slides of pathologic lung tissues, and compare them to the normal lung specimens. Record your observations in the Exercise 26 Review Sheets.

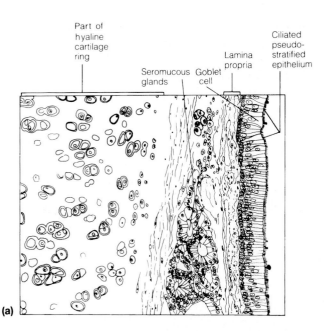

Part of
hyaline
cartilage
ring

Seromucous   Goblet
glands   cell

Lamina
propria

Ciliated
pseudo-
stratified
epithelium

(a)

Microscopic structure of the trachea, a bronchiole, and alveoli. (a) Cross section through the trachea. See corresponding Plate 31 in the Histology Atlas. **(b)** Cross-sectional view of a bronchiole. See corresponding Plate 32 in the Histology Atlas. **(c)** Alveoli. See corresponding Plate 30 in the Histology Atlas.

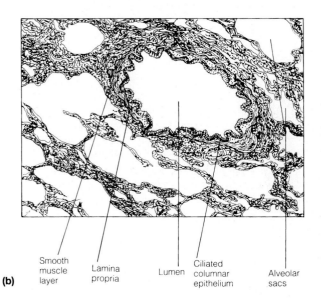

Smooth
muscle
layer

Lamina
propria

Lumen

Ciliated
columnar
epithelium

Alveolar
sacs

(b)

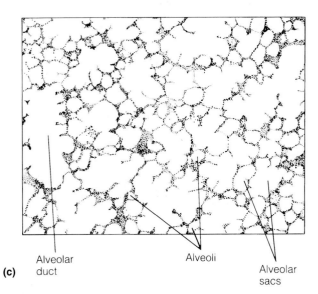

Alveolar
duct

Alveoli

Alveolar
sacs

(c)

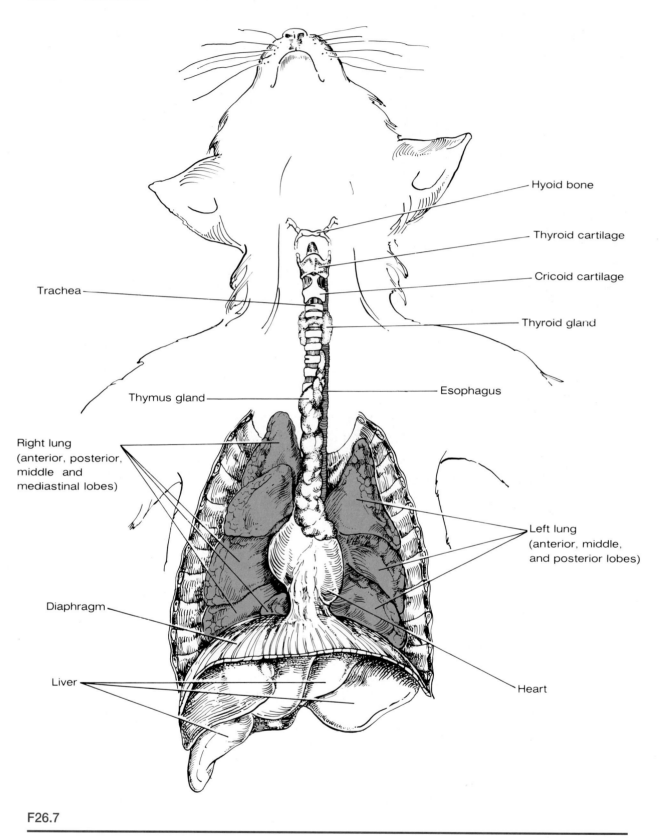

Hyoid bone

Thyroid cartilage

Cricoid cartilage

Thyroid gland

Trachea

Thyroid gland

Thymus gland

Esophagus

Right lung
(anterior, posterior,
middle and
mediastinal lobes)

Left lung
(anterior, middle,
and posterior lobes)

Diaphragm

Liver

Heart

F26.7

Respiratory system of the cat. See respiratory dissection, Plate D in the Cat Anatomy Atlas.

DISSECTION OF THE RESPIRATORY SYSTEM OF THE CAT

1.   Obtain your dissection animal. Examine the external nares, oral cavity, and oral pharynx. Use a probe to demonstrate the continuity between the oral pharynx and the nasal pharynx above. If desired, apply protective skin cream or don disposable gloves before beginning dissecting activities.

2.   After securing the animal to the dissecting tray, expose the respiratory structures by retracting the cut muscle and rib cage. Do not sever nerves and blood vessels located on either side of the trachea if these have not been studied. If you have not previously opened the thoracic cavity, make a medial longitudinal incision through the neck muscles and thoracic musculature to expose and view the thoracic organs.

3.   Using Figure 26.7 and Plate D in the Cat Anatomy Atlas as guides, identify the structures named in items 3 through 6. Examine the **trachea,** and determine by finger examination whether the cartilage rings are complete or incomplete posteriorly. Locate the *thyroid gland* inferior to the larynx on the trachea. Free the **larynx** from the attached muscle tissue, and pull the larynx anteriorly for ease of examination. Identify the **thyroid** and **cricoid cartilages** and the flaplike **epiglottis.** Find the hyoid bone located anterior to the larynx. Make a longitudinal incision through the ventral wall of the lar-ynx and locate the *true* and *false vocal cords* on the inner wall.

4.   Locate the large right and left common carotid arteries and the internal jugular veins on either side of the trachea. Also locate a conspicuous white band, the vagus nerve, which lies alongside the trachea, adjacent to the common carotid artery.

5.   Examine the contents of the thoracic cavity. Follow the trachea as it bifurcates into two **primary bronchi,** which plunge into the lungs. Note that there are two *pleural cavities* containing the lungs and that each lung is composed of many lobes. In humans there are three lobes in the right lung and two in the left. How does this compare to what is seen in the cat? Identify the pericardial sac containing the heart located in the mediastinum (if it is still present). Examine the pleura, and note its exceptionally smooth texture. Locate the **diaphragm** and the **phrenic nerve.** The phrenic nerve, clearly visible as a white "thread" running along the pericardium to the diaphragm, controls the activity of the diaphragm in breathing. Lift one lung and find the esophagus beneath the parietal pleura. Follow it through the diaphragm to the stomach.

6.   Make a longitudinal incision in the outer tissue of one lung lobe beginning at a primary bronchus. Attempt to follow part of the respiratory tree from this point down into the smaller subdivisions. Carefully observe the cut lung tissue (under a dissection scope, if one is available), noting the richness of the vascular supply and the irregular or spongy texture of the lung.

Anatomy of the Digestive System

OBJECTIVES

1. To state the overall function of the digestive system.

2. To identify on an appropriate diagram, torso model, or dissected animal the organs comprising the alimentary canal; and to name their subdivisions if any.

3. To name and/or identify the accessory digestive organs.

4. To describe the general functions of the digestive system organs or structures.

5. To describe the general histologic structure of the wall of the alimentary canal and/or label a cross-sectional diagram of the wall with the following terms: mucosa, submucosa, muscularis externa, and serosa or adventitia.

6. To list and explain the specializations of anatomical structure of the stomach and small intestine that contribute to their functional roles.

7. To list the major enzymes or enzyme groups produced by each of the following organs: salivary glands, stomach, small intestine, pancreas.

8. To name human deciduous and permanent teeth and describe the anatomy of the generalized tooth.

9. To recognize by microscopic inspection, or by viewing an appropriate diagram or photomicrograph, the histologic structure of the following organs:

small intestine	pancreas	stomach
salivary glands	tooth	liver

MATERIALS

Dissectible torso model
Anatomical chart of the human digestive system
Model of a villus and liver (if available)
Jaw model or human skull
Prepared microscope slides of the liver, pancreas, and mixed salivary glands; of longitudinal sections of the gastroesophageal junction and a tooth; and of cross sections of the stomach, duodenum, and ileum
Hand lens
Compound microscope
Colored pencils
Dissection animal, trays, instruments, and bone cutters
Protective hand cream or disposable gloves

 See Appendix B, Exercise 27 for links to A.D.A.M. Standard.

 See Appendix C, Exercise 27 for links to *Anatomy and PhysioShow: The Videodisc.*

The **digestive system** provides the body with the nutrients, water, and electrolytes essential for health. The organs of this system are responsible for food ingestion, digestion, absorption, and the elimination of the undigested remains as feces.

The digestive system consists of a hollow tube extending from the mouth to the anus, into which various accessory organs or glands empty their secretions (Figure 27.1). Food material within this tube, the *alimentary canal,* is technically outside the body, since it has contact only with the cells lining the tract. For ingested food to become available to the body cells, it must first be broken down *physically* (chewing, churning) and *chemically* (enzymatic hydrolysis) into its smaller diffusible molecules—a process called **digestion.** The digested end products can then pass through the epithelial cells lining the tract into the blood for distribution to body cells—a process termed **absorption.** In one sense, the digestive tract can be viewed as a disassembly line, in which food is carried from one stage of its digestive processing to the next by muscular activity, and its nutrients are made available to the cells of the body en route.

The organs of the digestive system are traditionally separated into two major groups: the **alimentary canal,** or **gastrointestinal (GI) tract,** and the **accessory digestive organs.** The alimentary canal is approximately 30 feet long in a cadaver but considerably less in a living person. It consists of the mouth, pharynx, esophagus, stomach, small and large intestines, rectum, and anus. The accessory structures include the salivary glands, gallbladder, liver, and pancreas, which secrete their products into the alimentary canal. These individual organs are described shortly.

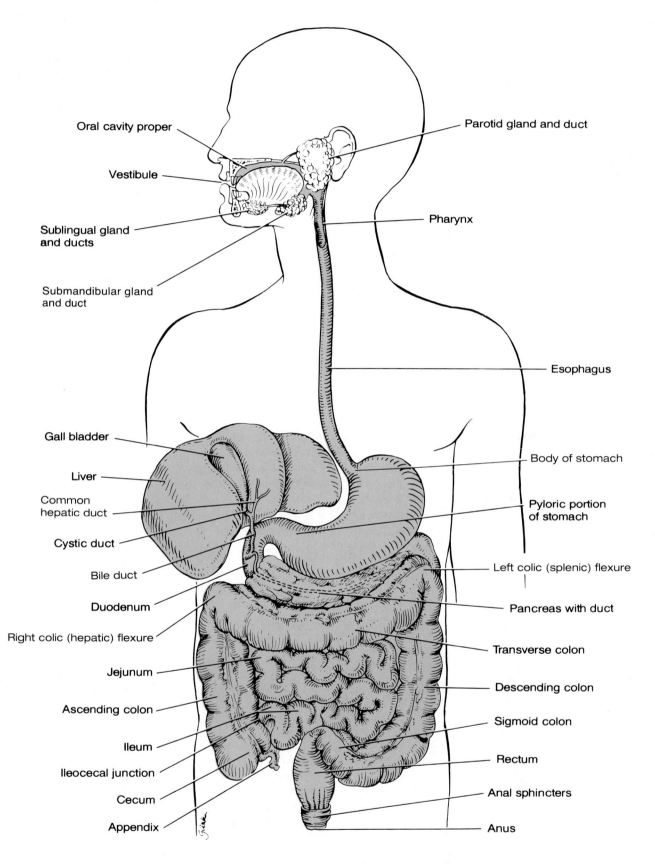

Oral cavity proper

Vestibule

Sublingual gland and ducts

Submandibular gland and duct

Parotid gland and duct

Pharynx

Esophagus

Gall bladder

Liver

Common hepatic duct

Cystic duct

Bile duct

Duodenum

Right colic (hepatic) flexure

Jejunum

Ascending colon

Ileum

Ileocecal junction

Cecum

Appendix

Body of stomach

Pyloric portion of stomach

Left colic (splenic) flexure

Pancreas with duct

Transverse colon

Descending colon

Sigmoid colon

Rectum

Anal sphincters

Anus

F27.1

The human digestive system: alimentary tube and accessory organs. (Liver and gallbladder are reflected superiorly and to the right.)

GENERAL HISTOLOGICAL PLAN OF THE ALIMENTARY CANAL

Because the alimentary canal has a shared basic structure (particularly from the esophagus to the anus), it makes sense to review that structure as we begin studying this group of organs. Once done, all that need be emphasized as the individual organs are described is their specializations for unique functions in the digestive process.

Essentially the alimentary canal walls have four basic **tunics** (layers). From the lumen outward, these are the *mucosa,* the *submucosa,* the *muscularis externa,* and the *serosa* or *adventitia* (Figure 27.2). Each of these tunics has a predominant tissue type and a specific function in the digestive process.

Mucosa (mucous membrane): The mucosa is the wet epithelial membrane abutting the alimentary canal lumen. It consists of a surface *epithelium* (in most cases, a simple columnar), a *lamina propria* (areolar connective tissue on which the epithelial layer rests), and a *muscularis mucosae* (a scant layer of smooth muscle fibers that enable local movements of the mucosa). The major functions of the mucosa are secretion (of enzymes, mucus, hormones, etc.), absorption of digested foodstuffs, and protection (against bacterial invasion). A particular mucosal region may be involved in one or all three functions.

Submucosa: Superficial to the mucosa, the submucosa is moderately dense connective tissue containing blood and lymphatic vessels, scattered lymph nodules, and nerve fibers. Its intrinsic nerve supply is called the *submucosal plexus.* Its major functions are nutrition and protection.

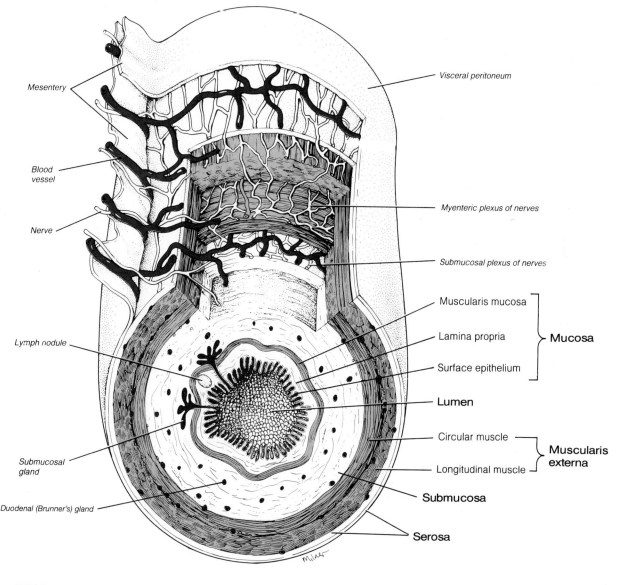

F27.2

Basic structural pattern of the alimentary canal wall.

Muscularis externa: The muscularis externa, also simply called the *muscularis,* typically is a bilayer of smooth muscle, with the deeper layer running circularly and the superficial layer running longitudinally. Another important intrinsic nerve plexus, the *myenteric plexus,* is associated with this tunic. By controlling the smooth muscle of the muscularis, this plexus is the major regulator of GI motility.

Serosa: The outermost serosa is equal to the *visceral peritoneum.* It consists of mesothelium associated with a thin layer of areolar connective tissue. In areas *outside* the abdominopelvic cavity, the serosa is replaced by an **adventitia,** a layer of coarse fibrous connective tissue that binds the organ to surrounding tissues. (This is the case with the esophagus.) The serosa reduces friction as the mobile GI tract organs work and slide across one another and the cavity walls. The adventitia anchors and protects the surrounded GI tract organ.

ORGANS OF THE ALIMENTARY CANAL

The sequential pathway and fate of food as it passes through the alimentary canal organs are described in the next sections. Identify each structure in Figure 27.1 and on the torso model as you work.

Oral Cavity or Mouth

Food enters the digestive tract through the **oral cavity,** or **mouth** (Figure 27.3). Within this mucous membrane–lined cavity are the gums, teeth, tongue, and openings of the ducts of the salivary glands. The **lips (labia)** protect the opening of the chamber anteriorly, the **cheeks** form its lateral walls, and the **palate,** its roof. The anterior portion of the palate is referred to as the **hard palate,** since bone (the palatine processes of the maxillae and the palatine bones) underlies it. The posterior **soft palate** is a fibromuscular structure that is unsupported by bone. The **uvula,** a fingerlike projection of the soft palate, extends inferiorly from its posterior margin. The soft palate rises to close off the oral cavity from the nasal and pharyngeal passages during swallowing. The floor of the oral cavity is occupied by the muscular **tongue,** which is largely supported by the **mylohyoid muscle** (Figure 27.4) and attaches to the hyoid bone, mandible, styloid processes, and pharynx. A membrane called the **lingual frenulum** secures the inferior midline of the tongue to the floor of the mouth. The space between the lips and cheeks and the teeth is the **vestibule;** the area that lies within the teeth and gums is the **oral cavity** proper.

On each side of the mouth at its posterior end are masses of lymphoid tissue, the **palatine tonsils** (Figure 27.4). Each lies in a concave area bounded anteriorly and posteriorly by membranes, the **palatoglossal arch** (anterior membrane) and the **palatopharyngeal arch** (posterior membrane). Another mass of lymphoid tissue, the **lingual tonsil,** covers the base of the tongue, posterior to the oral cavity proper. The tonsils, in common with other lymphoid tissues, are part of the body's defense system.

Very often in young children, the palatine tonsils become inflamed and enlarge, partially blocking the entrance to the pharynx posteriorly and making swallowing difficult and painful. This condition is called **tonsilitis.** ∎

Three pairs of salivary glands duct their secretion, saliva, into the oral cavity. One component of saliva, salivary amylase, begins the digestion of starchy foods within the oral cavity. (The salivary glands are discussed in more detail on pp. 307 and 308.)

As food enters the mouth, it is mixed with saliva and masticated (chewed). The cheeks and lips help hold the food between the teeth during mastication, and the highly mobile tongue manipulates the food during chewing and initiates swallowing. Thus the mechanical and chemical breakdown of food begins before the food has left the oral cavity. As noted in Exercise 20, the surface of the tongue is covered with papillae, many of which contain taste buds, receptors for taste sensation. So, in addition to its manipulative function, the tongue provides for the enjoyment and appreciation of the food ingested.

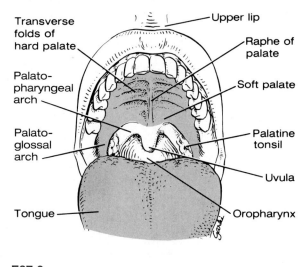

F27.3

Anterior view of the oral cavity.

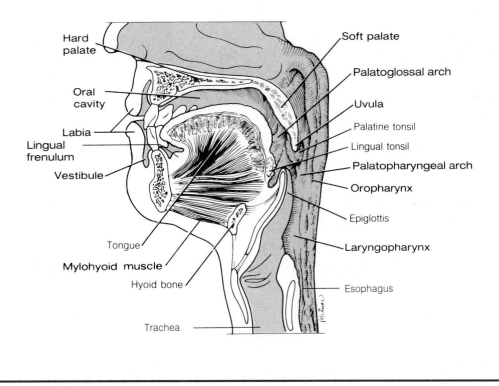

Hard palate
Oral cavity
Labia
Lingual frenulum
Vestibule
Tongue
Mylohyoid muscle
Hyoid bone
Trachea

Soft palate
Palatoglossal arch
Uvula
Palatine tonsil
Lingual tonsil
Palatopharyngeal arch
Oropharynx
Epiglottis
Laryngopharynx
Esophagus

F27.4

Sagittal view of the head showing oral, nasal, and pharyngeal cavities.

Pharynx

When the tongue initiates swallowing, the food passes posteriorly into the pharynx, a common passageway for food, fluid, and air (Figure 27.4). The pharynx is often subdivided anatomically into the **nasopharynx** (behind the nasal cavity), the **oropharynx** (behind the oral cavity extending from the soft palate to the epiglottis overlying the larynx), and the **laryngopharynx** (extending from the epiglottis to the base of the larynx), which is continuous with the esophagus.

The walls of the pharynx consist largely of two layers of skeletal muscles: an inner layer of longitudinal muscle (the levator muscles) and an outer layer of circular constrictor muscles, which initiate wavelike contractions that propel the food inferiorly into the esophagus. Its mucosa, like that of the oral cavity, contains a friction-resistant stratified squamous epithelium.

Esophagus

The **esophagus,** or gullet, extends from the pharynx through the diaphragm to the cardiac sphincter in the superior aspect of the stomach. It is approximately 10 inches long in humans and is essentially a food passageway that conducts food to the stomach in a wavelike peristaltic motion. The esophagus has no digestive or absorptive function. The walls at its superior end contain skeletal muscle, which is replaced by smooth muscle in the area nearing the stomach. Since the esophagus is located in the thoracic rather than the abdominal cavity, its outermost layer is an *adventitia,* rather than a serosa.

Stomach

The **stomach** (Figures 27.1 and 27.5) is on the left side of the abdominal cavity and is hidden by the liver and diaphragm. Different regions of the saclike stomach are the **cardiac region** (the area surrounding the **cardiac sphincter** through which food enters the stomach from the esophagus), the **fundus** (the expanded portion of the stomach, superolateral to the cardiac region), the **body** (midportion of the stomach, inferior to the fundus), and the **pyloric region** (the terminal part of the stomach, which is continuous with the small intestine through the **pyloric sphincter**).

The concave medial surface of the stomach is called the **lesser curvature;** its convex lateral surface is the **greater curvature.** Extending from these curvatures are two mesenteries, called *omenta.* The **lesser omentum** extends from the liver to attach to the lesser curvature of the stomach. The **greater omentum,** a saclike mesentery, extends from the greater curvature of the stomach, reflects downward over the abdominal contents to cover them in an apronlike fashion, and then blends with the **mesocolon** attaching the transverse colon to the posterior body wall. Figure 27.7 (p. 302) illustrates the omenta as well as the other peritoneal attachments of the abdominal organs.

The stomach is a temporary storage region for food as well as a site for mechanical and chemical breakdown of food. It contains a third *obliquely* oriented layer of smooth muscle in its muscularis externa that allows it to churn, mix, and pummel the food, physically reducing it to smaller fragments. Gastric glands of the mucosa

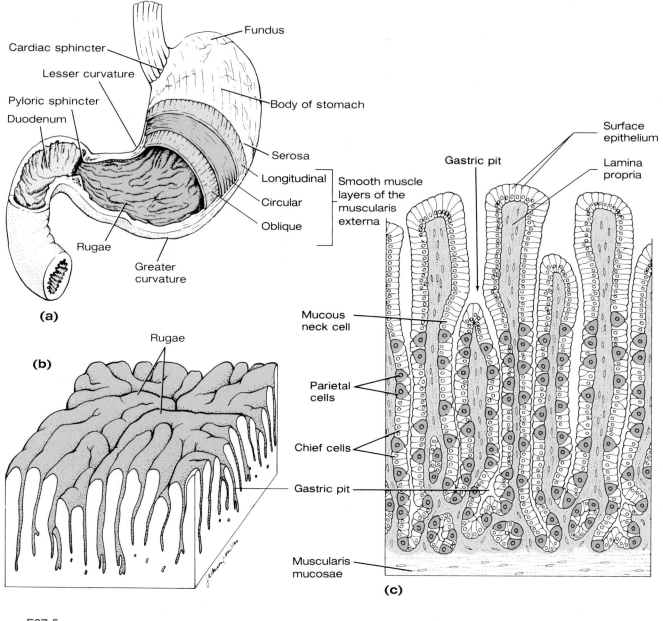

F27.5

Anatomy of the stomach. (a) Gross internal and external anatomy. **(b)** Section of the stomach wall showing rugae and gastric pits. **(c)** Enlarged view of gastric pits (longitudinal section).

secrete hydrochloric acid (HCl) and hydrolytic enzymes (primarily pepsinogen, the inactive form of *pepsin,* a protein-digesting enzyme), which begin the enzymatic, or chemical, breakdown of protein foods. The mucosal glands also secrete a viscous mucus that helps prevent the stomach itself from being digested by the proteolytic enzymes. Most digestive activity occurs in the pyloric region of the stomach. After the food has been processed in the stomach, it resembles a creamy mass (chyme), which enters the small intestine through the pyloric sphincter.

To prepare for the histological study you will be conducting now and later in the lab, obtain a microscope and the following slides: salivary glands (submandibular or sublingual); pancreas; liver; cross sections of the duodenum, ileum, and stomach; and longitudinal sections of a tooth and the gastroesophageal junction.

1. **Stomach:** The stomach slide will be viewed first. Refer to Figure 27.6a as you scan the tissue under low power to locate the muscularis externa; then move to high power to more closely examine this layer. Try to pick out the three smooth muscle layers. How does the extra (oblique) layer of smooth muscle found in the stomach correlate with the stomach's churning movements?

_____

_____

_____

Identify the gastric glands and the gastric pits (see Figures 27.5 and 27.6b). If the section is taken from the stomach fundus and is appropriately stained, you can identify, in the gastric glands, the blue-staining **chief,** or **zymogenic, cells,** which produce pepsinogen, and the red-staining **parietal cells,** which secrete HCl. Draw a small section of the stomach wall and label it appropriately.

2. **Gastroesophageal junction:** Scan the slide under low power to locate the junction between the end of the esophagus and the beginning of the stomach, the gastroesophageal junction. Compare your observations to Figure 27.6c. What is the functional importance of the epithelial differences seen in the two organs?

_____

_____

_____

F27.6

Histology of selected regions of the stomach and gastroesophageal junction. (a) Low-power view of the stomach wall. See corresponding Plate 33 in the Histology Atlas. **(b)** High-power view of gastric pits and glands. See corresponding Plate 34 in the Histology Atlas. **(c)** Gastroesophageal junction, longitudinal section. See corresponding Plate 35 in the Histology Atlas.

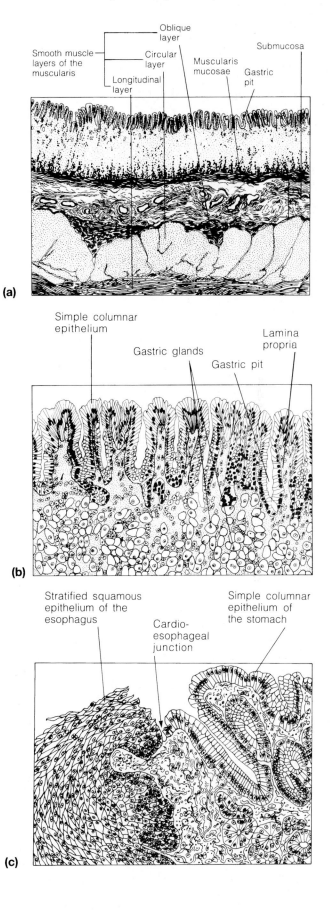

(a)

(b)

(c)

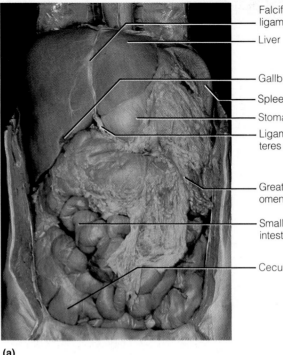

Falciform ligament

Liver

Gallbladder

Spleen

Stomach

Ligamentum teres

Greater omentum

Small intestine

Cecum

(a)

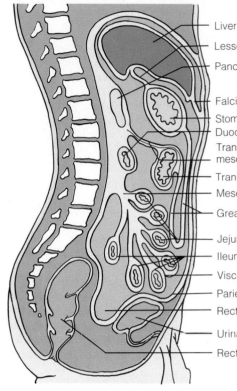

Liver

Lesser omentum

Pancreas

Falciform ligament

Stomach

Duodenum

Transverse mesocolon

Transverse colon

Mesentery

Greater omentum

Jejunum

Ileum

Visceral peritoneum

Parietal peritoneum

Rectovesical pouch

Urinary bladder

Rectum

(b)

F27.7

Peritoneal attachments of the abdominal organs.
(a) Superficial anterior view of abdominal cavity with omentum in place. **(b)** Sagittal view of a male torso. **(c)** Mesentery of the small intestine.

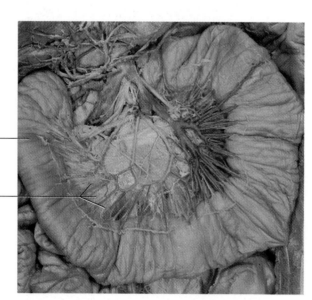

Small intestine

Spread mesentery

(c)

Small Intestine

The **small intestine** is a convoluted tube, 6 to 7 meters (about 20 feet) long in a cadaver but only about 2 m long during life because of its muscle tone. It extends from the pyloric sphincter to the ileocecal valve. The small intestine is suspended by a double layer of peritoneum, the fan-shaped **mesentery,** from the posterior abdominal wall (Figure 27.7), and it lies, framed laterally and superiorly by the large intestine, in the abdominal cavity. The small intestine has three subdivisions: (1) the **duodenum** extends from the pyloric sphincter for about 25 cm (10 inches) and curves around the head of the pancreas; most of the duodenum lies in a retroperitoneal position. (2) The **jejunum,** continuous with the duodenum, extends for 2.5 m (about 8 feet). Most of the jejunum occupies the umbilical region of the abdominal cavity. (3) The **ileum,** the terminal portion of the small intestine, is about 3.6 m (12 feet) long and joins the large intestine at the **ileocecal valve.** It is located inferiorly and somewhat to the right in the abdominal cavity, but its major portion lies in the hypogastric region.

Brush border enzymes, hydrolytic enzymes bound to the microvilli of the columnar epithelial cells, and, more importantly, enzymes produced by the pancreas and ducted into the duodenum via the **pancreatic duct** complete the enzymatic digestion process in the small intestine. Bile (formed in the liver) also enters the duodenum via the **bile duct** in the same area (see Figure 27.1). At the duodenum, the ducts join to form the bulb-like **hepatopancreatic ampulla** and empty their products into the duodenal lumen through the **major duodenal papilla,** an orifice controlled by a muscular valve called the **hepatopancreatic sphincter (sphincter of Oddi).**

302

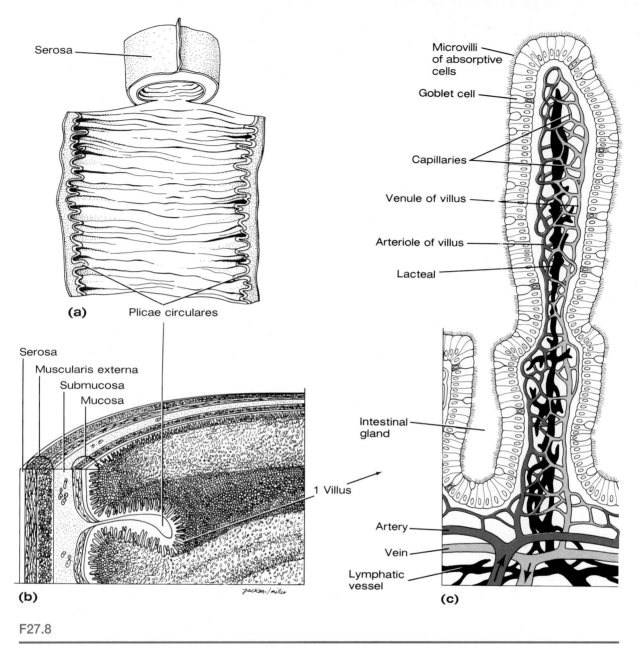

F27.8

Structural modifications of the small intestine. (a) Plicae circulares (circular folds) seen on the internal surface of the small intestine. (b) Enlargement of one plica circulare to show villi. (c) Detailed anatomy of a villus.

Nearly all nutrient absorption occurs in the small intestine, where three structural modifications that increase the mucosa absorptive area appear—the microvilli, villi, and plicae circulares (Figure 27.8). **Microvilli** are minute projections of the surface plasma membrane of the columnar epithelial lining cells of the mucosa; the **villi** are the fingerlike projections of the mucosa tunic that give it a velvety appearance and texture. The **plicae circulares** are deep folds of the mucosa and submucosa layers that force chyme to spiral through the intestine, mixing it and slowing its progress. These structural modifications, which increase the surface area, decrease in frequency and elaboration toward the end of the small intestine. Any residue remaining undigested and unabsorbed at the terminus of the small intestine enters the large intestine through the ileocecal valve. In contrast, the amount of lymphoid tissue in the submucosa of the small intestine (especially the aggregated lymphoid nodules called **Peyer's patches**) increases along the length of the small intestine and is very apparent in the ileum. This reflects the fact that the remaining undigested food residue contains large numbers of bacteria that must be prevented from entering the bloodstream.

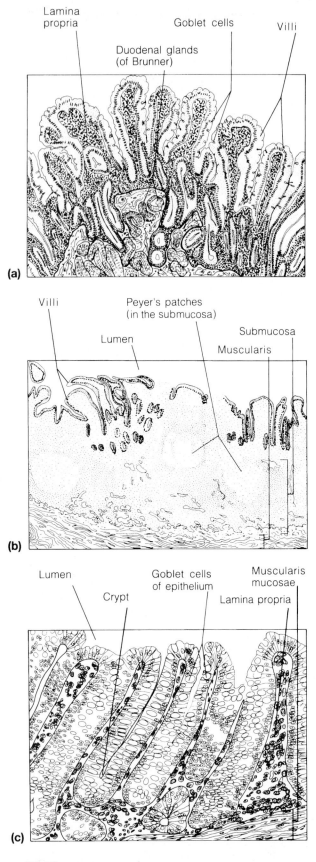

(a)

Lamina propria
Duodenal glands (of Brunner)
Goblet cells
Villi

(b)

Villi
Peyer's patches (in the submucosa)
Lumen
Submucosa
Muscularis

(c)

Lumen
Crypt
Goblet cells of epithelium
Muscularis mucosae
Lamina propria

F27.9

Histology of selected regions of the small and large intestines (cross-sectional views). **(a)** Duodenum of the small intestine. See corresponding Plate 36 in the Histology Atlas. **(b)** Ileum of the small intestine. See corresponding Plate 37 in the Histology Atlas. **(c)** Large intestine. See corresponding Plate 38 in the Histology Atlas.

1. **Duodenum:** Secure the slide of the duodenum (cross section) to the microscope stage. Observe the tissue under low power to identify the four basic tunics of the intestinal wall—that is, the **mucosa** lining (and its three sublayers), the **submucosa** (areolar connective tissue layer deep to the mucosa), the **muscularis externa** (composed of circular and longitudinal smooth muscle layers), and the **serosa** (the outermost layer, also called the *visceral peritoneum*). Consult Figure 27.9a to help you identify the scattered mucus-producing **duodenal glands** (Brunner's glands) in the submucosa.

What type of epithelium do you see here? _____

_____

Examine the large leaflike *villi,* which increase the surface area for absorption. Note also the *intestinal crypts* (crypts of Lieberkühn), invaginated areas of the mucosa between the villi containing the cells that produce intestinal juice, a watery mucus-containing mixture that serves as a carrier fluid for absorption of nutrients from the chyme. Sketch and label a small section of the duodenal wall, showing all layers and villi.

2. **Ileum:** The structure of the ileum resembles that of the duodenum, except that the villi are less elaborate (most of the absorption has occurred by the time the ileum is reached). Obtain and secure a slide of the ileum for viewing. Observe the villi, and identify the four layers of the wall and the large generally spherical Peyer's patches (Figure 27.9b). What tissue comprises Peyer's patches?

_____

3. If a *villus model* is available, identify the following cells or regions before continuing: absorptive epithelium, goblet cells, lamina propria, slips of the muscularis mucosae, capillary bed, and lacteal. If possible, also identify the intestinal crypts which lie between the villi.

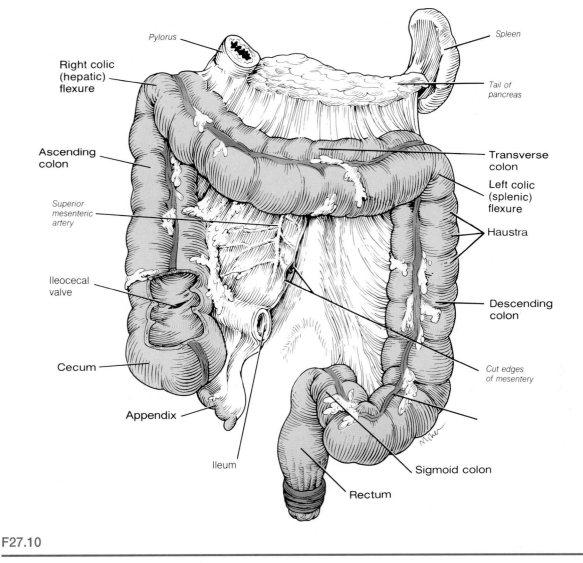

Pylorus

Right colic
(hepatic)
flexure

Spleen

*Tail of
pancreas*

Ascending
colon

Transverse
colon

Left colic
(splenic)
flexure

*Superior
mesenteric
artery*

Haustra

Ileocecal
valve

Descending
colon

Cecum

*Cut edges
of mesentery*

Appendix

Ileum

Sigmoid colon

Rectum

F27.10

The large intestine. (Section of the cecum removed to show the ileocecal valve.)

Large Intestine

The **large intestine** (see Figure 27.10) is about 1.5 m (5 feet) long and extends from the ileocecal valve to the anus. It encircles the small intestine on three sides and consists of the following subdivisions: the **cecum, appendix, colon, rectum,** and **anal canal.**

The blind tubelike appendix, which hangs from the cecum, is a trouble spot in the large intestine. Since it is generally twisted, it provides an ideal location for bacteria to accumulate and multiply. Inflammation of the appendix, or **appendicitis,** is the result. ■

The colon is divided into several distinct regions. The **ascending colon** travels up the right side of the abdominal cavity and makes a right-angle turn at the **right colic (hepatic) flexure** to cross the abdominal cavity as the **transverse colon.** It then turns at the **left colic (splenic) flexure** and continues down the left side of the abdominal cavity as the **descending colon,** where it

takes an S-shaped course as the **sigmoid colon.** The sigmoid colon, rectum, and the anal canal lie in the pelvis anterior to the sacrum and thus are not considered abdominal cavity structures. Except for the transverse and sigmoid colons, which are secured to the dorsal body wall by mesocolons (see Figure 27.7), the colon is retroperitoneal.

The anal canal terminates in the **anus,** the opening to the exterior of the body. The anus, which has an external sphincter of skeletal muscle (the voluntary sphincter) and an internal sphincter of smooth muscle (the involuntary sphincter), is normally closed except during defecation when the undigested remains of the food and bacteria are eliminated from the body as feces.

In the large intestine, the longitudinal muscle layer of the muscularis externa is reduced to three longitudinal muscle bands called the **teniae coli.** Since these bands are shorter than the rest of the wall of the large intestine, they cause the wall to pucker into small pocketlike sacs called **haustra.**

The major function of the large intestine is to consolidate and propel the unusable fecal matter toward the anus and eliminate it from the body. While it does that "chore," it (1) provides a site for the manufacture, by intestinal bacteria, of some vitamins (B and K), which it then absorbs into the bloodstream; and (2) reclaims most of the remaining water (and some of the electrolytes) from undigested food, thus conserving body water.

Watery stools, or diarrhea, result from any condition that rushes undigested food residue through the large intestine before it has had sufficient time to absorb the water (as in irritation of the colon by bacteria). Conversely, when food residue remains in the large intestine for extended periods (as with atonic colon or failure of the defecation reflex), excessive water is absorbed and the stool becomes hard and difficult to pass (constipation). ■

 Examine Figure 27.9c to compare the histology of the large intestine to that of the small intestine just studied.

ACCESSORY DIGESTIVE ORGANS

Teeth

By the age of 21, two sets of teeth have developed (Figure 27.11). The initial set, called the **deciduous** or **milk teeth,** normally appears between the ages of 6 months and 2½ years. The first of these to erupt are the lower central incisors. The child begins to shed the deciduous teeth around the age of 6, and a second set of teeth, the **permanent teeth,** gradually replace them. As the deeper permanent teeth progressively enlarge and develop, the roots of the deciduous teeth are resorbed, leading to their final shedding. During years 6–12, the child has mixed dentition—both permanent and deciduous teeth. Generally, by the age of 12, all of the deciduous teeth have been shed, or exfoliated.

Teeth are classified as **incisors, canines** (eye teeth), **premolars** (bicuspids), and **molars.** Teeth names reflect differences in relative structure and function. The incisors are chisel-shaped and exert a shearing action used in biting. Canines are cone shaped or fanglike, the latter description being much more applicable to the canines of animals whose teeth are used for the tearing of food. Incisors, canines, and premolars typically have single roots, though the first upper premolars may have two. The lower molars have two roots but the upper molars usually have three. The premolars have two cusps (grinding surfaces); the molars have broad crowns with rounded cusps specialized for the fine grinding of food.

Dentition is described by means of a **dental formula,** which designates the numbers, types, and position of the teeth in one side of the jaw. (Since tooth arrangement is bilaterally symmetrical, it is only necessary to designate one side of the jaw.) The complete dental formula for the deciduous teeth from the medial aspect of each jaw and proceeding posteriorly is as follows:

$$\frac{\text{Upper teeth: 2 incisors, 1 canine, 0 premolars, 2 molars}}{\text{Lower teeth: 2 incisors, 1 canine, 0 premolars, 2 molars}} \times 2$$

This formula is generally abbreviated to read as follows:

$$\frac{2,1,0,2}{2,1,0,2} \times 2 = 20 \text{ (number of deciduous teeth)}$$

The 32 permanent teeth are then described by the following dental formula:

$$\frac{2,1,2,3}{2,1,2,3} \times 2 = 32 \text{ (number of deciduous teeth)}$$

Although 32 is designated as the normal number of permanent teeth, not everyone develops a full complement. In many people, the No. 3 molars, commonly called *wisdom teeth,* never erupt.

 Identify the four types of teeth (incisors, canines, premolars, and molars) on the jaw model or human skull.

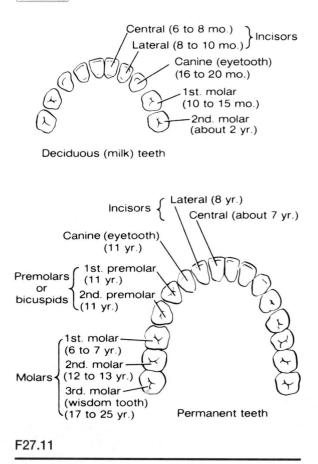

Deciduous (milk) teeth

F27.11

Human deciduous teeth and permanent teeth.
(Approximate time of teeth eruption shown in parentheses.)

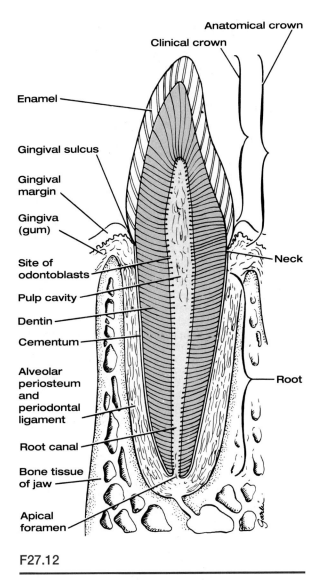

Anatomical crown
Clinical crown

Enamel

Gingival sulcus

Gingival margin

Gingiva (gum)

Site of odontoblasts

Pulp cavity

Dentin

Cementum

Alveolar periosteum and periodontal ligament

Root canal

Bone tissue of jaw

Apical foramen

Neck

Root

F27.12

Longitudinal section of human canine tooth.

A tooth consists of two major regions, the **crown** and the **root.** A longitudinal section made through a tooth shows the following basic anatomical plan (Figure 27.12). The crown is the superior portion of the tooth. The portion of the crown visible above the **gum,** or **gingiva,** is referred to as the **clinical crown.** The entire area covered by **enamel** is called the **anatomical crown.** Enamel is the hardest substance in the body and is fairly brittle. It consists of 95% to 97% inorganic calcium salts (chiefly $CaPO_4$) and thus is heavily mineralized. The crevice between the end of the anatomical crown and the upper margin of the gingiva is referred to as the **gingival sulcus** and its apical border is the **gingival margin.**

That portion of the tooth embedded in the alveolar portion of the jaw is the root, and the root and crown are connected by a slight constriction, the **neck.** The outermost surface of the root is covered by **cementum,** which is similar to bone in composition and less brittle than enamel. The cementum attaches the tooth to the **periodontal ligament** which holds the tooth in the alveolar socket and exerts a cushioning effect. **Dentin,** which comprises the bulk of the tooth, is the bonelike material medial to the enamel and cementum.

The **pulp cavity** occupies the central portion of the tooth. **Pulp,** connective tissue liberally supplied with blood vessels, nerves, and lymphatics, occupies this cavity and provides for tooth sensation and supplies nutrients to the tooth tissues. Specialized cells, called **odontoblasts,** which reside in the outer margins of the pulp cavity, produce the dentin. The pulp cavity extends into distal portions of the root and becomes the **root canal.** An opening at the root apex, the **apical foramen,** provides a route of entry into the tooth for the blood vessels, nerves, and other structures from the tissues beneath.

 Observe a slide of a longitudinal section of a tooth, and compare your observations with the structures detailed in Figure 27.12. Identify as many of these structures as possible.

Salivary Glands

Three pairs of major **salivary glands** (see Figure 27.1) empty their secretions into the oral cavity.

Parotid glands: large glands located anterior to the ear and ducting into the mouth over the second upper molar through the parotid duct

Submandibular glands: located inside the maxillary arch in the floor of the mouth and ducting under the tongue to the base of the lingual frenulum

Sublingual glands: small glands located most anteriorly in the floor of the mouth and emptying under the tongue via several small ducts

Food in the mouth and mechanical pressure (even chewing rubber bands or wax) stimulate the salivary glands to secrete saliva. Saliva consists primarily of mucin (a viscous glycoprotein), which moistens the food and helps to bind it together into a mass called a **bolus,** and a clear serous fluid containing the enzyme *salivary amylase.* Salivary amylase begins the digestion of starch (a large polysaccharide), breaking it down into disaccharides, or double sugars, and glucose. The secretion of the parotid glands is mainly serous, whereas the submandibular and sublingual glands are mixed glands that produce both mucin and serous components.

Examine salivary gland tissue under low power and then high power to become familiar with the appearance of a glandular tissue. Notice the clustered arrangement of the cells around their ducts. The cells are basically triangular, with their pointed ends facing the duct orifice. If possible, differentiate between the mucus-producing cells, which look hollow or have a clear cytoplasm, and the serous cells, which produce the clear, enzyme-containing fluid and have granules in their cytoplasm. The serous cells often form *demilune*s ("caps") around the more central mucous cells. Figure 27.13 may be helpful in this task. Draw your version of a small portion of the salivary gland tissue and label it appropriately.

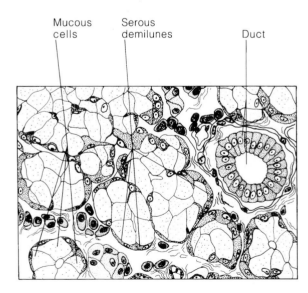

Mucous cells   Serous demilunes   Duct

F27.13

A mixed salivary gland. Corresponds to the photomicrograph in Plate 39 in the Histology Atlas.

Liver and Gallbladder

The **liver** (see Figure 27.1), the largest gland in the body, is located inferior to the diaphragm, more to the right than the left side of the body. As noted earlier, it hides the stomach from view in a superficial observation of abdominal contents. The human liver has four lobes and is suspended from the diaphragm and anterior abdominal wall by the **falciform ligament** (see Figure 27.7a).

The liver is one of the body's most important organs, and it performs many metabolic roles. However, its digestive function is to produce bile, which leaves the liver through the **common hepatic duct** and then enters the duodenum through the **bile duct.** Bile has no enzymatic action but emulsifies fats (spreads thin or breaks up large fat particles into smaller ones), thus creating a larger surface area for more efficient lipase activity. Without bile, very little fat digestion or absorption occurs.

When digestive activity is not occurring in the digestive tract, bile backs up the **cystic duct** and enters the **gallbladder,** a small, green sac on the inferior surface of the liver. It is stored there until needed for the digestive process. While in the gallbladder, bile is concentrated by the removal of water and some ions. When fat-rich food enters the duodenum, a hormonal stimulus causes the gallbladder to contract, releasing the stored bile and making it available to the duodenum.

If the common hepatic or bile duct is blocked (for example, by wedged gallstones), bile is prevented from entering the small intestine, accumulates, and eventually backs up into the liver. This exerts pressure on the liver cells, and bile begins to enter the bloodstream. As the bile circulates through the body, the tissues become yellow or **jaundiced.**

Blockage of the ducts is just one cause of jaundice. More often it results from actual liver problems such as **hepatitis** (an inflammation of the liver) or **cirrhosis,** a condition in which the liver is severely damaged, becoming hard and fibrous. Cirrhosis is almost guaranteed in those who drink excessive alcohol for many years. ■

As demonstrated by its highly organized anatomy, the liver (Figure 27.14) is very important in the initial processing of the nutrient-rich blood draining the digestive organs. Its structural and functional units are called **lobules.** Each lobule is a basically cylindrical structure consisting of cordlike arrays of parenchyma cells, which radiate outward from a central vein running upward in the longitudinal axis of the lobule. At each of the six corners of the lobule is a **portal triad,** so named because three basic structures are always present there: a branch of the *hepatic artery* (the functional blood supply of the liver), a branch of the *hepatic portal vein* (carrying nutrient-rich blood from the digestive viscera), and a *bile duct.* Between the liver parenchyma cells are blood-filled spaces, or **sinusoids,** through which blood from the hepatic portal vein and hepatic artery percolates past the parenchyma cells. Special phagocytic cells, **Kupffer cells,** line the sinusoids and remove debris such as bacteria from the blood as it flows past, while the paren-

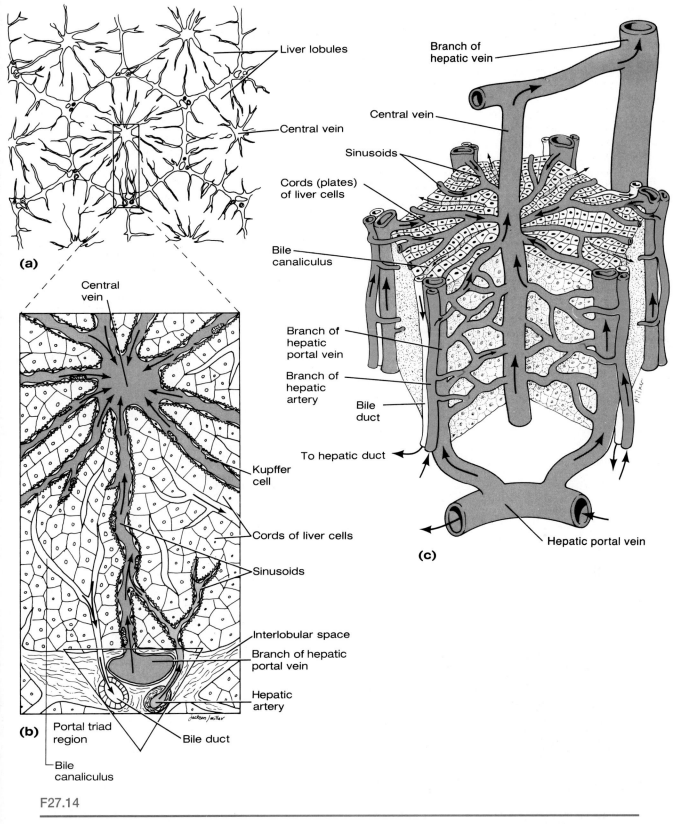

F27.14

Microscopic anatomy of the liver, diagrammatic view. (a) Several liver lobules (cross section). (b) Enlarged view of a portion of one liver lobule (cross section). (c) Portion of one liver lobule (three-dimensional representation). Arrows show direction of bile and blood flow.

chyma cells pick up oxygen and nutrients. Much of the glucose transported to the liver from the digestive system is stored as glycogen in the liver for later use, and amino acids are taken from the blood by the liver cells and utilized to make plasma proteins. The sinusoids empty into the central vein, and the blood ultimately drains from the liver via the *hepatic vein.*

Bile is continuously being made by the parenchyma cells. It flows through tiny canals, the **bile canaliculi,** which run between adjacent parenchyma cells toward the bile duct branches in the triad regions, where the bile eventually leaves the liver. Notice that the direction of blood and bile flow in the liver lobule is exactly opposite.

 Examine a slide of liver tissue and identify as many of the structural features illustrated in Figure 27.14 and Histology Atlas Plates 41 and 42 as possible. Also examine a three-dimensional model of the liver if this is available. Draw your observations below.

Pancreas

The **pancreas** is a soft, triangular gland that extends horizontally across the posterior abdominal wall from the spleen to the duodenum (see Figure 27.1). Like the duodenum, it is a retroperitoneal organ (see Figure 27.7). As noted in Exercise 21, the pancreas has both an endocrine function (it produces the hormones insulin and glucagon) and an exocrine (enzyme-producing) function. It produces a whole spectrum of hydrolytic enzymes, which it secretes in an alkaline fluid into the duodenum through the pancreatic duct. Pancreatic juice is very alkaline. Its high concentration of bicarbonate ion (HCO_3^-) neutralizes the acidic chyme entering the duodenum from the stomach, enabling the pancreatic and intestinal enzymes to operate at their optimal pH. (Optimal pH for digestive activity to occur in the stomach is very acidic and results from the presence of HCl; that for the small intestine is slightly alkaline.)

 Observe pancreas tissue under low power and then high power to distinguish between the lighter-staining, endocrine-producing clusters of cells (pancreatic islets) and the deeper-staining acinar cells, which produce the hydrolytic enzymes and form the major portion of the pancreatic tissue (see Figure 27.15). Notice the arrangement of the exocrine cells around their central ducts. Using colored pencils, appropriately color Figure 27.15 to match the slide you are viewing.

DISSECTION OF THE DIGESTIVE SYSTEM OF THE CAT

 1. Obtain your cat and secure it to the dissecting tray, dorsal surface down. Obtain all necessary dissecting instruments. If you have completed the dissection of the circulatory and respiratory systems, the abdominal cavity is already exposed and many of the digestive system structures have been previously identified. However, duplication of effort generally provides a good learning experience, so all of the digestive system structures will be traced and identified in this exercise.

2. To expose and identify the **salivary glands,** which secrete saliva into the mouth, remove the skin from one side of the head and clear the connective tissue away from the angle of the jaw, below the ear, and superior to the masseter muscle. Many lymph nodes are in this area and you should remove them if they obscure the salivary glands. The cat possesses five pairs of salivary glands, but only those glands described in humans are easily localized and identified (Figure 27.16). Locate the parotid gland on the cheek just inferior to the ear. Follow its duct over the surface of the masseter muscle to the angle of the mouth. The submandibular gland is posterior to

Acinar (exocrine) tissue

Connective tissue septum

Islet

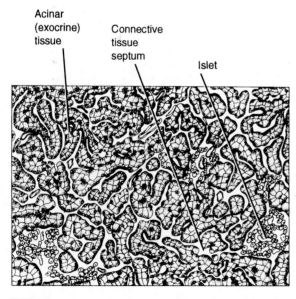

F27.15

Histology of the pancreas. The pancreatic islet cells produce insulin and glucagon (hormones). The acinar cells synthesize digestive enzymes for "export" to the duodenum. See also Plate 40 in the Histology Atlas.

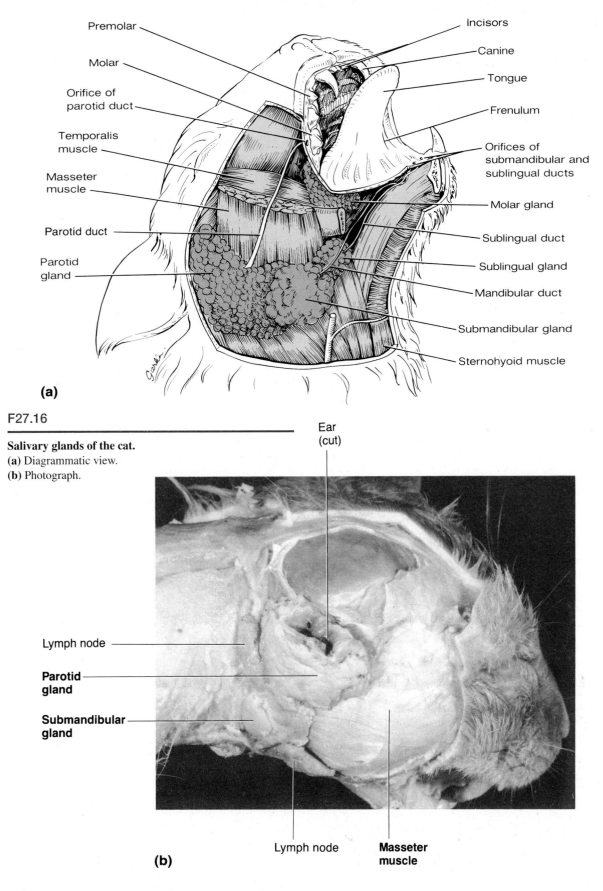

(a)

F27.16

Salivary glands of the cat.
(a) Diagrammatic view.
(b) Photograph.

Ear
(cut)

(b)

the parotid, near the angle of the jaw, and the sublingual gland is just anterior to the submandibular gland within the lower jaw. The ducts of the submandibular and sublingual glands run deep and parallel to each other and empty on the side of the frenulum of the tongue. These need not be identified on the cat.

3. To expose and identify the structures of the oral cavity, cut through the mandible with bone cutters just anterior to the angle to free the lower jaw from the maxilla. Observe the **teeth** of the cat. The dental formula for the adult cat is as follows: $\dfrac{3,1,3,1}{3,1,2,1} \times 2 = 30$

Identify the **hard** and **soft palates,** and use a probe to trace the hard palate to its posterior limits. Note the transverse ridges, or *rugae,* on the hard palate, which play a role in holding food in place while chewing.

Do these appear in humans?_____

Does the cat have a uvula?_____

Identify the **oropharynx** at the rear of the oral cavity and the palatine tonsils on the posterior walls at the junction between the oral cavity and oropharynx. Identify the **tongue** and rub your finger across its surface to feel the papillae. Some of the papillae, especially at the anterior end of the tongue, should feel sharp and bristly. These are the filiform papillae, which are much more numerous in the cat than in humans. What do you think their function is?

_____

Locate the **lingual frenulum** attaching the tongue to the floor of the mouth. Trace the tongue posteriorly until you locate the **epiglottis,** the flap of tissue that covers the entrance to the respiratory passageway when swallowing occurs. Identify the **esophageal opening** posterior to the epiglottis.

4.   Using Figure 27.17 and Plate F of the Cat Anatomy Atlas, locate the abdominal alimentary tube structures. (If the abdominal cavity has not been previously opened, make a midline incision from the rib cage to the pubic symphysis and then make four lateral cuts—two parallel to the rib cage and two at the inferior margin of the abdominal cavity so that the abdominal wall can be reflected back while you examine the abdominal contents. Observe the shiny membrane lining the inner surface of the abdominal wall, which is the **parietal peritoneum.**

5.   Identify the large reddish brown **liver** just beneath the diaphragm and the **greater omentum** covering the abdominal contents. The greater omentum assists in regulating body temperature and its phagocytic cells function in body protection. Notice that the greater omentum is riddled with fat deposits. Lift the greater omentum, noting its two-layered structure and attachments, and lay it to the side or remove it to make subsequent organ identifications easier. Does the liver of the cat have the same number of lobes as the human liver?

_____

6.   Lift the liver and examine its inferior surface to locate the **gallbladder,** a dark, greenish sac embedded in its ventral surface. Identify the **falciform ligament,** a delicate layer of mesentery separating the main lobes of the liver (right and left median lobes) and attaching the liver superiorly to the abdominal wall. Also identify the thickened area along the posterior edge of the falciform ligament, the *round ligament,* or *ligamentum teres,* a remnant of the umbilical vein of the embryo.

7.   Displace the left lobes of the liver to expose the **stomach.** Identify the cardiac, fundic, body, and pyloric regions of the stomach. What is the general shape of the stomach?

_____

Locate the **lesser omentum,** the serous membrane attaching the lesser curvature of the stomach to the liver. Make an incision through the stomach wall to expose its inner surface. Can you see the **rugae**? (When the stomach is empty, its mucosa is thrown into large folds called rugae. As the stomach fills, the rugae gradually disappear and are no longer visible.)

8.   Lift the stomach and locate the **pancreas,** which appears as a greyish or brownish diffuse glandular mass in the mesentery. It extends from the vicinity of the spleen and greater curvature of the stomach and wraps around the duodenum. Attempt to find the **pancreatic duct** as it empties into the duodenum at a swollen area referred to as the **hepatopancreatic ampulla.** Close to the pancreatic duct, locate the **bile duct** and trace its course superiorly to the point where it diverges into the **cystic duct** (gallbladder duct) and the **common hepatic duct** (duct from the liver). Notice that the duodenum assumes a looped position.

9.   Lift the **small intestine** to investigate the manner in which it is attached to the posterior body wall by the **mesentery.** Observe the mesentery closely. What types of structures do you see in this double peritoneal fold?

_____

Other than providing support for the intestine, what other functions does the mesentery have?

_____

_____

Trace the course of the small intestine from its proximal, or duodenal, end to its distal, or ileal, end. Can you see any obvious differences in the external anatomy of the small intestine from one end to the other?

_____

With a scalpel, slice open the distal portion of the ileum and flush out the inner surface with water. Feel the inner surface with your fingertip. How does it feel?

_____

Use a hand lens to see if you can see any **villi** and to locate the areas of lymphatic tissue called **Peyer's patches,** which appear as scattered white patches on the inner intestinal surface.

Return to the duodenal end of the small intestine. Make an incision into the duodenum. As before, flush

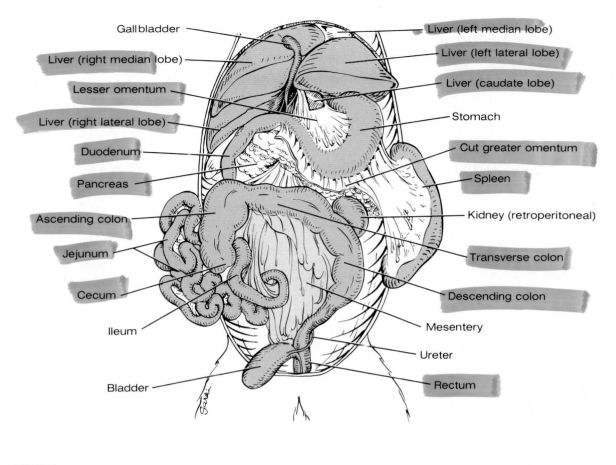

Gallbladder

Liver (right median lobe)

Lesser omentum

Liver (right lateral lobe)

Duodenum

Pancreas

Ascending colon

Jejunum

Cecum

Ileum

Bladder

Liver (left median lobe)

Liver (left lateral lobe)

Liver (caudate lobe)

Stomach

Cut greater omentum

Spleen

Kidney (retroperitoneal)

Transverse colon

Descending colon

Mesentery

Ureter

Rectum

F27.17

Digestive organs of the cat. Diagrammatic view including the left kidney (small intestine, liver, and greater omentum reflected). The greater omentum has been cut from its attachment to the stomach. (See also Plate F in the Cat Anatomy Atlas.)

the surface with water, and feel the inner surface. Does it feel any different than the ileal mucosa?

_____

If so, describe the difference. _____

_____

_____

Use the hand lens to observe the villi. What differences do you see between the villi in the two areas of the small intestine?

_____

_____

10.   Make an incision into the junction between the ileum and cecum to locate the **ileocecal valve.** Observe the **cecum,** the initial expanded part of the large in-

testine. (Lymph nodes may have to be removed from this area to observe it clearly.) Does the cat have an appendix?

_____

11.   Identify the short ascending, transverse, and descending portions of the **colon** and the **mesocolon,** a membrane that attaches the colon to the posterior body wall. Trace the descending colon to the **rectum,** which penetrates the body wall, and identify the **anus.**

Identify the two portions of the peritoneum, the parietal peritoneum lining the abdominal wall (identified previously) and the visceral peritoneum, which is the outermost layer of the wall of the abdominal organs (serosa).

12.   Prepare your cat for storage by wrapping it in paper towels wet with embalming fluid, return it to the plastic bag, and attach your name label. Wash the dissecting tray and instruments before continuing or leaving the laboratory.

Anatomy of the Urinary System

<table>
<tr><td>

OBJECTIVES

1. To describe the overall function of the urinary system.

2. To identify, on an appropriate diagram, torso model, or dissection specimen, the urinary system organs and to describe the general function of each.

3. To define *micturition,* and to explain pertinent differences in the control of the two bladder sphincters (internal and external).

4. To compare the course and length of the urethra in males and females.

5. To identify the following regions of the dissected kidney (longitudinal section): hilus, cortex, medulla, medullary pyramids, major and minor calyces, pelvis, renal columns, and capsule layers.

6. To trace the blood supply of the kidney from the renal artery to the renal vein.

7. To define the nephron as the physiological unit of the kidney and to describe its anatomy.

8. To define *glomerular filtration, tubular reabsorption,* and *tubular secretion,* and to indicate the nephron areas involved in these processes.

9. To recognize microscopic or diagrammatic views of the histologic structure of the kidney and bladder.

</td><td>

MATERIALS

Human dissectible torso model and/or anatomical chart of the human urinary system
3-dimensional model of the cut kidney and of a nephron (if available)
Dissection animal, tray, and instruments
Pig or sheep kidney, doubly or triply injected
Protective skin cream or disposable gloves
Prepared histologic slides of a longitudinal section of kidney and cross sections of the bladder
Compound microscope

 See Appendix B, Exercise 28 for links to A.D.A.M. Standard.

See Appendix C, Exercise 28 for links to *Anatomy and PhysioShow: The Videodisc.*

</td></tr>
</table>

Metabolism of nutrients by the body produces wastes (carbon dioxide, nitrogenous wastes, ammonia, and so on) that must be eliminated from the body if normal function is to continue. Although excretory processes involve several organ systems (the lungs excrete carbon dioxide and skin glands excrete salts and water), it is the **urinary system** that is primarily concerned with the removal of nitrogenous wastes from the body. In addition to this purely excretory function, the kidney maintains the electrolyte, acid-base, and fluid balances of the blood and is thus a major, if not *the* major, homeostatic organ of the body.

To perform its functions, the kidney acts first as a blood "filter," and then as a blood "processor." It allows toxins, metabolic wastes, and excess ions to leave the body in the urine, while simultaneously retaining needed substances and returning them to the blood. Malfunction of the urinary system, particularly of the kidneys, leads to a failure in homeostasis which, unless corrected, is fatal.

GROSS ANATOMY OF THE HUMAN URINARY SYSTEM

The urinary system (Figure 28.1) consists of the paired kidneys and ureters and the single urinary bladder and urethra. The kidneys perform the functions described above and manufacture urine in the process. The remaining organs of the system provide temporary storage reservoirs or transportation channels for urine.

 Examine the human torso model, a large anatomical chart, or a three-dimensional model of the urinary system to locate and study the anatomy and relationships of the urinary organs.

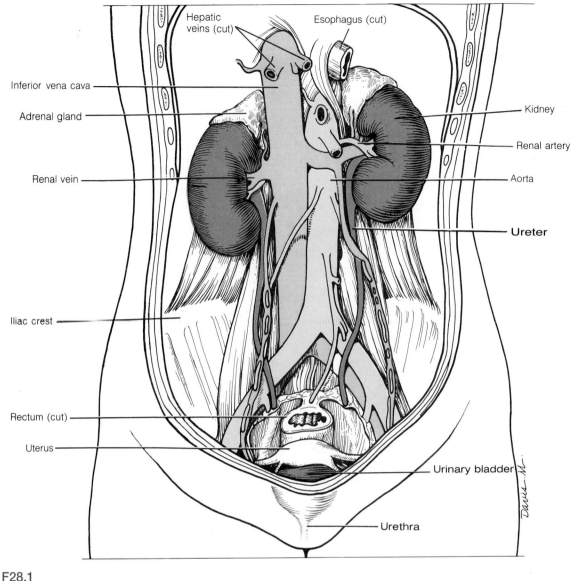

Hepatic veins (cut)

Esophagus (cut)

Inferior vena cava

Adrenal gland

Renal vein

Iliac crest

Rectum (cut)

Uterus

Kidney

Renal artery

Aorta

Ureter

Urinary bladder

Urethra

F28.1

Anterior view of the urinary organs of a human female. (Most unrelated abdominal organs have been removed.)

1. Locate the paired **kidneys** on the dorsal body wall in the superior lumbar region. Note that they are not positioned at exactly the same level. Because it is "crowded" by the liver, the right kidney is slightly lower than the left kidney. In a living person, fat deposits (the *adipose capsules*) hold the kidneys in place in a retroperitoneal position.

When the fatty material surrounding the kidneys is reduced or too meager in amount (in cases of rapid weight loss or in very thin individuals), the kidneys are less securely anchored to the body wall and may drop to a lower or more inferior position in the abdominal cavity. This phenomenon is called **ptosis.** ■

2. Observe the **renal arteries** as they diverge from the descending aorta and plunge into the indented medial region **(hilus)** of each kidney. Note also the **renal veins,** which drain the kidneys (circulatory drainage) and the

two **ureters,** which drain urine from the kidneys and conduct it by peristalsis to the bladder for temporary storage.

3. Locate the **urinary bladder,** and observe the point of entry of the two ureters into this organ. Also locate the single **urethra,** which drains the bladder. The triangular region of the bladder, which is delineated by these three openings (two ureteral and one urethral orifice), is referred to as the **trigone** (Figure 28.2). Although urine formation by the kidney is a continuous process, urine is usually removed from the body when voiding is convenient. In the meantime the bladder stores it temporarily.

Voiding, or **micturition,** is the process in which urine empties from the bladder. Two sphincter muscles or valves, the **internal urethral sphincter** (more superiorly located) and the **external urethral sphincter** (more inferiorly located) control the outflow of urine from the bladder. Ordinarily, the bladder continues to

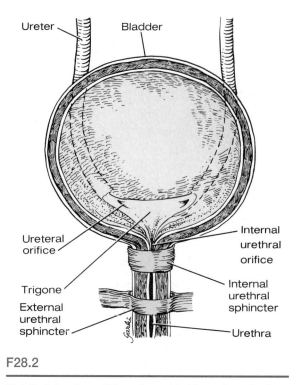

Ureter   Bladder

Ureteral orifice

Trigone

External urethral sphincter

Internal urethral orifice

Internal urethral sphincter

Urethra

F28.2

Detailed structure of the urinary bladder and urethral sphincters.

collect urine until about 200 ml have accumulated, at which time the stretching of the bladder wall activates stretch receptors. Impulses transmitted to the central nervous system subsequently produce reflex contractions of the bladder wall through parasympathetic nervous system pathways (i.e., via the pelvic splanchnic nerves). As the contractions increase in force and frequency, the stored urine is forced past the internal sphincter, which is a smooth muscle involuntary sphincter, into the superior part of the urethra. It is then that a person feels the urge to void. The inferior external sphincter consists of skeletal muscle and is reinforced and assisted by skeletal muscles of the pelvic diaphragm, which are also voluntarily controlled. If it is not convenient to void, the opening of this sphincter can be inhibited. Conversely, if the time is convenient, the sphincter may be relaxed and the stored urine flushed from the body. If voiding is inhibited, the reflex contractions of the bladder cease temporarily and urine continues to accumulate in the bladder. After another 200 to 300 ml of urine have been collected, the *micturition reflex* will again be initiated.

Lack of voluntary control over the external sphincter is referred to as **incontinence.** Incontinence is normal in children 2 years old or younger, as they have not yet gained control over the voluntary sphincter. In adults and older children, incontinence is generally a result of spinal cord injury, emotional problems, bladder irritability, or some other pathology of the urinary tract. ■

4.   Follow the course of the urethra to the body exterior. In the male, it is approximately 20 cm (8 inches) long, travels the length of the **penis,** and opens at its tip.

Its three named regions—the *prostatic, membranous,* and *spongy (penile) urethrae*—are described in more detail in Exercise 29 and illustrated in Figure 29.1 (pp. 325 and 326). The urethra of males has a dual function: it is a urine conduit to the body exterior, and it provides a passageway for the ejaculation of semen. Thus, in the male, the urethra is part of both the urinary and reproductive systems. In females, the urethra is very short, approximately 4 cm (1½ inches) long. There are no common urinary-reproductive pathways in the female, and the female's urethra serves only to transport urine to the body exterior. Its external opening, the **external urethral orifice,** lies anterior to the vaginal opening.

GROSS INTERNAL ANATOMY OF THE PIG OR SHEEP KIDNEY

1.   Obtain a preserved sheep or pig kidney, dissecting pan, and instruments. Observe the kidney to identify the **renal capsule,** a smooth transparent membrane that adheres tightly to the external aspect of the kidney.

2.   Find the ureter, renal vein, and renal artery at the hilus (indented) region. The renal vein has the thinnest wall and will be collapsed. The ureter is the largest of these structures and has the thickest wall.

3.   In preparation for dissection, don gloves or apply protective skin cream. Make a cut through the longitudinal axis (frontal section) of the kidney and locate the anatomical areas described below and depicted in Figure 28.3.

Kidney cortex: the superficial kidney region, which is lighter in color. If the kidney is doubly injected with latex, you will see a predominance of red and blue latex specks in this region indicating its rich vascular supply.

Medullary region: deep to the cortex; a darker, reddish-brown color. The medulla is segregated into triangular regions that have a striped, or striated, appearance—the **medullary (renal) pyramids.** The base of each pyramid faces toward the cortex. Its more pointed **apex,** or **papilla,** points to the innermost kidney region.

Renal columns: areas of tissue, more like the cortex in appearance, which segregate and dip inward between the pyramids.

Renal pelvis: medial to the hilus; a relatively flat, basinlike cavity that is continuous with the **ureter,** which exits from the hilus region. Fingerlike extensions of the pelvis should be visible. The larger, or primary, extensions are called the **major calyces;** subdivisions of the major calyces are the **minor calyces.** Notice that the minor calyces terminate in cuplike areas that enclose the apexes of the medullary pyramids and collect urine draining from the pyramidal tips into the pelvis.

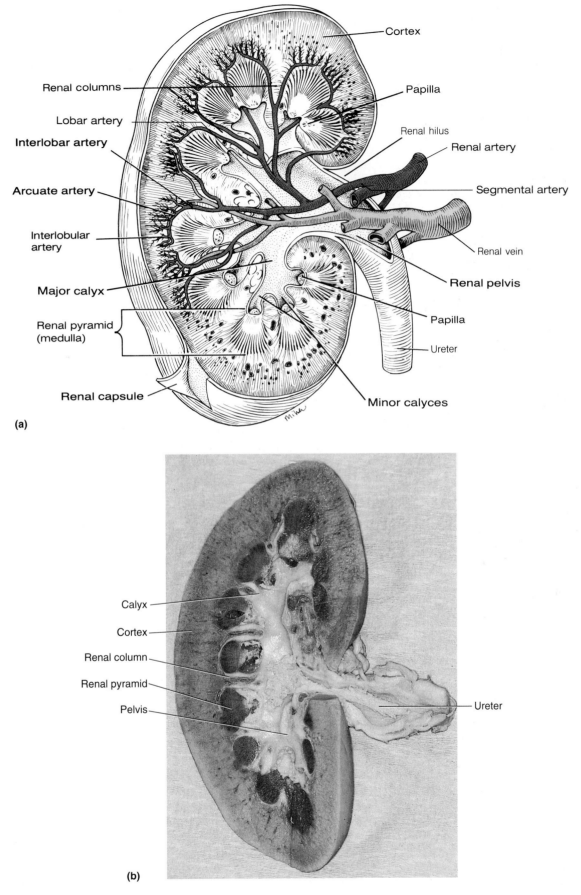

Cortex

Renal columns

Papilla

Lobar artery

Renal hilus

Interlobar artery

Renal artery

Segmental artery

Arcuate artery

Renal vein

Interlobular
artery

Renal pelvis

Major calyx

Papilla

Renal pyramid
(medulla)

Ureter

Renal capsule

Minor calyces

(a)

Calyx

Cortex

Renal column

Renal pyramid

Pelvis

Ureter

(b)

F28.3

Frontal section of a kidney. (a) Diagrammatic view, showing the larger arteries supplying the kidney tissue. (b) Photograph of a triple-injected pig kidney.

317

4.   If the preserved kidney is doubly or triply injected, follow the renal blood supply from the renal artery to the **glomeruli.** The glomeruli appear as little red and blue specks in the cortex region. (See Figures 28.3 and 28.4.)

Approximately a fourth of the total blood flow of the body is delivered to the kidneys each minute by the large **renal arteries.** As a renal artery approaches the kidney, it breaks up into five branches called **segmental arteries** which enter the hilus. Each segmental artery, in turn, divides into several **lobar arteries.** The lobar arteries branch to form **interlobar arteries,** which ascend toward the cortex in the renal column areas. At the top of the medullary region, these arteries give off arching branches, the **arcuate arteries,** which curve over the bases of the medullary pyramids. Small **interlobular arteries** branch off the arcuate arteries and ascend into the cortex, giving off the individual **afferent arterioles,** which provide the capillary networks (**glomeruli** and **peritubular capillary beds**) that supply the nephrons, or functional units, of the kidney. Blood draining from the nephron capillary networks in the cortex enters the **interlobular veins** and then drains through the **arcuate veins** and the **interlobar veins** to finally enter the **renal vein** in the pelvis region. (There are no lobar or segmental veins.)

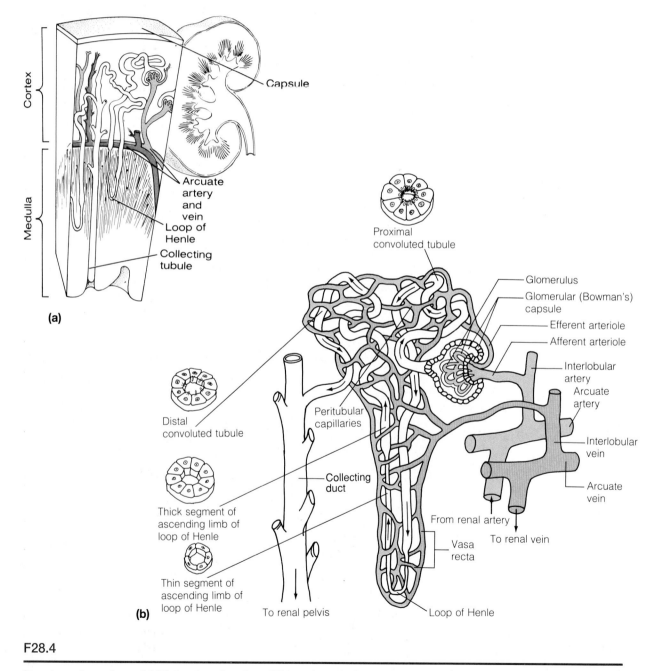

(a)

(b)

F28.4

Structure of a nephron. (a) Wedge-shaped section of kidney tissue, indicating the position of the nephrons in the kidney.
(b) Detailed nephron anatomy and associated blood supply.

MICROSCOPIC ANATOMY OF THE KIDNEY AND BLADDER

Obtain prepared slides of kidney and bladder tissue, and a compound microscope.

Kidney

Each kidney contains over a million nephrons, which are the anatomical units responsible for forming urine. Figure 28.4 depicts the detailed structure and the relative positioning of the nephrons in the kidney.

Each nephron consists of two major structures: a **glomerulus** (a capillary knot) and a **renal tubule.** During embryologic development, each renal tubule begins as a blind-ended tubule that gradually encloses an adjacent capillary cluster, or glomerulus. The enlarged end of the tubule encasing the glomerulus is the **glomerular (Bowman's) capsule,** and its inner, or visceral, wall consists of highly specialized cells called **podocytes.** Podocytes have long, branching processes (*foot processes*) that interdigitate with those of other podocytes and cling to the endothelial wall of the glomerular capillaries, thus forming a very porous epithelial membrane surrounding the glomerulus. The glomerulus-capsule complex is sometimes called the **renal corpuscle.**

The rest of the tubule is approximately 3 cm (1.25 inches) long. As it emerges from the glomerular capsule, it becomes highly coiled and convoluted, drops down into a long hairpin loop, and then again coils and twists before entering a collecting duct. In order from the glomerular capsule, the anatomical areas of the renal tubule are: the **proximal convoluted tubule, loop of Henle** (descending and ascending limbs), and the **distal convoluted tubule.** The wall of the renal tubule is composed almost entirely of cuboidal epithelial cells, with the exception of part of the descending limb (and sometimes part of the ascending limb) of the loop of Henle, which is simple squamous epithelium. The lumen surfaces of the cuboidal cells in the proximal convoluted tubule have dense microvilli (a cellular modification that greatly increases the surface area exposed to the lumen contents, or filtrate). Microvilli also occur on cells of the distal convoluted tubule but in greatly reduced numbers, revealing its less significant role in reclaiming filtrate contents.

Most nephrons, called **cortical nephrons,** are located entirely within the cortex. However, parts of the loops of Henle of the **juxtamedullary nephrons** (located close to the cortex-medulla junction) penetrate well into the medulla. The **collecting ducts,** each of which receives urine from many nephrons, run downward through the medullary pyramids, giving them their striped appearance. As the collecting ducts approach the renal pelvis, they fuse to form larger *papillary ducts,* which empty the final urinary product into the calyces and pelvis of the kidney.

The function of the nephron depends on several unique features of the renal circulation. The capillary vascular supply consists of two distinct capillary beds, the *glomerulus* and the *peritubular capillary bed.* Vessels leading to and from the **glomerulus,** the first capillary bed, are both arterioles: **the afferent arteriole** feeds the bed while the **efferent arteriole** drains it. The glomerular capillary bed has no parallel elsewhere in the body. It is a high-pressure bed along its entire length. Its high pressure is a result of two major factors: (1) the bed is *fed and drained* by arterioles (arterioles are high-resistance vessels as opposed to venules, which are low-resistance vessels), and (2) the afferent feeder arteriole is larger in diameter than the efferent arteriole draining the bed. The high hydrostatic pressure created by these two anatomical features forces out fluid and blood components smaller than proteins from the glomerulus into the glomerular capsule. That is, it forms the filtrate which is processed by the nephron tubule.

The **peritubular capillary bed** arises from the efferent arteriole draining the glomerulus. This set of capillaries cling intimately to the renal tubule and empty into the interlobular veins that leave the cortex. The peritubular capillaries are *low-pressure* very porous capillaries adapted for absorption rather than filtration and readily take up the solutes and water reabsorbed from the filtrate by the tubule cells. The juxtamedullary nephrons have additional looping vessels, called the **vasa recta** ("straight vessels"), that parallel their long loops of Henle in the medulla (see Figure 28.4). Hence, the two capillary beds of the nephron have very different, but complementary, roles: The glomerulus produces the filtrate and the peritubular capillaries reclaim most of that filtrate.

Urine formation is a result of three processes: *filtration, reabsorption,* and *secretion* (Figure 28.5). **Filtration,** the role of the glomerulus, is largely a passive process in which a portion of the blood passes from the glomerular bed into the glomerular capsule. This filtrate then enters the proximal convoluted tubule where tubular reabsorption and secretion begin. During **tubular reabsorption,** many of the filtrate components move through the tubule cells and return to the blood in the peritubular capillaries. Some of this reabsorption is passive, such as that of water which passes by osmosis, but the reabsorption of most substances depends on active transport processes and is highly selective. Which substances are reabsorbed at a particular time depends on the composition of the blood and needs of the body at that time. Substances that are almost entirely reabsorbed from the filtrate include water, glucose, and amino acids. Various ions are selectively reabsorbed or allowed to go out in the urine according to what is required to maintain appropriate blood pH and electrolyte composition. Waste products (urea, creatinine, uric acid, and drug metabolites) are reabsorbed to a much lesser degree or not at all. Most (75% to 80%) of tubular reabsorption occurs in the proximal convoluted tubule; the balance occurs in other areas, especially the distal convoluted tubules and collecting ducts.

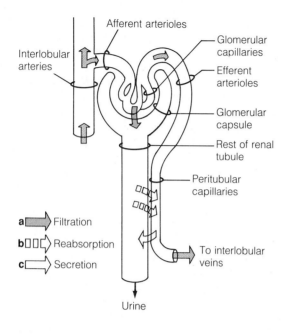

F28.5

The kidney depicted as a single, large nephron. A kidney actually has millions of nephrons acting in parallel. The three major mechanisms by which the kidneys adjust the composition of plasma are (**a**) glomerular filtration, (**b**) tubular reabsorption, and (**c**) tubular secretion. Solid pink arrows show the path of blood flow through the renal microcirculation.

Tubular secretion is essentially the reverse process of tubular reabsorption. Substances such as hydrogen and potassium ions and creatinine move either from the blood of the peritubular capillaries through the tubular cells or from the tubular cells into the filtrate to be disposed of in the urine. This process is particularly important for the disposal of substances not already in the filtrate (such as drug metabolites), and as a device for controlling blood pH.

Observe a model of the nephron before continuing on with the microscope study of the kidney.

1.   Hold the longitudinal section of the kidney up to the light to identify cortical and medullary areas. Then secure the slide on the microscope stage and scan the slide under low power.

2.   Move the slide so that you can see the cortical area. Identify a glomerulus, which appears as a ball of tightly packed material containing many small nuclei (Figure 28.6). It is usually delineated by a vacant-appearing region (corresponding to the space between the visceral and parietal layers of the glomerular capsule) that surrounds it.

3.   Notice that the renal tubules are cut at various angles. Also try to differentiate between the thin-walled loop of Henle portion of the tubules and the cuboidal epithelium of the proximal convoluted tubule, which has dense microvilli.

Bladder

1.   Scan the bladder tissue. Identify its three layers—mucosa, muscular layer, and fibrous adventitia.

2.   Study the mucosa with its highly specialized transitional epithelium. The plump, transitional epithelial cells have the ability to slide over one another, thus decreasing the thickness of the mucosa layer as the bladder fills and stretches to accommodate the increased urine volume. Depending on the degree of stretching of the bladder, the mucosa may be three to eight cell layers thick. Compare your observations of the transitional epithelium of the mucosa to that shown in Figure 5.3h (p. 38).

3.   Examine the heavy muscular wall (detrusor muscle), which consists of three irregularly arranged muscular layers. The innermost and outermost muscle layers are arranged longitudinally; the middle layer is arranged circularly. Attempt to differentiate the three muscle layers.

4.   Draw a small section of the bladder wall, and label all regions or tissue areas.

5.   Compare your sketch of the bladder wall to the structure of the ureter wall shown in Plate 45 in the Histology Atlas. How are the two organs similar histologically?

_____

_____

_____

What is/are the most obvious differences?

_____

_____

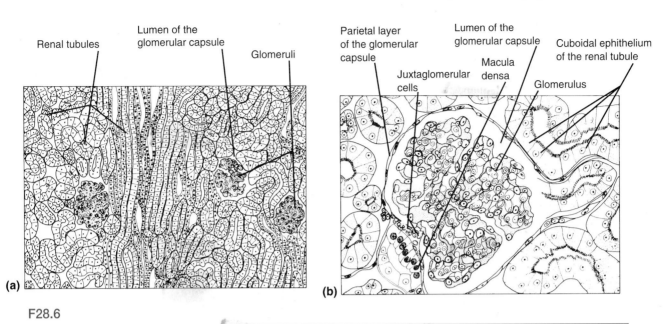

F28.6

Microscopic structure of kidney tissue. (a) Low-power view of the renal cortex. (b) Detailed structure of the glomerulus. (See corresponding Plates 43 and 44 in the Histology Atlas.)

DISSECTION OF THE CAT URINARY SYSTEM

The structures of the reproductive and urinary systems are often considered together as the *urogenital system,* since they have common embryologic origins. However, the emphasis in this dissection is on identifying the structures of the urinary tract (Figures 28.7 and 28.8) with only a few references to contiguous reproductive structures. The anatomy of the reproductive system is studied in Exercise 29.

1. Don gloves or apply skin cream. Obtain your dissection specimen, and pin or tie its limbs to the dissection tray. Reflect the abdominal viscera (most importantly the small intestine) to locate the kidneys high on the dorsal body wall. Note that the kidneys in the cat, as well as in the human, are retroperitoneal (behind the peritoneum).

2. Carefully remove the peritoneum, and clear away the bed of fat that invests the kidneys. Locate the adrenal (suprarenal) glands lying superiorly and medial to the kidneys.

3. Identify the renal artery (red latex injected), the renal vein (blue latex injected), and the ureter at the hilus region of the kidney. (You may find two renal veins leaving one kidney in the cat but not in humans.)

4. Trace the ureters to the urinary bladder, a smooth muscular sac located superiorly to the small intestine. If your cat is a female, be careful not to confuse the ureters

with the uterine tubes, which lie superior to the bladder in the same general region. (See Figure 28.8 and Plate B in the Cat Anatomy Atlas.) Observe the sites where the ureters enter the bladder. How would you describe the entrance point anatomically?

_____

_____

5. Cut through the bladder wall, and examine the region of the urethral exit to see if you can discern any evidence of the internal sphincter.

6. If your cat is a male, identify the prostate gland (part of the male reproductive system), which encircles the neck of the bladder (see Figure 28.7). Notice that the urinary bladder is somewhat fixed in position by ligaments.

7. Using a probe, trace the urethra as it exits from the bladder to its terminus in the **urogenital sinus*** (opening into the vagina) in the female cat or into the penis of the male. Dissection to expose the urethra along its entire length should not be done at this time because of possible damage to the reproductive structures, which you will study in Exercise 29.

8. Before cleaning up the dissection materials, observe a cat of the opposite sex.

* In the human female, the urethra does not empty into the vagina but has a separate external opening located above the vaginal orifice.

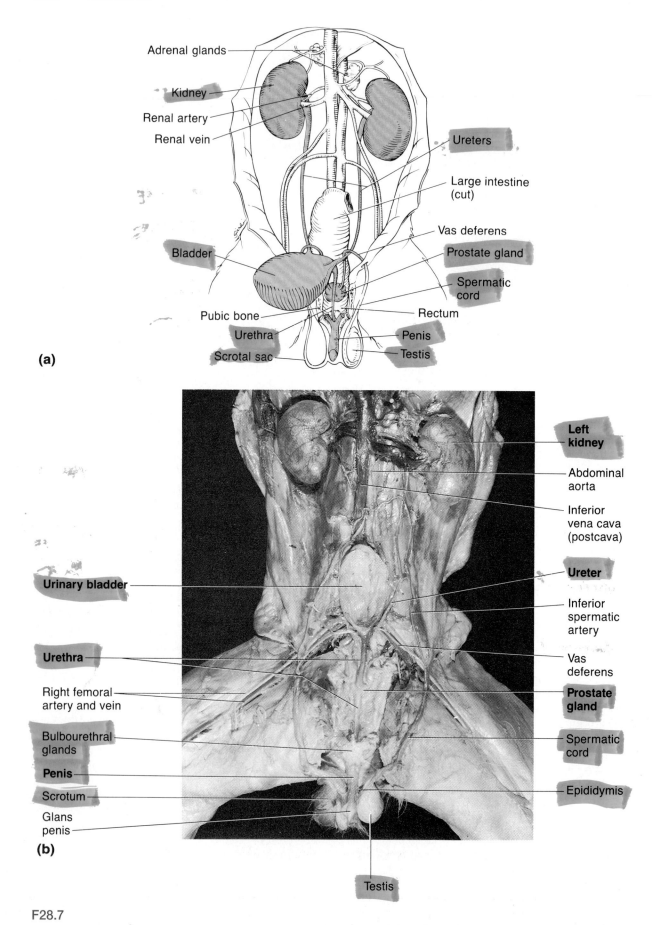

Adrenal glands

Kidney

Renal artery

Renal vein

Ureters

Large intestine (cut)

Vas deferens

Prostate gland

Spermatic cord

Bladder

Pubic bone

Rectum

Urethra

Penis

Scrotal sac

Testis

(a)

Left kidney

Abdominal aorta

Inferior vena cava (postcava)

Ureter

Inferior spermatic artery

Urinary bladder

Urethra

Vas deferens

Right femoral artery and vein

Prostate gland

Bulbourethral glands

Spermatic cord

Penis

Epididymis

Scrotum

Glans penis

Testis

(b)

F28.7

Urinary system of the male cat. (Reproductive structures also indicated.) **(a)** Diagrammatic view. **(b)** Photograph of male urogenital system. See dissection photo, Plate A in the Cat Anatomy Atlas.

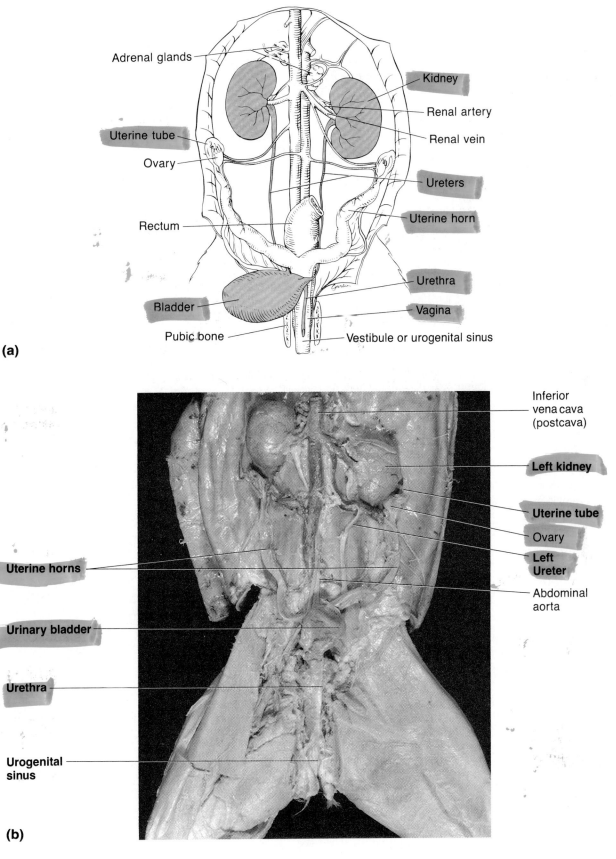

(a)

(b)

F28.8

Urinary system of the female cat. (Reproductive structures are also indicated.) **(a)** Diagrammatic view. **(b)** Photograph of female urogenital system. See dissection photo, Plate B in the Cat Anatomy Atlas.

Anatomy of the Reproductive System

OBJECTIVES

1. To discuss the general function of the reproductive system.

2. To identify and name the structures of the male and female reproductive systems when provided with an appropriate model or diagram, and to discuss the general function of each.

3. To define *semen,* discuss its composition, and name the organs involved in its production.

4. To trace the pathway followed by a sperm from its site of formation to the external environment.

5. To name the exocrine and endocrine products of the testes and ovaries, indicating the cell types or structures responsible for the production of each.

6. To identify homologous structures of the male and female systems.

7. To discuss the microscopic structure of the penis, epididymis, uterine tube, and uterus, and to relate structure to function.

8. To define *ejaculation, erection,* and *gonad.*

9. To discuss the function of the fimbriae and ciliated epithelium of the uterine (fallopian) tubes.

10. To identify the fundus, body, and cervical regions of the uterus.

11. To define *endometrium, myometrium,* and *ovulation.*

12. To identify the major reproductive structures of the male and female dissection animal, and to recognize and discuss pertinent differences between the reproductive structures of the human and the dissection animal.

MATERIALS

Models or large laboratory charts of the male and female reproductive tracts

Prepared microscope slides of cross sections of the penis, epididymis, uterine tube, and uterus showing endometrium (proliferative phase); also longitudinal sections of the testis and ovary and a sperm smear

Compound microscope

Dissection animal, tray, and instruments

Bone clippers

Protective skin cream or plastic gloves

Small metric ruler

 See Appendix B, Exercise 29 for links to A.D.A.M. Standard.

See Appendix C, Exercise 29 for links to *Anatomy and PhysioShow: The Videodisc.*

Most simply stated, the biologic function of the **reproductive system** is to perpetuate the species. Thus the reproductive system is unique, since the other organ systems of the body function primarily to sustain the existing individual.

The essential organs of reproduction are those that produce the germ cells—the testes and the ovaries. The reproductive role of the male is to manufacture sperm and to deliver them to the female reproductive tract. The female, in turn, produces eggs. If the time is suitable, the combination of sperm and egg produces a fertilized egg, which is the first cell of a new individual. Once fertilization has occurred, the female uterus provides a nurturing, protective environment in which the embryo, later called the fetus, develops until birth.

Although the drive to reproduce is strong in all animals, in humans this drive is also intricately related to nonbiologic factors. Emotions and social considerations often enhance or thwart its expression.

GROSS ANATOMY OF THE HUMAN MALE REPRODUCTIVE SYSTEM

The primary reproductive organs of the male are the **testes,** the male *gonads* which have both an exocrine (sperm production) and an endocrine (testosterone production) function. All other reproductive structures are conduits or sources of secretions, which aid in the safe delivery of the sperm to the body exterior or female reproductive tract.

 As the following organs and structures are described, locate them on Figure 29.1, and then identify them on a three-dimensional model of the male reproductive system or on a large laboratory chart.

The paired oval testes lie in the **scrotal sac** outside the abdominopelvic cavity. The temperature there (approximately 94°F, or 34°C) is slightly lower than body temperature, a requirement for producing viable sperm.

The accessory structures forming the *duct system* are the epididymis, the ductus deferens, the ejaculatory duct, and the urethra. The **epididymis** is an elongated structure running up the posterolateral aspect of the testis and capping its superior aspect. The epididymis forms the first portion of the duct system and provides a site for immature sperm that enter it from the testis to complete their maturation process. The **ductus deferens** (sperm duct) arches superiorly from the epididymis, passes through the inguinal canal into the pelvic cavity, and courses over the superior aspect of the urinary bladder. In life, the ductus deferens (also called the *vas deferens*) is enclosed along with blood vessels and nerves in a connective tissue sheath called the **spermatic cord.** The terminus of the ductus deferens enlarges to form the region called the **ampulla,** which empties into the **ejaculatory duct.** Contraction of the ejaculatory duct propels the sperm through the prostate gland to the **prostatic urethra,** which in turn empties into the **membranous urethra** and then into the **penile urethra,** which runs through the length of the penis to the body exterior.

The spermatic cord is easily palpated through the skin of the scrotum. When a *vasectomy* is performed, a small incision is made in each side of the scrotum, and each ductus deferens is cut through or cauterized. Although sperm are still produced, they can no longer reach the body exterior; thus a man is sterile after this procedure (and 12 to 15 ejaculations to clear the conducting tubules).

The *accessory glands* include the prostate gland, the paired seminal vesicles, and bulbourethral glands. These glands produce **seminal fluid,** the liquid medium in which sperm leave the body. The **seminal vesicles,** which produce about 60% of seminal fluid, lie at the posterior wall of the urinary bladder close to the terminus of the ductus deferens. They produce a viscous alkaline secretion containing fructose (a simple sugar) and other substances that nourish the sperm passing through the tract or promote the fertilizing capability of sperm in some way. The duct of each seminal vesicle merges with a ductus deferens to form the ejaculatory duct (mentioned above); thus sperm and seminal fluid enter the urethra together.

The **prostate gland** encircles the urethra just inferior to the bladder. It secretes a milky fluid into the urethra, which plays a role in activating the sperm.

Hypertrophy of the prostate gland, a troublesome condition commonly seen in elderly men, constricts the urethra so that urination is difficult. ■

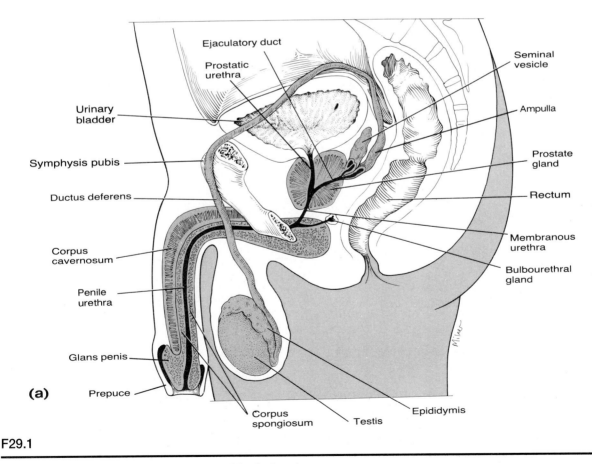

(a)

Ejaculatory duct

Prostatic urethra

Urinary bladder

Symphysis pubis

Ductus deferens

Corpus cavernosum

Penile urethra

Glans penis

Prepuce

Corpus spongiosum

Testis

Epididymis

Seminal vesicle

Ampulla

Prostate gland

Rectum

Membranous urethra

Bulbourethral gland

F29.1

Reproductive system of the human male. (a) Midsagittal section.

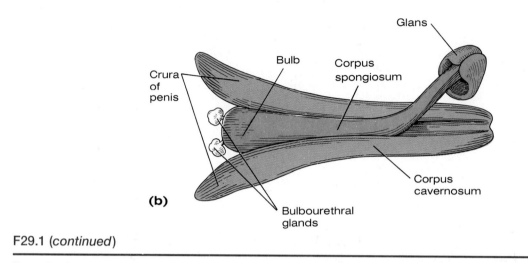

(b)

F29.1 *(continued)*

Reproductive system of the human male. (b) Inferior view of the penis with corpus spongiosum partially reflected.

The **bulbourethral glands** are tiny, pea-shaped glands inferior to the prostate. They produce a thick, clear, alkaline mucus that drains into the membranous urethra. This secretion acts to wash residual urine out of the urethra when ejaculation of **semen** (sperm plus seminal fluid) occurs. The relative alkalinity of seminal fluid also buffers the sperm against the acidity of the female reproductive tract.

The **penis,** part of the external genitalia of the male along with the scrotal sac, is the copulatory organ of the male and is designed to deliver sperm into the female reproductive tract. It consists of a shaft, which terminates in an enlarged tip, the **glans** (see Figure 29.1b). The skin covering the penis is loosely applied, and it reflects downward to form a circular fold of skin, the **prepuce,** or **foreskin,** around the proximal end of the glans. (The foreskin is removed in the surgical procedure called *circumcision.*) Internally, the penis consists primarily of three elongated cylinders of erectile tissue, which engorge with blood during sexual excitement. This causes the penis to become rigid and enlarged so that it may more adequately serve as a penetrating device. This event is called **erection.** The paired dorsal cylinders are the **corpora cavernosa.** The single ventral **corpus spongiosum** surrounds the penile urethra.

MICROSCOPIC ANATOMY OF SELECTED MALE REPRODUCTIVE ORGANS

Testis

Each testis is covered by a dense connective tissue capsule called the **tunica albuginea** (literally, "white tunic"). Extensions of this sheath enter the testis, dividing it into a number of lobes, each of which houses one to four highly coiled **seminiferous tubules,** the sperm-forming factories (Figure 29.2). The seminiferous tubules of each lobe converge to empty the sperm into another set of tubules, the **rete testis,** at the mediastinum of the testis. Sperm traveling through the rete

testis then enter the epididymis, located on the exterior aspect of the testis, as previously described. Lying between the seminiferous tubules and softly padded with connective tissue are the **interstitial cells,** which produce testosterone, the hormonal product of the testis.

1.   Obtain a slide of the testis and a microscope. Examine the slide under low power to identify the cross-sectional views of the cut seminiferous tubules. Then rotate the high-power lens into position and observe the wall of one of the cut tubules. As you work, refer to Plate 49 in the Histology Atlas to make the following identifications.

2.   Scrutinize the cells at the periphery of the tubule. The cells in this area are the **spermatogonia,** which undergo frequent mitoses to increase their number and maintain their population. About half of the spermatogonia's "offspring" become **spermatocytes,** spermatogenic cells that undergo meiosis, which leads to the formation of spermatids having half the usual genetic composition. The remaining daughter cells resulting from mitotic divisions of spermatogonia remain at the tubule periphery to maintain the germ cell line.

3.   Observe the cells in the middle of the tubule wall. There you should see a large number of spermatocytes that are obviously undergoing a nuclear division process. Look for the chromosomes, visible only during nuclear division, which have the appearance of coiled springs.

4.   Examine the cells at the tubule lumen. Identify the small round-nucleated spermatids, many of which may appear lopsided and look as though they are starting to lose their cytoplasm. See if you can find a spermatid embedded in an elongated cell type, a **sustentacular (Sertoli) cell,** which extends inward from the periphery of the tubule. The sustentacular cells nourish the spermatids as they begin their transformation into sperm. Also in the adluminal area, locate sperm, which can be

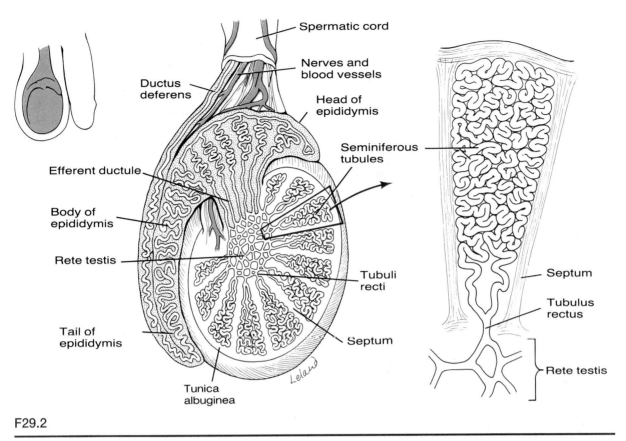

F29.2

Longitudinal-section view of the testis showing seminiferous tubules. (Epididymis and part of the ductus deferens also shown.)

identified by their tails. The sperm develop directly from the spermatids by the loss of extraneous cytoplasm and the development of a propulsive tail.

5. Identify the testosterone-producing *interstitial cells* lying external to and between the seminiferous tubules.

6. Obtain a prepared slide of human sperm and view it under high power or with the oil immersion lens. See the photograph of sperm in Plate 50 in the Histology Atlas. Identify the head (essentially the nucleus of the spermatid), acrosome, and tail regions. (The acrosome, which caps the nucleus anteriorly, contains enzymes involved in sperm penetration of the egg.)

Epididymis

Obtain a cross section of the epididymis. Notice the abundant tubule cross sections resulting from the fact that the coiling epididymis tubule has been cut through many times in the specimen. Look for sperm in the lumen of the tubule. Using Figure 29.3a as a guide, examine the composition of the tubule wall carefully. Identify the *stereocilia* of the pseudo-

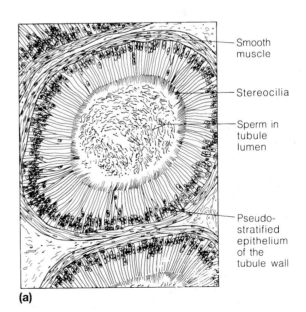

(a)

F29.3

Microscopic anatomy of selected organs of the male reproductive duct system. (a) Epididymis (see corresponding Plate 46 in the Histology Atlas).

stratified columnar epithelial lining. These nonmotile microvilli absorb excess fluid and pass nutrients to the sperm in the lumen. Now identify the smooth muscle layer. What do you think the function of the smooth muscle is?

_____

_____

_____

Penis

Obtain a cross section of the penis. Scan the tissue under low power to identify the urethra and the cavernous bodies. Compare your observations to Figure 29.3b. Observe the lumen of the urethra carefully. What type of epithelium do you see?

_____

Explain the function of this type of epithelium.

_____

_____

_____

GROSS ANATOMY OF THE HUMAN FEMALE REPRODUCTIVE SYSTEM

The **ovaries** (female gonads) are the primary reproductive organs of the female. Like the testes of the male, the ovaries produce both an exocrine product (the eggs, or ova) and endocrine products (estrogens and progesterone). The other accessory structures of the female reproductive system transport, house, nurture, or otherwise serve the needs of the reproductive cells and/or the developing fetus.

The reproductive structures of the female are generally considered in terms of internal organs and external organs, or external genitalia.

As you read the descriptions of these structures, locate them on Figures 29.4 and 29.5 and then on the female reproductive system model or large laboratory chart.

The **external genitalia (vulva)** consist of the mons pubis, the labia majora and minora, the clitoris, the urethral and vaginal orifices, the hymen, and the greater vestibular glands. The **mons pubis** is a rounded fatty eminence overlying the pubic symphysis. Running inferiorly and posteriorly from the mons pubis are two elongated, pigmented, hair-covered skin folds, the **labia majora,** which are homologous to the scrotum of the male. These enclose two smaller hair-free folds, the **labia minora.** (Terms indicating only one of the two folds in each case are *labium majus* and *minus,* respectively.) The labia minora, in turn, enclose a region called

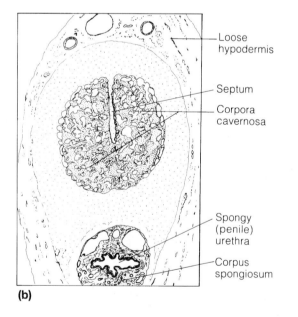

(b)

F29.3 *(continued)*

Microscopic anatomy of selected organs of the male reproductive duct system. (b) Cross-sectional view of the penis (see also Plate 47 in the Histology Atlas).

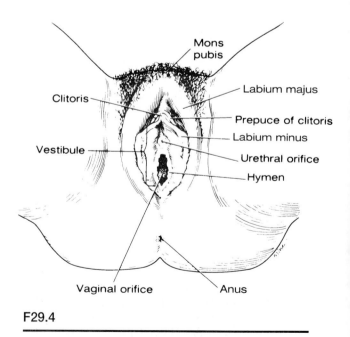

F29.4

External genitalia of the human female.

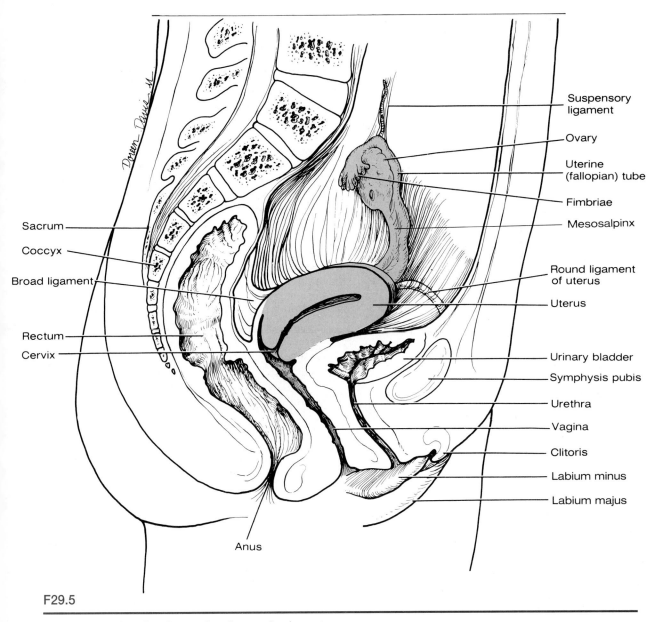

Suspensory ligament

Ovary

Uterine (fallopian) tube

Fimbriae

Mesosalpinx

Round ligament of uterus

Uterus

Urinary bladder

Symphysis pubis

Urethra

Vagina

Clitoris

Labium minus

Labium majus

Sacrum

Coccyx

Broad ligament

Rectum

Cervix

Anus

F29.5

Midsagittal section of the human female reproductive system.

the **vestibule,** which contains many structures—the clitoris, most anteriorly, followed by the urethral orifice and the vaginal orifice. The diamond-shaped region between the anterior end of the labial folds, the ischial tuberosities laterally, and the anus posteriorly is called the **perineum.**

The **clitoris** is a small protruding structure, homologous to the male penis. Like its counterpart, it is composed of highly sensitive, erectile tissue. It is hooded by skin folds of the anterior labia minora, referred to as the **prepuce of the clitoris.** The urethral orifice, which lies posterior to the clitoris, is the outlet for the urinary system and has no reproductive function in the female. The vaginal opening is partially closed by a thin fold of mucous membrane called the **hymen** and is flanked by the pea-sized, mucus-secreting **greater vestibular glands.**

These glands (not depicted in the illustrations) lubricate the distal end of the vagina during coitus.

The internal female organs include the vagina, uterus, uterine tubes, ovaries, and the ligaments and supporting structures that suspend these organs in the pelvic cavity. The **vagina** extends for approximately 10 cm (4 inches) from the vestibule to the uterus superiorly. It serves as a copulatory organ and birth canal, and permits passage of the menstrual flow. The pear-shaped **uterus,** situated between the bladder and the rectum, is a highly muscular organ with its narrow end, the **cervix,** directed inferiorly. The major portion of the uterus is referred to as the **body;** its superior rounded region above the entrance of the uterine tubes is called the **fundus.** A fertilized egg is implanted in the uterus, which houses the embryo or fetus during its development.

In some cases, the fertilized egg may implant in a uterine tube or even on the abdominal viscera, creating an **ectopic pregnancy.** Such implantations are usually unsuccessful and may even endanger the mother's life, because the uterine tubes cannot accommodate the increasing size of the fetus. ■

The apical region of the **endometrium,** the thick mucosal lining of the uterus, sloughs off periodically (about every 28 days) in response to cyclic changes in the levels of ovarian hormones in the woman's blood. This sloughing-off process, which is accompanied by bleeding, is referred to as **menstruation,** or **menses.**

The **uterine,** or **fallopian, tubes** enter the superior region of the uterus and extend laterally for about 10 cm (4 inches) toward the **ovaries** in the peritoneal cavity. The distal ends of the tubes are funnel-shaped and have fingerlike projections called **fimbriae.** Unlike the male duct system, there is no actual contact between the female gonad and the initial part of the female duct system—the uterine tube.

Because of this open passageway between the female reproductive organs and the peritoneal cavity, reproductive system infections, such as gonorrhea, can spread to cause widespread inflammations of the pelvic viscera, a condition called **pelvic inflammatory disease** or **PID.** ■

Within the ovaries, the female gametes (eggs) develop in sac-like structures called *follicles.* The growing follicles also produce *estrogens.* When a developing egg has reached the appropriate stage of maturity, it is ejected from the ovary in an event called **ovulation.** The ruptured follicle is then converted to a second type of endocrine gland, called a *corpus luteum,* which secretes progesterone (and some estrogens).

The flattened almond-shaped ovaries lie adjacent to the uterine tubes but are not connected to them; consequently, an ovulated egg* enters the pelvic cavity. The waving fimbriae of the uterine tubes create fluid currents that, if successful, draw the egg into the lumen of the uterine tube, where it begins its passage to the uterus, propelled by the cilia of the tubule walls. The usual and most desirable site of fertilization is the uterine tube, because the journey to the uterus takes about 3 to 4 days and an egg is viable for up to 24 hours after it is expelled

from the ovary. Thus, sperm must swim upward through the vagina and uterus, and into the uterine tubes to reach the egg. This must be an arduous journey, because they must swim against the downward current created by ciliary action—rather like swimming against the tide!

The internal female organs are all retroperitoneal, except the ovaries. They are supported and suspended somewhat freely by ligamentous folds of peritoneum. The peritoneum takes an undulating course. From the pelvic cavity floor it moves superiorly over the top of the bladder, reflects over the anterior and posterior surfaces of the uterus, and then over the rectum, and up the posterior body wall. The fold that encloses the uterine tubes and uterus and secures them to the lateral body walls is the **broad ligament.** The part of the broad ligament specifically anchoring the uterus is called the **mesometrium** and that anchoring the uterine tubes, the **mesosalpinx.** The **round ligaments,** fibrous cords that run from the uterus to the labia majora, also help attach the uterus to the body wall. The ovaries are supported medially by the **ovarian ligament** (extending from the uterus to the ovary), laterally by the **suspensory ligaments,** and posteriorly by a fold of the broad ligament, the **mesovarium.**

MICROSCOPIC ANATOMY OF SELECTED FEMALE REPRODUCTIVE ORGANS

Uterine Tube

Obtain a prepared slide of a cross-sectional view of a uterine tube for examination. Notice the highly folded mucosa (the folds nearly fill the tubule lumen) as illustrated in Figure 29.6. Then switch to high power to examine the ciliated secretory epithelium. Draw your observations of the tubule mucosa below.

---

* To simplify this discussion, the ovulated cell is called an egg. What is actually expelled from the ovary is an earlier stage of development called a secondary oocyte.

Serosa   Smooth   Highly   Lumen
          muscle   folded
                    mucosa

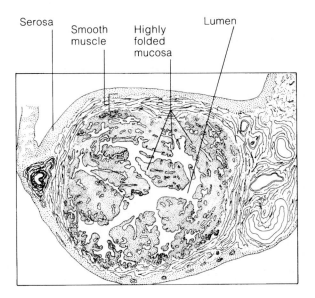

F29.6

Cross-sectional view of a uterine tube. Notice its highly folded mucosa. (See also corresponding Plate 48 in the Histology Atlas.)

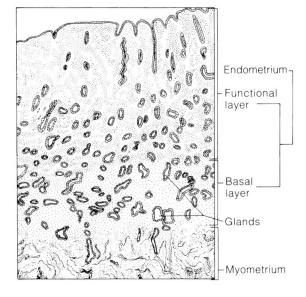

Endometrium
Functional layer
Basal layer
Glands
Myometrium

F29.7

Structure of the uterine endometrium (proliferative phase). Corresponds to Plate 52 in the Histology Atlas.

Wall of the Uterus

Obtain a cross-sectional view of the uterine wall. Identify the three layers of the uterine wall—the endometrium, myometrium, and serosa. Also identify the two strata of the endometrium: the **functional layer** (stratum functionalis) and the **basal layer** (stratum basalis), which forms a new functional layer each month. Figure 29.7 of the proliferative endometrium may be of some help in this study.

As you study the slide, notice that the bundles of smooth muscle are oriented in several different directions. What is the function of the myometrium (smooth muscle layer) during the birth process?

_____

_____

_____

What is the importance of the abundant endometrial glands?

_____

_____

_____

_____

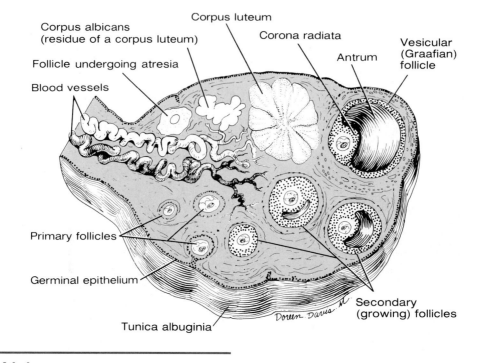

F29.8

Anatomy of the human ovary.

Ovary

Because many different stages of ovarian development exist within the ovary at any one time, a single microscopic preparation will contain follicles at many different stages of development. Obtain a cross section of ovary tissue, and identify the following structures. Refer to Figures 29.8 and 29.9 as you work.

Germinal epithelium: Outermost layer of the ovary.

Primary follicle: One or a few layers of cuboidal follicle cells surrounding the larger central developing ovum.

Secondary (growing) follicles: Follicles consisting of several layers of follicle (granulosa) cells surrounding the central developing ovum, and beginning to show evidence of fluid accumulation and **antrum** (central cavity) formation.

Vesicular (Graafian) follicle: At this stage of development, the follicle has a large antrum containing fluid produced by the granulosa cells. The developing ovum is pushed to one side of the follicle and is surrounded by a capsule of several layers of granulosa cells called the **corona radiata** (radiating crown). When the immature ovum (secondary oocyte) is released, it enters the uterine tubes with its corona radiata intact. The connective tissue stroma (background tissue) adjacent to the mature

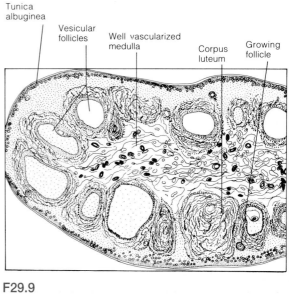

F29.9

Line drawing of a photomicrograph of the human ovary. (See corresponding Plate 51 in the Histology Atlas.)

follicle forms a capsule that encloses the follicle and is called the **theca.** (See also Plate 24 in the Histology Atlas.)

Corpus luteum: A solid glandular structure or a structure containing a scalloped lumen that develops from the ovulated follicle. (See Plate 25 in the Histology Atlas.)

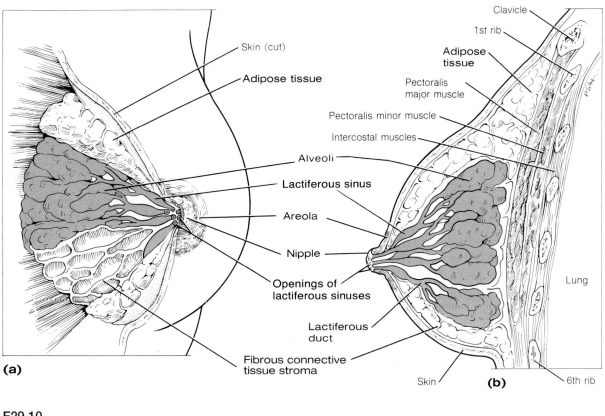

(a)

Skin (cut)
Adipose tissue
Alveoli
Lactiferous sinus
Areola
Nipple
Openings of lactiferous sinuses
Lactiferous duct
Fibrous connective tissue stroma

Clavicle
1st rib
Adipose tissue
Pectoralis major muscle
Pectoralis minor muscle
Intercostal muscles
Lung
Skin
(b)
6th rib

F29.10

Female mammary gland. (a) Anterior view. (b) Sagittal section.

THE MAMMARY GLANDS

The **mammary glands** or breasts exist, of course, in both sexes, but they have a reproduction-related function only in females. Since the function of the mammary glands is to produce milk to nourish the newborn infant, their importance is more closely associated with events that occur when reproduction has already been accomplished. Periodic stimulation by the female sex hormones, especially estrogens, increases the size of the female mammary glands at puberty. During this period, the duct system becomes more elaborate, and fat is deposited—fat deposition being the more important contributor to increased breast size.

The rounded, skin-covered mammary glands lie anterior to the pectoral muscles of the thorax, attached to them by connective tissue. Slightly below the center of each breast is a pigmented area, the **areola,** which surrounds a centrally protruding **nipple** (Figure 29.10).

Internally each mammary gland consists of 15 to 20 **lobes** which radiate around the nipple and are separated by fibrous connective tissue and adipose, or fatty, tissue. Within each lobe are smaller chambers called **lobules,** containing the glandular **alveoli** that produce milk during lactation. The alveoli of each lobule pass the milk into a number of **lactiferous ducts,** which join to form an expanded storage chamber, the **lactiferous sinus,** as they approach the nipple. The sinuses open to the outside at the nipple.

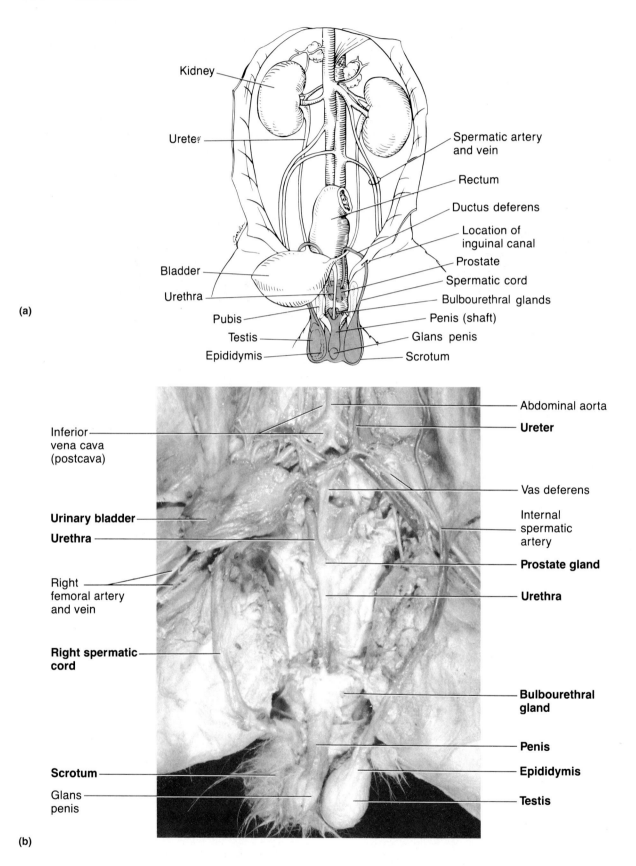

(a)

(b)

F29.11

Reproductive system of the male cat. (a) Diagrammatic view. (b) Photograph. See dissection photo, Plate A in the Cat Anatomy Atlas.

DISSECTION OF THE REPRODUCTIVE SYSTEM OF THE CAT

Obtain your cat, a dissection tray, the necessary dissection instruments, and, if desired, protective skin cream or plastic gloves. After you have completed the study of the reproductive structures of your specimen, observe a cat of the opposite sex. (The following instructions assume that the abdominal cavity has been opened in previous dissection exercises.)

Male Reproductive System

Refer to Figure 29.11 and Plate A in the Cat Anatomy Atlas as you identify the male structures.

 1.   Identify the **penis** and notice the prepuce covering the glans. Carefully cut through the skin overlying the penis to expose the cavernous tissue beneath, then cross section the penis to see the relative positioning of the three cavernous bodies.

2.   Identify the **scrotal sac,** and then carefully make a shallow incision through the scrotum to expose the testes. Note that the scrotum is divided internally.

3.   Lateral to the medial aspect of the scrotal sac, locate the **spermatic cord,** which contains the spermatic artery, vein, and nerve, as well as the ductus deferens, and follow it up through the inguinal canal into the abdominal cavity. (It is not necessary to cut through the pelvic bone; a slight tug on the spermatic cord in the scrotal sac region will reveal its position in the abdominal cavity.)

Carefully loosen the spermatic cord from the connective tissue investing it, and follow its course as it travels superiorly in the pelvic cavity, loops over the ureter,* and then courses posterior to the bladder and enters the prostate gland.

4.   Notice that the **prostate gland** is comparatively smaller in the cat than in the human, and it is more distal to the bladder. (In the human, the prostate is immediately adjacent to the base of the bladder.) Carefully slit open the prostate gland to follow the **ductus deferens** to the urethra, which exits from the bladder midline. The male cat urethra, like that of the human, serves as both a urinary and sperm duct. In the human, the ductus deferens is joined by the duct of the seminal vesicle to form the ejaculatory duct, which enters the prostate. Seminal vesicles are not present in the cat.

5.   Trace the **urethra** to the proximal end of the cavernous tissue of the penis. Carefully split the proximal portion of the penis along a sagittal plane to reveal the **bulbourethral glands** lying beneath it.

6.   Once again, turn your attention to the testis. Cut it from its attachment to the spermatic cord and carefully slit open the **tunica vaginalis** capsule enclosing it. Identify the **epididymis** running along one side of the testis. Make a longitudinal cut through the testis and epididymis. Can you see the tubular nature of the epididymis and the rete testis portion of the testis with the naked eye?

---

* This position of the spermatic cord and ductus deferens is due to the fact that during fetal development, the testis was in the same relative position as the ovary is in the female. In its descent, it passes laterally and ventrally to the ureter.

Female Reproductive System

Refer to Figure 29.12 and Plate B in the Cat Anatomy Atlas showing a dissection of the urogenital system of the female cat as you identify the structures described below.

1.   Unlike the pear-shaped simplex, or one-part, uterus of the human, the uterus of the cat is Y-shaped (bipartite or bicornuate) and consists of a **uterine body** from which two **uterine horns** (cornua) diverge. Such an enlarged uterus enables the animal to produce litters. Examine the abdominal cavity and identify the bladder and the body of the uterus lying just dorsal to it.

2.   Follow one of the uterine horns as it travels superiorly in the body cavity. Identify the thin mesentery (the *broad ligament*), which helps anchor it and the other reproductive structures to the body wall. Approximately halfway up the length of the uterine horn, it should be possible to identify the more important *round ligament,* a cord of connective tissue extending laterally and posteriorly from the uterine horn to the region of the body wall that would correspond to the inguinal region of the male.

3.   Examine the **uterine tube** and **ovary** at the distal end of the uterine horn just caudal to the kidney. Observe how the funnel-shaped end of the uterine tube curves around the ovary. As in the human, the distal end of the tube is fimbriated, or fringed, and the tube is lined with ciliated epithelium. The uterine tubes of the cat are tiny and relatively much shorter than in the human. Identify the *ovarian ligament,* a short thick cord that extends from the uterus to the ovary and anchors the ovary to the body wall. Also observe the ovarian artery and vein passing through the mesentery to the ovary and uterine structures.

4.   Return to the body of the uterus and follow it caudad to the bony pelvis. Use bone clippers to cut through the median line of the pelvis (the pubic symphysis), cutting carefully so you do not damage the urethra deep to it. Expose the pelvic region by pressing the thighs dorsally. Follow the uterine body caudally to the vagina, and note the point where the urethra draining the bladder and the **vagina** enter a common chamber, the **urogenital sinus.** How does this anatomical arrangement compare to that seen in the human female?

_____

_____

5.   Observe the **vulva** of the cat, which is similar to the human vulva. Identify the raised **labia majora** surrounding the urogenital opening.

6.   To determine the length of the vagina, which is difficult to ascertain by external inspection, slit through the vaginal wall just superior to the urogenital sinus and cut toward the body of the uterus with scissors. Reflect the cut edges, and identify the muscular cervix of the uterus. Approximately how long is the vagina of the cat? (Measure the distance between the urogenital sinus and the cervix.)

_____

7.   When you have completed your observations of both male and female cats, clean your dissecting instruments and tray and properly wrap the cat for storage.

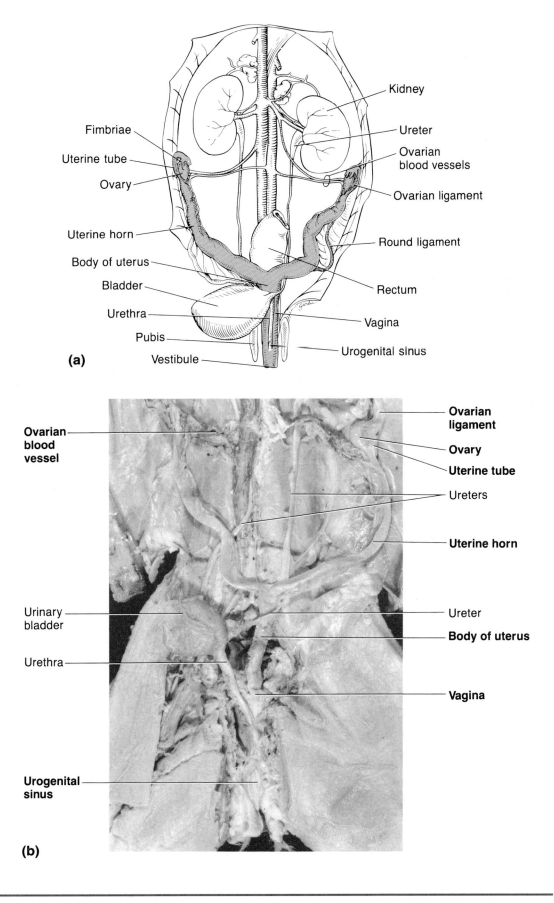

(a)

Kidney

Fimbriae

Uterine tube

Ovary

Ureter

Ovarian blood vessels

Ovarian ligament

Uterine horn

Body of uterus

Round ligament

Bladder

Urethra

Rectum

Pubis

Vagina

Vestibule

Urogenital sinus

(b)

Ovarian blood vessel

Ovarian ligament

Ovary

Uterine tube

Ureters

Urinary bladder

Uterine horn

Urethra

Ureter

Body of uterus

Vagina

Urogenital sinus

F29.12

Reproductive system of the female cat. (a) Diagrammatic view. **(b)** Photograph. See dissection of female urogenital system, Plate B in the Cat Anatomy Atlas.

Review Sheet

EXERCISE 1

The Language of Anatomy

Surface Anatomy

1. Match each of the following descriptions with a key equivalent, and record the key letter or term in front of the description.

    Key: a. buccal   c. deltoid   e. patellar
            b. calcaneal   d. digital   f. scapular

    _____ 1. cheek   _____ 4. anterior aspect of knee

    _____ 2. pertaining to the fingers   _____ 5. heel of foot

    _____ 3. shoulder blade region   _____ 6. curve of shoulder

2. Indicate the following body areas on the accompanying diagram by placing the correct key letter at the end of each line.

    Key:

    a. abdominal
    b. antecubital
    c. axillary
    d. brachial
    e. cervical
    f. femoral
    g. gluteal
    h. inguinal
    i. lumbar
    j. occipital
    k. oral
    l. popliteal
    m. pubic
    n. sural
    o. thoracic
    p. umbilical

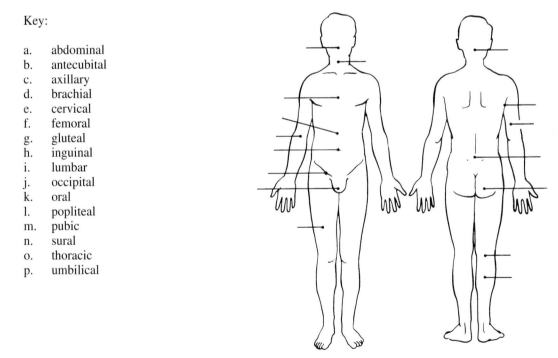

3. Classify each of the surface anatomy terms in the key of question 2 above, into one of the body regions indicated below. Insert the appropriate key letters on the answer blanks.

    _____ 1. Appendicular   _____ 2. Axial

Body Orientation, Direction, Planes, and Sections

1. Describe completely the standard human anatomical position. _____

    _____

2. Define *section:* _____

3. Several incomplete statements are listed below. Correctly complete each statement by choosing the appropriate anatomical term from the key. Record the key letters and/or terms on the correspondingly numbered blanks below.

    Key: a. anterior   e. lateral   i. sagittal
          b. distal   f. medial   j. superior
          c. frontal   g. posterior   k. transverse
          d. inferior   h. proximal

    In the anatomical position, the face and palms are on the __1__ body surface; the buttocks and shoulder blades are on the __2__ body surface; and the top of the head is the most __3__ part of the body. The ears are __4__ and __4__ to the shoulders and __5__ to the nose. The heart is __6__ to the vertebral column (spine) and __7__ to the lungs. The elbow is __8__ to the fingers but __9__ to the shoulder. The abdominopelvic cavity is __10__ to the thoracic cavity and __11__ to the spinal cavity. In humans, the dorsal surface can also be called the __12__ surface; however, in quadruped animals, the dorsal surface is the __13__ surface.

    If an incision cuts the heart into right and left parts, the section is a __14__ section; but if the heart is cut so that superior and inferior portions result, the section is a __15__ section. You are told to cut a dissection animal along two planes so that the kidneys are observable in both sections. The two sections that meet this requirement are the __16__ and __17__ sections.

    1. _____   7. _____   13. _____

    2. _____   8. _____   14. _____

    3. _____   9. _____   15. _____

    4. _____   10. _____   16. _____

    5. _____   11. _____   17. _____

    6. _____   12. _____

4. Correctly identify each of the nine areas of the abdominal surface by inserting the appropriate term for each of the letters indicated in the drawing on the next page.

    a. _____   d. _____

    b. _____   e. _____

    c. _____   f. _____

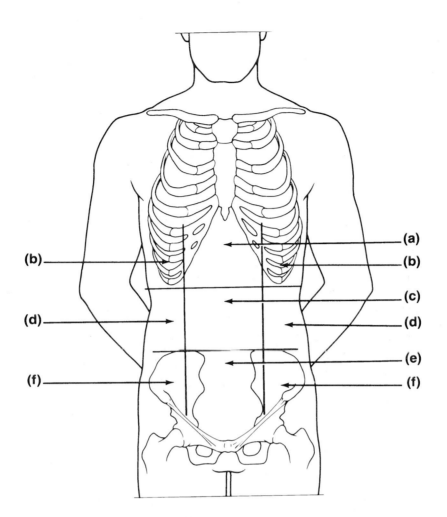

Body Cavities

1. Which body cavity would have to be opened for the following types of surgery? (Insert letter of key choice in same-numbered blank.)

Key: a. abdominopelvic   c. dorsal   e. thoracic

       b. cranial   d. spinal   f. ventral

1. surgery to remove a cancerous lung lobe
2. removal of the uterus or womb
3. removal of a brain tumor
4. appendectomy
5. stomach ulcer operation

The abdominopelvic and thoracic cavities are subdivisions of the __6__ body cavity, while the cranial and spinal cavities are subdivisions of the __7__ body cavity. The __8__ body cavity is totally surrounded by bone, and thus affords its contained structures very good protection.

1. _____
2. _____
3. _____
4. _____
5. _____
6. _____
7. _____
8. _____

RS3

2. Name the serous membranes covering the lungs (#1), the heart (#2), and the organs of the abdominopelvic cavity (#3), and insert your responses in the blanks on the right.

1. _____

2. _____

3. _____

3. Name the muscle that subdivides the ventral body cavity. _____

4. Which of the following organ systems are represented in all three subdivisions of the ventral body cavity? (Circle all appropriate responses.)

respiratory   circulatory   reproductive   lymphatic

nervous   excretory (urinary)   muscular   integumentary

5. Which organ system would not be represented in any of the body cavities? _____

6. What are the bony landmarks of the abdominopelvic cavity? _____

_____

7. Which body cavity affords the least protection to its internal structures? _____

8. What is the function of the serous membranes of the body? _____

_____

_____

9. A nurse informs you that she is about to take blood from the antecubital region. What portion of your body

should you present to her? _____

10. What do peritonitis, pleurisy, and pericarditis (pathologic conditions) have in common? _____

_____

11. Why are these conditions accompanied by a great deal of pain? _____

_____

12. The mouth, or buccal cavity, and its extension, which stretches through the body inside the digestive system, is

not listed as an internal body cavity. Why is this so? _____

_____

_____

Review Sheet

EXERCISE 2

Organ Systems Overview

1. Use the key below to indicate the body systems that perform the following functions for the body.

Key: a. cardiovascular   d. integumentary   g. nervous   j. skeletal
       b. digestive   e. lymphatic   h. reproductive   k. urinary
       c. endocrine   f. muscular   i. respiratory

_____ 1. rids the body of nitrogen-containing wastes

_____ 2. is affected by removal of the thyroid gland

_____ 3. provides support and levers on which the muscular system acts

_____ 4. includes the heart

_____ 5. causes the onset of the menstrual cycle

_____ 6. protects underlying organs from drying out and from mechanical damage

_____ 7. protects the body; destroys bacteria and tumor cells

_____ 8. breaks down ingested food into its building blocks

_____ 9. removes carbon dioxide from the blood

_____ 10. delivers oxygen and nutrients to the tissues

_____ 11. moves the limbs; facilitates facial expression

_____ 12. conserves body water or eliminates excesses

_____ and _____ 13. facilitate conception and childbearing

_____ 14. controls the body by means of chemical molecules called hormones

_____ 15. is damaged when you cut your finger or get a severe sunburn

2. Using the above key, choose the *organ system* to which each of the following sets of organs or body structures belong:

_____ 1. thymus, spleen, lymphatic vessels

_____ 2. bones, cartilages, tendons

_____ 3. pancreas, pituitary, adrenals

_____ 4. trachea, bronchi, alveoli

_____ 5. kidneys, bladder, ureters

_____ 6. testis, vas deferens, urethra

_____ 7. esophagus, large intestine, rectum

_____ 8. arteries, veins, heart

3. Using the key below, place the following organs in their proper body cavity.

Key: a.   abdominopelvic   b.   cranial   c.   spinal   d.   thoracic

_____ 1. stomach   _____ 6.   urinary bladder

_____ 2. esophagus   _____ 7.   heart

_____ 3. large intestine   _____ 8.   trachea

_____ 4. liver   _____ 9.   brain

_____ 5. spinal cord   _____ 10.   rectum

4. Using the organs listed in item 3 above, record, by number, which would be found in the abdominal regions listed below:

_____ 1.   hypogastric region   _____ 4.   epigastric region

_____ 2.   right lumbar region   _____ 5.   left iliac region

_____ 3.   umbilical region   _____ 6.   left hypochondriac region

5. The five levels of organization of a living body are cell, _____ ,

_____ , _____ , and organism.

6. Define *organ.* _____

_____

_____

7. During the course of this laboratory exercise, a rat was dissected. What is the *value* of observing the anatomy of a rat (or any other small mammal) when *human anatomy* is the actual topic of study?

_____

_____

_____

RS6

Review Sheet

EXERCISE 3

The Microscope

Care and Structure of the Compound Microscope

1. Label all indicated parts of the microscope.

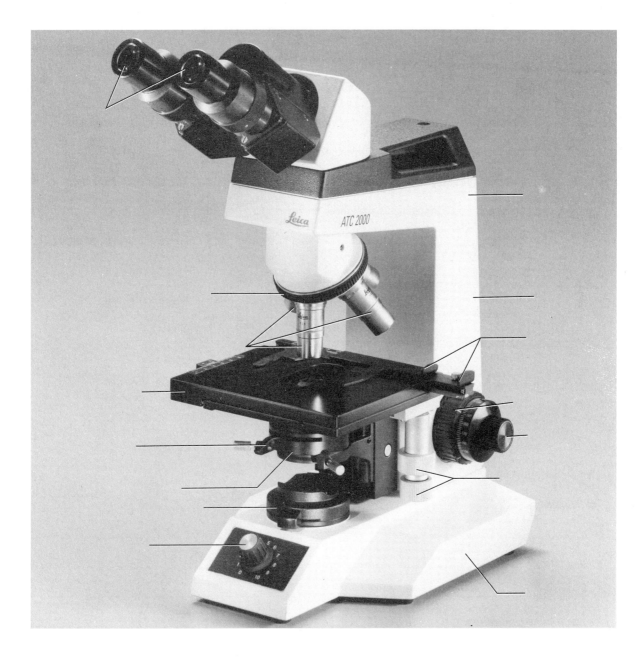

2. The following statements are true or false. If true, write *T* on the answer blank. If false, correct the statement by writing on the blank the proper word or phrase to replace the one that is <u>underlined</u>.

_____ 1. The microscope lens may be cleaned <u>with any soft tissue.</u>

_____ 2. The coarse adjustment knob may be used in focusing <u>with all three objectives.</u>

_____ 3. The microscope should be stored with the <u>oil immersion</u> lens in position over the stage.

_____ 4. When beginning to focus, the <u>low-power</u> lens should be used.

_____ 5. In low power, always focus <u>toward</u> the specimen.

_____ 6. A coverslip should always be used <u>with the high-power and oil lenses.</u>

_____ 7. The greater the amount of light delivered to the objective lens, the <u>less</u> the resolution.

3. Match the microscope structures given in column B with the statements in column A that identify or describe them:

Column A

_____ 1. platform on which the slide rests for viewing

_____ 2. lens located at the superior end of the body tube

_____ 3. secure(s) the slide to the stage

_____ 4. delivers a concentrated beam of light to the specimen

_____ 5. used for precise focusing once initial focusing has been done

_____ 6. carries the objective lenses; rotates so that the different objective lenses can be brought into position over the specimen

_____ 7. used to increase the amount of light passing through the specimen

Column B

a. coarse adjustment knob

b. condenser

c. fine adjustment knob

d. iris diaphragm

e. mechanical stage or spring clips

f. movable nosepiece

g. objective lenses

h. ocular

i. stage

4. Explain the proper technique for transporting the microscope.

_____

_____

5. Define the following terms.

real image: _____

_____

virtual image: _____

_____

6. Define *total magnification:* _____

_____

7. Define *resolution:* _____ ___

Viewing Objects Through the Microscope

1. Complete, or respond to, the following statements:

    _____ 1. The distance from the bottom of the objective lens in use to the specimen is called the _____.

    _____ 2. The resolution of the human eye is approximately _____ μm.

    _____ 3. The area of the specimen seen when looking through the microscope is the _____.

    _____ 4. If a microscope has a 10× ocular and the total magnification at a particular time is 950×, the objective lens in use at that time is _____ ×.

    _____ 5. If, after focusing in low power, only the fine adjustment need be used to focus the specimen at the higher powers, the microscope is said to be _____.

    _____ 6. If, when using a 10× ocular and a 15× objective, the field size is 1.5 mm, the approximate field size with a 30× objective is _____ mm.

    _____ 7. If the size of the high-power field is 1.2 mm, an object that occupies approximately a third of that field has an estimated diameter of _____ mm.

    _____ 8. Assume there is an object on the left side of the field that you want to bring to the center (that is, toward the apparent right). In what direction would you move your slide?

    _____ 9. If the object is in the top of the field and you want to move it downward to the center, you would move the slide _____.

2. You have been asked to prepare a slide with letter *k* on it (as below). In the circle below, draw the *k* as seen in the low-power field.

3. The numbers for the field sizes below are too large to represent the typical compound microscope lens system, but the relationships depicted are accurate. Figure out the magnification of fields 1 and 3, and the field size of 2. (*Hint:* Use your ruler.)

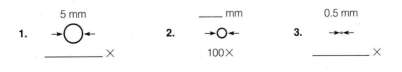

4. Say you are observing an object in the low-power field. When you switch to high-power, it is no longer in your field of view. Why might this occur? _____

_____

_____

What should be done initially to prevent this from happening? _____

_____

5. Do the following factors increase or decrease as one moves to higher magnifications with the microscope?

resolution _____   amount of light needed _____

working distance _____   depth of field _____

6. A student has the high-power lens in position and appears to be intently observing the specimen. The instructor, noting a working distance of about 1 cm, knows the student isn't actually seeing the specimen.

How so? _____

7. Why is it important to be able to use your microscope to perceive depth when studying slides?

_____

_____

8. If you are observing a slide of tissue two cell layers thick, how can you determine which layer is superior?

_____

_____

_____

9. Describe the proper procedure for preparing a wet mount.

_____

_____

_____

10. Give two reasons why the light should be dimmed when viewing living or unstained material.

_____

_____

11. Indicate the probable cause of the following situations arising during use of a microscope.

   a. Only half of field illuminated: _____

_____

   b. Field does not change as mechanical stage is moved: _____

_____

Review Sheet

EXERCISE 4

The Cell—
Anatomy and Division

Anatomy of the Composite Cell

1. Define the following:

   organelle: _____

   _____

   cell: _____

2. Although cells have differences that reflect their specific functions in the body, what functional capabilities do all

   cells exhibit? _____

3. Identify the following cell parts:

   _____ 1. external boundary of cell; regulates flow of materials into and out of the cell; site of cell signaling

   _____ 2. contains digestive enzymes of many varieties; "suicide sac" of the cell

   _____ 3. scattered throughout the cell; major site of ATP synthesis

   _____ 4. slender extensions of the plasma membrane that increase its surface area

   _____ 5. stored glycogen granules, crystals, pigments, and so on

   _____ 6. membranous system consisting of flattened sacs and vesicles; packages proteins for export

   _____ 7. control center of the cell; necessary for cell division and cell life

   _____ 8. two rod-shaped bodies near the nucleus; direct formation of the mitotic spindle

   _____ 9. dense, darkly staining nuclear body; packaging site for ribosomes

   _____ 10. contractile elements of the cytoskeleton

   _____ 11. membranous system; involved in intracellular transport of proteins and synthesis of membrane lipids

   _____ 12. attached to membrane systems or scattered in the cytoplasm; synthesize proteins

   _____ 13. threadlike structures in the nucleus; contain genetic material (DNA)

   _____ 14. site of free radical detoxification

4. In the following diagram, label all parts provided with a leader line.

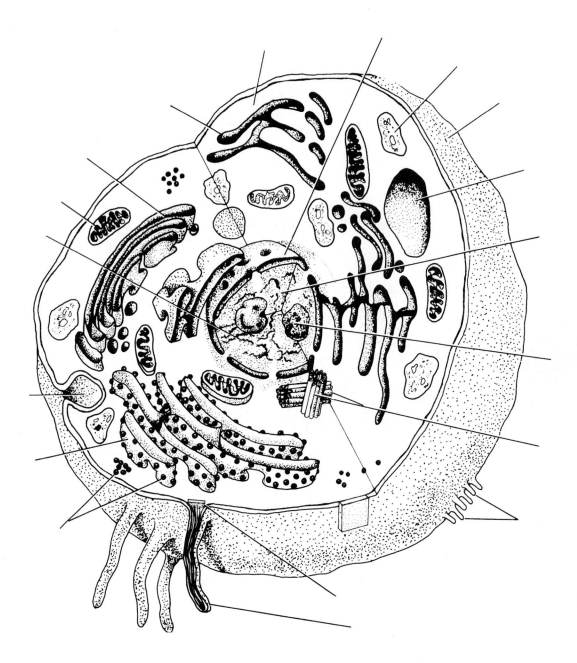

Observing Differences and Similarities in Cell Structure

1. For each of the following cell types, list (a) *one* important structural characteristic observed in the laboratory, and (b) the function that structure complements or ensures.

    1. squamous epithelium   a. _____

                               b. _____

    2. sperm   a. _____

                               b. _____

3. smooth muscle   a. _____

                                  b. _____

4. red blood cells   a. _____

                                  b. _____

2. What is the significance of the red blood cell being anucleate (without a nucleus)? _____

_____

Did it ever have a nucleus? _____ When? _____

Cell Division: Mitosis and Cytokinesis

1. Identify the three phases of mitosis in the following photomicrographs.

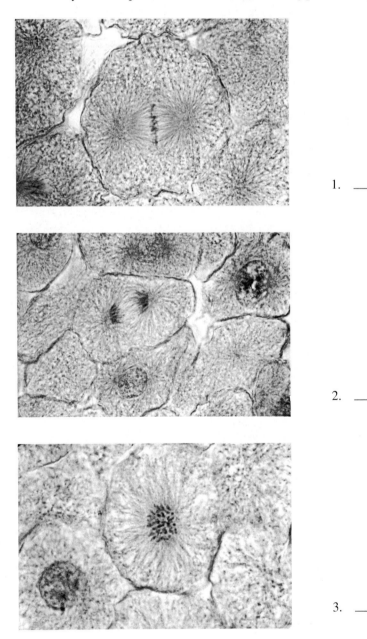

1. _____

2. _____

3. _____

2. What is the importance of mitotic cell division? _____

_____

3. Complete or respond to the following statements:

Division of the __1__ is referred to as mitosis. Cytokinesis is division of the __2__ . The major structural difference between chromatin and chromosomes is that the latter is __3__ . Chromosomes
attach to the spindle fibers by undivided structures called __4__ . If
a cell undergoes mitosis but not cytokinesis, the product is __5__ .
The structure that acts as a scaffolding for chromosomal attachment and movement is called the __6__ . __7__ is the period of cell
life when the cell is not involved in division. Two cell populations
in the body that do not undergo cell division are __8__ and __9__ .
The implication of an inability of a cell population to divide is that
when some of its members die, they are replaced by __10__ .

1. _____

2. _____

3. _____

4. _____

5. _____

6. _____

7. _____

8. _____

9. _____

10. _____

4. Using the key, categorize each of the events described below according to the phase in which it occurs.

Key: a. prophase   b. anaphase   c. telophase   d. metaphase   e. none of these

_____ 1.   Chromatin coils and condenses, forming chromosomes.

_____ 2.   The chromosomes (chromatids) are V-shaped.

_____ 3.   The nuclear membrane re-forms.

_____ 4.   Chromosomes stop moving toward the poles.

_____ 5.   Chromosomes line up in the center of the cell.

_____ 6.   The nuclear membrane fragments.

_____ 7.   The mitotic spindle forms.

_____ 8.   DNA synthesis occurs.

_____ 9.   Centrioles replicate.

_____ 10. Chromosomes first appear to be duplex structures.

_____ 11. Chromosomal centromeres are attached to the kinetochore fibers.

_____ 12. Cleavage furrow forms.

_____ 13. The nuclear membrane(s) is absent.

5. What is the physical advantage of the chromatin coiling and condensing to form short chromosomes at the onset

of mitosis? _____

_____

Review Sheet

EXERCISE 5

Classification of Tissues

Tissue Structure and Function—General Review

1. Define *tissue:* _groups of cells that are similar in structure and function_ _____

2. Use the key choices to identify the major tissue types described below.

Key: a. connective tissue   b. epithelium   c. muscle   d. nervous tissue

epithelium _____ 1. lines body cavities and covers the body's external surface

connective _____ 2. pumps blood, flushes urine out of the body, allows one to swing a bat

nervous _____ 3. transmits electrochemical impulses

epithelium _____ 4. anchors, packages, and supports body organs

epithelium _____ 5. cells may absorb, secrete, and protect

nervous _____ 6. most involved in regulating and controlling body functions

muscle _____ 7. major function is to contract

nervous _____ 8. synthesizes hormones

muscle _____ 9. the most durable tissue type

connective ___ 10. abundant nonliving extracellular matrix

connective ___ 11. most widespread tissue in the body

nervous ___ 12. forms nerves and the brain

Epithelial Tissue

1. Describe the general characteristics of epithelial tissue. _found where organs encounter_ _the ambient environment, cellular, polar, basement membrane,_ _avascular, regenerates_ _____

2. On what bases are epithelial tissues classified? _shape of cells, # of layers_ _____

3. What are the major functions of epithelium in the body? (Give examples.) *protection, absorption, control permeability, move fluids, secretions, sensations*

4. How is the function of epithelium reflected in its arrangement? _____

5. Where is ciliated epithelium found? *stomach lining, respiratory lining*

   What role does it play? *moving fluids across the cells*

6. Transitional epithelium is actually stratified squamous epithelium, but there is something special about it.

   How does it differ structurally from other stratified squamous epithelia? *it can be stretched because it is tall*

   How does this reflect its function in the body? *it is in organs (bladder) which have to stretch*

7. How do the endocrine and exocrine glands differ in structure and function? *endocrine cells pinch of to allow the secretion to be carried throughout the body, exocrine excrete through ducts*

8. Respond to the following with the key choices.

   Key: a. pseudostratified ciliated columnar   c. simple cuboidal   e. stratified squamous
         b. simple columnar   d. simple squamous   f. transitional

   simple squamous _____ 1. lining of the esophagus

   simple columnar _____ 2. lining of the stomach

   pseudostratified ciliated columnar 3. lung tissue, alveolar sacs

   simple cuboidal _____ 4. tubules of the kidney

   stratified squamous _____ 5. epidermis of the skin

   transitional _____ 6. lining of bladder; peculiar cells that have the ability to slide over each other

   simple squamous _____ 7. forms the thin serous membranes; a single layer of flattened cells

RS16

Connective Tissue

1. What are the general characteristics of connective tissues? _____

    _____

    _____

    _____

    _____

2. What functions are performed by connective tissue? _____

    _____

3. How are the functions of connective tissue reflected in its structure? _____

    _____

    _____

4. Using the key, choose the best response to identify the connective tissues described below.

    Key: a. adipose connective tissue   e. fibrocartilage
          b. areolar connective tissue   f. hematopoietic tissue
          c. dense connective tissue   g. hyaline cartilage
          d. elastic cartilage   h. osseous tissue

    _____ 1. attaches bones to bones and muscles to bones

    _____ 2. acts as a storage depot for fat

    _____ 3. the dermis of the skin

    _____ 4. makes up the intervertebral discs

    _____ 5. forms your hip bone

    _____ 6. composes basement membranes; a soft packaging tissue
                                          with a jellylike matrix

    _____ 7. forms the larynx, the costal cartilages of the ribs, and the
                                          embryonic skeleton

    _____ 8. provides a flexible framework for the external ear

    _____ 9. firm, structurally amorphous matrix heavily invaded with
                                          fibers; appears glassy and smooth

    _____ 10. matrix hard owing to calcium salts; provides levers for
                                          muscles to act on

    _____ 11. insulates against heat loss

5. Why are adipose cells called "signet ring" cells? _____

    _____

Muscle Tissue

The three types of muscle tissue exhibit similarities as well as differences. Check the appropriate space in the chart below to indicate which muscle types exhibit each characteristic.

Characteristic	Skeletal	Cardiac	Smooth
Voluntarily controlled			
Involuntarily controlled			
Striated			
Has a single nucleus in each cell			
Has several nuclei per cell			
Found attached to bones			
Allows you to direct your eyeballs			
Found in the walls of the stomach, uterus, and arteries			
Contains spindle-shaped cells			
Contains branching cylindrical cells			
Contains long, nonbranching cylindrical cells			
Has intercalated discs			
Concerned with locomotion of the body as a whole			
Changes the internal volume of an organ as it contracts			
Tissue of the heart			

Nervous Tissue

1. What two physiologic characteristics are highly developed in nervous tissue? _____

    _____

2. In what ways are nerve cells similar to other cells? _____

    _____

    How are they different? _____

3. Sketch a neuron, recalling in your diagram the most important aspects of its structure. Below the diagram, describe how its particular structure relates to its function in the body.

_____

_____

For Review
Label the following tissue types here and on the next pages, and identify all major structures.

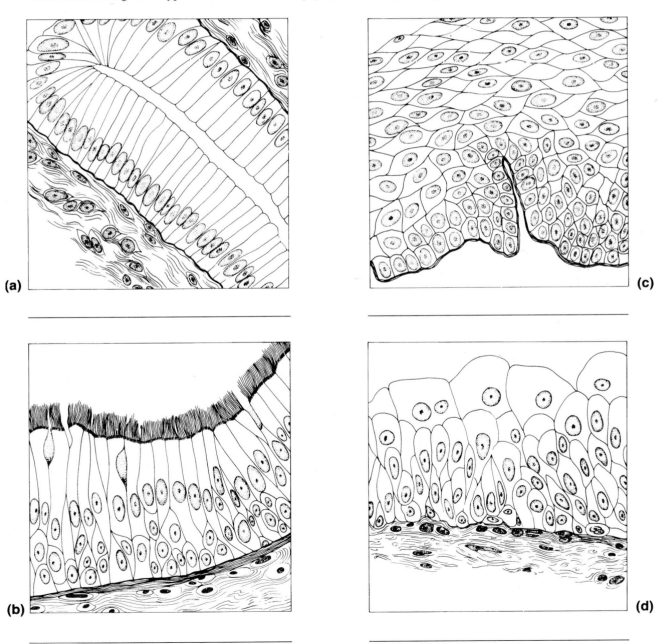

(a)

(c)

(b)

(d)

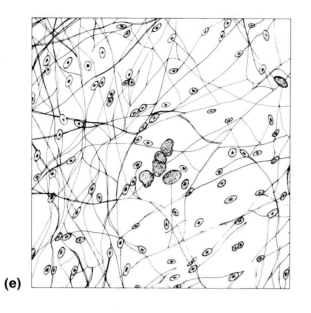

(e)

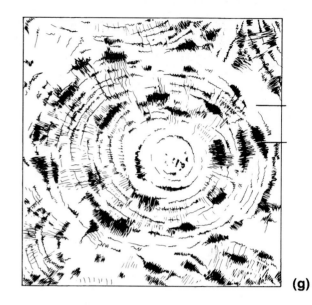

(g)

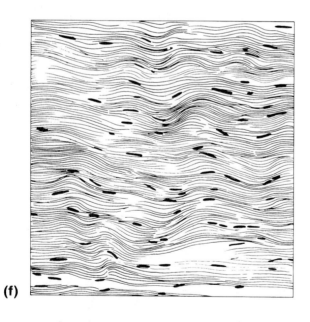

(f)

(h)

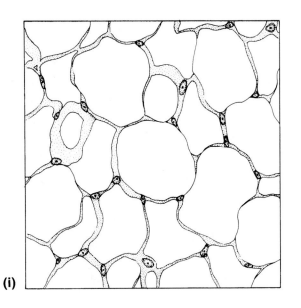

(i)

_____

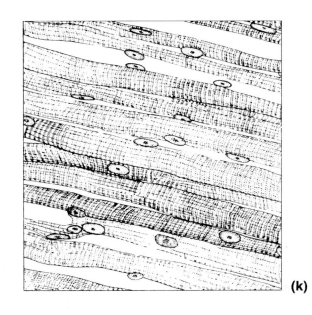

(k)

_____

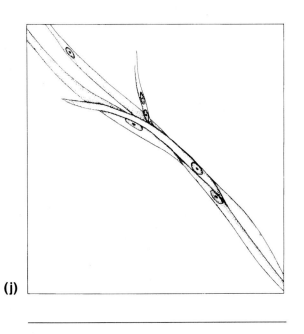

(j)

_____

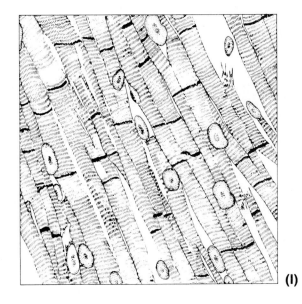

(l)

_____

Review Sheet

EXERCISE 6

The Integumentary System

Basic Structure of the Skin

1. Complete the following statements by writing the appropriate word or phrase on the correspondingly numbered blanks:

The two basic tissues of which the skin is composed are dense irregular connective tissue, which makes up the dermis, and __1__ , which forms the epidermis. The tough water-repellent protein found in the epidermal cells is called __2__ . The pigments, melanin and __3__ , contribute to skin color. A localized concentration of melanin is referred to as a __4__ .

1. _____

2. _____

3. _____

4. _____

2. Four protective functions of the skin are _____

_____ .

3. Using the key choices, choose all responses that apply to the following descriptions.

Key: a. stratum basale
      b. stratum corneum
      c. stratum granulosum

      d. stratum lucidum
      e. stratum spinosum
      f. papillary layer

      g. reticular layer
      h. epidermis as a whole
      i. dermis as a whole

_____ 1. translucent cells containing keratin fibrils

_____ 2. dead cells

_____ 3. dermis layer responsible for fingerprints

_____ 4. vascular region

_____ 5. major skin area where derivatives (nails and hair) arise

_____ 6. epidermal region exhibiting rapid cell division

_____ 7. scalelike dead cells, full of keratin, that constantly slough off

_____ 8. also called the stratum germinativum

_____ 9. has abundant elastic and collagenic fibers

_____ 10. site of melanin formation

_____ 11. area where tonofilaments first appear

4. Label the skin structures and areas indicated in the accompanying diagram of skin. Then, complete the statements that follow.

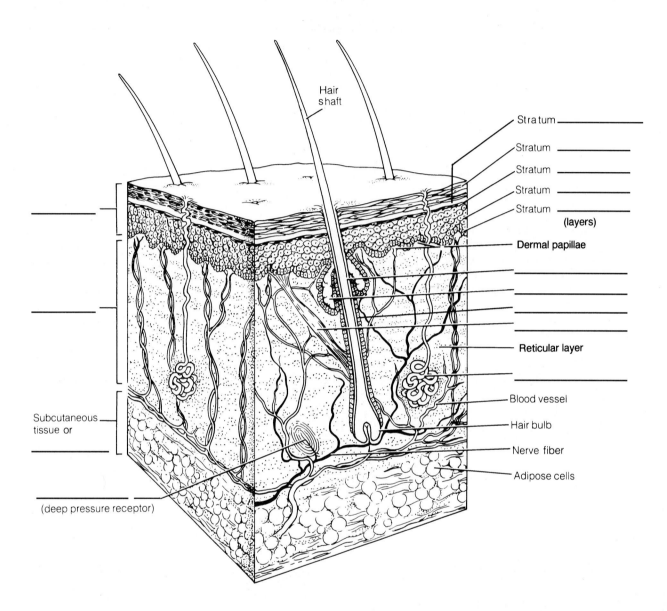

Hair
shaft

Stratum _____

Stratum _____

Stratum _____

Stratum _____

Stratum _____
(layers)

Dermal papillae

_____

_____

_____

Reticular layer

_____

Blood vessel

Hair bulb

Nerve fiber

Adipose cells

Subcutaneous
tissue or
_____

(deep pressure receptor)

1. _____ granules extruded from the keratinocytes prevent water loss by diffusion

   through the epidermis.

2. The origin of the fibers in the dermis is _____.

3. Glands that respond to rising androgen levels are the _____ glands.

4. Phagocytic cells that occupy the epidermis are called _____.

5. A unique touch receptor formed from a stratum basale cell and a nerve fiber is a

   _____.

5. What substance is manufactured in the skin (but is not a secretion) to play a role elsewhere in the body?

_____

6. What sensory receptors are found in the skin? _____

_____

_____

7. A nurse tells a doctor that a patient is cyanotic. What is cyanosis? _____

    What does its presence imply? _____

8. What is the mechanism of a suntan? _____

_____

9. What is a decubitus ulcer? _____

    Why does it occur? _____

_____

10. Some injections hurt more than others. On the basis of what you have learned about skin structure, can you

    determine why this is so? _____

_____

Appendages of the Skin

1. Using key choices, respond to the following descriptions.

    Key: a. arrector pili   d. hair follicle   g. sweat gland—apocrine
          b. cutaneous receptors   e. nail   h. sweat gland—eccrine
          c. hair   f. sebaceous glands

    _____ 1. A blackhead is an accumulation of oily material that is produced by

                                        _____.

    _____ 2. Tiny muscles, attached to hair follicles, that pull the hair upright during
                                    fright or cold.

    _____ 3. More numerous variety of perspiration gland.

    _____ 4. Sheath formed of both epithelial and connective tissues.

    _____ 5. Less numerous type of perspiration-producing gland; found mainly in
                                    the pubic and axillary regions.

    _____ 6. Found everywhere on body except palms of hands and soles of feet.

    _____ 7. Primarily dead/keratinized cells.

    _____ 8. Specialized nerve endings that respond to temperature, touch, etc.

    _____ 9. Its secretion contains cell fragments.

    _____ 10. "Sports" a lunula and a cuticle.

2. How does the skin help in regulating body temperature? (Describe two different mechanisms.) _____

_____

_____

_____

_____

Plotting the Distribution of Sweat Glands

1. With what substance in the bond paper does the iodine painted on the skin react? _____

2. Which skin area—the forearm or palm of hand—has more sweat glands? _____

   Which other body areas would, if tested, prove to have a high density of sweat glands? _____

   _____

3. What organ system controls the activity of the eccrine sweat glands? _____

STUDENT NAME ____________________

LAB TIME/DATE ____________________

Review Sheet

EXERCISE 7

Classification of Body Membranes

Classification of Body Membranes

1. Complete the following chart:

Membrane	Tissue type (epithelial/connective)	Common locations	Functions
cutaneous			
mucous			
serous			
synovial			

2. Respond to the following statements by choosing an answer from the key.

Key: a. cutaneous b. mucous c. serous d. synovial

__________ 1. membrane type associated with skeletal system structures

__________ 2. always formed of simple squamous epithelium

__________, __________ 3. membrane types *not* found in the ventral body cavity

__________ 4. the only membrane type in which goblet cells are found

__________ 5. the only *external* membrane

__________ 6. "wet" membranes

__________ 7. adapted for absorption and secretion

__________ 8. has parietal and visceral layers

STUDENT NAME ____________________

LAB TIME/DATE ____________________

Review Sheet

EXERCISE 8 Overview of the Skeleton: Classification and Structure of Bones and Cartilages

Bone Markings

1. Match the terms in column B with the appropriate description in column A:

	Column A	Column B
________	1. sharp, slender process*	a. condyle
________	2. small rounded projection*	b. crest
________	3. narrow ridge of bone*	c. epicondyle
________	4. large rounded projection*	d. fissure
________	5. structure supported on neck†	e. foramen
________	6. armlike projection†	f. fossa
________	7. rounded, convex projection†	g. head
________	8. narrow depression or opening‡	h. meatus
________	9. canal-like structure‡	i. ramus
________	10. opening through a bone‡	j. sinus
________	11. shallow depression†	k. spine
________	12. air-filled cavity	l. trochanter
________	13. large, irregularly shaped projection*	m. tubercle
________	14. raised area of a condyle*	n. tuberosity

* A site of muscle attachment.

† Takes part in joint formation.

‡ A passageway for nerves or blood vessels.

Classification of Bones

1. The four major anatomical classifications of bones are long, short, flat, and irregular. Which category has the least amount of spongy bone relative to its total volume? ____________________

2. Classify each of the bones in the next chart into one of the four major categories by checking the appropriate column. Use appropriate references as necessary.

	Long	Short	Flat	Irregular
humerus				
phalanx				
parietal				
calcaneus				
rib				
vertebra				
ulna				

Gross Anatomy of the Typical Long Bone

1. Use the terms below to identify the structures marked by leader lines and braces in the diagrams (some terms are used more than once). After labeling the diagrams, use the listed terms to characterize the statements following the diagrams.

Key:
a. articular cartilage
b. compact bone
c. diaphysis
d. endosteum

e. epiphyseal line
f. epiphysis
g. medullary cavity
h. periosteum

i. red marrow cavity
j. trabeculae of spongy bone
k. yellow marrow

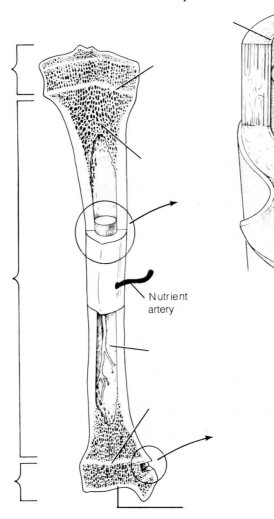

Nutrient artery

_____ 1. contains spongy bone in adults

_____ 2. made of compact bone

_____ 3. site of blood cell formation

_____ 4. major submembranous site of osteoclasts

_____ 5. scientific term for bone shaft

_____ 6. contains fat in adult bones

_____ 7. growth plate remnant

_____ 8. major submembranous site of osteoblasts

2. What differences between compact and spongy bone can be seen with the naked eye? _____

_____

3. What is the function of the periosteum? _____

_____

Microscopic Structure of Compact Bone

1. Trace the route taken by nutrients through a bone, starting with the periosteum and ending with an osteocyte in a

   lacuna. Periosteum __→_____ →_____

   _____ →_____ →__ osteocyte

2. Several descriptions of bone structure are given in column B. Identify the structure involved by choosing the appropriate term from column A and placing its letter in the blank.

Column A

a. canaliculi

b. central canal

c. concentric lamellae

d. lacunae

Column B

_____ 1. layers of bony matrix around a central canal

_____ 2. site of osteocytes

_____ 3. longitudinal canal carrying blood vessels, lymphatics, and nerves

_____ 4. minute canals connecting osteocytes of an osteon

3. On the photomicrograph of bone below (208×), identify all structures named in column A in question 2 above.

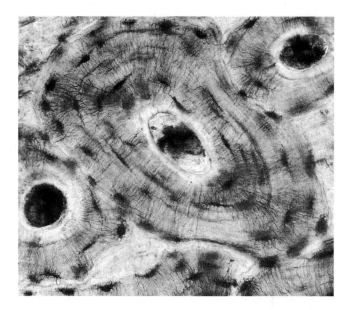

Chemical Composition of Bone

1. What is the function of the organic matrix in bone?_____

2. Name the important organic bone components._____

3. Calcium salts form the bulk of the inorganic material in bone. What is the function of the calcium salts?

   _____

4. Which is responsible for bone structure? (circle the appropriate response)

   inorganic portion   organic portion   both contribute

Cartilages of the Skeleton

Using key choices, identify each type of cartilage described (in terms of its body location or function) below.

Key: a. elastic   b. fibrocartilage   c. hyaline

_____ 1. supports the external ear

_____ 2. between the vertebrae

_____ 3. forms the walls of the voice box (larynx)

_____ 4. the epiglottis

_____ 5. articular cartilages

_____ 6. meniscus in a knee joint

_____ 7. connects the ribs to the sternum

_____ 8. most effective at resisting compression

_____ 9. most springy and flexible

_____ 10. most abundant

Review Sheet

EXERCISE 9

The Axial Skeleton

The Skull

1. The skull is one of the major components of the axial skeleton. Name the other two:

   _____ and _____

   What structures do each of these areas protect? _____

   _____

   _____

2. Define *suture:* _____

3. With one exception, the skull bones are joined by sutures. Name the exception. _____

   _____

4. What are the four major sutures of the skull, and what bones do they connect? _____

   _____

   _____

   _____

   _____

5. Name the eight bones composing the cranium.

   _____ _____ _____ _____

   _____ _____ _____ _____

6. Give two possible functions of the sinuses. _____

   _____

7. What is the orbit? _____

   What bones contribute to the formation of the orbit? _____

   _____

8. Why can the sphenoid bone be called the keystone of the cranial floor? _____

   _____

   _____

9. What is a cleft palate? _____

_____

10. Match the bone names in column B with the descriptions in column A.

Column A		Column B
_____ 1. forehead bone		a. ethmoid
_____ 2. cheekbone		b. frontal
_____ 3. lower jaw		c. hyoid
_____ 4. bridge of nose		d. lacrimal
_____ 5. posterior bones of the hard palate		e. mandible
_____ 6. much of the lateral and superior cranium		f. maxilla
_____ 7. most posterior part of cranium		g. nasal
_____ 8. single, irregular, bat-shaped bone forming part of the cranial floor		h. occipital
_____ 9. tiny bones bearing tear ducts		i. palatine
_____ 10. anterior part of hard palate		j. parietal
_____ 11. superior and medial nasal conchae formed from its projections		k. sphenoid
_____ 12. site of mastoid process		l. temporal
_____ 13. site of sella turcica		m. vomer
_____ 14. site of cribriform plate		n. zygomatic
_____ 15. site of mental foramen		
_____ 16. site of styloid processes		

_____, _____, _____, and

_____ 17. four bones containing paranasal sinuses

_____ 18. condyles here articulate with the atlas

_____ 19. foramen magnum contained here

_____ 20. small U-shaped bone in neck, where many tongue muscles attach

_____ 21. middle ear found here

_____ 22. nasal septum

_____ 23. bears an upward protrusion, the "cock's comb," or crista galli

_____, _____ 24. contain alveoli bearing teeth

11. Using choices from column B in question 10 and from the numbered key to the right, identify all bones and bone markings provided with leader lines in the two diagrams below.

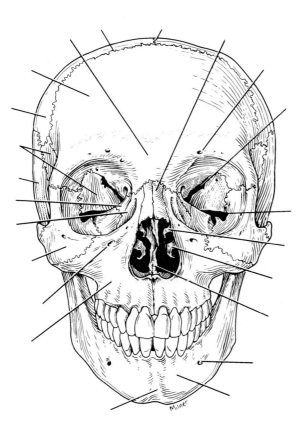

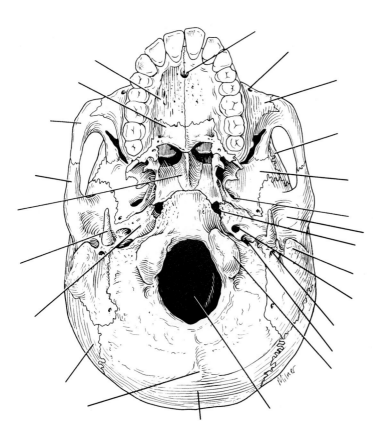

1. carotid canal

2. coronal suture

3. external occipital protuberance

4. foramen lacerum

5. foramen magnum

6. foramen ovale

7. glabella

8. incisive foramen

9. inferior nasal concha

10. inferior orbital fissure

11. infraorbital foramen

12. jugular foramen

13. mandibular fossa

14. mandibular symphysis

15. mastoid process

16. mental foramen

17. middle nasal concha of ethmoid

18. occipital condyle

19. palatine process of maxilla

20. sagittal suture

21. styloid process

22. stylomastoid foramen

23. superior orbital fissure

24. supraorbital foramen

25. zygomatic process of temporal bone

The Vertebral Column

1. Using the key, correctly identify the vertebral parts/areas described below. (More than one choice may apply in some cases.) Also use the key letters to correctly identify the vertebral areas in the diagram.

Key: a. body
      b. intervertebral foramina
      c. lamina

     d. pedicle
     e. spinous process
     f. superior articular process

     g. transverse process
     h. vertebral arch
     i. vertebral foramen

_____ 1. cavity enclosing the nerve cord

_____ 2. weight-bearing portion of the vertebra

_____ 3. provide levers against which muscles pull

_____ 4. provide an articulation point for the ribs

_____ 5. openings providing for exit of spinal nerves

_____ 6. structures that form an enclosure for the spinal cord

2. The distinguishing characteristics of the vertebrae composing the vertebral column are noted below. Correctly identify each described structure/region by choosing a response from the key.

Key: a. atlas
      b. axis
      c. cervical vertebra—typical

     d. coccyx
     e. lumbar vertebra

     f. sacrum
     g. thoracic vertebra

_____ 1. vertebral type containing foramina in the transverse processes, through which the vertebral arteries ascend to reach the brain

_____ 2. dens here provides a pivot for rotation of the first cervical vertebra (C_1)

_____ 3. transverse processes faceted for articulation with ribs; spinous process pointing sharply downward

_____ 4. composite bone; articulates with the hip bone laterally

_____ 5. massive vertebrae; weight-sustaining

_____ 6. "tail bone"; vestigial fused vertebrae

_____ 7. supports the head; allows a rocking motion in conjunction with the occipital condyles

_____ 8. seven components; unfused

_____ 9. twelve components; unfused

3. Name two factors/structures that allow for flexibility of the vertebral column.

_____ and _____

4. Describe how a spinal nerve exits from the vertebral column. _____ _____

_____

5. Which two spinal curvatures are obvious at birth? _____ and _____

Under what conditions do the secondary curvatures develop?_____

_____

_____

6. Diagram the normal spinal curvatures and the abnormal spinal curvatures named below. (Use posterior or lateral views as necessary.)

Normal curvature   Lordosis   Scoliosis   Kyphosis

7. What kind of tissue comprises the intervertebral discs?_____

8. What is a herniated disc?_____

What problems might it cause?_____

_____

The Bony Thorax

1. The major components of the thorax (excluding the vertebral column) are the_____

and the _____ .

2. Differentiate between a true rib and a false rib. _____

_____

_____

Is a floating rib a true or a false rib? _____

3. What is the general shape of the thoracic cage?_____

4. Using the terms at the right, identify the regions and landmarks of the bony thorax.

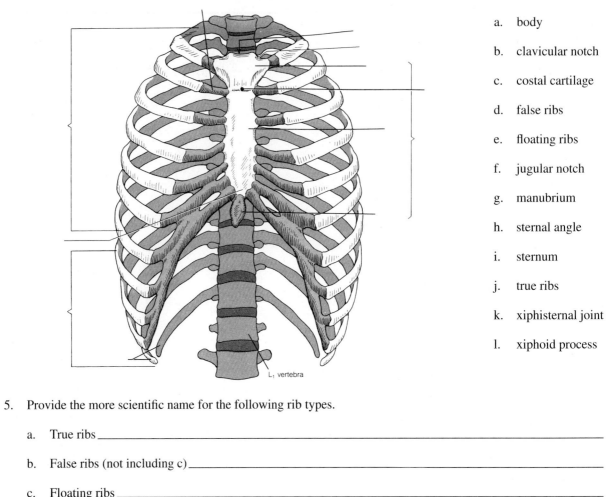

L₁ vertebra

a. body

b. clavicular notch

c. costal cartilage

d. false ribs

e. floating ribs

f. jugular notch

g. manubrium

h. sternal angle

i. sternum

j. true ribs

k. xiphisternal joint

l. xiphoid process

5. Provide the more scientific name for the following rib types.

a. True ribs _____

b. False ribs (not including c) _____

c. Floating ribs _____

The Ap

Bones of the Shoulder Girdle and Upper Limb

1. Match the bone names or markings in column B with the descriptions in column A.

Column A

_____ 1. raised area on lateral surface of humerus to which deltoid muscle attaches

_____ 2. arm bone

_____ , _____ 3. bones of the shoulder girdle

_____ , _____ 4. forearm bones

_____ 5. scapular region to which the clavicle connects

_____ 6. shoulder girdle bone that is unattached to the axial skeleton

_____ 7. shoulder girdle bone that transmits forces from the upper limb to the bony thorax

_____ 8. depression in the scapula that articulates with the humerus

_____ 9. process above the glenoid cavity that permits muscle attachment

_____ 10. the "collarbone"

_____ 11. distal condyle of the humerus that articulates with the ulna

_____ 12. medial bone of forearm in anatomical position

_____ 13. rounded knob on the humerus; adjoins the radius

_____ 14. anterior depression, superior to the trochlea, which receives part of the ulna when the forearm is flexed

_____ 15. forearm bone involved in formation of the elbow joint

_____ 16. wrist bones

_____ 17. finger bones

_____ 18. heads of these bones form the knuckles

_____ , _____ 19. bones that articulate with the clavicle

Column B

a. acromion

b. capitulum

c. carpals

d. clavicle

e. coracoid process

f. coronoid fossa

g. deltoid tuberosity

h. glenoid cavity

i. humerus

j. metacarpals

k. olecranon fossa

l. olecranon process

m. phalanges

n. radial tuberosity

o. radius

p. scapula

q. sternum

r. styloid process

s. trochlea

t. ulna

the clavicle at risk to fracture when a person falls on his or her shoulder? _____

_____

Why is there generally no problem in the arm clearing the widest dimension of the thoracic cage?

_____

_____

4.   What is the total number of phalanges in the hand? _____

5.   What is the total number of carpals in the wrist? _____

     In the proximal row, the carpals are (medial to lateral) _____

     In the distal row, they are (medial to lateral) _____

6.   Using items from the list at the right, identify the anatomical landmarks and regions of the scapula.

a.   acromion

b.   coracoid process

c.   glenoid cavity

d.   inferior angle

e.   infraspinous fossa

f.   lateral border

g.   medial border

h.   spine

i.   superior angle

j.   superior border

k.   suprascapular notch

l.   supraspinous fossa

Bones of the Pelvic Girdle and Lower Limb

1. Compare the pectoral and pelvic girdles by choosing appropriate descriptive terms from the key.

   Key: a. flexibility most important
        b. massive
        c. lightweight
        d. insecure axial and limb attachments
        e. secure axial and limb attachments
        f. weightbearing most important

   Pectoral: _____, _____, _____   Pelvic: _____, _____, _____

2. What organs are protected, at least in part, by the pelvic girdle? _____

   _____

3. Distinguish between the true pelvis and the false pelvis. _____

   _____

   _____

4. Name five differences between the male and female pelves. _____

   _____

   _____

   _____

   _____

5. Deduce why the pelvic bones of a four-legged animal such as the cat or pig are much less massive than those of

   the human. _____

   _____

6. A person instinctively curls over the abdominal area in times of danger. Why? _____

   _____

7. For what anatomical reason do many women appear to be slightly knock-kneed? _____

   _____

   _____

   _____

8. What does *fallen arches* mean? _____

   _____

9. Match the bone names and markings in column B with the descriptions in column A.

Column A

_____, _____, and

_____ 1. fuse to form the coxal bone

_____ 2. inferoposterior "bone" of the coxal bone

_____ 3. point where the coxal bones join anteriorly

_____ 4. superiormost margin of the coxal bone

_____ 5. deep socket in the coxal bone that receives the head of the thigh bone

_____ 6. joint between the axial skeleton and the pelvic girdle

_____ 7. longest, strongest bone in body

_____ 8. thin lateral leg bone

_____ 9. heavy medial leg bone

_____, _____ 10. bones forming the knee joint

_____ 11. point where the patellar ligament attaches

_____ 12. kneecap

_____ 13. shin bone

_____ 14. medial ankle projection

_____ 15. lateral ankle projection

_____ 16. largest tarsal bone

_____ 17. ankle bones

_____ 18. bones forming the instep of the foot

_____ 19. opening in hip bone formed by the pubic and ischial rami

_____ and _____ 20. sites of muscle attachment on the proximal femur

_____ 21. tarsal bone that "sits" on the calcaneus

Column B

a. acetabulum

b. calcaneus

c. femur

d. fibula

e. gluteal tuberosity

f. greater sciatic notch

g. greater and lesser trochanters

h. iliac crest

i. ilium

j. ischial tuberosity

k. ischium

l. lateral malleolus

m. lesser sciatic notch

n. linea aspera

o. medial malleolus

p. obturator foramen

q. metatarsals

r. patella

s. pubic symphysis

t. pubis

u. sacroiliac joint

v. talus

w. tarsals

x. tibia

y. tibial tuberosity

Summary of Skeleton

1. Identify all indicated bones (or groups of bones) in the diagram of the articulated skeleton.

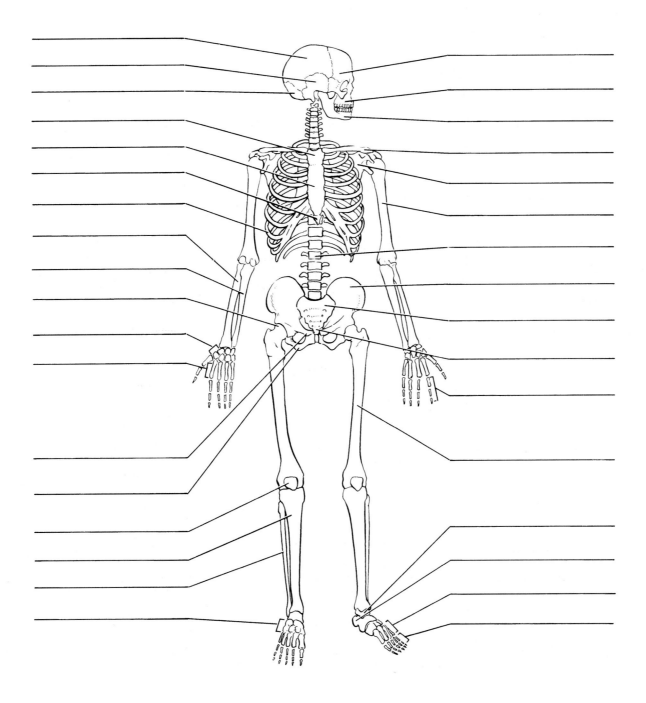

2. Three bones of the appendicular skeleton are diagrammed below. Using the choices from below, correctly identify all bone markings provided with leader lines.

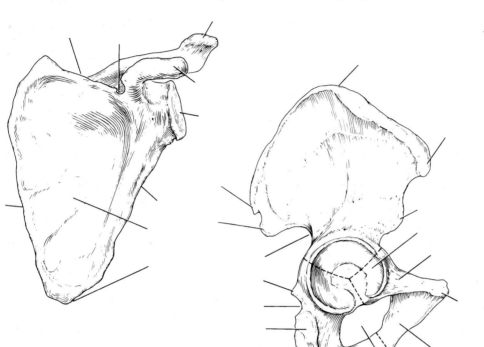
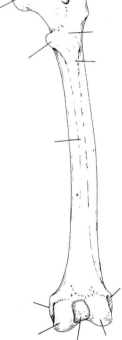

Choices for girdle bones

a. acetabulum

b. acromion

c. anterior inferior iliac spine

d. anterior superior iliac spine

e. coracoid process

f. glenoid cavity

g. greater sciatic notch

h. iliac crest

i. inferior angle of scapula

j. ischial ramus

k. ischial spine

l. ischial tuberosity

m. lateral border of scapula

n. lesser sciatic notch

o. medial border of scapula

p. obturator foramen

q. posterior inferior iliac spine

r. posterior superior iliac spine

s. pubic tubercle

t. pubis—inferior ramus

u. pubis—superior ramus

v. subscapular fossa

w. superior border of scapula

x. suprascapular notch

Choices for limb bone

1. fovea capitis

2. gluteal tuberosity

3. greater trochanter

4. head

5. intercondylar fossa

6. intertrochanteric crest

7. lateral condyle

8. lateral epicondyle

9. lesser trochanter

10. linea aspera

11. medial condyle

12. medial epicondyle

Review Sheet

EXERCISE 11

The Fetal Skeleton

1. Are the same skull bones seen in the adult found in the fetal skull? _____

2. How does the size of the fetal face compare to its cranium? _____

   _____

   How does this compare to the adult skull? _____

   _____

   _____

3. What are the outward conical projections in some of the fetal cranial bones? _____

   _____

4. What is a fontanel? _____

   What is its fate? _____

   What is the function of the fontanels in the fetal skull? _____

   _____

5. Describe how the fetal skeleton compares with the adult skeleton in the following areas:

   vertebrae _____

   _____

   os coxae _____

   _____

   carpals and tarsals _____

   sternum _____

   frontal bone _____

   patella _____

   rib cage _____

6. How does the size of the fetus's head compare to the size of its body? _____

   _____

7. Using the terms listed, identify each of the fontanels shown on the fetal skull below.

a. anterior fontanel   b. mastoid fontanel   c. posterior fontanel   d. sphenoidal fontanel

Parietal bone

Frontal bone

Occipital bone

Temporal bone

Review Sheet

EXERCISE 12	Articulations and Body Movements

Types of Joints

1. Use key responses to identify the joint types described below.

Key: a. cartilaginous   b. fibrous   c. synovial

_____ 1. typically allows a slight degree of movement

_____ 2. includes the pubic symphysis and joints between the vertebral bodies

_____ 3. essentially immovable joints

_____ 4. sutures are the most remembered examples

_____ 5. characterized by cartilage connecting the bony portions

_____ 6. all have a fibrous capsule lined with a synovial membrane surrounding a joint cavity

_____ 7. all are freely movable or diarthrotic

_____ 8. bone regions are united by fibrous connective tissue

_____ 9. include the hip, knee, and elbow joints

2. Match the joint subcategories in column B with their descriptions in column A, and place an asterisk (*) beside all choices that are examples of synovial joints.

Column A		Column B	
_____ 1. joint between skull bones		a.	ball and socket
_____ 2. joint between the axis and atlas		b.	condyloid
_____ 3. hip joint		c.	gliding
_____ 4. intervertebral joints (between articular processes)		d.	hinge
_____ 5. joint between forearm bones and wrist		e.	pivot
_____ 6. elbow		f.	saddle
_____ 7. interphalangeal joints		g.	suture
_____ 8. intercarpal joints		h.	symphysis
_____ 9. joint between tarsus and tibia/fibula		i.	synchondrosis
_____ 10. joint between skull and vertebral column		j.	syndesmosis
_____ 11. joint between jaw and skull			
_____ 12. joints between proximal phalanges and metacarpal bones			
_____ 13. epiphyseal plate of a child's long bone			
_____ 14. a multiaxial joint			
_____ , _____ 15. biaxial joints			
_____ , _____ 16. uniaxial joints			

3. What characteristics do all joints have in common? _____

_____

4. Describe the structure and function of the following structures or tissues in relation to a synovial joint and label the structures indicated by leader lines in the diagram.

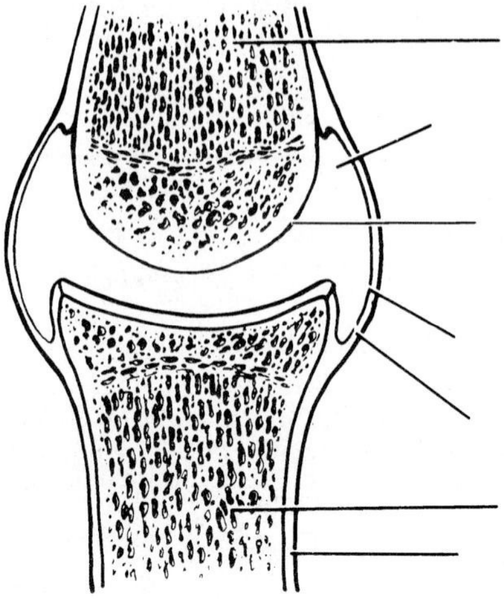

ligament _____

_____

tendon _____

_____

hyaline cartilage _____

_____

synovial membrane _____

_____

_____

bursa _____

5. Which joint, the hip or the knee, is more stable? _____

Name two important factors that contribute to the stability of the hip joint.

_____ and _____

Name two important factors that contribute to the stability of the knee.

_____ and _____

Body Movements

1. Complete the following statements:

The movable attachment of a muscle is called its __1__, and its stationary attachment is called the __2__. Winding up for a pitch (as in baseball) can properly be called __3__. To keep your seat when riding a horse, the tendency is to __4__ your thighs. In running, the action at the hip joint is __5__ in reference to the leg moving forward and __6__ in reference to the leg in the posterior position. In kicking a football, the action at the knee is __7__. In climbing stairs, the hip and knee of the forward leg are both __8__. You have just touched your chin to your chest. This is __9__ of the neck. Using a screwdriver with your arm straight requires __10__ of the arm. Consider all the movements of which the arm is capable. One often used for strengthening all the upper arm and shoulder muscles is __11__. Movement of the head that signifies "no" is __12__. Standing on your toes, as in ballet, requires __13__ of the foot. Action that moves the distal end of the radius across the ulna is __14__. Raising the arms laterally away from the body is called __15__ of the arms. Walking on one's heels is __16__.

1. _____

2. _____

3. _____

4. _____

5. _____

6. _____

7. _____

8. _____

9. _____

10. _____

11. _____

12. _____

13. _____

14. _____

15. _____

16. _____

Joint Disorders

1. What structural joint changes are common to the elderly? _____

_____

_____

2. Define:

sprain _____

_____

dislocation _____

_____

STUDENT NAME ______________________

LAB TIME/DATE ______________________

Review Sheet

EXERCISE 13 Microscopic Anatomy, Organization, and Classification of Skeletal Muscle

Skeletal Muscle Cells and Their Packaging into Muscles

1. What capability is most highly expressed in muscle tissue? ______________________

2. Use the items on the right to correctly identify the structures described on the left.

________ 1.	connective tissue ensheathing a bundle of muscle cells	a. endomysium
________ 2.	bundle of muscle cells	b. epimysium
________ 3.	contractile unit of muscle	c. fascicle
________ 4.	a muscle cell	d. fiber
________ 5.	thin reticular connective tissue investing each muscle cell	e. myofilament
________ 6.	plasma membrane of the muscle fiber	f. myofibril
________ 7.	a long filamentous organelle with a banded appearance found within muscle cells	g. perimysium
		h. sarcolemma
________ 8.	actin- or myosin-containing structure	i. sarcomere
________ 9.	cord of collagen fibers that attaches a muscle to a bone	j. sarcoplasm
		k. tendon

3. Why are the connective tissue wrappings of skeletal muscle important? (Give at least three reasons.)

____

____

____

____

4. Why are indirect—that is, tendinous—muscle attachments to bone seen more often than direct attachments?

____

____

____

5. How does an aponeurosis differ from a tendon? _____

_____

_____

6. The diagram illustrates a small portion of a muscle myofibril. Using letters from the key, correctly identify each structure indicated by a leader line or a bracket. Below the diagram make a sketch of how this segment of the myofibril would look if contracted.

Key: a. actin filament   d. myosin filament
      b. A band   e. sarcomere
      c. I band   f. Z line

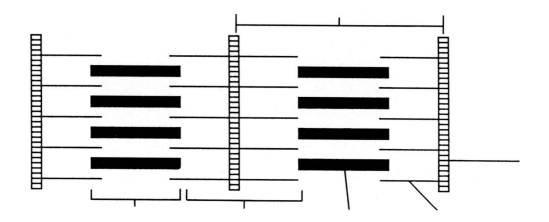

The Neuromuscular Junction

Complete the following statements:

The junction between a motor neuron's axon and the muscle cell membrane is called a neuromuscular junction or a __1__ junction. A motor neuron and all of the skeletal muscle cells it stimulates is called a __2__ . The axonal terminals of each motor axon have numerous projections called __3__ . The actual gap between the axonal terminal and the muscle cell is called a __4__ . Within the axonal terminal are many small vesicles containing a neurotransmitter substance called __5__ . When the __6__ reaches the ends of the axon, the neurotransmitter is released and diffuses to the muscle cell membrane to combine with receptors there. The combining of the neurotransmitter with the muscle membrane receptors causes the membrane to become permeable to sodium, which results in the influx of sodium ions and __7__ of the membrane. Then contraction of the muscle cell occurs. Before a muscle cell can be stimulated to contract again, __8__ must occur.

1. _____

2. _____

3. _____

4. _____

5. _____

6. _____

7. _____

8. _____

Classification of Skeletal Muscles

1. Several criteria were given relative to the naming of muscles. Match the criteria (column B) to the muscle cell names (column A). Note that more than one criterion may apply in some cases.

Column A	Column B
_____ 1. gluteus maximus	a. action of the muscle
_____ 2. adductor magnus	b. shape of the muscle
_____ 3. biceps femoris	c. location of the origin and/or insertion of the muscle
_____ 4. abdominis transversus	d. number of origins
_____ 5. extensor carpi ulnaris	e. location of muscle relative to a bone or body region
_____ 6. trapezius	f. direction in which the muscle fibers run relative to some imaginary line
_____ 7. rectus femoris	g. relative size of the muscle
_____ 8. external oblique	

2. When muscles are discussed relative to the manner in which they interact with other muscles, the terms shown in the key are often used. Match the key terms with the appropriate definitions.

Key: a. antagonist   b. fixator   c. prime mover   d. synergist

_____ 1. agonist

_____ 2. postural muscles, for the most part

_____ 3. reverses and/or opposes the action of a prime mover

_____ 4. stabilizes a joint so that the prime mover may act at more distal joints

_____ 5. performs the same movement as the prime mover

_____ 6. immobilizes the origin of a prime mover

Review Sheet

EXERCISE 14

Gross Anatomy of the Muscular System

Muscles of the Head and Neck

1. Using choices from the list at the right, correctly identify muscles provided with leader lines on the diagram.

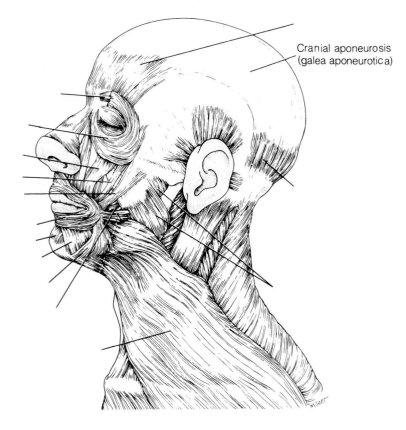

Cranial aponeurosis
(galea aponeurotica)

a.   buccinator

b.   corrugator supercilii

c.   depressor anguli oris

d.   depressor labii inferioris

e.   epicranius frontalis

f.   epicranius occipitalis

g.   levator labii superioris

h.   masseter

i.   mentalis

j.   platysma

k.   orbicularis oculi

l.   orbicularis oris

m.   zygomaticus

2. Using the terms provided above, identify the muscles described next.

_____ 1.   used in smiling

_____ 2.   used to suck in your cheeks

_____ 3.   used in blinking and squinting

_____ 4.   used to pout (pulls the corners of the mouth downward)

_____ 5.   raises your eyebrows for a questioning expression

_____ 6.   used to form the vertical frown crease on the forehead

_____ 7.   your "kisser"

_____ 8.   prime mover to raise the lower jawbone

_____ 9.   tenses skin of the neck during shaving

Muscles of the Trunk

1. Correctly identify both intact and transected (cut) muscles depicted in the diagram, using the terms given at the right. (Not all terms will be used in this identification.)

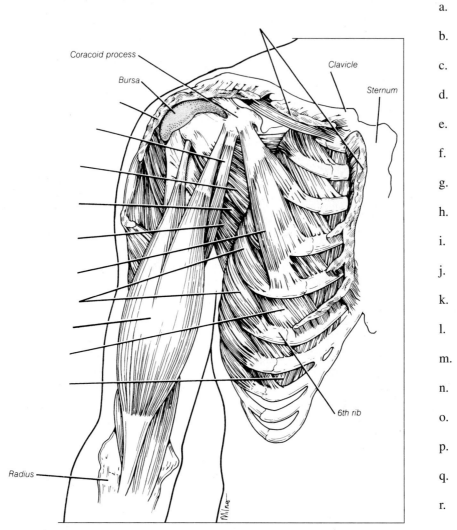

a. biceps brachii

b. coracobrachialis

c. deltoid (cut)

d. external intercostals

e. external oblique

f. internal intercostals

g. internal oblique

h. latissimus dorsi

i. pectoralis major (cut)

j. pectoralis minor

k. rectus abdominis

l. rhomboids

m. serratus anterior

n. subscapularis

o. teres major

p. teres minor

q. transversus abdominis

r. trapezius

2. Using choices from the terms provided in question 1 above, identify the major muscles described next:

_____ 1. a major spine flexor

_____ 2. prime mover for pulling the arm posteriorly

_____ 3. prime mover for shoulder flexion

_____ 4. assume major responsibility for forming the abdominal girdle (three pairs of muscles)

_____ 5. pulls the shoulder backward and downward

_____ 6. prime mover of shoulder abduction

_____ 7. important in shoulder adduction; antagonists of the shoulder abductor (two muscles)

_____ 8. moves the scapula forward and downward

_____ 9. small, inspiratory muscles between the ribs; elevate the ribs

_____ 10. extends the head

_____ 11. pull the scapulae medially

RS56

Muscles of the Upper Limb

1. Using terms from the list on the right, correctly identify all muscles provided with leader lines in the diagram. Note that not all the listed terms will be used in this exercise.

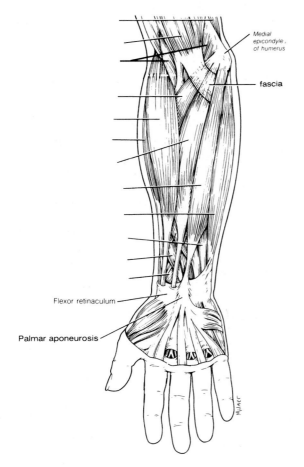

a. biceps brachii

b. brachialis

c. brachioradialis

d. extensor carpi radialis longus

e. extensor digitorum

f. flexor carpi radialis

g. flexor carpi ulnaris

h. flexor digitorum superficialis

i. flexor pollicis longus

j. palmaris longus

k. pronator quadratus

l. pronator teres

m. supinator

n. triceps brachii

2. Use the terms provided in question 1 to identify the muscles described next.

_____ 1. places the palm upward (two muscles)

_____ 2. flexes the forearm and supinates the hand

_____ 3. forearm flexors; no role in supination (two muscles)

_____ 4. elbow extensor

_____ 5. power wrist flexor and abductor

_____ 6. flexes wrist and distal phalanges

_____ 7. pronate the hand (two muscles)

_____ 8. flexes the thumb

_____ 9. extends and abducts the wrist

_____ 10. extends the wrist and digits

_____ 11. flat muscle that is a weak wrist flexor

Muscles of the Lower Limb

1. Using the terms listed to the right, correctly identify all muscles provided with leader lines in the diagram below. Not all listed terms will be used in this exercise.

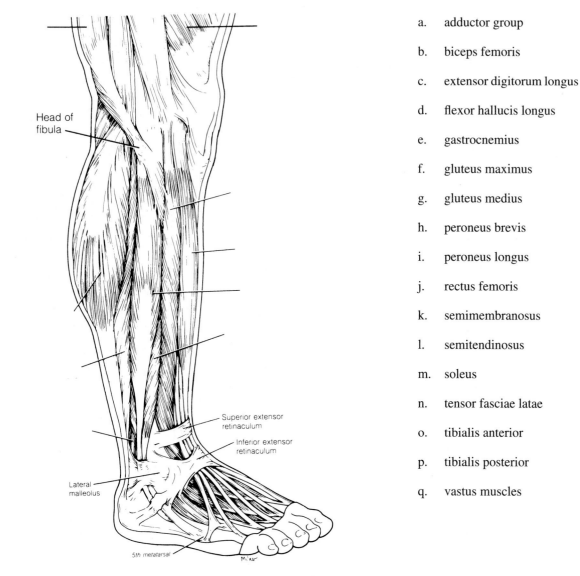

a. adductor group

b. biceps femoris

c. extensor digitorum longus

d. flexor hallucis longus

e. gastrocnemius

f. gluteus maximus

g. gluteus medius

h. peroneus brevis

i. peroneus longus

j. rectus femoris

k. semimembranosus

l. semitendinosus

m. soleus

n. tensor fasciae latae

o. tibialis anterior

p. tibialis posterior

q. vastus muscles

2. Use the key terms in exercise 1 to respond to the descriptions below.

_____ 1. flexes the great toe and inverts the ankle

_____ 2. lateral compartment muscles that plantar flex and evert the ankle (two muscles)

_____ 3. move the thigh laterally to take the "at ease" stance (two muscles)

_____ 4. used to extend the hip when climbing stairs

_____ 5. prime movers of ankle plantar flexion (two muscles)

_____ 6. major foot inverter

_____ 7. prime mover of ankle dorsiflexion

_____ 8. allow you to draw your legs to the midline of your body, as when standing at attention

_____ 9. extends the toes

_____ 10. extend thigh and flex knee (three muscles)

_____ 11. extends knee and flexes thigh

Muscle Recognition: General Review

1. Identify the lettered muscles in the diagram of the human anterior superficial musculature by matching the letter with one of the following muscle names:

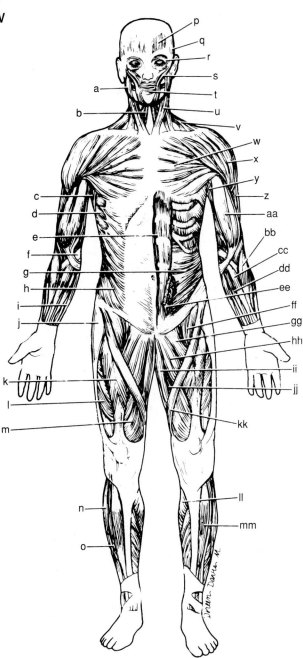

_____ 1. orbicularis oris

_____ 2. pectoralis major

_____ 3. external oblique

_____ 4. sternocleidomastoid

_____ 5. biceps brachii

_____ 6. deltoid

_____ 7. vastus lateralis

_____ 8. brachioradialis

_____ 9. frontalis

_____ 10. rectus femoris

_____ 11. pronator teres

_____ 12. rectus abdominis

_____ 13. sartorius

_____ 14. gracilis

_____ 15. flexor carpi ulnaris

_____ 16. adductor longus

_____ 17. palmaris longus

_____ 18. flexor carpi radialis

_____ 19. latissimus dorsi

_____ 20. orbicularis oculi

_____ 21. gastrocnemius

_____ 22. masseter

_____ 23. trapezius

_____ 24. tibialis anterior

_____ 25. extensor digitorum longus

_____ 26. tensor fasciae latae

_____ 27. pectineus

_____ 28. sternohyoid

_____ 29. serratus anterior

_____ 30. adductor magnus

_____ 31. vastus medialis

_____ 32. transversus abdominis

_____ 33. peroneus longus

_____ 34. iliopsoas

_____ 35. temporalis

_____ 36. zygomaticus

_____ 37. coracobrachialis

_____ 38. triceps brachii

_____ 39. internal oblique

2. Identify each of the lettered muscles in this diagram of the human posterior superficial musculature by matching its letter to one of the following muscle names:

_____ 1.   gluteus maximus

_____ 2.   semimembranosus

_____ 3.   gastrocnemius

_____ 4.   latissimus dorsi

_____ 5.   deltoid

_____ 6.   iliotibial tract (tendon)

_____ 7.   teres major

_____ 8.   semitendinosus

_____ 9.   trapezius

_____ 10. biceps femoris

_____ 11. triceps brachii

_____ 12. external oblique

_____ 13. gluteus medius

_____ 14. gracilis

_____ 15. flexor carpi ulnaris

_____ 16. extensor carpi ulnaris

_____ 17. extensor digitorum communis

_____ 18. extensor carpi radialis longus

_____ 19. occipitalis

_____ 20. extensor carpi radialis brevis

_____ 21. sternocleidomastoid

_____ 22. adductor magnus

_____ 23. anconeus

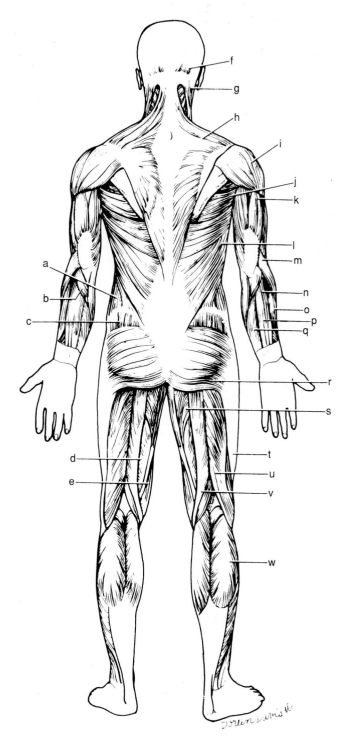

Review Sheet

EXERCISE 15

Histology of Nervous Tissue

1. The cellular unit of the nervous system is the neuron. What is the major function of this cell type?

   _____

2. Name four types of neuroglia and list at least four functions of these cells. (You will need to consult your textbook for this.)

   Types

   _____

   _____

   _____

   _____

   Functions

   _____

   _____

   _____

   _____

   _____

   _____

   _____

3. Match each statement with a response chosen from the key.

   Key: a. afferent neuron   e. ganglion   i. nuclei
         b. association neuron   f. neuroglia   j. peripheral nervous system
         c. central nervous system   g. neurotransmitters   k. synapse
         d. efferent neuron   h. nerve   l. tract

   _____ 1. the brain and spinal cord collectively

   _____ 2. specialized supporting cells in the CNS

   _____ 3. junction or point of close contact between neurons

   _____ 4. a bundle of nerve processes inside the central nervous system

   _____ 5. neuron serving as part of the conduction pathway between sensory and motor neurons

   _____ 6. spinal and cranial nerves and ganglia

   _____ 7. collection of nerve cell bodies found outside the CNS

   _____ 8. neuron that conducts impulses away from the CNS to muscles and glands

   _____ 9. neuron that conducts impulses toward the CNS from the body periphery

   _____ 10. chemicals released by neurons that stimulate or inhibit other neurons or effectors

Neuron Anatomy

1. Match the following anatomical terms (column B) with the appropriate description or function (column A).

Column A	Column B
_____ 1. region of the cell body from which the axon originates	a. axon
_____ 2. secretes neurotransmitters	b. axonal terminal
_____ 3. receptive region of a neuron	c. axon hillock
_____ 4. insulates the nerve fibers	d. dendrite
_____ 5. is site of the nucleus and the most important metabolic area	e. myelin sheath
_____ 6. may be involved in the transport of substances within the neuron	f. neuronal cell body
_____ 7. essentially rough endoplasmic reticulum, important metabolically	g. neurofibril
_____ 8. impulse generator and transmitter	h. Nissl bodies

2. Draw a "typical" neuron in the space below. Include and label the following structures on your diagram: cell body, nucleus, Nissl bodies, dendrites, axon, axon collaterals, myelin sheath, and nodes of Ranvier.

3. How is one-way conduction at synapses assured? _____

_____

4. What anatomical characteristic determines whether a particular neuron is classified as unipolar, bipolar, or multi-

polar? _____

Make a simple line drawing of each type here.

Unipolar neuron   Bipolar neuron   Multipolar neuron

5. Describe how the Schwann cells form the myelin sheath and the neurilemma encasing the nerve processes. (You may want to diagram the process.) _____

_____

_____

_____

Structure of a Nerve

1. What is a nerve? _____

_____

2. State the location of each of the following connective tissue coverings:

   endoneurium _____

   perineurium _____

   epineurium _____

3. What is the value of the connective tissue wrappings found in a nerve? _____

_____

4. Define *mixed nerve:* _____

_____

5. Identify all indicated parts of the nerve section.

Review Sheet

EXERCISE 16   Gross Anatomy of the Brain and Cranial Nerves

The Human Brain

1. Match the letters on the diagram of the human brain (right lateral view) to the appropriate terms listed at the left:

    _____ 1.   frontal lobe

    _____ 2.   parietal lobe

    _____ 3.   temporal lobe

    _____ 4.   precentral gyrus

    _____ 5.   parieto-occipital
                   sulcus

    _____ 6.   postcentral gyrus

    _____ 7.   lateral sulcus   _____ 10.   medulla

    _____ 8.   central sulcus   _____ 11.   occipital lobe

    _____ 9.   cerebellum   _____ 12.   pons

2. In which of the cerebral lobes would the following functional areas be found?

    auditory area _____   olfactory area _____

    primary motor area _____   visual area _____

    primary sensory area _____   Broca's area _____

3. Which of the following structures are *not* part of the brain stem? (Circle the appropriate response or responses.)

    cerebral hemispheres   pons   midbrain   cerebellum   medulla   diencephalon

4. Complete the following statements by writing the proper word or phrase on the corresponding blanks at the right.

    A(n) __1__ is an elevated ridge of cerebral tissue. The convolutions seen in the cerebrum are important because they increase the __2__ . Gray matter is composed of __3__ . White matter is composed of __4__ . A fiber tract that provides for communication between different parts of the same cerebral hemisphere is called a(n) __5__ , whereas one that carries impulses to and from the cerebrum from and to lower CNS areas is called a(n) __6__ tract. The lentiform nucleus along with the amygdaloid and caudate nuclei are collectively called the __7__ .

    1. _____

    2. _____

    3. _____

    4. _____

    5. _____

    6. _____

    7. _____

5. Identify the structures on the following sagittal view of the human brain by matching the lettered areas to the proper terms at the left:

_____ 1. cerebellum

_____ 2. cerebral aqueduct

_____ 3. cerebral hemisphere

_____ 4. cerebral peduncle

_____ 5. choroid plexus

_____ 6. corpora quadrigemina

_____ 7. corpus callosum

_____ 8. fornix

_____ 9. fourth ventricle

_____ 10. hypothalamus

_____ 11. mammillary bodies

_____ 12. massa intermedia

_____ 13. medulla oblongata

_____ 14. optic chiasma

_____ 15. pineal body

_____ 16. pituitary gland

_____ 17. pons

_____ 18. septum pellucidum

_____ 19. thalamus

6. Using the letters from the diagram in item 5, match the appropriate structures with the descriptions given below:

_____ 1. site of regulation of body temperature and water balance; most important autonomic center

_____ 2. consciousness depends on the function of this part of the brain

_____ 3. located in the midbrain; contains reflex centers for vision and audition

_____ 4. responsible for regulation of posture and coordination of complex muscular movements

_____ 5. important synapse site for afferent fibers traveling to the sensory cortex

_____ 6. contains autonomic centers regulating blood pressure, heart rate, and respiratory rhythm, as well as coughing, sneezing, and swallowing centers

_____ 7. large commissure connecting the cerebral hemispheres

_____ 8. fiber tract involved with olfaction

_____ 9. connects the third and fourth ventricles

_____ 10. encloses the third ventricle

7. Embryologically, the brain arises from the rostral end of a tubelike structure that quickly becomes divided into three major regions. Groups of structures that develop from the embryonic brain are listed below. Designate the embryonic origin of each group as the hindbrain, midbrain, or forebrain.

_____ 1. the diencephalon, including the thalamus, optic chiasma, and hypothalamus

_____ 2. the medulla, pons, and cerebellum

_____ 3. the cerebral hemispheres

8. What is the function of the basal nuclei? _____

_____

9. What is the corpus striatum, and how is it related to the fibers of the internal capsule? _____

_____

_____

10. A brain hemorrhage within the region of the right internal capsule results in paralysis of the left side of the body.

Explain why the left side (rather than the right side) is affected. _____

_____

11. Explain why trauma to the base of the brain is often much more dangerous than trauma to the frontal lobes. (*Hint:* Think about the relative functioning of the cerebral hemispheres and the brain stem structures. Which contain centers more vital to life?)

_____

_____

_____

12. In "split brain" experiments, the main commissure connecting the cerebral hemispheres is cut. Name this commissure: _____

Meninges of the Brain

Identify the meningeal (or associated) structures described below:

_____ 1. outermost meninx covering the brain; composed of tough fibrous connective tissue

_____ 2. innermost meninx covering the brain; delicate and highly vascular

_____ 3. structures instrumental in returning cerebrospinal fluid to the venous blood in the dural sinuses

_____ 4. structure that forms the cerebrospinal fluid

_____ 5. middle meninx; like a cobweb in structure

_____ 6. its outer layer forms the periosteum of the skull

_____ 7. a dural fold that attaches the cerebrum to the crista galli of the skull

_____ 8. a dural fold separating the cerebrum from the cerebellum

Cerebrospinal Fluid

Fill in the following flowchart by delineating the circulation of cerebrospinal fluid from its formation site (assume that this is one of the lateral ventricles) to the site of its reabsorption into the venous blood:

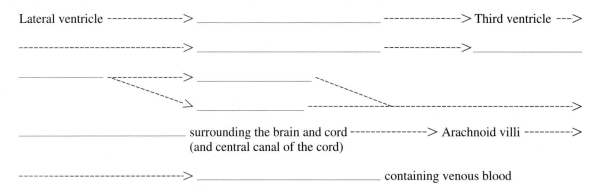

Lateral ventricle -------------> _____ --------------> Third ventricle --->

-----------------------------> _____ -------------->_____

_____ <-----------> _____ ` ` `

-------------> _____ --->

_____ surrounding the brain and cord -------------> Arachnoid villi --------->
(and central canal of the cord)

-----------------------------> _____ containing venous blood

Cranial Nerves

1. Using the terms below, correctly identify all structures indicated by leader lines on the diagram below.

a. abducens nerve (I)

b. accessory nerve (XI)

c. cerebellum

d. cerebral peduncle

e. decussation of the pyramids

f. facial nerve (VII)

g. frontal lobe of cerebral hemisphere

h. glossopharyngeal nerve (IX)

i. hypoglossal nerve (XII)

j. longitudinal fissure

k. mammillary body

l. medulla oblongata

m. oculomotor nerve (III)

n. olfactory bulb

o. olfactory tract

p. optic chiasma

q. optic nerve (II)

r. optic tract

s. pituitary gland

t. pons

u. spinal cord

v. temporal lobe of cerebral hemisphere

w. trigeminal nerve (V)

x. trochlear nerve (IV)

y. vagus nerve (X)

z. vestibulocochlear nerve (VIII)

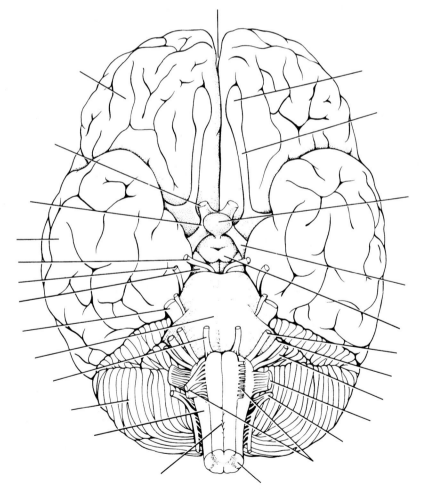

2. Provide the name and number of the cranial nerves involved in each of the following activities, sensations, or disorders:

_____ 1. shrugging the shoulders

_____ 2. smelling a flower

_____ 3. raising the eyelids; focusing the lens of the eye for accommodation; and pupillary constriction

_____ 4. slows the heart; increases the mobility of the digestive tract

_____ 5. involved in Bell's palsy (facial paralysis)

_____ 6. chewing food

_____ 7. listening to music; seasickness

_____ 8. secretion of saliva; tasting well-seasoned food

_____ 9. involved in "rolling" the eyes (three nerves—provide numbers only)

_____ 10. feeling a toothache

_____ 11. reading *Playgirl* or *Playboy* magazine

_____ 12. purely sensory in function (three nerves—provide numbers only)

Dissection of the Sheep Brain

1. In your own words, describe the relative hardness of the sheep brain tissue as observed when cutting into it.

_____

Because formalin hardens all tissue, what conclusions might you draw about the relative hardness and texture of

living brain tissue? _____

2. How does the relative size of the cerebral hemispheres compare in sheep and human brains? _____

_____

What is the significance of this difference? _____

_____

3. What is the significance of the fact that the olfactory bulbs are relatively much larger in the sheep brain than in the

human brain? _____

_____

Review Sheet
_____

EXERCISE 17

Spinal Cord, Spinal Nerves, and the Autonomic Nervous System

Anatomy of the Spinal Cord

1. Match the descriptions given below to the proper anatomical term:

   a. cauda equina   b. conus medullaris   c. filium terminale   d. foramen magnum

   _____ 1. most superior boundary of the spinal cord

   _____ 2. meningeal extension beyond the spinal cord terminus

   _____ 3. spinal cord terminus

   _____ 4. collection of spinal nerves traveling in the vertebral canal below the terminus of the spinal cord

2. Using the terms below, correctly identify on the diagram all structures provided with leader lines.

   a. anterior (ventral) horn   f. dorsal root of spinal nerve   k. posterior (dorsal) horn

   b. arachnoid mater   g. dura mater   l. spinal nerve

   c. central canal   h. gray commissure   m. ventral ramus of spinal nerve

   d. dorsal ramus of spinal nerve   i. lateral horn   n. ventral root of spinal nerve

   e. dorsal root ganglion   j. pia mater   o. white matter

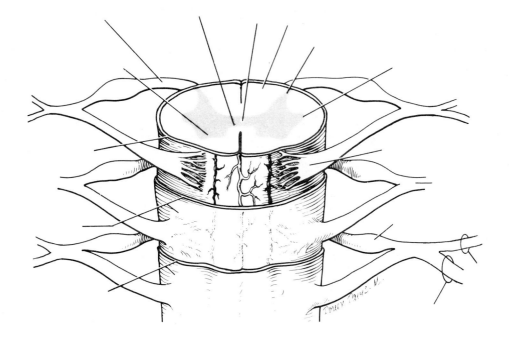

3. Choose the proper answer from the following key to respond to the descriptions relating to spinal cord anatomy.

Key: a. afferent   b. efferent   c. both afferent and efferent   d. association

_____ 1. neuron type found in dorsal horn   _____ 4. fiber type in ventral root

_____ 2. neuron type found in ventral horn   _____ 5. fiber type in dorsal root

_____ 3. neuron type in dorsal root ganglion   _____ 6. fiber type in spinal nerve

4. The spinal cord is enlarged in two regions, the _____ and the _____

regions. What is the significance of these enlargements? _____

_____

5. How does the position of the gray and white matter differ in the spinal cord and the cerebral hemispheres?

_____

_____

6. Choose the name of the tract, from the following key, that might be damaged when the following conditions are observed. (More than one choice may apply.)

_____ 1. uncoordinated movement   Key:

_____ 2. lack of voluntary movement   a. fasciculus gracilis
     b. fasciculus cuneatus
_____ 3. tremors, jerky movements   c. lateral corticospinal tract
     d. ventral corticospinal tract
_____ 4. diminished pain perception   e. tectospinal tract
     f. rubrospinal tract
_____ 5. diminished sense of touch   g. lateral spinothalamic tract
     h. ventral spinothalamic tract
     i. dorsal spinocerebellar tract
     j. vestibulospinal tract
     k. olivospinal tract
     l. ventral spinocerebellar tract

7. Use an appropriate reference to describe the functional significance of an upper motor neuron and a lower motor neuron:

upper motor neuron _____

_____

lower motor neuron _____

_____

Will contraction of a muscle occur if the lower motor neurons serving it have been destroyed? _____ If the upper motor neurons serving it have been destroyed? _____ Using an appropriate reference, differentiate between flaccid and spastic paralysis and note the possible causes of each. _____

_____

_____

_____

_____

Spinal Nerves and Nerve Plexuses

1. In the human, there are 31 pairs of spinal nerves named according to the region of the vertebral column from which they issue. The spinal nerves are named below; note, by number, the vertebral level at which they emerge:

cervical nerves _____   sacral nerves _____

lumbar nerves _____   thoracic nerves _____

2. The ventral rami of spinal nerves C_1 through T_1 and T_{12} through S_4 take part in forming _____,

which serve the _____ of the body. The ventral rami of T_2 through T_{12} run

between the ribs to serve the _____. The dorsal rami of the spinal nerves

serve _____ .

3. What would happen if the following structures were damaged or transected? (Use key choices for responses.)

Key: a. loss of motor   b. loss of sensory   c. loss of both motor and
            function   function   sensory function

_____ 1. dorsal root of a spinal nerve   _____ 3. anterior ramus of a spinal nerve

_____ 2. ventral root of a spinal nerve

4. Define *plexus:* _____

_____

5. Name the major nerves that serve the following body areas:

_____ 1. head, neck, shoulders (name plexus only)

_____ 2. diaphragm

_____ 3. posterior thigh

_____ 4. leg and foot (name two)

_____ 5. most anterior forearm muscles

_____ 6. arm muscles (name two)

_____ 7. abdominal wall (name plexus only)

_____ 8. anterior thigh

_____ 9. medial side of the hand

The Autonomic Nervous System

1. For the most part, sympathetic and parasympathetic fibers serve the same organs and structures. How can they exert antagonistic effects? (After all, nerve impulses are nerve impulses—aren't they?)

   _____

2. Name three structures that receive sympathetic but not parasympathetic innervation.

   _____

   _____

3. A pelvic splanchnic nerve contains (circle one):

   a. preganglionic sympathetic fibers   c. preganglionic parasympathetic fibers

   b. postganglionic sympathetic fibers   d. postganglionic parasympathetic fibers

4. The following chart states a number of conditions. Use a check mark to show which division of the autonomic nervous system is involved in each.

Sympathetic division	Condition	Parasympathetic division
	Secretes norepinephrine; adrenergic fibers	
	Secretes acetylcholine; cholinergic fibers	
	Long preganglionic axon; short postganglionic axon	
	Short preganglionic axon; long postganglionic axon	
	Arises from cranial and sacral nerves	
	Arises from spinal nerves T_1 through L_3	
	Normally in control	
	"Fight or flight" system	
	Has more specific control (Look it up!)	

5. You are alone in your home late in the evening, and you hear an unfamiliar sound in your backyard. List four events promoted by the sympathetic nervous system that would aid you in coping with this rather frightening situation:

   _____

   _____

   _____

   _____

   _____

6. Often after surgery, people are temporarily unable to urinate, and bowel sounds are absent. What division of the

   ANS is affected by the anesthesia? _____

Review Sheet

EXERCISE 18

Special Senses: Vision

Anatomy of the Eye

1. Three accessory eye structures contribute to the formation of tears and/or aid in lubrication of the eyeball. Name each and then name its major secretory product. Indicate which has antibacterial properties by circling the correct secretory product.

Accessory structures	Product

2. The eyeball is wrapped in adipose tissue within the orbit. What is the function of the adipose tissue?

_____

What seven bones form the bony orbit? (Think! If you can't remember, check a skull or your text.)

_____   _____   _____

_____   _____

_____   _____

3. Why does one often have to blow one's nose after having a good cry?_____

_____

4. Identify the extrinsic eye muscle predominantly responsible for the actions described below.

_____ 1.   turns the eye laterally

_____ 2.   turns the eye medially

_____ 3.   turns the eye up and laterally

_____ 4.   turns the eye inferiorly

_____ 5.   turns the eye superiorly

_____ 6.   turns the eye down and laterally

5. What is a sty?_____

_____

Conjunctivitis?_____

6. Using the terms listed on the right, correctly identify all structures provided with leader lines in the diagram.

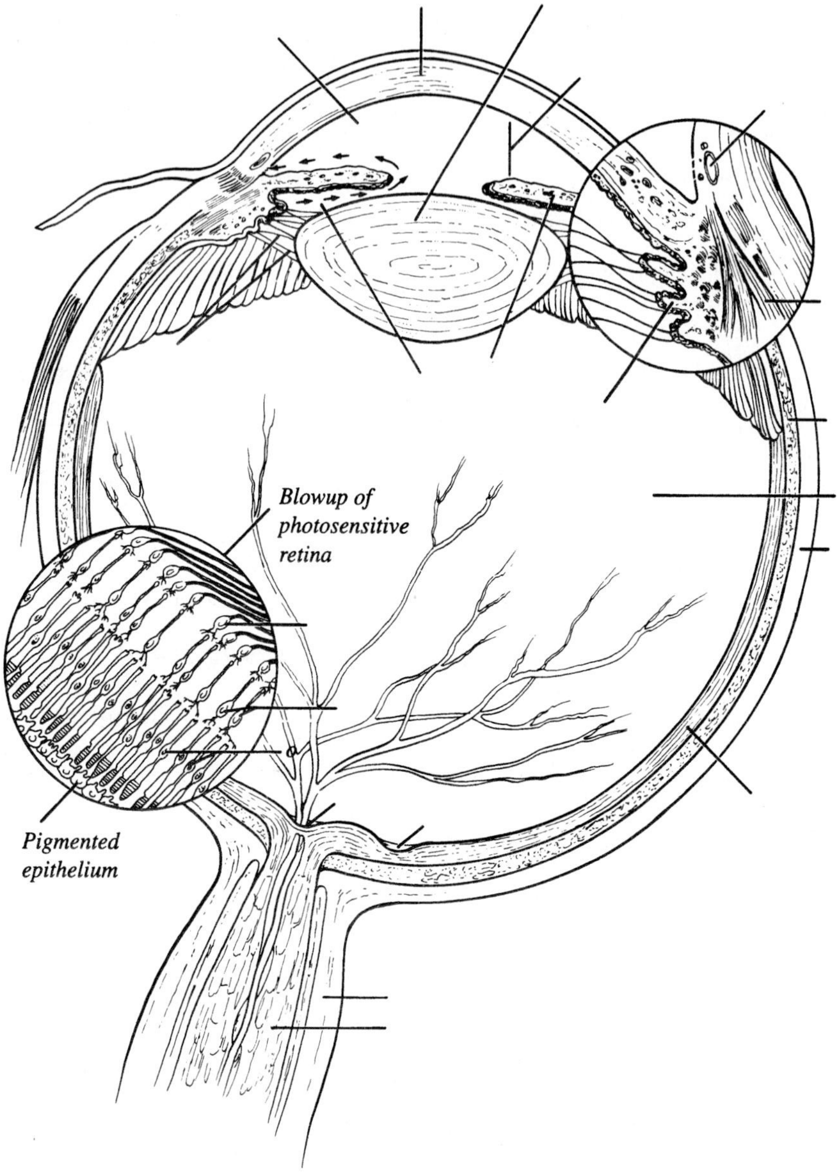

a. anterior chamber

b. anterior segment containing aqueous humor

c. bipolar neurons

d. ciliary body and processes

e. ciliary muscle

f. choroid

g. cornea

h. dura mater

i. fovea centralis

j. ganglion cells

k. iris

l. lens

m. optic disc

n. optic nerve

o. photoreceptors

p. posterior chamber

q. retina

r. sclera

s. scleral venous sinus

t. suspensory ligaments

u. vitreous body in posterior segment

Blowup of photosensitive retina

Pigmented epithelium

Notice the arrows drawn close to the left side of the iris in the diagram above. What do they indicate?

_____

_____

_____

7. Match the key responses with the descriptive statements that follow.

Key: a.   aqueous humor   e.   cornea   j.   retina
         b.   choroid   f.   fovea centralis   k.   sclera
         c.   ciliary body   g.   iris   l.   scleral venous sinus
         d.   ciliary processes of   h.   lens   m.   suspensory ligament
             the ciliary body   i.   optic disc   n.   vitreous humor

_____ 1.   attaches the lens to the ciliary body

_____ 2.   fluid filling the anterior segment of the eye

_____ 3.   the "white" of the eye

_____ 4.   part of the retina that lacks photoreceptors

_____ 5.   modification of the choroid that controls the shape of the crystalline lens

_____ 6.   contains the ciliary muscle

_____ 7.   drains the aqueous humor from the eye

_____ 8.   tunic containing the rods and cones

_____ 9.   substance occupying the posterior segment of the eyeball

_____ 10.   forms the bulk of the heavily pigmented vascular tunic

_____ , _____ 11.   smooth muscle structures

_____ 12.   area of critical focusing and discriminatory vision

_____ 13.   form (by filtration) the aqueous humor

_____ , _____ , _____ ,

_____ 14.   light-bending media of the eye

_____ 15.   anterior continuation of the sclera—your "window on the world"

_____ 16.   composed of tough, white, opaque, fibrous connective tissue

8. The iris is composed primarily of two smooth muscle layers, one arranged radially and the other circularly.

Which of these dilates the pupil? _____

9. You would expect the pupil to be dilated in which of the following circumstances? Circle the correct response(s).

a.   in brightly lit surroundings   c.   during focusing for near vision

b.   in dimly lit surroundings   d.   in observing distant objects

10. The intrinsic eye muscles are under the control of which of the following? (Circle the correct response.)

                   autonomic nervous system   somatic nervous system

Dissection of the Cow (Sheep) Eye

1. What modification of the choroid that is not present in humans is found in the cow eye? _____

   _____

   What is its function? _____

   _____

   _____

2. What is the anatomical appearance of the retina? _____

   _____

   At what point is it attached to the posterior aspect of the eyeball? _____

Microscopic Anatomy of the Retina

1. The two major layers of the retina are the epithelial and nervous layers. In the nervous layer, the neuron populations are arranged as follows from the epithelial layer to the vitreous humor. (Circle all proper responses.)

   bipolar cells, ganglion cells, photoreceptors   photoreceptors, ganglion cells, bipolar cells

   ganglion cells, bipolar cells, photoreceptors   photoreceptors, bipolar cells, ganglion cells

2. The axons of the _____ cells form the optic nerve, which exits from the eyeball.

3. Complete the following statements by writing either *rods* or *cones* on each blank:

   The dim light receptors are the _____. Only _____ are found

   in the fovea centralis, whereas mostly _____ are found in the periphery of the retina.

   _____ are the photoreceptors that operate best in bright light and allow for color vision.

Visual Pathways to the Brain

1. The visual pathway to the occipital lobe of the brain consists most simply of a chain of five neurons. Beginning with the photoreceptor cell of the retina, name them and note their location in the pathway.

   (1) *Photoreceptor cell, retina*   (4) _____

   (2) _____   (5) _____

   (3) _____

2. Visual field tests are done to reveal destruction along the visual pathway from the retina to the optic region of the brain. Note where the lesion is likely to be in the following cases:

   Normal vision in left eye visual field; absence of vision in right eye visual field: _____

   Normal vision in both eyes for right half of the visual field; absence of vision in both eyes for left half of the

   visual field: _____

3. How is the right optic *tract* anatomically different from the right optic *nerve*? _____

_____

_____

_____

4. Why is the ophthalmoscopic examination an important diagnostic tool? _____

_____

Visual Tests and Experiments

1. Match the terms in column B with the descriptions in column A:

Column A	Column B
_____ 1. light bending	a. accommodation
_____ 2. ability to focus for close (under 20 ft) vision	b. astigmatism
_____ 3. normal vision	c. emmetropia
_____ 4. inability to focus well on close objects (farsightedness)	d. hyperopia
_____ 5. nearsightedness	e. myopia
_____ 6. blurred vision due to unequal curvatures of the lens or cornea	f. refraction

2. Complete the following statements:

In farsightedness, the light is focused __1__ the retina. The lens required to treat myopia is a __2__ lens. The "near point" increases with age because the __3__ of the lens decreases as we get older. A convex lens, like that of the eye, produces an image that is upside down and reversed from left to right. Such an image is called a __4__ image.

1. _____

2. _____

3. _____

4. _____

3. Use terms from the key to complete the statements concerning near and distance vision.

   Key: a. contracted   b. decreased   c. increased   d. relaxed   e. taut

   During distance vision: The ciliary muscle is ____, the suspensory ligament is ____, the convexity of the

   lens is ____, and light refraction is ____. During close vision: The ciliary muscle is ____, the suspensory liga-

   ment is ____, lens convexity is ____, and light refraction is ____.

4. Explain why vision is lost when light hits the blind spot. _____

5. Record your Snellen eye test results below:

Left eye (without glasses) _____ (with glasses) _____

Right eye (without glasses) _____ (with glasses) _____

Is your visual acuity normal, less than normal, or better than normal? _____

Explain. _____

_____

Explain why each eye is tested separately when using the Snellen eye chart. _____

_____

Explain 20/40 vision. _____

_____

Explain 20/10 vision. _____

_____

6. Define *astigmatism:* _____

_____

How can it be corrected? _____

7. Record the distance of your near point of accommodation as tested in the laboratory:

right eye _____ left eye _____

Is your near point within the normal range for your age? _____

8. Define *presbyopia:* _____

_____

What causes it? _____

9. To which wavelengths of light do the three cone types of the retina respond maximally?

_____, _____, and _____

10. How can you explain the fact that we see a great range of colors even though only three cone types exist?

_____

_____

11. From what condition does color blindness result? _____

_____

12. Many college students struggling through mountainous reading assignments are told that they need glasses for "eyestrain." Why is it more of a strain on the extrinsic and intrinsic eye muscles to look at close objects than at far objects? _____

_____

Review Sheet

EXERCISE 19

Special Senses: Hearing and Equilibrium

Anatomy of the Ear

1. Select the terms from column B that apply to the column A descriptions. Some terms are used more than once.

Column A

_____, _____,

_____ 1. structures comprising the outer or external ear

_____, _____,

_____ 2. structures composing the inner ear

_____, _____,

_____ 3. collectively called the ossicles

_____, _____

_____ 4. ear structures not involved with audition

_____ 5. involved in equalizing the pressure in the middle ear with atmospheric pressure

_____ 6. vibrates at the same frequency as sound waves hitting it; transmits the vibrations to the ossicles

_____, _____

_____ 7. contain receptors for the sense of balance

_____ 8. transmits the vibratory motion of the stirrup to the fluid in the scala vestibuli of the inner ear

_____ 9. acts as a pressure relief valve for the increased fluid pressure in the scala tympani; bulges into the tympanic cavity

_____ 10. passage between the throat and the tympanic cavity

_____ 11. fluid contained within the membranous labyrinth

_____ 12. fluid contained within the osseous labyrinth and bathing the membranous labyrinth

Column B

a.   auditory (pharyngo-tympanic) tube

b.   anvil (incus)

c.   cochlea

d.   endolymph

e.   external auditory canal

f.   hammer (malleus)

g.   oval window

h.   perilymph

i.   pinna

j.   round window

k.   semicircular canals

l.   stirrup (stapes)

m.   tympanic membrane

n.   vestibule

2. Sound waves hitting the eardrum initiate its vibratory motion. Trace the pathway through which vibrations and fluid currents are transmitted to finally stimulate the hair cells in the organ of Corti. (Name the appropriate ear structures in their correct sequence.) Eardrum → _____

_____

_____

3. Identify all indicated structures and ear regions in the following diagram.

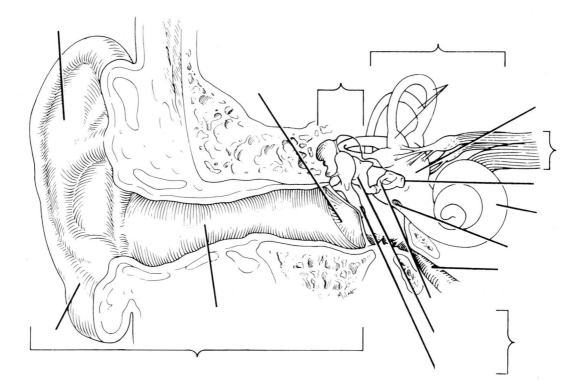

4. Match the membranous labyrinth structures listed in column B with the descriptive statements in column A:

Column A

_____, _____ 1. sacs found within the vestibule

_____ 2. contains the organ of Corti

_____, _____ 3. sites of the maculae

_____ 4. positioned in all spatial planes

_____ 5. hair cells of organ of Corti rest on this membrane

_____ 6. gelatinous membrane overlying the hair cells of the organ of Corti

_____ 7. contains the crista ampullaris

_____, _____, _____, _____ 8. function in static equilibrium

_____, _____, _____, _____ 9. function in dynamic equilibrium

_____ 10. carries auditory information to the brain

_____ 11. gelatinous cap overlying hair cells of the crista ampullaris

_____ 12. grains of calcium carbonate in the maculae

Column B

a. ampulla

b. basilar membrane

c. cochlear duct

d. cochlear nerve

e. cupula

f. otoliths

g. saccule

h. semicircular ducts

i. tectorial membrane

j. utricle

k. vestibular nerve

5. Describe how sounds of different frequency (pitch) are differentiated in the cochlea. _____

_____

_____

_____

6. Explain the role of the endolymph of the semicircular canals in activating the receptors during angular motion.

_____

_____

_____

7. Explain the role of the otoliths in perception of static equilibrium (head position). _____

_____

_____

_____

Laboratory Tests

1. Was the auditory acuity measurement made during the experiment on page 210 the same or different for both

   ears? _____ What factors might account for a difference in the acuity of the two ears?

   _____

   _____

2. During the sound localization experiment on page 210, in which position(s) was the sound least easily located?

   _____

   How can this phenomenon be explained? _____

   _____

3. In the experiment on page 211, which tuning fork was the most difficult to hear? _____ Hz

   What conclusion can you draw? _____

   _____

4. When the tuning fork handle was pressed to your forehead during the Weber test, where did the sound seem to

   originate? _____

   Where did it seem to originate when one ear was plugged with cotton? _____

   How do sound waves reach the cochlea when conduction deafness is present? _____

   _____

5. Indicate whether the following conditions relate to conduction deafness (C) or sensorineural (central) deafness (S):

_____ 1. can result from the fusion of the ossicles

_____ 2. can result from a lesion on the cochlear nerve

_____ 3. sound heard in one ear but not in the other during bone and air conduction

_____ 4. can result from otitis media

_____ 5. can result from impacted cerumen or a perforated eardrum

_____ 6. can result from a blood clot in the auditory cortex

6. The Rinne test evaluates an individual's ability to hear sounds conducted by air or bone. Which is more indicative of normal hearing? _____

7. Define *nystagmus:* _____

_____

8. What is the usual reason for conducting the Romberg test? _____

_____

Was the degree of sway greater with the eyes open or closed? _____

Why? _____

9. Normal balance, or equilibrium, depends on input from a number of sensory receptors. Name them.

_____

_____

Review Sheet

EXERCISE 20

Special Senses: Taste and Olfaction

Localization and Anatomy of Taste Buds

1. Name three sites where receptors for taste are found, and circle the predominant site:

   _____ , _____ , and

   _____

2. Describe the cellular makeup and arrangement of a taste bud. (Use a diagram, if helpful.) _____

   _____

   _____

Localization and Anatomy of the Olfactory Receptors

1. Describe the cellular composition and the location of the olfactory epithelium. _____

   _____

   _____

2. How and why does sniffing improve your sense of smell? _____

   _____

Laboratory Experiments

1. Taste and smell receptors are both classified as _____ , because they both re-

   spond to _____

2. Why is it impossible to taste substances with a dry tongue? _____

   _____

3. State the most important sites of the taste-specific receptors, as determined during the plotting exercise in the laboratory:

   salt _____   sour _____

   bitter _____   sweet _____

4. The basic taste sensations are elicited by specific chemical substances or groups. Name them:

   salt _____   sour _____

   bitter _____   sweet _____

5. Babies tend to favor bland foods, whereas adults tend to like highly seasoned foods. What is the basis for this phenomenon? _____

_____

_____

6. How palatable is food when you have a cold? _____

Explain. _____

_____

Review Sheet

EXERCISE 21

Functional Anatomy of the Endocrine Glands

Gross Anatomy and Basic Function of the Endocrine Glands

1. Both the endocrine and nervous systems are major regulating systems of the body; however, the nervous system has been compared to an airmail delivery system and the endocrine system to the pony express. Briefly explain this comparison.

    _____

    _____

    _____

2. Define *hormone:* _____

    _____

3. Chemically, hormones belong chiefly to two molecular groups, the _____

    and the _____.

4. What do all hormones have in common? _____

    _____

5. Define *target organ:* _____

    _____

6. Why don't all tissues respond to all hormones? _____

    _____

7. Identify the endocrine organ described by the following statements:

    _____ 1. located in the throat; bilobed gland connected by an isthmus

    _____ 2. found close to the kidney

    _____ 3. a mixed gland, located close to the stomach and small intestine

    _____ 4. paired glands suspended in the scrotum

    _____ 5. ride "horseback" on the thyroid gland

    _____ 6. found in the pelvic cavity of the female, concerned with ova and female hormone production

    _____ 7. found in the upper thorax overlying the heart; large during youth

    _____ 8. found in the roof of the third ventricle

8. For each statement describing hormonal effects, identify the hormone(s) involved by choosing a number from key A, and note the hormone's site of production with a letter from key B. More than one hormone may be involved in some cases.

Key A:

1. ACTH
2. ADH
3. aldosterone
4. cortisone
5. epinephrine
6. estrogens
7. FSH
8. glucagon
9. GH
10. insulin
11. LH
12. melatonin
13. MSH
14. oxytocin
15. progesterone
16. prolactin
17. PTH
18. serotonin
19. testosterone
20. thymosin
21. thyrocalcitonin/calcitonin
22. T_4 / T_3
23. TSH

Key B:

a. adrenal cortex
b. adrenal medulla
c. anterior pituitary
d. hypothalamus
e. ovaries
f. pancreas
g. parathyroid glands
h. pineal gland
i. posterior pituitary
j. testes
k. thymus gland
l. thyroid gland

_____, _____ 1. basal metabolism hormone

_____, _____ 2. programming of T lymphocytes

_____, _____ and _____, _____ 3. regulate blood calcium levels

_____, _____ and _____, _____ 4. released in response to stressors

_____, _____ and _____, _____ 5. drives development of secondary sexual characteristics

_____, _____; _____, _____; _____, _____; and _____, _____ 6. regulate the function of another endocrine gland

_____, _____ 7. mimics the sympathetic nervous system

_____, _____ and _____, _____ 8. regulate blood glucose levels; produced by the same "mixed" gland

_____, _____ and _____, _____ 9. directly responsible for regulation of the menstrual cycle

_____, _____ and _____, _____ 10. regulate the ovarian cycle

_____, _____ and _____, _____ 11. maintenance of salt and water balance in the ECF

_____, _____ and _____, _____ 12. directly involved in milk production and ejection

_____, _____ 13. questionable function; may stimulate the melanocytes of the skin

9. Although the pituitary gland is often referred to as the master gland of the body, the hypothalamus exerts some control over the pituitary gland. How does the hypothalamus control both anterior and posterior pituitary functioning?

_____

_____

_____

_____

_____

10. Indicate whether the release of the hormones listed below is stimulated by (a), another hormone; (b), the nervous system (neurotransmitters, or releasing factors); or (c), humoral factors (the concentration of specific nonhormonal substances in the blood or extracellular fluid):

_____ 1. T_4 / T_3 _____ 4. parathyroid hormone   _____ 7. ADH

_____ 2. insulin   _____ 5. testosterone   _____ 8. TSH, FSH

_____ 3. estrogens   _____ 6. norepinephrine   _____ 9. aldosterone

11. Name the hormone that would be produced in *inadequate* amounts under the following conditions. (Use your textbook as necessary.)

_____ 1. sexual immaturity

_____ 2. tetany

_____ 3. excessive diuresis without high blood glucose levels

_____ 4. polyurea, polyphagia, and polydipsia

_____ 5. abnormally small stature, normal proportions

_____ 6. miscarriage

_____ 7. lethargy, hair loss, low BMR, obesity

12. Name the hormone that is produced in *excessive* amounts in the following conditions. (Use your textbook as necessary.)

_____ 1. lantern jaw and large hands and feet in the adult

_____ 2. bulging eyeballs, nervousness, increased pulse rate

_____ 3. demineralization of bones, spontaneous fractures

Microscopic Anatomy of Selected Endocrine Glands (optional)

1. Choose a response from the key below to name the hormone(s) produced by the cell types listed:

Key: a. insulin   d. calcitonin   g. glucagon
      b. GH, prolactin   e. TSH, ACTH, FSH, LH   h. PTH
      c. T_4 / T_3 f. mineralocorticoids   i. glucocorticoids

_____ 1. parafollicular cells of the thyroid   _____ 6. zona fasciculata cells

_____ 2. follicular epithelial cells of the thyroid   _____ 7. zona glomerulosa cells

_____ 3. beta cells of the islets of Langerhans   _____ 8. chief cells

_____ 4. alpha cells of the islets of Langerhans   _____ 9. acidophil cells of the anterior pituitary

_____ 5. basophil cells of the anterior pituitary

2. Five diagrams of the microscopic structures of the endocrine glands are presented here. Identify each and name all indicated structures.

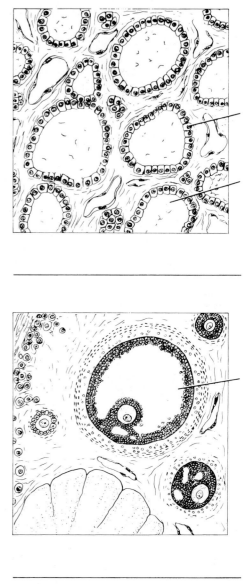

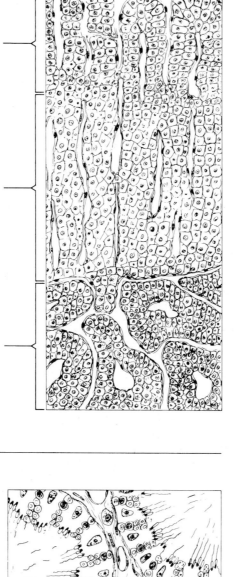

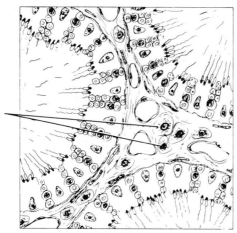

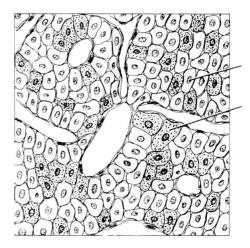

Review Sheet

EXERCISE 22   # Blood

Composition of Blood

1. What is the blood volume of an average-size adult? _____ liters

2. What determines whether blood is bright red or a dull brick-red? _____

_____

3. Use the key to identify the cell type(s) or blood elements that fit the following descriptive statements.

Key: a.   red blood cell   d.   basophil   g.   lymphocyte
        b.   megakaryocyte   e.   monocyte   h.   formed elements
        c.   eosinophil   f.   neutrophil   i.   plasma

_____ 1.   most numerous leukocyte

_____ , _____ , and

_____ 2.   granulocytes

_____ 3.   also called an erythrocyte; anucleate formed element

_____ , _____ 4.   actively phagocytic leukocytes

_____ , _____ 5.   agranulocytes

_____ 6.   ancestral cell of platelets

_____ 7.   (a) through (g) are all examples of these

_____ 8.   number rises during parasite infections

_____ 9.   releases histamine; promotes inflammation

_____ 10.   many formed in lymphoid tissue

_____ 11.   transports oxygen

_____ 12.   primarily water, noncellular; the fluid matrix of blood

_____ 13.   increases in number during prolonged infections

_____ , _____ , _____ ,

_____ , _____ 14.   also called white blood cells

4. List three classes of nutrients normally found in plasma: _____ ,

_____ , and _____

Name two gases. _____ and _____

Name three ions. _____ , _____ , and _____

5. Describe the consistency and color of the plasma you observed in the laboratory. _____

_____

6. What is the average life span of a red blood cell? How does its anucleate condition affect this life span?

_____

_____

_____

7. From memory, describe the structural characteristics of each of the following blood cell types as accurately as possible, and note the percentage of each in the total white blood cell population.

eosinophils _____

_____

neutrophils _____

_____

lymphocytes _____

_____

basophils _____

_____

monocytes _____

_____

8. Correctly identify the blood pathologies described in column A by matching them with selections from column B:

Column A		Column B
_____ 1. abnormal increase in the number of WBCs		a. anemia
_____ 2. abnormal increase in the number of RBCs		b. leukocytosis
_____ 3. condition of too few RBCs or of RBCs with hemoglobin deficiencies		c. leukopenia
		d. polycythemia
_____ 4. abnormal decrease in the number of WBCs		

Hematologic Tests

1. Broadly speaking, why are hematologic studies of blood so important in the diagnosis of disease?

   _____

   _____

2. In the chart below, record information from the blood tests you conducted. Complete the chart by recording values for healthy male adults and indicating the significance of high or low values for each test.

Test	Student test results	Normal values (healthy male adults)	Significance High values	Low values
Total WBC count	*(not conducted)*			
Total RBC count	*(not conducted)*			
Hematocrit				
Hemoglobin determination				
Coagulation time				

3. Why is a differential WBC count more valuable than a total WBC count when trying to pin down the specific source of pathology? _____

   _____

   _____

4. What name is given to the process of RBC production? _____ _____

   What acts as a stimulus for this process? _____

   What organ provides this stimulus and under what conditions? _____

   _____

   _____

5. Discuss the effect of each of the following factors on RBC count. Consult an appropriate reference as necessary, and explain your reasoning.

athletic training (for example, running 4 to 5 miles per day over a period of 6 to 9 months) _____

_____

_____

a permanent move from sea level to a high-altitude area _____

_____

_____

6. Define *hematocrit:* _____

7. If you had a high hematocrit, would you expect your hemoglobin determination to be high or low?

_____ Why? _____

_____

8. What is an anticoagulant? _____

Name two anticoagulants used in conducting the hematologic tests. _____

and _____

What is the body's natural anticoagulant? _____

9. If your blood clumped with both anti-A and anti-B sera, your ABO blood type would be _____

To what ABO blood groups could you give blood? _____

From which ABO donor types could you receive blood? _____

Which ABO blood type is most common? _____ Least common? _____

10. Explain why an Rh-negative person does not have a transfusion reaction on the first exposure to Rh-positive

blood but *does* have a reaction on the second exposure. _____

_____

What happens when an ABO blood type is mismatched for the first time? _____

_____

11. Record your observations of the five demonstration slides viewed.

a. Macrocytic hypochromic anemia: _____

b. Microcytic hypochromic anemia: _____

_____

c. Sickle-cell anemia: _____

d. Lymphocytic leukemia (chronic): _____

_____

e. Eosinophilia: _____

Which of slides a through e above corresponds with the following conditions?

_____ 1. iron-deficient diet   _____ 4. lack of vitamin B_{12}

_____ 2. a type of bone marrow cancer   _____ 5. a tapeworm infestation in the body

_____ 3. genetic defect that causes hemoglobin   _____ 6. a bleeding ulcer
           to become sharp/spiky

Review Sheet

EXERCISE 23

Anatomy of the Heart

Gross Anatomy of the Human Heart

1. An anterior view of the heart is shown here. Identify each numbered structure by writing its name on the correspondingly numbered line:

1. _____

2. _____

3. _____

4. _____

5. _____

6. _____

7. _____

8. _____

9. _____

10. _____

11. _____

12. _____

13. _____

14. _____

15. _____

16. _____

17. _____

18. _____

19. _____

20. _____

21. _____

2. What is the function of the fluid that fills the pericardial sac? _____

_____

3. Match the terms in the key to the descriptions provided below.

_____ 1. location of the heart in the thorax

_____ 2. superior heart chambers

_____ 3. inferior heart chambers

_____ 4. visceral pericardium

_____ 5. "anterooms" of the heart

_____ 6. equals cardiac muscle

_____ 7. provide nutrient blood to the heart muscle

_____ 8. lining of the heart chambers

_____ 9. actual "pumps" of the heart

_____ 10. drains blood into the right atrium

Key:

a. atria

b. coronary arteries

c. coronary sinus

d. endocardium

e. epicardium

f. mediastinum

g. myocardium

h. ventricles

4. What is the function of the valves found in the heart? _____

_____

5. Can the heart function with leaky valves? (Think! Can a water pump function with leaky valves?) _____

6. What is the role of the chordae tendineae? _____

_____

7. Define:

angina pectoris _____

_____

pericarditis _____

Pulmonary, Systemic, and Cardiac Circulations

1. A simple schematic of a so-called general circulation is shown below. What part of the circulation is missing from

   this diagram? _____

   Add to the diagram as best you can to make it depict a complete systemic/pulmonary circulation and reidentify "general circulation" as the correct subcirculation.

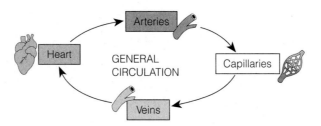

2. Differentiate clearly between the roles of the pulmonary and systemic circulations. _____

_____

_____

3. Complete the following scheme of circulation in the human body:

Right atrium through the tricuspid valve to the _____ through the _____

_____ valve to the pulmonary trunk to the _____

to the capillary beds of the lungs to the _____ to the _____

of the heart through the _____ valve to the _____ through the

_____ valve to the _____ to the systemic arteries to the

_____ of the tissues to the systemic veins to the _____ and

_____ entering the right atrium of the heart.

4. If the mitral valve does not close properly, which circulation is affected? _____

5. Why might a thrombus (blood clot) in the anterior descending branch of the left coronary artery cause sudden

death? _____

_____

Microscopic Anatomy of Cardiac Muscle

1. How would you distinguish cardiac muscle from skeletal muscle? _____

_____

_____

_____

_____

2. Add the following terms to the diagram of cardiac muscle at the right:

   a. intercalated disc

   b. nucleus of cardiac fiber

   c. striations

   d. cardiac muscle fiber

3. What role does the unique structure of cardiac muscle play in its function? (Note: before attempting a response, *describe* the unique anatomy.) _____

_____

_____

_____

Dissection of the Sheep Heart

1. During the sheep heart dissection, you were asked initially to identify the right and left ventricles without cutting into the heart. During this procedure, what differences did you observe between the two chambers?

_____

_____

Knowing that structure and function are related, how would you say this structural difference reflects the relative

functions of these two heart chambers? _____

_____

_____

_____

2. Semilunar valves prevent backflow into the _____ ; AV valves prevent backflow into the

_____ . Using your own observations, explain how the operation of the semilunar valves

differs from that of the AV valves. _____

_____

_____

_____

_____

3. Differentiate clearly between the location and appearance of pectinate muscle and trabeculae carneae. _____

_____

_____

4. Two remnants of fetal structures are observable in the heart—the ligamentum arteriosum and the fossa ovalis. What were they called in the fetal heart, where was each located, and what common purpose did they serve as functioning fetal structures?

_____

_____

_____

_____

RS102

Review Sheet

EXERCISE 24

Anatomy of Blood Vessels

Microscopic Structure of the Blood Vessels

1. Use key choices to identify the blood vessel tunic described.

   Key: a. tunica intima   b. tunica media   c. tunica externa

   _____ 1. most internal tunic

   _____ 2. bulky middle tunic contains smooth muscle and elastin

   _____ 3. its smooth surface decreases resistance to blood flow

   _____ 4. tunic(s) of capillaries

   _____, _____, _____ 5. tunic(s) of arteries and veins

   _____ 6. is especially thick in elastic arteries

   _____ 7. most superficial tunic

2. Servicing the capillaries is the essential function of the organs of the circulatory system. Explain this statement.

   _____

   _____

   _____

3. Cross-sectional views of an artery and of a vein are shown here. Identify each; and on the lines beneath, note the structural details that enabled you to make these identifications:

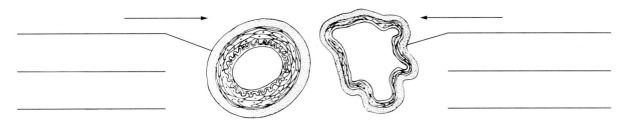

4. Why are valves present in veins but not in arteries? _____

   _____

   _____

5. Name two events *occurring within the body* that aid in venous return:

   _____

   and _____

6. Why are the walls of arteries proportionately thicker than those of the corresponding veins? _____

_____

_____

Major Systemic Arteries and Veins of the Body

1. Use the key on the right to identify the arteries or veins described on the left.

    _____ 1. the arterial system has one of these; the venous system has two

    _____ 2. these arteries supply the myocardium

    _____, _____ 3. two paired arteries serving the brain

    _____ 4. longest vein in the lower limb

    _____ 5. artery on the dorsum of the foot checked after leg surgery

    _____ 6. serves the posterior thigh

    _____ 7. supplies the diaphragm

    _____ 8. formed by the union of the radial and ulnar veins

    _____, _____ 9. two superficial veins of the arm

    _____ 10. artery serving the kidney

    _____ 11. veins draining the liver

    _____ 12. artery that supplies the distal half of the large intestine

    _____ 13. drains the pelvic organs

    _____ 14. what the external iliac artery becomes on entry into the thigh

    _____ 15. major artery serving the arm

    _____ 16. supplies most of the small intestine

    _____ 17. join to form the inferior vena cava

    _____ 18. an arterial trunk that has three major branches, which run to the liver, spleen, and stomach

    _____ 19. major artery serving the tissues external to the skull

    _____, _____, _____ 20. three veins serving the leg

    _____ 21. artery generally used to take the pulse at the wrist

Key:
a. anterior tibial
b. basilic
c. brachial
d. brachiocephalic
e. celiac trunk
f. cephalic
g. common carotid
h. common iliac
i. coronary
j. deep femoral
k. dorsalis pedis
l. external carotid
m. femoral
n. greater saphenous
o. hepatic
p. inferior mesenteric
q. internal carotid
r. internal iliac
s. peroneal
t. phrenic
u. posterior tibial
v. radial
w. renal
x. subclavian
y. superior mesenteric
z. vertebral

2. The human arterial and venous systems are diagrammed on the next two pages. Identify all indicated blood vessels.

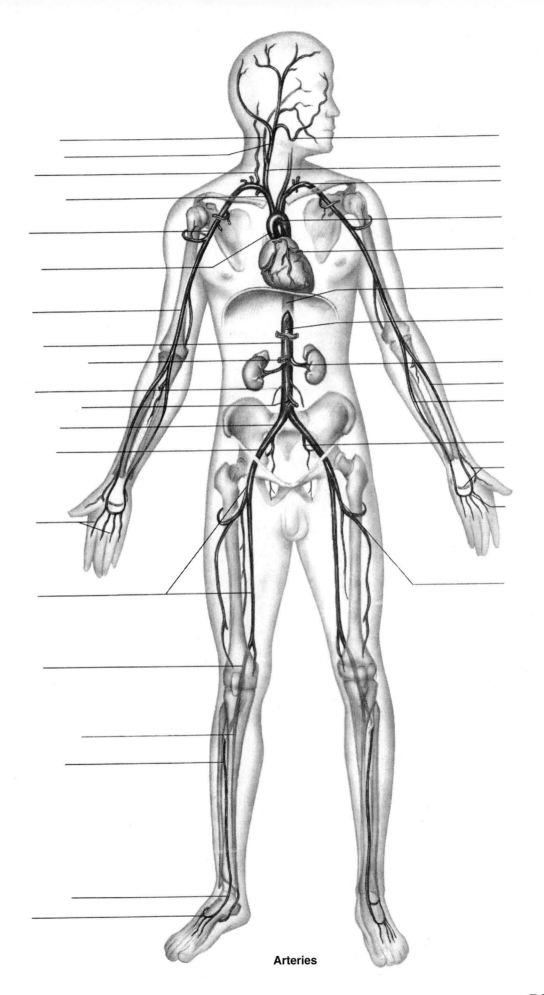

Arteries

RS105

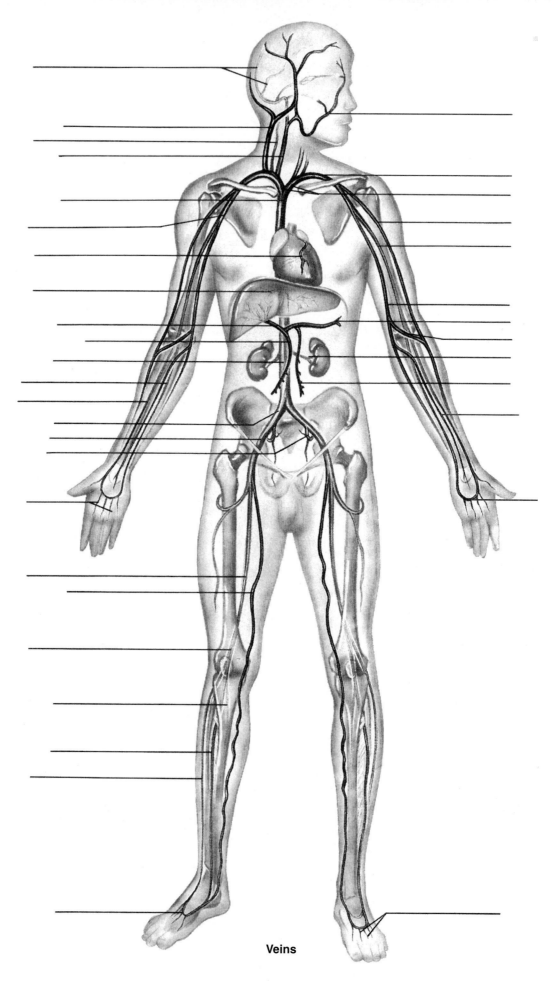

Veins

RS106

3. Trace the blood flow for the following situations:

a. From the capillary beds of the left thumb to the capillary beds of the right thumb _____

_____

_____

_____

_____

_____

_____

b. From the bicuspid valve to the tricuspid valve by way of the great toe _____

_____

_____

_____

_____

_____

c. From the pulmonary vein to the pulmonary artery by way of the right side of the brain _____

_____

_____

_____

_____

_____

_____

Special Circulations

Pulmonary circulation:

1. Trace the pathway of a carbon dioxide gas molecule in the blood from the inferior vena cava until it leaves the bloodstream. Name all structures (vessels, heart chambers, and others) passed through en route.

_____

_____

_____

2. Trace the pathway of an oxygen gas molecule from an alveolus of the lung to the right atrium of the heart. Name all structures through which it passes. _____

_____

_____

3. Most arteries of the adult body carry oxygen-rich blood, and the veins carry oxygen-depleted, carbon dioxide–rich blood. What is different about the pulmonary arteries and veins? _____

_____

_____

4. How do the arteries of the pulmonary circulation differ structurally from the systemic arteries? What condition is indicated by this anatomical difference? _____

_____

_____

Arterial supply of the brain:

1. What two paired arteries enter the skull to supply the brain?

_____ and _____

2. Branches of the paired arteries just named cooperate to form a ring of blood vessels encircling the pituitary gland, at the base of the brain. What name is given to this communication network? _____

What is its function? _____

_____

3. What portion of the brain is served by the anterior and middle cerebral arteries? _____

_____

Both the anterior and middle cerebral arteries arise from the _____ arteries.

4. Trace the pathway of a drop of blood from the aorta to the left occipital lobe of the brain, noting all structures through which it flows. _____

_____

Hepatic portal circulation:

1. Complete the labeling in the diagram of vessels of the hepatic portal system.

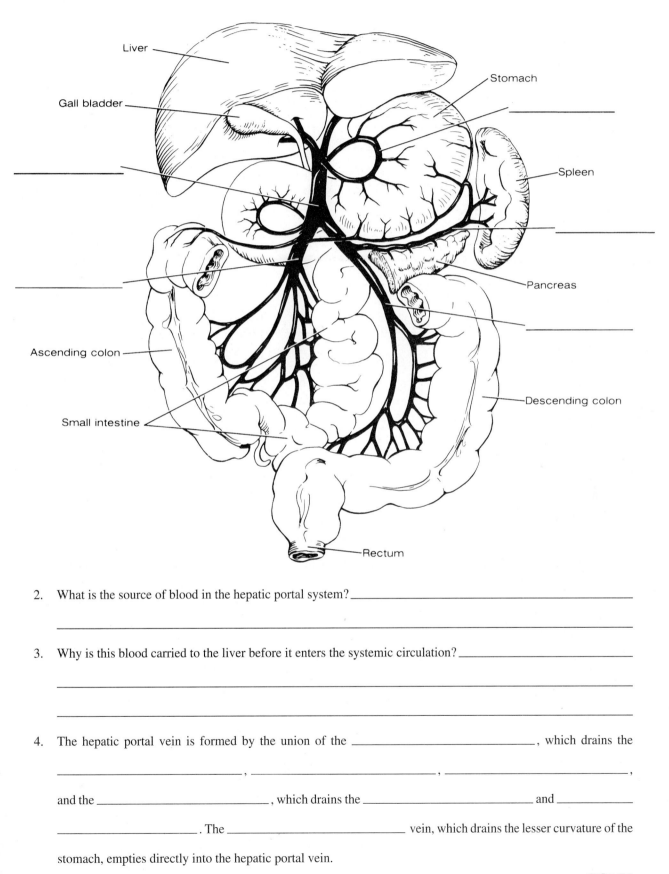

2. What is the source of blood in the hepatic portal system? _____

_____

3. Why is this blood carried to the liver before it enters the systemic circulation? _____

_____

_____

4. The hepatic portal vein is formed by the union of the _____ , which drains the

_____ , _____ , _____ ,

and the _____ , which drains the _____ and _____

_____ . The _____ vein, which drains the lesser curvature of the

stomach, empties directly into the hepatic portal vein.

5. Trace the flow of a drop of blood from the small intestine to the right atrium of the heart, noting all structures encountered or passed through on the way. _____

_____

_____

Fetal circulation:

1. The failure of two of the fetal bypass structures to become obliterated after birth can cause congenital heart disease, in which the youngster would have improperly oxygenated blood. Which two structures are these?

    _____ and _____

2. For each of the following structures, first indicate its function in the fetus; and then note what happens to it or what it is converted to after birth. Circle the blood vessel that carries the most oxygen-rich blood.

Structure	Function in fetus	Fate
Umbilical artery		
Umbilical vein		
Ductus venosus		
Ductus arteriosus		
Foramen ovale		

3. What organ serves as a respiratory/digestive/excretory organ for the fetus? _____

Dissection of the Blood Vessels of the Cat

1. What differences did you observe between the origin of the left common carotid arteries in the cat and in the human? _____

   _____

   _____

   Between the origin of the internal and external iliac arteries? _____

   _____

   _____

   _____

2. How do the relative sizes of the external and internal jugular veins differ in the human and the cat? _____

   _____

   _____

3. In the cat the inferior vena cava is called the _____,

   and the superior vena cava is referred to as the _____.

Review Sheet

EXERCISE 25

The Lymphatic System

The Lymphatic System

1. Identify all structures that have leader lines on the diagram below. Use the key for your selection.

Key:
a. bone marrow
b. cisterna chyli
c. lymph nodes

d. lymphatic vessels
e. right lymphatic duct
f. spleen

g. thoracic duct
h. thymus
i. tonsils

1. _____

2. _____

3. _____

4. _____

5. _____

6. _____

7. _____

8. _____

9. _____

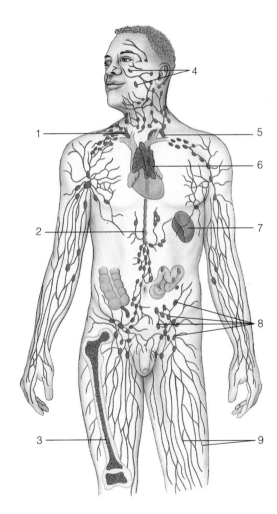

2. Explain why the lymphatic system is a one-way system, whereas the blood vascular system is a two-way system.

_____

_____

3. How do lymphatic vessels resemble veins?_____

   How do lymphatic capillaries differ from blood capillaries?_____

4. What is the function of the lymphatic vessels?_____

   _____

5. What is lymph?_____

6. What factors are involved in the flow of lymphatic fluid?_____

   _____

7. What name is given to the terminal duct draining most of the body?_____

8. What is the cisterna chyli?_____

   _____

   How does the composition of lymph in the cisterna chyli differ from that in the general lymphatic stream?

   _____

9. Which portion of the body is drained by the right lymphatic duct?_____

10. Note three areas where lymph nodes are densely clustered: _____,

    _____ , and _____

11. What are the two major functions of the lymph nodes?_____

    _____

12. The radical mastectomy is an operation in which a cancerous breast, surrounding tissues, and the underlying muscles of the anterior thoracic wall, plus the axillary lymph nodes, are removed. After such an operation, the arm usually swells, or becomes edematous, and is very uncomfortable—sometimes for months. Why?

    _____

Microscopic Anatomy of a Lymph Node

1. In the space to the right, make a rough drawing of the structure of a lymph node. Identify the cortex area, germinal centers, and medulla. For each identified area, note the cell type (T cell, B cell, or macrophage) most likely to be found there.

2. What structural characteristic ensures a *slow* flow of lymph through a lymph node and why is this desirable?

   _____

   _____

Review Sheet

EXERCISE 26

Anatomy of
the Respiratory System

Upper and Lower Respiratory Structures

1. Complete the labeling of the diagram of the upper respiratory structures (sagittal section).

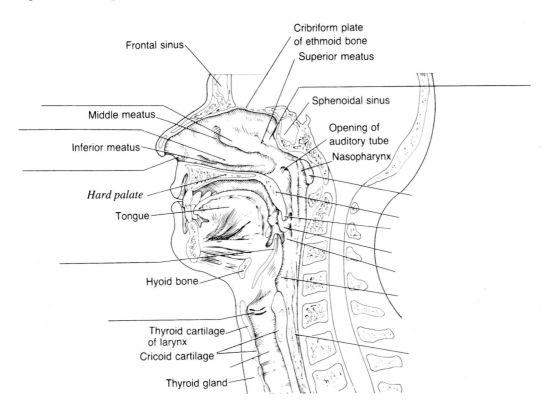

Frontal sinus

Cribriform plate
of ethmoid bone

Superior meatus

Sphenoidal sinus

Middle meatus

Opening of
auditory tube

Inferior meatus

Nasopharynx

Hard palate

Tongue

Hyoid bone

Thyroid cartilage
of larynx

Cricoid cartilage

Thyroid gland

2. Two pairs of vocal folds are found in the larynx. Which pair are the true vocal cords (superior or inferior)?

_____

3. What is the significance of the fact that the human trachea is reinforced with cartilage rings?

_____

Of the fact that the rings are incomplete posteriorly? _____

_____

4. Name the specific cartilages in the larynx that correspond to the following descriptions:

    1. forms the Adam's apple _____   3. shaped like a signet ring _____

    2. a "lid" for the larynx _____   4. vocal cord attachment _____

5. Trace a molecule of oxygen from the external nares to the pulmonary capillaries of the lungs: External nares →

_____

_____

_____

6. What is the function of the pleural membranes? _____

_____

_____

7. Name two functions of the nasal cavity mucosa: _____

8. The following questions refer to the primary bronchi:

   Which is longer? _____   Larger in diameter? _____   More horizontal? _____

   The more common site for lodging of a foreign object that had entered the respiratory passageways? _____

9. Match the terms in column B to those in column A.

   Column A

   _____ 1. nerve that activates the diaphragm during inspiration

   _____ 2. "floor" of the nasal cavity

   _____ 3. food passageway posterior to the trachea

   _____ 4. flaps over the glottis during swallowing of food

   _____ 5. contains the vocal cords

   _____ 6. part of the conducting pathway between the larynx and the primary bronchi

   _____ 7. pleural layer lining the walls of the thorax

   _____ 8. site from which oxygen enters the pulmonary blood

   _____ 9. autonomic nervous system nerve serving the thoracic region

   _____ 10. opening between the vocal folds

   _____ 11. fleshy lobes in the nasal cavity

   Column B

   a. alveolus

   b. bronchiole

   c. conchae

   d. epiglottis

   e. esophagus

   f. glottis

   g. larynx

   h. palate

   i. parietal pleura

   j. phrenic nerve

   k. primary bronchi

   l. trachea

   m. vagus nerve

   n. visceral pleura

10. What portions of the respiratory system are referred to as anatomical dead space? _____

_____

_____

   Why? _____

11. Define *external respiration:* _____

_____

internal respiration: _____

_____

12. On the diagram below identify alveolar epithelium, capillary endothelium, alveoli, and red blood cells, and bracket the respiratory membrane.

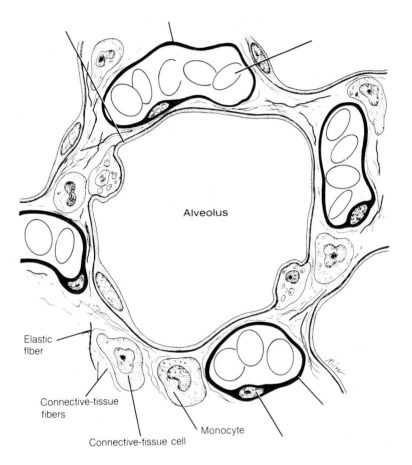

Alveolus

Elastic fiber

Connective-tissue fibers

Connective-tissue cell

Monocyte

Sheep Pluck Demonstration

1. Does the lung inflate part by part or as a whole, like a balloon? _____

   What happened when the pressure was released? _____

   What type of tissue ensures this phenomenon? _____

Examination of Prepared Slides of Lung and Trachea Tissue

1. The tracheal epithelium is ciliated and has goblet cells. What is the function of each of these modifications?

   Cilia? _____

   Goblet cells? _____

2. The tracheal epithelium is said to be pseudostratified. Why?_____

_____

_____

3. What structural characteristics of the alveoli make them an ideal site for the diffusion of gases?

_____

Why does oxygen move from the alveoli into the pulmonary capillary blood?_____

_____

_____

4. If you observed pathologic lung sections, what were the responsible conditions and how did the tissue differ from normal lung tissue?

Slide Type   **Observations**

Dissection of the Respiratory System of the Cat

1. Are the cartilaginous rings in the cat trachea complete or incomplete?_____

2. How does the number of lung lobes in the cat compare with the number in humans?_____

_____

3. Describe the appearance of lung tissue under the dissection microscope._____

_____

STUDENT NAME ______________________

LAB TIME/DATE ______________________

Review Sheet

EXERCISE 27

Anatomy of the Digestive System

General Histological Plan of the Alimentary Canal

1. The general anatomical features of the digestive tube have been presented. Fill in the table below to complete the information listed.

Wall layer	Subdivisions of the layer	Major functions
mucosa		
submucosa		
muscularis externa		
serosa or adventitia		

Organs of the Alimentary Canal

1. The tubelike digestive system canal that extends from the mouth to the anus is the ______________ canal.

2. How is the muscularis externa of the stomach modified? ______________________

____

How does this modification relate to the function of the stomach? ______________________

____

3. Match the items in column B with the descriptive statements in column A.

Column A

_____ 1. structure that suspends the small intestine from the posterior body wall

_____ 2. fingerlike extensions of the intestinal mucosa that increase the surface area for absorption

_____ 3. large collections of lymphoid tissue found in the submucosa of the small intestine

_____ 4. deep folds of the mucosa and submucosa that extend completely or partially around the circumference of the small intestine

_____, _____ 5. regions that break down foodstuffs mechanically

_____ 6. mobile organ that manipulates food in the mouth and initiates swallowing

_____ 7. conduit for both air and food

_____, _____, _____ 8. three structures continuous with and representing modifications of the peritoneum

_____ 9. the "gullet"; no digestive/absorptive function

_____ 10. folds of the gastric mucosa

_____ 11. sacculations of the large intestine

_____ 12. projections of the plasma membrane of a mucosal epithelial cell

_____ 13. valve at the junction of the small and large intestines

_____ 14. primary region of food and water absorption

_____ 15. membrane securing the tongue to the floor of the mouth

_____ 16. absorbs water and forms feces

_____ 17. area between the teeth and lips/cheeks

_____ 18. wormlike sac that outpockets from the cecum

_____ 19. initiates protein digestion

_____ 20. structure attached to the lesser curvature of the stomach

_____ 21. organ distal to the stomach

_____ 22. valve controlling food movement from the stomach into the duodenum

_____ 23. posterosuperior boundary of the oral cavity

_____ 24. location of the hepatopancreatic sphincter through which pancreatic secretions and bile pass

_____ 25. serous lining of the abdominal cavity wall

_____ 26. principal site for the synthesis of vitamin K by microorganisms

_____ 27. region containing two sphincters through which feces are expelled from the body

_____ 28. bone-supported anterosuperior boundary of the oral cavity

Column B

a. anus

b. appendix

c. esophagus

d. frenulum

e. greater omentum

f. hard palate

g. haustra

h. ileocecal valve

i. large intestine

j. lesser omentum

k. mesentery

l. microvilli

m. oral cavity

n. parietal peritoneum

o. Peyer's patches

p. pharynx

q. plicae circulares

r. pyloric valve

s. rugae

t. small intestine

u. soft palate

v. stomach

w. tongue

x. vestibule

y. villi

z. visceral peritoneum

4.   Correctly identify all organs depicted in the diagram below.

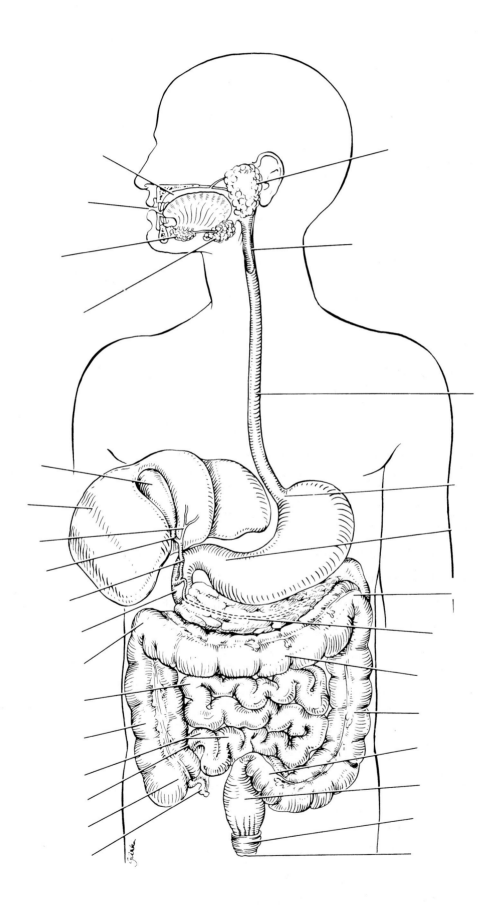

5. You have studied the histologic structure of a number of organs in this laboratory. Three of these are diagrammed below. Identify each.

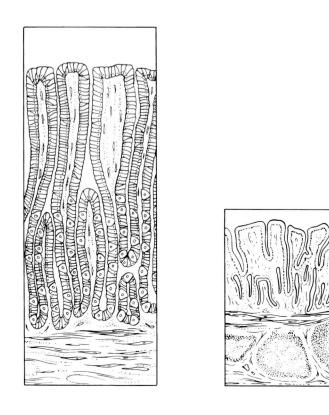 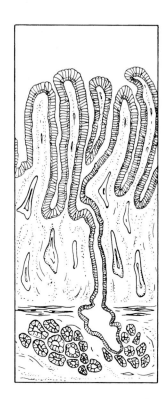

_____   _____   _____

6. What transition in epithelium type exists at the cardio-esophageal junction?_____

_____

How do the epithelia of these two organs relate to their specific functions?_____

_____

_____

7. What cells of the stomach produce HCl? _____ Pepsinogen?_____

8. Name three structures always found in the portal triad regions of the liver._____,

_____ , and_____.

9. Where would you expect to find the Kupffer cells of the liver?_____

What is their function?_____

10. Why is the liver so dark red in the living animal?_____

_____

Accessory Digestive Organs

1. Various types of glands form a part of the alimentary tube wall or duct their secretions into it. Match the glands listed in column B with the function/locations described in column A.

Column A

Column B

_____ 1. produce(s) mucus; found in the submucosa of the small intestine

a. duodenal glands

_____ 2. produce(s) a product containing amylase that begins starch breakdown in the mouth

b. gastric glands

c. intestinal crypts

_____ 3. produce(s) a whole spectrum of enzymes and an alkaline fluid that is secreted into the duodenum

d. liver

_____ 4. produce(s) bile that it secretes into the duodenum via the bile duct

e. pancreas

_____ 5. produce(s) HCl and pepsinogen

f. salivary glands

_____ 6. found in the mucosa of the small intestine; produce(s) intestinal juice

2. Which of the salivary glands produces a secretion that is mainly serous?_____

3. What is the role of the gallbladder? _____

4. Use the key to identify each tooth area described below.   Key: a. anatomical crown

_____ 1. visible portion of the tooth *in situ*

b. cementum

_____ 2. material covering the tooth root

c. clinical crown

_____ 3. hardest substance in the body

d. dentin

_____ 4. attaches the tooth to bone and surrounding alveolar structures

e. enamel

_____ 5. portion of the tooth embedded in bone

f. gingiva

_____ 6. forms the major portion of tooth structure; similar to bone

g. odontoblasts

_____ 7. form the dentin

h. periodontal ligament

_____ 8. site of blood vessels, nerves, and lymphatics

i. pulp

_____ 9. entire portion of the tooth covered with enamel

j. root

5. In the human, the number of deciduous teeth is _____; the number of permanent teeth is_____.

6. The dental formula for permanent teeth is $\dfrac{2, 1, 2, 3}{2, 1, 2, 3}$

Explain what this means: _____

_____

What is the dental formula for the deciduous teeth?

7. What teeth are the "wisdom teeth"? _____

Dissection of the Digestive System of the Cat

1. Several differences between cat and human digestive anatomy should have become apparent during the dissection. Note the pertinent differences between the human and the cat relative to the following structures:

Structure	Cat	Human
tongue papillae		
number of liver lobes		
appendix		

Review Sheet

EXERCISE 28

Anatomy of the Urinary System

Gross Anatomy of the Human Urinary System

1. Complete the following statements:

   The kidney is referred to as an excretory organ because it excretes __1__ wastes. It is also a major homeostatic organ because it maintains the electrolyte, __2__ , and __3__ balance of the blood.
   Urine is continuously formed by the __4__ and is routed down the __5__ by the mechanism of __6__ to a storage organ called the __7__ . Eventually, the urine is conducted to the body __8__ by the urethra. In the male, the urethra is __9__ inches long and transports both urine and __10__ . The female urethra is __11__ inches long and transports only urine.
   Voiding or emptying the bladder is called __12__ . Voiding has both voluntary and involuntary components. The voluntary sphincter is the __13__ sphincter. An inability to control this sphincter is referred to as __14__ .

   1. _____
   2. _____
   3. _____
   4. _____
   5. _____
   6. _____
   7. _____
   8. _____
   9. _____
   10. _____
   11. _____
   12. _____
   13. _____
   14. _____

2. What is the function of the fat cushion that surrounds the kidneys in life? _____

   _____

3. Define *ptosis.* _____

   _____

4. Why is incontinence a normal phenomenon in the child under 1½ to 2 years old? _____

   _____

   What events may lead to its occurrence in the adult? _____

   _____

   _____

5. Complete the labeling of the diagram to correctly identify the urinary system organs.

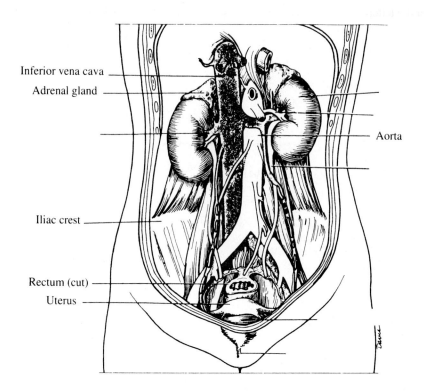

Inferior vena cava

Adrenal gland

Aorta

Iliac crest

Rectum (cut)

Uterus

Gross Internal Anatomy of the Pig or Sheep Kidney

Match the appropriate structure in column B to its description in column A.

Column A

_____ 1. smooth membrane, tightly adherent to the kidney surface

_____ 2. portion of the kidney containing mostly collecting ducts

_____ 3. portion of the kidney containing the bulk of the nephron structures

_____ 4. superficial region of kidney tissue

_____ 5. basinlike area of the kidney, continuous with the ureter

_____ 6. a cup-shaped extension of the pelvis that encircles the apex of a pyramid

_____ 7. area of cortical tissue running between the medullary pyramids

Column B

a. cortex

b. medulla

c. minor calyx

d. renal capsule

e. renal column

f. renal pelvis

Microscopic Anatomy of the Kidney and Bladder

1. Match each of the lettered structures on the diagram of the nephron (and associated renal blood supply) on the left with the terms on the right:

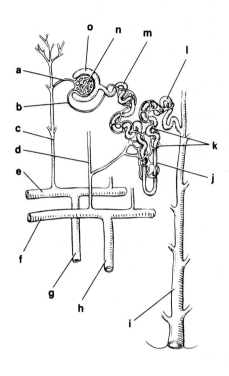

_____ 1. collecting duct

_____ 2. glomerulus

_____ 3. peritubular capillaries

_____ 4. distal convoluted tubule

_____ 5. proximal convoluted tubule

_____ 6. interlobar artery

_____ 7. interlobular artery

_____ 8. arcuate artery

_____ 9. interlobular vein

_____ 10. efferent arteriole

_____ 11. arcuate vein

_____ 12. loop of Henle

_____ 13. afferent arteriole

_____ 14. interlobar vein

_____ 15. glomerular capsule

2. Using the terms provided in item 1, identify the following:

_____ 1. site of filtrate formation

_____ 2. primary site of tubular reabsorption

_____ 3. secondarily important site of tubular reabsorption

_____ 4. structure that conveys the processed filtrate (urine) to the renal pelvis

_____ 5. blood supply that directly receives substances from the tubular cells

_____ 6. its inner (visceral) membrane forms part of the filtration membrane

3. Explain *why* the glomerulus is such a high-pressure capillary bed. _____

_____

How does its high pressure condition aid its function of filtrate formation? _____

_____

4. What structural modification of certain tubule cells enhances their ability to reabsorb substances from the filtrate?

_____

5. Explain the mechanism of tubular secretion and explain its importance in the urine formation process. _____

_____

_____

_____

6. Compare and contrast the composition of blood plasma and glomerular filtrate. _____

_____

7. Trace a drop of blood from the time it enters the kidney in the renal artery until it leaves the kidney through the

renal vein. Renal artery $\rightarrow$ _____

_____

_____ $\rightarrow$ renal vein

8. Trace the anatomical pathway of a molecule of creatinine (metabolic waste) from the glomerular capsule to the
urethra. Note each microscopic and/or gross structure it passes through in its travels. Name the subdivisions of the

renal tubule. Glomerular capsule $\rightarrow$ _____

_____

_____ $\rightarrow$ urethra

9. What is important functionally about the specialized epithelium (transitional epithelium) in the bladder?

_____

_____

Dissection of the Cat Urinary System

1. How does the position of the kidneys in the cat differ from their position in humans? _____

_____

_____

2. How does the site of urethral emptying in the female cat differ from its termination point in the human female?

_____

_____

_____

3. What gland encircles the neck of the bladder in the male? _____ Is this part of the urinary

system? _____ What is its function? _____

_____

Review Sheet

EXERCISE 29

Anatomy of the Reproductive System

Gross Anatomy of the Human Male Reproductive System

1. List the two principal functions of the testis: _____

2. Identify all indicated structures or portions of structures on the diagrammatic view of the male reproductive system below.

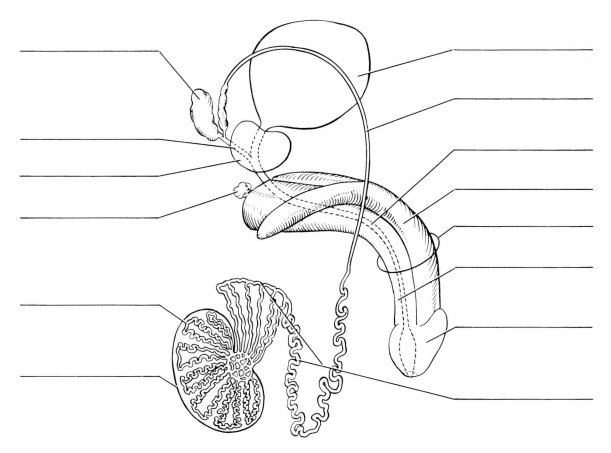

3. A common part of any physical examination of the male is palpation of the prostate gland. How is this accomplished? (Think!) _____

4. How might enlargement of the prostate gland interfere with urination or the reproductive ability of the male?

_____

_____

5. Match the terms in column B to the descriptive statements in column A.

Column A

_____ 1. copulatory organ/penetrating device

_____ 2. site of sperm/androgen production

_____ 3. muscular passageway conveying sperm to the ejaculatory duct; in the spermatic cord

_____ 4. transports both sperm and urine

_____ 5. sperm maturation site

_____ 6. location of the testis in adult males

_____ 7. loose fold of skin encircling the glans penis

_____ 8. portion of the urethra between the prostate gland and the penis

_____ 9. empties a secretion into the prostatic urethra

_____ 10. empties a secretion into the membranous urethra

Column B

a. bulbourethral glands

b. epididymis

c. glans penis

d. membranous urethra

e. penile urethra

f. penis

g. prepuce

h. prostate gland

i. prostatic urethra

j. seminal vesicles

k. scrotum

l. testes

m. vas (ductus) deferens

6. Why are the testes located in the scrotum? _____

_____

7. Describe the composition of semen and name all structures contributing to its formation. _____

_____

_____

8. Of what importance is the fact that seminal fluid is alkaline? _____

_____

9. What structures comprise the spermatic cord? _____

_____

Where is it located? _____

_____

10. Using the following terms, trace the pathway of sperm from the testes to the urethra: rete testis, epididymis, seminiferous tubule, ductus deferens.

_____ → _____ → _____ → _____

11. Using an appropriate reference, define *cryptorchidism* and discuss its significance.

_____

_____

Gross Anatomy of the Human
Female Reproductive System

1. On the diagram of a frontal section of a portion of the female reproductive system seen below, identify all indicated structures.

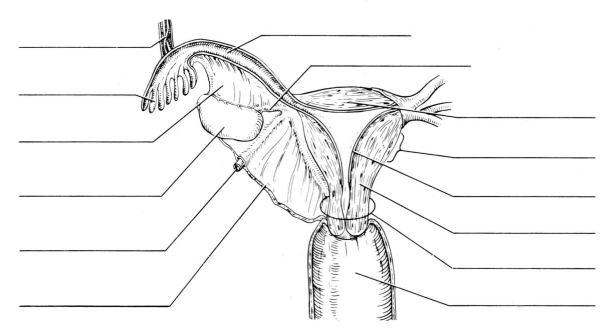

2. Identify the female reproductive system structures described below:

_____ 1. site of fetal development

_____ 2. copulatory canal

_____ 3. "fertilized egg" typically formed here

_____ 4. becomes erectile during sexual excitement

_____ 5. duct extending superolaterally from the uterus

_____ 6. partially closes the vaginal canal; a membrane

_____ 7. produces eggs, estrogens, and progesterone

_____ 8. fingerlike ends of the fallopian tube

3. Do any sperm enter the pelvic cavity of the female? Why or why not? _____

_____

4. What is an ectopic pregnancy, and how can it happen? _____

_____

_____

_____

5. Name the structures composing the external genitalia, or vulva, of the female. _____

_____

_____

6. Put the following vestibular-perineal structures in their proper order from the anterior to the posterior aspect: vaginal orifice, anus, urethral opening, and clitoris.

   Anterior limit: _____ → _____ → _____ → _____

7. Name the male structure that is homologous to the female structures named below.

   labia majora _____   clitoris _____

8. Assume a couple has just consummated the sex act and the male's sperm have been deposited in the woman's vagina. Trace the pathway of the sperm through the female reproductive tract.

_____

9. Define *ovulation:* _____

The Mammary Glands

1. To describe breast function, complete the following sentences:

   Milk is formed by _____ within the _____ of the breast. Milk

   is then excreted into enlarged storage regions called _____ and then finally through the

_____.

2. Describe the procedure for self-examination of the breasts. (Men are not exempt from breast cancer, you know!)

_____

_____

_____

_____

_____

_____

Microscopic Anatomy of Selected Male and Female Reproductive Organs

1. The testis is divided into a number of lobes by connective tissue. Each of these lobes contains one to four

    _____, which converge on a tubular region at the testis hilus called the

    _____.

2. What is the function of the cavernous bodies seen in the male penis? _____

    _____

    _____

3. Name the three layers of the uterine wall from the inside out.

    _____, _____, _____

    Which of these is sloughed during menses? _____

    Which contracts during childbirth? _____

4. What is the function of the stereocilia exhibited by the epithelial cells of the mucosa of the epididymis? _____

    _____

5. On the diagram showing the sagittal section of the human testis, correctly identify all structures provided with leader lines.

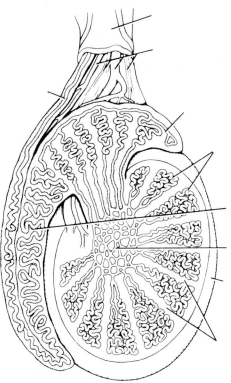

Dissection of the Reproductive System of the Cat

1. The female cat has a _____ uterus; that of the human female is _____ .

   Explain the difference in structure of these two uterine types. _____

   _____

   _____

2. What reproductive advantage is conferred by the feline uterine type? _____

   _____

3. Cite differences noted between the cat and the human relative to the following structures:

   uterine tubes or oviducts _____

   site of entry of ductus deferens into the urethra _____

   _____

   location of the prostate gland _____

   _____

   seminal vesicles _____

   urethral and vaginal openings in the female _____

   _____

   _____

The Metric System

Measurement	Unit and abbreviation	Metric equivalent	Metric to English conversion factor	English to metric conversion factor
Length	1 kilometer (km)	= 1000 (10^3) meters	1 km = 0.62 mile	1 mile = 1.61 km
	1 meter (m)	= 100 (10^2) centimeters = 1000 millimeters	1 m = 1.09 yards 1 m = 3.28 feet 1 m = 39.37 inches	1 yard = 0.914 m 1 foot = 0.305 m
	1 centimeter (cm)	= 0.01 (10^{-2}) meter	1 cm = 0.394 inch	1 foot = 30.5 cm 1 inch = 2.54 cm
	1 millimeter (mm)	= 0.001 (10^{-3}) meter	1 mm = 0.039 inch	
	1 micrometer (µm) [formerly micron (µ)]	= 0.000001 (10^{-6}) meter		
	1 nanometer (nm) [formerly millimicron (mµ)]	= 0.000000001 (10^{-9}) meter		
	1 angstrom (Å)	= 0.0000000001 (10^{-10}) meter		
Area	1 square meter (m²)	= 10,000 square centimeters	1 m² = 1.1960 square yards 1 m² = 10.764 square feet	1 square yard = 0.8361 m² 1 square foot = 0.0929 m²
	1 square centimeter (cm²)	= 100 square millimeters	1 cm² = 0.155 square inch	1 square inch = 6.4516 cm²
Mass	1 metric ton (t)	= 1000 kilograms	1 t = 1.103 ton	1 ton = 0.907 t
	1 kilogram (kg)	= 1000 grams	1 kg = 2.205 pounds	1 pound = 0.4536 kg
	1 gram (g)	= 1000 milligrams	1 g = 0.0353 ounce 1 g = 15.432 grains	1 ounce = 28.35 g
	1 milligram (mg)	= 0.001 gram	1 mg = approx. 0.015 grain	
	1 microgram (µg)	= 0.000001 gram		
Volume (solids)	1 cubic meter (m³)	= 1,000,000 cubic centimeters	1 m³ = 1.3080 cubic yards 1 m³ = 35.315 cubic feet	1 cubic yard = 0.7646 m³ 1 cubic foot = 0.0283 m³
	1 cubic centimeter (cm³ or cc)	= 0.000001 cubic meter = 1 milliliter	1 cm³ = 0.0610 cubic inch	1 cubic inch = 16.387 cm³
	1 cubic millimeter (mm³)	= 0.000000001 cubic meter		
Volume (liquids and gases)	1 kiloliter (kl or kL)	= 1000 liters	1 kL = 264.17 gallons	1 gallon = 3.785 L 1 quart = 0.946 L
	1 liter (l or L)	= 1000 milliliters	1 L = 0.264 gallons 1 L = 1.057 quarts	
	1 milliliter (ml or mL)	= 0.001 liter = 1 cubic centimeter	1 ml = 0.034 fluid ounce 1 ml = approx. $\frac{1}{4}$ teaspoon 1 ml = approx. 15–16 drops (gtt.)	1 quart = 946 ml 1 pint = 473 ml 1 fluid ounce = 29.57 ml 1 teaspoon = approx. 5 ml
	1 microliter (µl or µL)	= 0.000001 liter		
Time	1 second (s)	= $\frac{1}{60}$ minute		
	1 millisecond (ms)	= 0.001 second		
Temperature	Degrees Celsius (°C)		$°F = \frac{9}{5}°C + 32$	$°C = \frac{5}{9}(°F - 32)$

A.D.A.M. Correlations

Appendix B lists correlations to *A.D.A.M. Standard* or *A.D.A.M. Comprehensive* for appropriate exercises in this manual. When you choose a system from the "View" menu of the main menu bar, the selected system will be highlighted and the rest of the anatomy will be dimmed. You will also see a text overview of the selected system in the right corner. When in "Full Anatomy" in *A.D.A.M. Standard* or *A.D.A.M. Comprehensive*, "View" refers to the "Anterior," "Lateral," "Posterior," or "Medial" view in the primary window. "Layer" refers to one or more of the layers listed in the lower right corner of the computer screen. Note that the layer name in *A.D.A.M. Standard* or *A.D.A.M. Comprehensive* may be different from the key structure you are studying, but using the indicated layer will bring the structure into full view.

Exercise 1
The Language of Anatomy

Go to "Full Anatomy" to illustrate a coronal plane
View: Anterior
Layer: Skull—coronal section

Go to "Full Anatomy" to illustrate a midsagittal plane
View: Medial
Layer: Skin

Go to the "Cross Sections" book to illustrate a transverse plane
Choose the eye button for the location of the cross-sectional illustration or MRI image of your choice*

Go to "Full Anatomy"
View: Medial
Layer: Central nervous system and sacral plexus
Key Structure: Dorsal body cavity

Go to "Full Anatomy"
View: Anterior
Layer: Diaphragm
Key Structure: Ventral body cavity

Go to "Full Anatomy"
View: Anterior
Layers: Pericardial sac, Parietal peritoneum, Peritoneum
Key Structures: Serous membranes of the ventral body cavity

Go to "Full Anatomy"
View: Medial
Layer: Parietal pleura
Key Structure: Parietal pleura

Exercise 2
Organ Systems Overview

Go to "View" on the Menu Bar
Choose each body system and review text and illustrations.

Exercise 7
Classification of Body Membranes

Go to "Full Anatomy"
View: Anterior
Layer: Lungs
Key Structures: Parietal pleural membranes

Go to "Full Anatomy"
View: Medial
Layer: Parietal pleura
Key Structures: Peritoneal membranes

Exercise 8
Overview of the Skeleton: Classification and Structure of Bones and Cartilages

Go to "View" on Menu Bar
Choose "Skeletal System"

Exercise 9
The Axial Skeleton

Go to "View," choose "Skeletal System"

Go to "Full Anatomy"
View: Anterior
Layer: Skull
Key Structures: Bones of the cranium

Go to "Full Anatomy"
View: Posterior
Layer: Bones—coronal section
Key Structures: Bones of the cranium

Go to "Full Anatomy"
View: Lateral
Layer: Skull
Key Structures: Bones of the cranium and face

Go to "Full Anatomy"
View: Anterior
Layer: Skull
Key Structures: Bones of the face

Go to "Full Anatomy"
View: Lateral
Layer: Hyoid bone
Key Structure: Hyoid bone

Go to "Full Anatomy"
View: Anterior
Layer: Skull—coronal section
Key Structures: Paranasal sinuses

Go to "Full Anatomy"
View: Anterior
Layer: Atlas
Key Structure: Vertebral column

Go to "Full Anatomy"
View: Posterior
Layer: Posterior vertebral column
Key Structure: Posterior vertebral column

Go to "Full Anatomy"
View: Medial
Layer: Central nervous system and sacral plexus
Key Structures: Vertebral column and intervertebral discs

Go to "Full Anatomy"
View: Posterior
Layer: Bones—coronal section
Key Structures: Sacrum and lumbar vertebrae

Go to "Full Anatomy"
View: Anterior
Layer: Ribs
Key Structure: Bony thorax

---

*Open the Cross Sections book in the Library window, and then choose the eye button as requested. "Level 1" refers to the cross section at the top of the image and "Level 7" refers to the bottom cross section.

Exercise 10
The Appendicular Skeleton

Go to "View," choose "Skeletal System"

Go to "Full Anatomy"
View: Anterior
Layer: Clavicle
Key Structure: Pectoral girdle

Go to "Full Anatomy"
View: Posterior
Layer: Scapula
Key Structure: Pectoral girdle

Go to "Full Anatomy"
View: Anterior
Layer: Humerus
Key Structure: Humerus

Go to "Full Anatomy"
View: Anterior
Layer: Ulna
Key Structure: Humerus

Go to "Full Anatomy"
View: Anterior
Layer: Bones—coronal section
Key Structures: Bones of the wrist and hand

Go to "Full Anatomy"
View: Anterior
Layer: Femur
Key Structures: Bones of the pelvic girdle

Using "Options," choose "Female"
Go to "Full Anatomy"
View: Anterior
Layer: Bones—coronal section
Key Structure: Female pelvis

Using "Options," choose "Male"
Go to "Full Anatomy"
View: Anterior
Layer: Bones—coronal section
Key Structure: Male pelvis

Go to "Full Anatomy"
View: Anterior
Layer: Femur
Key Structure: Femur

Go to "Full Anatomy"
View: Anterior
Layer: Tibia
Key Structure: Tibia

Go to "Full Anatomy"
View: Anterior
Layer: Bones of the foot
Key Structures: Bones of the foot

Exercise 11
Articulations and Body Movements

Go to "Full Anatomy"
View: Lateral
Layer: Synovial joint of the knee
Key Structure: Synovial joint of the knee

Go to "Full Anatomy"
View: Anterior
Layer: Synovial joint capsule of the hip
Key Structure: Synovial joint capsule of the hip

Exercise 14
Gross Anatomy of the Muscular System

Go to "Full Anatomy"
View: Anterior
Layer: Orbicularis oculi
Key Structures: Muscles of the face

Go to "Full Anatomy"
View: Anterior
Layer: Masseter muscle
Key Structures: Muscles of mastication

Go to "Full Anatomy"
View: Lateral
Layers: Platysma muscle, Sternocleidomastoid muscle
Key Structures: Superficial muscles of the neck

Go to "Full Anatomy"
View: Lateral
Layer: Strap muscles
Key Structures: Deep muscles of the neck

Go to "Full Anatomy"
View: Lateral
Layers: Pectoralis major muscle, Serratus anterior muscle
Key Structures: Thorax and shoulder muscles

Go to "Full Anatomy"
View: Anterior
Layers: External intercostal muscles, Internal intercostal muscle
Key Structures: Thorax muscles

Go to "Full Anatomy"
View: Anterior
Layers: Rectus abdominis muscle, External abdominal oblique muscle, Internal abdominal oblique muscle
Key Structures: Abdominal wall muscles

Go to "Full Anatomy"
View: Lateral
Layer: Latissimus dorsi muscle
Key Structures: Thorax muscles

Go to "Full Anatomy"
View: Posterior
Layer: Rhomboideus muscle
Key Structures: Posterior muscles of the trunk

Go to "Full Anatomy"
View: Posterior
Layers: Semispinalis muscle, Splenius muscle
Key Structures: Muscles associated with the vertebral column

Go to "Full Anatomy"
View: Posterior
Layer: Brachialis muscle
Key Structures: Muscles of the humerus that act on the forearm

Go to "Full Anatomy"
View: Anterior
Layers: Flexor carpi radialis muscle, Palmaris longus muscle, Flexor carpi ulnaris muscle, Flexor digitorum superficialis muscle
Key Structures: Anterior muscles of the forearm that act on the hand and fingers

Go to "Full Anatomy"
View: Anterior
Layer: Abductor pollicis longus muscle
Key Structures: Deep muscles of the forearm that act on the hand and fingers

Go to "Full Anatomy"
View: Anterior
Layers: Sartorius muscle
Key Structures: Muscles acting on the thigh

Go to "Full Anatomy"
View: Anterior
Layers: Rectus femoris muscle, Vastus lateralis muscle, Vastus medialis muscle, Vastus intermedius muscle, Tensor fasciae latae
Key Structures: Quadriceps

Go to "Full Anatomy"
View: Posterior
Layers: Gluteus maximus muscle,
Gluteus medius muscle, Gluteus
minimus muscle
Key Structures: Muscles acting on the
thigh and originating on the pelvis

Go to "Full Anatomy"
View: Posterior
Layers: Long head of the biceps
femoris muscle, Semitendinosus
muscle, Semimembranosus muscle
Key Structures: Hamstrings

Go to "Full Anatomy"
View: Posterior
Layers: Gastrocnemius muscle, Soleus
muscle, popliteus muscle, Tibialis
posterior
Key Structures: Superficial muscles
acting on the foot and ankle

Go to "Full Anatomy"
View: Anterior
Layers: Tibialis anterior muscle,
Extensor digitorum longus muscle,
Extensor hallucis longus muscle
Key Structures: Muscles acting on the
foot and ankle

Exercise 16
Gross Anatomy of the Brain and Cranial Nerves

Go to "View," choose "Nervous
System"

Go to "Full Anatomy"
View: Anterior
Layer: Skull—coronal section
Key Structure: Cerebrum

Go to "Full Anatomy"
View: Lateral
Layer: Brain
Key Structure: Cerebrum

Go to the "Cross Sections" book,
choose Level 1 (in the cerebral
region)*
Key Structure: Cerebrum

Go to "Full Anatomy"
View: Medial
Layer: Central nervous system and
sacral plexus
Key Structures: Diencephalon and
pituitary gland

Go to the "Cross Sections" book,
choose Level 2*
Key Structures: Cerebellum,
cerebrum, and cranial nerves V and VI

Go to "Full Anatomy"
View: Lateral
Layer: Falx cerebri
Key Structures: Internal anatomy of
the cerebrum and meningeal extensions

Go to the "Cross Sections" book,
choose Level 1 (in the cerebral region)
Key Structures: Internal anatomy of
the cerebrum

Go to "Full Anatomy"
View: Lateral
Layer: Cranial nerves
Key Structures: Cranial nerves

Go to "Full Anatomy"
View: Lateral
Layer: Right vagus nerve
Key Structure: Vagus nerve

Go to "Full Anatomy"
View: Anterior
Layer: Phrenic and vagus nerves
Key Structures: Phrenic and vagus
nerves

Exercise 17
Spinal Cord, Spinal Nerves, and the Autonomic Nervous System

Go to "View," choose "Nervous
System"

Go to "Full Anatomy"
View: Posterior
Layer: Spinal Cord
Key Structure: Spinal cord

Go to the "Cross Sections" book,
choose Level 3 (in the cervical region)
Key Structure: Spinal cord

Go to "Full Anatomy"
View: Anterior
Layer: Deep nerve plexuses and inter-
costal nerves
Key Structures: Deep nerve plexuses
and intercostal nerves

Go to "Full Anatomy"
View: Anterior
Layer: Brachial plexus and branches
Key Structures: Brachial plexus and
branches

Go to "Full Anatomy"
View: Medial
Layer: Central nervous system and
sacral plexus
Key Structures: Central nervous
system and sacral plexus

Go to "Full Anatomy"
View: Medial
Layer: Sciatic nerve and branches
Key Structures: Sciatic nerve and
branches

Go to "Full Anatomy"
View: Posterior
Layer: Spinal cord
Key Structure: Spinal cord

Go to "Full Anatomy"
View: Posterior
Layer: Tibial nerve and branches
Key Structure: Tibial nerve

Go to "Full Anatomy"
View: Medial
Layer: Autonomic nerve plexuses
Key Structures: Autonomic nerve
plexuses

Go to "Full Anatomy"
View: Anterior
Layer: Sympathetic trunk
Key Structure: Sympathetic trunk

Exercise 18
Special Senses: Vision

Go to "Full Anatomy"
View: Anterior
Layer: Muscles of the eye
Key Structures: Muscles and external
anatomy of the eye

Go to "Full Anatomy"
View: Lateral
Layer: Eye muscles
Key Structures: Eye muscles

Go to the "Cross Sections" book,
choose the eye button at Level 2 (in
the cephalic region)*
Key Structure: Eye

---

*Open the Cross Sections book in the Library window, and then choose the eye button as requested. "Level 1" refers to the cross section at the top of the image and "Level 7" refers to the bottom cross section.

Exercise 21
Functional Anatomy of the Endocrine Glands

Go to "Options," choose "Female"
Go to "View," choose "Endocrine System"

Go to "Options," choose "Male"
Go to "View," choose "Endocrine System"

Go to "Full Anatomy"
View: Lateral
Layer: Tongue
Key Structure: Pituitary gland

Go to "Full Anatomy"
View: Anterior
Layer: Thyroid gland
Key Structure: Thyroid gland

Go to "Full Anatomy"
View: Anterior
Layer: Suprarenal gland
Key Structure: Suprarenal gland

Go to "Full Anatomy"
View: Anterior
Layer: Pancreas
Key Structure: Pancreas

Go to "Options," choose "Female"
View: Medial
Layer: Ovarian artery
Key Structure: Ovaries

Go to "Options," choose "Male"
View: Anterior
Layer: Testes
Key Structure: Testes

Go to "Full Anatomy"
View: Anterior
Layer: Remnant of thymus gland
Key Structure: Remnant of thymus gland

Go to "Full Anatomy"
View: Anterior
Layer: Testes
Key Structure: Testes

Exercise 23
Anatomy of the Heart

Go to "Full Anatomy"
View: Anterior
Layers: Ribs, Cardiac veins, Heart—cut section
Key Structure: Position of the heart and valves in the thoracic cavity

Go to "Full Anatomy"
View: Anterior
Layer: Heart
Key Structure: Heart

Go to "Full Anatomy"
View: Anterior
Layer: Pericardiacophrenic vein
Key Structure: Fibrous pericardium

Go to "Full Anatomy"
View: Anterior
Layer: Epicardium
Key Structure: Visceral pericardium

Go to "Full Anatomy"
View: Anterior
Layer: Heart—cut section
Key Structures: Heart chambers and heart valves

Go to "Full Anatomy"
View: Anterior
Layer: Coronary arteries
Key Structure: Coronary arteries

Exercise 24
Anatomy of Blood Vessels

Go to "View," choose "Cardiovascular System"

Go to "Full Anatomy"
View: Anterior
Layer: Aortic arch and branches
Key Structures: Aortic arch and branches

Go to "Full Anatomy"
View: Anterior
Layer: Common carotid artery and branches
Key Structures: Branches of the aortic arch

Go to "Full Anatomy"
View: Anterior
Layer: Celiac trunk and branches
Key Structures: Major branches of the descending aorta

Go to "Full Anatomy"
View: Anterior
Layer: Descending thoracic aorta
Key Structure: Descending thoracic aorta

Go to "Full Anatomy"
View: Anterior
Layer: Abdominal aorta and branches
Key Structures: Abdominal aorta and branches

Go to "Options," choose "Male"
View: Anterior
Layer: Gonadal veins
Key Structures: Abdominal aorta and branches

Go to "Full Anatomy"
View: Lateral
Layer: Aortic arch and branches
Key Structures: Lateral of the aorta and vena cava

Go to "Full Anatomy"
View: Anterior
Layer: Superior vena cava and tributaries
Key Structures: Veins draining into the vena cava

Go to "Full Anatomy"
View: Anterior
Layers: Radial artery, Deep femoral artery
Key Structure: Pulse points

Go to "Full Anatomy"
View: Anterior
Layer: Pulmonary arteries
Key Structure: Pulmonary circulation

Go to "Full Anatomy"
View: Lateral
Layer: Aortic arch and branches
Key Structure: Arterial supply of the brain

Go to "Full Anatomy"
View: Anterior
Layer: Portal vein and tributaries
Key Structure: Hepatic portal circulation

Exercise 25
The Lymphatic System

Go to "View," choose "Lymphatic System"
Key Structures: Lymphatic vessels and lymph nodes

Go to "Full Anatomy"
View: Anterior
Layer: Thoracic duct
Key Structure: Thoracic duct

Go to "Full Anatomy"
View: Anterior
Layer: Axillary and cervical lymph nodes
Key Structures: Axillary and cervical lymph nodes

Go to "Full Anatomy"
View: Anterior
Layer: Remnant of thymus gland
Key Structure: Remnant of thymus
gland
(View histology also)*

Go to "Full Anatomy"
View: Anterior
Layer: Spleen
Key Structure: Spleen
(View histology also)*

Go to "Full Anatomy"
View: Anterior
Layer: Palatine glands and tonsils
Key Structures: Palatine glands and
tonsils
(View histology also)*

Go to "Full Anatomy"
View: Anterior
Layer: Axillary and cervical lymph
nodes
Key Structures: Axillary and cervical
lymph nodes
(View histology also)*

Exercise 26
Anatomy of the
Respiratory System

Go to "View," choose "Respiratory
System"

Go to "Full Anatomy"
View: Medial
Layer: Nasal conchae
Key Structure: Upper respiratory tract

Go to "Full Anatomy"
View: Lateral
Layer: Mediastinal parietal pleura of
left pleural cavity
Key Structure: Upper respiratory tract

Go to "Full Anatomy"
View: Anterior
Layer: Thyroid gland
Key Structure: Larynx

Go to "Full Anatomy"
View: Anterior
Layer: Epiglottis
Key Structure: Epiglottis

Go to "Full Anatomy"
View: Medial
Layer: Structures in the hilum of the
right lung
Key Structure: Lower respiratory
structures

Go to "Full Anatomy"
View: Anterior
Layer: Trachea
Key Structure: Lower respiratory
structures

Go to "Full Anatomy"
View: Anterior
Layer: Lungs
Key Structures: Lungs

Go to "Full Anatomy"
View: Anterior
Layer: Lungs—coronal section
Key Structure: Lungs—coronal
section

Go to "Full Anatomy"
View: Lateral
Layer: Right lung
Key Structure: Right lung

Go to "Full Anatomy"
View: Lateral
Layer: Parietal pleura
Key Structure: Parietal pleura

Go to "Full Anatomy"
View: Anterior
Layer: Diaphragm
Key Structure: Diaphragm

Go to the "Cross sections" book,
choose Level 5 and Level 6†
Key Structure: Lungs

Go to "Full Anatomy"
View: Lateral
Layer: Trachea
Key Structure: Trachea
(View histology also)*

Exercise 27
Anatomy of the Digestive
System

Go to "View," choose "Digestive
System"

Go to "Full Anatomy"
View: Lateral
Layer: Esophagus
Key Structure: Oral cavity

Go to "Full Anatomy"
View: Anterior
Layer: Oral mucosa
Key Structures: Tonsils

Go to "Full Anatomy"
View: Lateral
Layers: Parotid gland and duct, Deep
salivary glands
Key Structures: Salivary glands

Go to "Full Anatomy"
View: Lateral
Layer: Esophagus
Key Structure: Esophagus
(View histology also)*

Go to "Full Anatomy"
View: Anterior
Layers: Stomach, Stomach—coronal
section
Key Structure: Stomach

Go to "Full Anatomy"
View: Medial
Layer: Stomach
Key Structure: Stomach

Go to "Full Anatomy"
View: Lateral
Layer: Esophagus
Key Structure: Esophagus

Go to "Full Anatomy"
View: Anterior
Layer: Small intestine
Key Structure: Ileum

Go to "Full Anatomy"
View: Anterior
Layer: Duodenum
Key Structure: Duodenum

Go to "Full Anatomy"
View: Anterior
Layer: Transverse colon
Key Structure: Transverse colon

Go to "Full Anatomy"
View: Anterior
Layer: Ascending and descending
colon
Key Structure: Ascending and
descending colon

*In "Full Anatomy," select the structure with the Identify tool, and then select "Histology" from the pop-up submenu.
†Open the Cross Sections book in the Library window, and then choose the eye button as requested. "Level 1" refers to the cross section at the top of the image and "Level 7" refers to the bottom cross section.

Go to "Full Anatomy"
View: Medial
Layer: Colon
Key Structure: Colon

Go to "Full Anatomy"
View: Anterior
Layer: Skull
Key Structures: Teeth

Go to "Full Anatomy"
View: Anterior
Layers: Common bile duct, Liver,
Liver—coronal section, Gallbladder
Key Structures: Common bile duct,
liver, gallbladder

Go to "Full Anatomy"
View: Anterior
Layer: Pancreas
Key Structure: Pancreas
(View histology also)*

Exercise 28
Anatomy of the Urinary System

Go to "Options," choose "Female"
Go to "View," choose "Urinary System"

Go to "Options," choose "Male"
Go to "View" on the Menu Bar
Choose Urinary System

Go to "Full Anatomy"
View: Anterior
Layers: Kidney, Kidney—
longitudinal section
Key Structure: Kidneys

Go to "Full Anatomy"
View: Anterior
Layer: Ureter
Key Structure: Ureter

Go to "Full Anatomy"
View: Anterior
Layer: Urinary bladder
Key Structure: Urinary bladder

Go to "Options," choose "Female"
Go to "Full Anatomy"
View: Medial
Layer: Urinary bladder
Key Structure: Urethra

Go to "Options," choose "Male"
View: Medial
Layer: Urinary bladder and prostate gland
Key Structure: Urethra

Exercise 29
Anatomy of the Reproductive System

Go to "Options," choose "Male"
Go to "View," choose "Reproductive System"

Go to "Options," choose "Male"
View: Medial
Layer: Urinary bladder and prostate gland
Key Structure: Urethra

Go to "Options," choose "Male"
View: Anterior
Layer: Testes
Key Structure: Testes
(View histology also)*

Go to "Options," choose "Male"
View: Anterior
Layer: Penis
Key Structure: Penis

Go to "Options," choose "Female"
Go to "View," choose "Reproductive System"

Go to "Options," choose "Female"
View: Medial
Layer: Uterus
Key Structure: Uterus

Go to "Options," choose "Female"
View: Anterior
Layer: Uterus
Key Structure: Uterus
(View histology also)*

Go to "Options," choose "Female"
View: Anterior
Layer: Breast
Key Structure: Mammary glands

*In "Full Anatomy," select the structure with the Identify tool, and then select "Histology" from the pop-up submenu.

Videodisc Correlations

Appendix C includes barcodes for images from *Anatomy and PhysioShow: The Videodisc* for *Human Anatomy and Physiology*, third edition by Elaine Marieb for appropriate exercises in this lab manual. The barcodes are listed in the order that topics are presented in each exercise.

Exercise 1
The Language of Anatomy

Planes of the body

Search frame 37–39

Dorsal and ventral body cavities and their subdivisions

Search frame 40–42

Serous membrane relationships

Search frame 4428–4431

Exercise 4
The Cell: Anatomy and Division

Structure of the generalized cell

Search frame 5970

The endoplasmic reticulum

Search frame 6310

Role of the Golgi apparatus in packaging proteins for cellular use and for secretion

Search frame 6315

Cytoskeleton

Search frame 6319

Cilia structure and function

Search frame 7810–7811

The cell cycle

Search frame 8983

The stages of mitosis

Search frame 9444

Human blood smear

Search frame 53018

Simple squamous epithelium

Search frame 53024

Exercise 5
Classification of Tissues

Epithelial tissue: simple squamous epithelium

Search frame 14022–14043

Areolar connective tissue: a prototype (model) connective tissue

Search frame 14047

Connective tissues: embryonic connective tissue

Search frame 14049–14069

Muscle tissues: skeletal muscle

Search frame 14074–14078

Nervous tissue

Search frame 14080

Alveoli, alveolar ducts and sacs

Search frame 53000

Cross section of the trachea

Search frame 53001

Simple squamous epithelium

Search frame 53024

Simple cuboidal epithelium

Search frame 53025

Simple columnar epithelium

Search frame 53026

Pseudostratified epithelium

Search frame 53027

Transitional epithelium

Search frame 53028

Loose areolar connective tissue

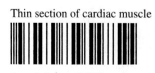

Search frame 53029

Reticular tissue of the spleen

Search frame 53030

Adipose tissue

Search frame 53031

Dense regular connective tissue

Search frame 53032

Dense irregular connective tissue

Search frame 53033

Hyaline cartilage

Search frame 53034

Elastic cartilage

Search frame 53035

Fibrocartilage

Search frame 53036

Skeletal muscle

Search frame 53037

Visceral muscle

Search frame 53038

Thin section of cardiac muscle

Search frame 53039

Skeletal muscle

Search frame 52977

Multipolar neuron

Search frame 52979

Exercise 6
The Integumentary System

Skin structure

Search frame 14089

Diagram showing the main features—layers and relative distribution of the different cell types—in epidermis of thin skin

Search frame 14092

Structure of a hair and hair follicle

Search frame 14093

Scalp tissue

Search frame 53040

Exercise 7
Classification of
Body Membranes

Classes of epithelial membranes

Search frame 14071–14073

Goblet cells, example of unicellular exocrine glands

Search frame 14046

General structure of a synovial joint

Search frame 14240

Serous membrane relationships

Search frame 4428–4431

Exercise 8
Overview of the Skeleton: Classification and Structure of Bones and Cartilages

The human skeleton

Search frame 14128–14132

The structure of a long bone (humerus of arm)

Search frame 14104

Microscopic structure of compact bone

Search frame 14107

Connective tissues: others

Search frame 14067

Cartilage in the adult skeleton and body

Search frame 14101

Connective tissues: cartilage

Search frame 14063

Epiphyseal plate of a bone

Search frame 53041

Hyaline cartilage

Search frame 53034

Elastic cartilage

Search frame 53035

Fibrocartilage

Search frame 53036

Exercise 9
The Axial Skeleton

Anatomy of the anterior and posterior aspects of the skull

Search frame 14134–14137

Anatomy of the lateral aspects of the skull

Search frame 14139–14142

Anatomy of the inferior portion of the skull

Search frame 14145–14151

Detailed anatomy of some isolated facial bones

Search frame 14152–14156

Special anatomical characteristics of the orbits

Search frame 14158

Special anatomical characteristics of the nasal cavity

Search frame 14163–14164

Paranasal sinuses

Search frame 14165–14166

The vertebral column

Search frame 14167

Structure of a typical vertebra

Search frame 14170

The first and second cervical vertebrae

Search frame 14171–14173

Posterolateral views of articulated vertebrae

Search frame 14174–14176

The sacrum and coccyx

Search frame 14177–14178

The bony thorax

Search frame 14179–14181

Exercise 10
The Appendicular Skeleton

The human skeleton

Search frame 14128–14132

Bones of the pectoral girdle

Search frame 14182

The humerus of the arm

Search frame 14188–14191

Bones of the forearm

Search frame 14194–14197

Bones of the hand

Search frame 14200–14201

Bones of the pelvic girdle

Search frame 14202–14204

Bones of the right thigh and knee

Search frame 14205–14210

The tibia and fibula of the right leg

Search frame 14211–14216

Bones of the right foot

Search frame 14217–14218

Exercise 11
Fetal Skeleton

Differences in the growth rates for some parts of the body compared to others determine body proportions

Search frame 14221

Exercise 12
Articulations and Body Movements

Fibrous joints

Search frame 14235–14236

Cartilaginous joints

Search frame 14237–14239

General structure of a synovial joint

Search frame 14240

Types of synovial joints

Search frame 14242–14245

Knee joint relationships

Search frame 14248–14251

Photograph of a hand deformed by rheumatoid arthritis

Search frame 14253

Exercise 13
Microscopic Anatomy, Organization, and Classification of Skeletal Muscle

Connective tissue wrappings of skeletal muscle

Search frame 14256

Microscopic anatomy of a skeletal muscle fiber

Search frame 14257

Myofilament composition in skeletal muscle

Search frame 14261–14264

Relationship of the sarcoplasmic reticulum and T tubules to the myofibrils of skeletal muscle

Search frame 14265

The neuromuscular junction

Search frame 16254

Skeletal muscle

Search frame 52977

Part of a motor unit

Search frame 52978

Skeletal muscle

Search frame 53037

**Exercise 14
Gross Anatomy of the
Muscular System**

Anterior view of superficial muscles of the body

Search frame 21296

Posterior view of superficial muscles of the body

Search frame 21308

Lateral view of muscles of the scalp, face, and neck

Search frame 21315

Muscles promoting mastication and tongue movement

Search frame 21322–21323

Muscles of the anterior neck and throat that promote swallowing

Search frame 21326–21329

Muscles of the neck and vertebral column causing movements of the head and trunk

Search frame 21331–21335

Superficial muscles of the thorax and shoulder acting on the scapula and arm

Search frame 21347–21349

Muscles crossing the shoulder and elbow joint, respectively causing movements of the arm and forearm

Search frame 21351–21355

Muscles of the abdominal wall

Search frame 21336–21340

Muscles of the anterior fascial compartment of the forearm acting on the wrist and fingers

Search frame 21356

Muscles of the posterior fascial compartment of the forearm acting on the wrist and fingers

Search frame 21362

Anterior and medial muscles promoting movements of the thigh and leg

Search frame 21366–21370

Posterior muscles of the right hip and thigh

Search frame 21371–21374

Muscles of the anterior compartment of the right leg

Search frame 21375–21379

Muscles of the lateral compartment of the right leg

Search frame 21380–21383

Muscles of the posterior compartment of the right leg

Search frame 21384–21392

Exercise 15
Histology of Nervous Tissue

Supporting cells of nervous tissue

Search frame 21395–21399

Structure of a motor neuron

Search frame 22301

The neuromuscular junction

Search frame 16254

Relationship of Schwann cells to axons in the peripheral nervous system

Search frame 22303–22306

Structure of a nerve

Search frame 27597

Teased myelinated axons

Search frame 52981

Multipolar neuron

Search frame 52979

Neurons differentially stained

Search frame 52980

General sensory receptors classified by structure and function

Search frame 27588

Exercise 16
Gross Anatomy of the Brain and Cranial Nerves

Embryonic development of the human brain

Search frame 27536

Three-dimensional views of the ventricles of the brain

Search frame 27540–27541

Regions of the brain

Search frame 27539

Lobes and fissures of the cerebral hemispheres

Search frame 27542

Functional areas of the left cerebral cortex

Search frame 27549

Diencephalon and brain stem structures

Search frame 27554

Basal nuclei

Search frame 27552–27553

Meninges of the brain

Search frame 27560

Formation, location, and circulation of CSF

Search frame 27563–27564

Summary of location and function of the cranial nerves

Search frame 27602

Exercise 17
Spinal Cord, Spinal Nerves, and the Autonomic Nervous System

Anatomy of the spinal cord

Search frame 27567–27569

Major ascending (sensory) and descending (motor) tracts of the spinal cord

Search frame 27573

Formation and branches of spinal nerves

Search frame 27606–27607

The cervical plexus

Search frame 27610

The brachial plexus

Search frame 27612–27616

The lumbar plexus

Search frame 27619–27622

The sacral plexus

Search frame 27625–27627

Comparison of the somatic and autonomic nervous systems

Search frame 27646

Sympathetic pathways

Search frame 27649–27651

Exercise 18
Special Senses: Vision

Accessory structures of the eye

Search frame 27668–27670

Extrinsic muscles of the eye

Search frame 27671–27672

Internal structure of the eye

Search frame 27673–27677

Schematic view of the major three neuronal populations composing the neural layer of the retina

Search frame 27678

Photoreceptors of the retina

Search frame 27682

Visual fields of the eyes and visual pathway to the brain

Search frame 27686

Focusing for distant and close vision

Search frame 27680–27681

Retina of the eye

Search frame 52988

Exercise 19
Special Senses: Hearing and Equilibrium

Structure of the ear

Search frame 27687–27690

Anatomy of the cochlea

Search frame 28262–28264

Localization and structure of a crista ampullaris

Search frame 28269–28271

One turn of the cochlea

Search frame 52989

The organ of Corti

Search frame 52990

Exercise 20
Special Senses: Taste and Olfaction

Location and structure of taste buds

Search frame 27662–27664

Olfactory receptors

Search frame 27665

Taste buds

Search frame 52991

Olfactory epithelium

Search frame 52992

Exercise 21
Functional Anatomy of the Endocrine Glands

Location of the major endocrine organs of the body

Search frame 28276

Structural and functional relationships of the pituitary and the hypothalamus

Search frame 29556–29559

Gross and microscopic anatomy of the thyroid gland

Search frame 29560

Summary of parathyroid hormone's effects on bone, the intestines, and the kidneys

Search frame 29568

Microscopic structure of the adrenal gland

Search frame 29571

Regulation of blood sugar levels by insulin and glucagon

Search frame 29576

The thyroid gland

Search frame 52993

Pancreatic islet

Search frame 52994

Posterior pituitary

Search frame 52995

Adrenal gland

Search frame 52996

Exercise 22
Blood

Major components of whole blood

Search frame 29583

Life cycle of red blood cells

Search frame 30216

Structure of hemoglobin

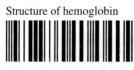

Search frame 30211–30212

Human blood smear

Search frame 53018

Monocyte

Search frame 53019

Neutrophils

Search frame 53020

Eosinophil

Search frame 53021

Lymphocyte

Search frame 53022

Basophil

Search frame 53023

Exercise 23
Anatomy of the Heart

Location of the heart in the mediastinum of the thorax

Search frame 30228–30230

The pericardial layers and layers of the heart wall

Search frame 30232

Gross anatomy of the heart

Search frame 30235–30241

Beating heart

Search frame 30245

Heart valves

Search frame 34828

Operation of the atrioventricular valves of the heart

Search frame 38381–38383

Operation of the semilunar valves

Search frame 39433–39436

Microscopic anatomy of cardiac muscle

Search frame 39437–39438

Thin section of cardiac muscle

Search frame 53039

Exercise 24
Anatomy of Blood Vessels

Structure of arteries, veins, and capillaries

Search frame 40246–40249

Anatomy of a capillary bed

Search frame 40250

Arteries of the abdomen

Search frame 43333–43340

Venous drainage of the head, neck, and brain

Search frame 43344–43347

Pulmonary circulation

Search frame 43324

Arteries of the head, neck, and brain

Search frame 43327–43330

Veins of the abdomen

Search frame 43348–43350

Circulation in fetus and newborn

Search frame 52952–52955

Exercise 25
The Lymphatic System

The lymphatic system

Search frame 43359–43361

Lymphoid organs

Search frame 43364

Structure of the spleen

Search frame 43367–43368

Spleen

Search frame 52998

Palatine tonsil

Search frame 52999

Exercise 26
Anatomy of the Respiratory System

Basic anatomy of the upper respiratory tract

Search frame 45725

Anatomical relationships of organs in the thoracic cavity

Search frame 46428

The major respiratory organs shown in relation to surrounding structures

Search frame 45724

Respiratory zone structures

Search frame 46423

Anatomy of the respiratory membrane

Search frame 46425–46426

Tissue composition of the tracheal wall

Search frame 46421

Alveoli, alveolar ducts and sacs

Search frame 53000

Cross section of the trachea

Search frame 53001

Exercise 27
Anatomy of the Digestive System

Organs of the alimentary canal and related accessory organs

Search frame 46449

The basic structure of the wall of the alimentary canal consists of four layers: the mucosa, submucosa, muscularis externa, and serosa

Search frame 47808

Anatomy of the oral cavity (mouth)

Search frame 47811–47813

Major salivary glands

Search frame 47815–47816

Microscopic structure of the wall of the esophagus

Search frame 47820

Anatomy of the stomach

Search frame 47826–47828

Microscopic anatomy of the stomach

Search frame 47829–47831

The duodenum of the small intestine, and related organs

Search frame 47832

Structural modifications of the small intestine that increase its surface area for digestion and absorption

Search frame 47834

Scanning electron micrographs of the villi and microvilli of the small intestine

Search frame 47836

Gross anatomy of the large intestine

Search frame 47857–47859

Longitudinal section of a canine tooth within its bony alveolus

Search frame 47817

Major salivary glands

Search frame 47815–47816

Microscopic anatomy of the liver

Search frame 47840

Stomach wall showing three tunics

Search frame 53002

Palatine tonsil

Search frame 52999

Gastroesophageal junction

Search frame 53004

Gastric glands and pits

Search frame 53003

Villi and duodenal glands

Search frame 53005

Ileum showing Peyer's patches

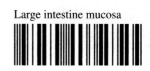

Search frame 53006

Large intestine mucosa

Search frame 53007

Pig liver

Search frame 53009

Pancreas

Search frame 53008

Pancreatic islet

Search frame 52994

Exercise 28
Anatomy of the Urinary System

Organs of the urinary system

Search frame 50752

Internal anatomy of the kidney

Search frame 50753

Location and structure of nephrons

Search frame 50755–50759

Detailed anatomy of nephrons and their vasculature

Search frame 50760–50763

The kidney depicted schematically as a single, large, uncoiled nephron

Search frame 50764

The filtration membrane

Search frame 52822–52824

Summary of nephron functions

Search frame 52828

Structure of the urinary bladder and urethra

Search frame 52834–52836

Glomerulus

Search frame 53046

Urinary bladder wall

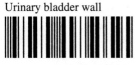

Search frame 53047

Exercise 29
Anatomy of the Reproductive System

Reproductive organs of the male

Search frame 52870

Structure of the penis

Search frame 52875

Internal structure of the testis

Search frame 52873–52874

Internal organs of the female reproductive system

Search frame 52903–52907

Structure of an ovary

Search frame 52910–52913

Anterior view of the internal reproductive organs from a female cadaver

Search frame 52914

Epididymis

Search frame 53010

The penis (transverse section)

Search frame 53011

Uterus: proliferative phase

Search frame 53015

Credits

PHOTOGRAPH CREDITS

9.5c, 14.3c, 14.4c, 14.5c, 14.6a, 14.10b, 14.11b, 14.12c, 16.2c, 16.4b, 16.6a,b, 16.7c, 17.1b, 26.1b, 27.7: From *A Stereoscopic Atlas of Human Anatomy* by David L. Bassett, M.D.

2.1, 2.2, 2.3b, 18.6, 22.4, 22.5, 22.7, 27.16b, 27.17b: © The Benjamin/ Cummings Publishing Company. Photos by Jack Scanlon, Holyoke Community College.

Histology Atlas

Plates 1–60: © The Benjamin/Cummings Publishing Company. Photos by Victor Eroschenko, University of Idaho, except as noted.

Plate 10: Courtesy of Churchill-Livingstone.

Plate 20: © Carolina Biological Supply Company.

Plate 30: Courtesy of Marian Rice.

Beauchene skull photographs: © Somso Modelle.

Cat Anatomy Atlas

Plates A–F: © Paul Waring/BioMed Arts Associates, Inc.

Exercise 1

1.1: © Paul Waring/BioMed Arts Associates, Inc.

Exercise 2

2.5: © Carolina Biological Supply Company/Phototake.

Exercise 3

3.1: Courtesy of Leica Inc., Deerfield, IL. 3.5: © The Benjamin/Cummings Publishing Company. Photo by Victor Eroschenko.

Exercise 4

4.1b: © Don W. Fawcett/Photo Researchers, Inc. 4.3a–f: © Ed Reschke.

Exercise 5

5.3, 5.5a,c–j, 5.6a–c: © The Benjamin/ Cummings Publishing Company. Photo by Allen Bell, University of New England. 5.5b: Courtesy of Marian Rice. 5.7c: © Ed Reschke. 5.5k: © The Benjamin/ Cummings Publishing Company. Photo by Victor Eroschenko.

Exercise 6

6.3: B. Kozier and G. Erb, *Fundamentals of Nursing,* 5th ed. (Redwood City, CA: Addison-Wesley Nursing, a division of Benjamin/Cummings, 1995). © 1995 Addison-Wesley Publishing Company. 6.6: © The Benjamin/Cummings Publishing Company. Photo by Victor Eroschenko.

Exercise 8

8.3b: Courtesy of Marian Rice.

Exercise 9

9.8c: A. P. Spence, E. B. Mason, *Human Anatomy and Physiology,* 3rd ed. (Redwood City, CA: Benjamin/ Cummings, 1987). © 1987 The Benjamin/Cummings Publishing Company.

Exercise 10

10.5b: Courtesy of the Department of Anatomy and Histology, University of California, San Francisco. 10.6d: © The Benjamin/Cummings Publishing Company. Photo by San Francisco State University. 10.8c, 10.9b, 10.10b: Courtesy of Richard Hambert, Stanford University.

Exercise 11

11.2: © The Benjamin/Cummings Publishing Company. Photo by San Francisco State University.

Exercise 12

12.2: Courtesy of University of California, San Francisco, Department of Anatomy and Histology.

Exercise 13

13.2b: Courtesy of Marian Rice.

Exercise 14

14.8f: Courtesy of University of Kansas Medical Center. Photo by Stephen Spector. 14.14–14.25: © Paul Waring/BioMed Arts Associates, Inc.

Exercise 15

15.2a: © Manfred Kage/Peter Arnold, Inc. 15.3g: Courtesy of G. L. Scott, J. A. Feilbach, T. A. Duff.

Exercise 16

16.10c, 16.12b: © The Benjamin/Cummings Publishing Company. Photo by Dr. Sharon Cummings, University of California, Davis. 16.11: © The Benjamin/Cummings Publishing Company. Photo by Wally Cash, Kansas State University.

Exercise 17

17.10b, 17.11b: © Paul Waring/BioMed Arts Associates, Inc.

Exercise 18

18.7: © Elena Dorfman. 18.8: Courtesy of Christopher S. Sherman, Swedish Hospital Medical Center.

Exercise 22

22.6: © Boehringer Ingelheim International GmbH. Photo by Lennart Nilsson.

Exercise 23

23.3b: © Lennart Nilsson, *Behold Man,* Little, Brown and Company. 23.6: © The Benjamin/Cummings Publishing Company. Photo by Allen Bell, University of New England. 23.7, 23.8: © The Benjamin/Cummings Publishing Company. Photo by Wally Cash, Kansas State University.

Exercise 24

24.5: © Science Photo Library/Photo Researchers. 24.9: L. Gartner and J. Hiatt, *Color Atlas of Histology* (Baltimore, Maryland: Williams and Wilkins). © 1990 Williams and Wilkins. 24.16b, 24.19b: © Paul Waring/BioMed Arts Associates, Inc.

Exercise 26

26.3c: © SPL, Custom Medical Stock Photo.

Exercise 28

28.3b: © The Benjamin/Cummings Publishing Company. Photo by Wally Cash, Kansas State University. 28.7b, 28.8b: © Paul Waring/BioMed Arts Associates, Inc.

Exercise 29

29.11b, 29.12b: © Paul Waring/BioMed Arts Associates, Inc.

Review Sheets

Page RS7: Courtesy of Leica, Inc., Deerfield, IL. Page RS13: © The Benjamin/Cummings Publishing Company. Photos by Victor Eroschenko. Page RS31: © The Benjamin/Cummings Publishing Company. Photo by Allen Bell, University of New England. Page RS101: Courtesy of Marian Rice.

TRADEMARK ACKNOWLEDGEMENTS

Betadine is a registered trademark of The Purdue Frederick Company.

Landau is a registered trademark of Landau Uniform.

Novocain is a registered trademark of Sterling Drug, Inc.

Seal-Ease is a registered trademark of Becton, Dickinson, and Company.

Index